Modeling Forest Trees and Stands

Harold E. Burkhart • Margarida Tomé

Modeling Forest Trees and Stands

Springer

Harold E. Burkhart
Department of Forestry
Virginia Tech
Cheatham Hall 319
Blacksburg, VA 24061
USA

Margarida Tomé
Universidade Técnica de Lisboa
Instituto Superior de Agronomia
DRAT
Tapada da Ajuda
Lisboa, Portugal

ISBN 978-94-007-1597-4 (hardcover)
ISBN 978-90-481-3169-3 (softcover)
ISBN 978-90-481-3170-9 (eBook)
DOI 10.1007/978-90-481-3170-9
Springer Dordrecht Heidelberg New York London

Library of Congress Control Number: 2012937653

Printed on acid-free paper

Springer is part of Springer Science+Business Media (www.springer.com)

Preface

This book is intended to serve as a text for graduate-level courses in forest modeling and as a reference for researchers working in growth and yield modeling. Estimating tree volumes and stand yields has a long history in forestry, starting with tabulations and progressing to use of advanced statistical analyses and electronic computers for implementation. Continuing advances in statistical science and computing technology have fueled increasingly sophisticated approaches to modeling of forest trees and stands. The focus of this volume is on contemporary methods (meaning generally from the 1960s forward) for modeling individual tree characteristics and forest stand dynamics, growth and yield. It is our hope that this summary and synthesis of past work on empirical modeling of forest trees and stands will provide a useful platform from which future researchers can build and move the field forward.

After an introduction (Chap. 1), the first part of this book (Chaps. 2, 3, 4, and 5) covers quantification of tree form, taper, volume, biomass, and crown characteristics. Part 2 provides a transition to stand modeling with chapters on growth functions (Chap. 6), quantification of site quality (Chap. 7), and measures of stand and point density (Chaps. 8 and 9, respectively). The third part contains an introductory chapter (Chap. 10) to modeling forest stand development, which is followed by Chaps. 11, 12, 13, and 14 on whole-stand, diameter-distribution, size-class, and individual-tree model structures for even-aged stands. Special considerations and modeling methods for uneven-aged stands are described in Chap. 15. Extensions to commonly-applied stand modeling approaches are given in part 4, with summary information on modeling response to silvicultural treatments (Chap. 16), incorporating wood quality information (Chap. 17), and implementing and evaluating models (Chap. 18).

Following an introduction to the topic, each chapter provides increasingly advanced information on the principal modeling approaches that have been applied. Although an in-depth background in statistical analysis is required for comprehension of the more advanced techniques presented, the material is sequenced such that it should be suitable for use in courses at the masters or doctoral level.

The scope of this volume is limited to empirical modeling of forest trees and stands; coverage of process models and related approaches is not provided. No geographic restrictions were placed on what was included, but due to differences in contributions to empirical modeling of forests that have been made in different regions and for different species, some areas and species are inevitably more prominently represented than others. Citations are largely limited to literature published in English; where English language summaries of work published in other languages are available, the English versions are cited. Articles from the peer-reviewed literature are cited, where possible, but in some instances papers from proceedings, research bulletins and the like are cited because they are the primary source for the material. The primary publications for the statistical techniques employed in various modeling applications are generally not cited in this book, but they can be located by consulting references listed in the forest modeling papers cited at the end of each chapter. The list of references at the end of each chapter consists of literature that is cited therein plus selected citations of additional papers relevant to the topics covered. Metric units are used throughout, and to the extent possible, standard symbols of the International Union of Forest Research Organization are used.

Finally, we wish to acknowledge, with deep gratitude, the critical role that our mentors, colleagues, and especially our students have played in developing our appreciation for and understanding of modeling forest trees and stands. A number of professional associates provided ideas, review comments and suggestions for various portions of this treatise. Special thanks go to Professor Timothy Gregoire for his contributions to the planning and early stages of development of the book proposal. Numerous individuals provided helpful review comments on selected portions of the manuscript as it was being developed, including Ralph Amateis, Finto Anthony, Clara Antón Fernández, Quang Cao, Klaus von Gadow, Oscar Garcia, Samantha Gill, Steven Knowe, Tony Kozak, Thomas Lynch, Douglas Maguire, Arne Pommerening, Charles Sabatia, Mahadav Sharma, Mike Strub, Guillermo Trincado, Curtis VanderSchaaf, and Boris Zeide. Graduate assistants at Virginia Tech, namely Micky Allen, Gavin Corral, Nabin Gyawali, and Ram Thapa, provided valuable support in obtaining reference material, proof reading, and performing other tasks associated with preparation of this volume. To all who contributed to this effort in numerous ways we say "thanks", while affirming that any remaining errors or shortcomings in this work are ours alone.

Blacksburg, Virginia Harold E. Burkhart
Lisboa, Portugal Margarida Tomé

Contents

Chapter 1
Introduction

1.1 Quantifying Forest Trees and Stands

Quantitative information on trees and stands is required for assessment and management of forests. The basic management entity is generally the forest stand, which, from a forestry standpoint, can be defined as "a contiguous group of trees sufficiently uniform in age-class distribution, composition, and structure, and growing on a site of sufficiently uniform quality, to be a distinguishable unit" (Helms 1998). Forest stands are often described as being pure (single species) or mixed-species, even-aged or uneven-aged. Production forestry commonly involves establishment of pure, even-aged stands. Due to the substantial investments required for intensive management for timber production, highly accurate models of tree attributes and stand development are required. Models for various stand types are included in this volume; however, much research has been focused on pure, even-aged stands and that focus is reflected in the overview of forest tree and stand modeling that constitutes this book.

While the bulk of this volume is devoted to modeling forest stand dynamics, stands are composed of trees and models of tree attributes such as stem taper, volume and weight, as well as tree biomass and crown characteristics, are central components in a comprehensive growth and yield prediction system. Managers need information not only on volume but also on value, thus necessitating a detailed quantitative description of individual tree characteristics. In addition, tree and hence stand response to silvicultural treatments and environmental influences is related to tree characteristics such as crown size; accordingly increasingly complete quantitative descriptions of individual trees are required.

A number of approaches, including process, hybrid, and empirical models, have been taken to modeling tree and stand development. Empirical forest growth and yield models are developed using statistical techniques and calibrated with comprehensive data sets. They are aimed at describing growth for a range of silvicultural practices and site conditions. Their relatively simple data input requirements and accuracy in predicting growth have made empirical models an essential tool for

H.E. Burkhart and M. Tomé, *Modeling Forest Trees and Stands*,
DOI 10.1007/978-90-481-3170-9_1,

forest management. Process-based model include a description of the behavior of a system in terms of a set of functional relationships (e.g. plant growth relationships) and their interactions with each other and the system (Landsberg and Sands 2011). However, both types of models have some empiricism, i.e. are based on data that describe some process or part of the system. The essential difference between the so-called empirical models, such as conventional growth and yield models, and process-based models is that empirical models essentially describe the observed data at a given organizational level (e.g. tree or stand level) in terms of attributes at the same level while in process models the empiricism is at lower levels than in single tree or stand models (Thornley and Johnson 1990). When developing process-based models the primary units of simulation are the processes. These models deal, essentially, with dry mass production processes (at the leaf or canopy level) and its distribution to plant parts, resulting in tree or stand growth. In tree and stand models the primary unit of simulation is the tree or the stand, respectively. This means that process-based models are based on causal processes at a hierarchical level below the entire system while empirical models directly model the tree or stand variables. Models that combine process-based and empirical components are often designated as hybrid models. Mäkela et al. (2000) defined hybrid models as those that contain both causal and empirical elements at the same hierarchical level. In practice process-based models are used when modeling is undertaken for the purpose of understanding while growth and yield models are widely used when the objective is prediction. Data inputs for process models are often lacking in forest inventories, which limits their use in wide-scale applications. Here we assume the objective is primary prediction and we emphasize empirical modeling.

Statistical techniques are essential tools in development of empirical models and the variety and complexity of the techniques used has greatly increased in the last decades. Mixed-effects models, simultaneous fitting of systems of equations, generalized linear models, and many other techniques, are now commonly applied in model development. It is beyond the scope of this book to present underlying statistical techniques in detail. References on forest modeling, which are included with each chapter, provide an entry point to books on statistical theory and methods as well as to key journal papers reporting advances in statistical science.

1.2 Modeling Approaches

In the preface to his book *Empirical Model Building* Thompson (1989) states:

> Empirical model building refers to a mindset that lends itself to constructing practical models useful in describing and coping with real-world situations.

The empirical models described in the chapters that follow are indeed practical models that are useful for describing real-world situations – and hence helpful in solving real-world problems. The objectives of the modeling endeavor and the level of understanding of the phenomenon being modeled often dictate an empirical approach. Empirical models have been applied with successful results for a host of forestry applications.

Thompson (1989, pp. 2–3) describes the various ways in which scientists approach the concept of models and categorizes modelers in three groups that appear to have the greatest numbers of adherents, namely:

Idealists are not really data oriented. They are rather concerned with theory as a mental process that takes a cavalier attitude toward the "real world." Their attitude can be summed up as follows: "If facts do not conform to theory, then so much the worse for the facts." For them, the "model" is all.
Radical Pragmatists are the opposite end of the spectrum from that of the Idealists. The Radical Pragmatists hold that data are everything. Every situation is to be treated more or less uniquely. There is no "truth." All models are false. Instead of model building, Radical Pragmatists fit curves to data. They do not look on the fitted curve as something of general applicability, but rather as an empirical device for coping with a particular situation.
Realists occupy a ground intermediate to that of the Idealists and that of the Radical Pragmatists. They hold that the universe is governed by rational and consistent laws. Models, to the Realist, are approximations to bits and pieces of these laws. The Realist knows the model is not quite right but hopes it is incomplete rather than false. The collection of data is useful in testing a model and enabling the Realist to modify it in an appropriate fashion. It is this truth seeking, interactive procedure between mind and data which might be termed *empirical model building*.

Empirical model building by the Realist approach follows the process that is commonly referred to as the "scientific method." In this interactive method scientists proceed to understand portions of the real world by proposing theoretical mechanisms, testing these against observations, and revising theory where it does not conform to data. Much model building in forestry falls within this definition of the Realist approach. Some modelers lean more towards the Idealists end of the spectrum while others tend more toward the Radical Pragmatists. But the main distinctions among forest models and modelers are a matter of modeling objective – that is whether modeling for understanding or for prediction (Burkhart 1999).

1.3 Empirical Modeling of Forests

In classical growth and yield research, the emphasis has been on the development of prediction tools to provide decision support for practical forest management. The approach involves taking field measurements of the relevant variables such as age, tree diameters and heights, and site quality, and fitting regression models that allow reproduction of the essential characteristics of the field data set. In its most data-intensive form, this modeling approach can be regarded as descriptive (following the Radical Pragmatists School) essentially with the aim of summarizing and reproducing field data. As more data became available, sophisticated analytical methods were developed, and more powerful computing technology was obtained, this approach evolved to produce rather elaborate yield models, with increasing detail on individual trees and stand structure. Over the past decades, whole-stand, diameter-distribution, size-class, and individual-tree models have been developed for forest management purposes. At the same time increasing detail on crown structure and branching characteristics, in addition to stem volume, has been included in the models (Mohren and Burkhart 1994).

Provided the conditions under which these models are applied are the same (or very similar) as those under which the basic data have been collected, such models produce accurate predictions of future yield. To the extent that appropriate underlying functional forms are specified, these empirical models can also be used confidently for limited extrapolation beyond the range of the field data. Growth and yield models are used for inventory updating, estimating stand structure and stand productivity, evaluating silvicultural alternatives, and in general for decision support in forest management and planning (Burkhart 1990). To develop such models requires extensive databases from long-term permanent field plots, covering various management regimes and site productivities. Data from designed experiments are also often incorporated to extend the range of conditions for which reliable predictions can be made.

1.4 Organization of Book Contents

The material in this book is organized into four parts. The first part deals with quantifying attributes of individual trees. Tree form and taper are covered in Chap. 2, tree-stem volume equations are addressed in Chap. 3, tree weight and biomass estimation is the topic of Chap. 4, and Chap. 5 deals with characterizing tree crowns.

The second part provides a transition towards forest stand modeling, starting with background information on growth functions in Chap. 6. Chapter 7 addresses quantifying site quality, with the primary focus being on the height-age or site index concept. Stand density measures is the topic of Chap. 8, and Chap. 9 deals with measures of point density.

Forest stand modeling constitutes a major focus of this text and reference book. Chapter 10 provides context and introductory information for the chapters on commonly-used model structures that follow. Chapter 11 describes whole-stand models for even-aged forests. Due to the large amount of effort that researchers have devoted to using continuous distributions to produce stand tables (numbers of trees by diameter classes) and yield estimates from whole stand characteristics, Chap. 12 is devoted to this area, commonly referred to as diameter-distribution models. Size-class models are the subject of Chap. 13 and individual-tree models are described in

Chap. 14. This suite of stand models, from whole-stand to individual-tree levels, is developed in Chaps. 11, 12, 13, and 14 in the context of even-aged stands. Chapter 15 deals with special considerations involved with modeling uneven-aged stands and extends and expands the modeling techniques described in Chaps. 11, 12, 13, and 14 to the age-indeterminate situation. Chapter 15 also covers modeling techniques, most notably matrix models, that have been found highly useful for uneven-aged management.

The final part of the book extends forest stand modeling approaches by describing methods to incorporate silvicultural treatments (Chap. 16) and wood quality attributes (Chap. 17) in growth and yield models. Chapter 18 provides a discussion of model evaluation and implementation.

1.5 Abbreviations and Symbols Used

Certain abbreviations (such as *dbh* for diameter at breast height) and symbols are used in multiple chapters; others are defined when they appear in the text. For clarification and ease of reference, a description of common symbols is given here. The symbols used are based on the International Union of Forest Research Organizations (IUFRO) standardization recommended in 1959 and republished in 1965 (van Soest et al. 1965). Capital letters are used to denote totals per unit area (most of the stand variables) while lowercase letters generally refer to individual or mean values for tree variables. Whenever appropriate, the same symbol is used for the tree (lowercase letters) and for the stand (capital letters). For instance v symbolizes tree volume, whereas V designates stand volume.

The most referred to tree and stand variables are listed in Table 1.1. Additional variables are defined as needed in the appropriate chapters. For convenience, most symbols and variables are defined within each chapter when they are first used.

Conventions followed to develop notation and symbols include:

- Arithmetic mean values of tree variables are indicated by the same symbol used for the corresponding tree variable with a superimposed bar (for instance d for tree dbh and $\bar{d}$ for the mean diameter);
- Mean values, other than arithmetic means, are defined as needed;
- Increment in one variable is indicated by the letter i or the symbol Δ followed by the symbol of the variable (for example, tree volume increment is iv or Δv and stand volume increment is iV or ΔV).

We have strived for consistency in notation within this volume by defining the most commonly used symbols (Table 1.1). Conventions vary across the primary literature that is synthesized and presented herein, however, and we have incorporated certain notation that is typically used in the various subject areas covered to facilitate augmenting our presentation with more in-depth material.

Table 1.1 Abbreviations and symbols

Tree and stand variables

Symbol	Units	Description
Age variables for trees and even-aged stands		
t	Years	Age (independently of the definition: from seed, from planting, at dbh); also used to denote elapsed time
t_b	Years	Base age (for S determination)
Diameter and diameter related variables		
d	cm	Diameter at breast height (1.3 m, 1.4 m, 4.5 ft = 1.37 m or other)
$\bar{d}$	cm	Arithmetic mean diameter
$\overline{d}_g$	cm	Quadratic mean diameter
d_{hi}	cm	Diameter at height h_i above ground
g	cm^2	Tree basal area at breast height
g_{hi}	cm^2	Cross-sectional area at height h_i above ground
Height and height related variables		
h	m	Total tree height
h_m	m	Merchantable tree height
$\overline{h}$	m	Arithmetic mean height
h_{di}	m	Height from the ground to point diameter d_i on the tree stem (corresponding to diameter d_i or some other limit of measurement on utilization)
h_d	m	Height to *dbh*
h_{cb}	m	Height to the base of the crown
Crown and crown related variables		
c_w	m	Crown width, crown diameter
c_l	m	Crown length
c_a	m^2	Crown area
c_v	m^3	Crown volume
c_r	–	Crown ratio (crown length divided by total tree height)
Stem volume and weight variables		
v	m^3	Total volume (to the tip of the tree)
v_m	m^3	Merchantable volume of stem
v_{di}	m^3	Volume to top diameter d_i (cm)
v_{hi}	m^3	Volume to height h_i (m)
w	kg	Total weight (green or dry) to tip of tree
w_m	kg	Merchantable weight (green or dry) of stem
w_{di}	kg	Weight (green or dry) to top diameter d_i (cm)
w_{hi}	kg	Weight (green or dry) to height h_i (m)
Dominant height and site index variables		
h_{dom}	m	Dominant height (independently of the definition)
S	m	Site index for a specified base age
Stand density variables		
G	m^2ha^{-1}	Basal area
N	ha^{-1}	Number of trees
Stand volume variables		
V	m^3ha^{-1}	Total volume (to the tip of the tree)
V_m	m^3ha^{-1}	Merchantable volume stand
V_{di}	m^3ha^{-1}	Volume to top diameter d_i (cm)
V_{hi}	m^3ha^{-1}	Volume to height h_i (m)

References

Bunnell FL (1989) Alchemy and uncertainty: what good are models? USDA Forest Service, Pacific Northwest Research Station, Portland, General Technical Report, PNW-GTR-232

Burkhart HE (1987) Data collection and modeling approaches for forest growth and yield prediction. In Chappell HN, Maguire DA (eds) Predicting forest growth and yield: current issues, future prospects, Institute of Forest Resources, University of Washington. Contribution No. 58, pp 3–16

Burkhart HE (1990) Status and future of growth and yield models. In LaBau VJ, Cunia T (eds) State-of-the-art methodology of forest inventory: a symposium proceedings. USDA Forest Service, Pacific Northwest Forest Experiment Station, Portland, pp 409–414, General Technical Report, PNW-GTR-263

Burkhart HE (1999) Development of empirical growth and yield models. In: Amaro A, Tomé M (eds) Proceedings IUFRO conference on empirical and process-based models for forest tree and stand growth simulation. Edicoes Salamandra, Lisboa, pp 53–60

Burkhart H (2003) Suggestions for choosing an appropriate level for modeling forest stands. In: Amaro A, Reed D, Soares P (eds) Modelling forest systems. CAB International, Wallingford, pp 3–10

Dennis B, Brown BE, Stage AR, Burkhart HE, Clark S (1985) Problems of modeling growth and yield of renewable resources. Am Stat 39:374–383

Helms JA (ed) (1998) The dictionary of forestry. Society of American Foresters, Bethesda

Landsberg JJ, Sands P (2011) Physiological ecology of forest production: principles, processes, and models. Academic, London

Mäkela A, Landsberg J, Ek AR, Burk TE, Ter-Mikaelian M, Agren GI, Oliver CD, Puttonen P (2000) Process-based models for forest ecosystem management: current state of the art and challenges for practical implementation. Tree Physiol 20:289–298

Mohren GMJ, Burkhart HE (1994) Contrasts between biologically-based process models and management-oriented growth and yield models. For Ecol Manage 69:1–5

Thompson JR (1989) Empirical model building. Wiley, New York

Thornley JHM, Johnson IR (1990) Plant and crop modeling. A mathematical approach to plant and crop physiology. Clarendon, Oxford

van Soest J, Ayral P, Schober R, Hummel FC (1965) The standardization of symbols in forest mensuration. University of Maine Agricultural Experiment Station, Technical Bulletin 15

Chapter 2
Tree Form and Stem Taper

2.1 Form and Taper

Although sometimes used as synonyms, the terms tree form and stem taper have specific connotations in a forestry context. Form refers to the characteristic shape of the tree, whereas taper is the rate of decrease in stem diameter with increasing height from ground level to the tree tip. The general form of trees can be divided into three basic classes: excurrent, decurrent, or shrub. Species exhibit an excurrent or conical crown shape when terminal growth exceeds branch lateral growth; excurrent crowns are typical of many conifers and some deciduous trees (Fig 2.1a). Because trees of excurrent form generally have a single main stem they have been preferred for production of solid wood products, in particular, and with their often-times high commercial value these species have been the focus of much of the research aimed at accurate quantitative descriptions of the main stem.

Decurrent or spreading crowns, which are typical of many deciduous species, result when lateral branches grow nearly as fast as or faster than the terminal leader. This crown form often produces repeated forking of the main stem (Fig. 2.1b). Quantitative description of such stems is more difficult than with single-stemmed trees, but, with appropriate definitions and predictor variables, stem taper functions for decurrent forms have been developed. Fowler and Rennie (1988), for instance, presented a method to use merchantable height in lieu of total height when developing stem profile equations for hardwood trees in Tennessee, USA.

Shrubs are woody perennial plants that generally lack a well-defined main stem (Fig. 2.1c). Although not of primary interest from a wood products and forest management perspective, the volume or biomass of shrubs is sometimes needed to evaluate overall site occupancy and productivity, amounts of fuel for fires, and for other purposes. While stem taper functions are not of general interest for shrubs, volume and/or biomass equations may be needed. Additional measurements, and special definitions, have been employed when developing volume and biomass prediction equations for shrubs (see Chap. 3).

H.E. Burkhart and M. Tomé, *Modeling Forest Trees and Stands*,
DOI 10.1007/978-90-481-3170-9_2,

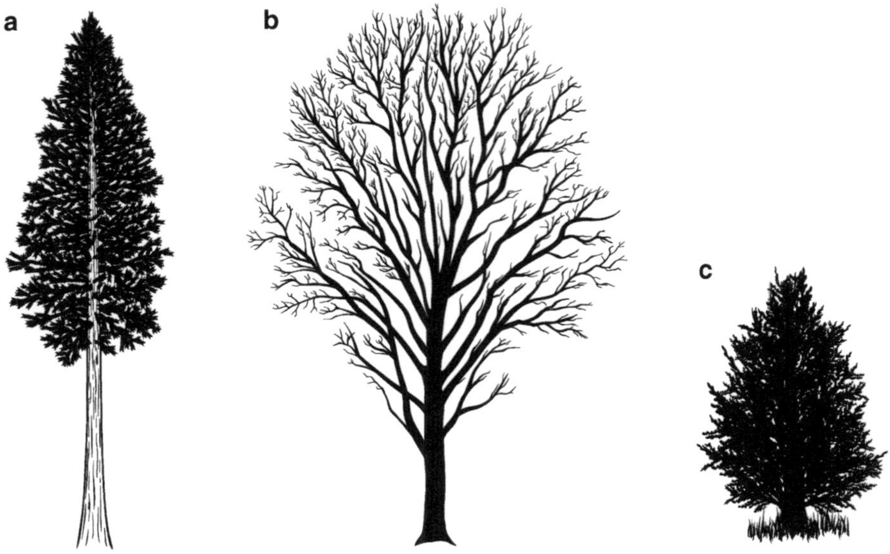

Fig. 2.1 Excurrent (**a**), decurrent (**b**), and shrub (**c**) tree forms

The stem of excurrent trees resembles a solid of revolution, namely a paraboloid. A more detailed analysis shows that it is more realistic to consider three parts in the stem that can be approximated by different solids of revolution: neiloid for the bottom, paraboloid for the central part of the stem and cone for the top (Fig. 2.2). If stem form (that is, the shape of the stem or some portion thereof) is approximately that of a regular geometric solid, the volume can be estimated from measurement of the appropriate widths and length.

Tree stems, however, have multiple inflexion points along their length resulting in multiple geometric shapes being approximated, and resulting in a mathematical description of the entire bole being very difficult to obtain. The functions used to describe tree stem shape by predicting diameter along the bole as a function of the height from the ground as well as of tree diameter and height are generally designated as taper functions. Stem taper functions are exceedingly flexible and can provide estimates of diameter at any point along the stem, total stem volume, merchantable volume, and merchantable height to any top diameter and from any stump height, as well as individual log volumes of any length at any height from ground. Consequently, a great deal of effort has been devoted to modeling stem taper, especially for single-stemmed species of high commercial value.

2.2 Stem Taper Functions

Many different forms of taper functions have been developed for various tree species. In general terms a taper function takes the form

$$d_i = f(d, h, h_i) \tag{2.1}$$

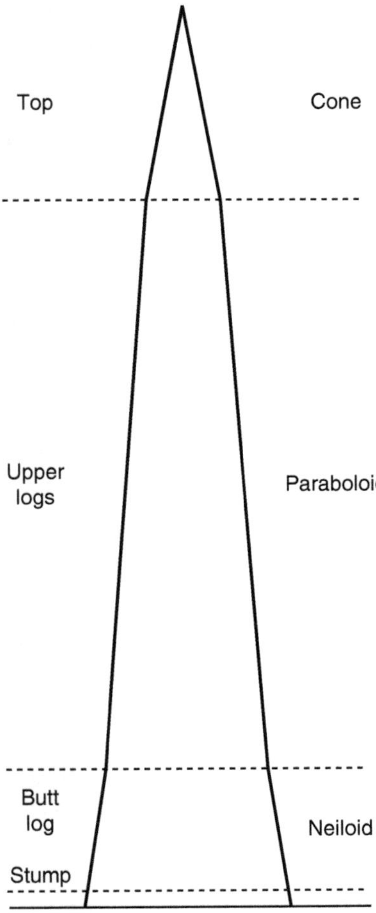

Fig. 2.2 Geometric shapes assumed by different portions of tree boles (Adapted from Avery and Burkhart 2002)

where d_i is the stem diameter at height h_i above ground for a tree with diameter at breast height d and total height h. Diameter inside bark or outside bark may be used as the dependent variable. Taper curves can also be used to estimate the height at a specified diameter by inverting Eq. 2.1. Such equations allow estimation of merchantable heights to specified top diameters and computation of length of quantities for specific products.

To obtain stem volume for any desired portion of the tree bole, the expression for stem cross-sectional area is integrated over the length desired. If one assumes that tree cross-sections are circular in shape, then the cross-sectional area g_i in square units for a given d_i is:

$$g_i = \frac{\pi}{4} {d_i}^2 = \mathrm{k}{d_i}^2$$

where k is $\pi/4$ multiplied by a constant depending on the units of d_i and g_i.

Integrating the expression for area in square units over the length desired in units gives the volume in cubic units for that segment:

$$v_{h_1-h_2}(cubic\ units) = \mathrm{k}\int_{h_1}^{h_2} {d_i}^2 dh_i$$

where $v_{h_1-h_2}$ is volume for the tree segment between h_1 and h_2, k is a constant depending on the units and h_1 and h_2 denote the limits of integration.

Taper functions can be grouped according to various categories. For purposes of this discussion three broad categories will be examined: simple taper functions, segmented functions and variable-exponent models.

Simple taper functions depict the entire tree profile with a single equation. Tree boles are highly variable, however, with multiple inflection points and it is difficult to describe their shapes over entire lengths without rather complex functions. Segmented taper functions consist of sub-models to describe different portions of the tree bole; these sub-models are then joined for a description of the entire stem. Variable exponent equations consist of a continuous function for describing the shape of the bole from ground to tip by using a changing exponent to describe the lower (neiloid), mid (paraboloid), and upper (conic) forms of the stem.

Integrating for volume estimation and inverting the equation for estimating height at a specified stem diameter becomes more computationally complex (and sometimes requires numerical approximations) when complex functions are employed. Fortunately, computing technology has largely eliminated the computational difficulties associated with fitting taper equations to data and applying them for estimating stem diameters and volumes.

2.2.1 Simple Equations

The parabolic function of Kozak et al. (1969) provides an example of a simple taper function:

$$ {d_i}^2/d^2 = \mathrm{b}_0 + \mathrm{b}_1\,(h_i/h) + \mathrm{b}_2\left({h_i}^2/h^2\right) \tag{2.2}$$

where d_i is diameter at any given height h_i above ground, and b_0, b_1 and b_2 are regression coefficients. Kozak et al. (1969) developed the equation for the prediction of the tree profile inside bark but the model has been used either to predict d_i or du_i. If d_i and d are in the same units and h_i and h are also expressed in the same units, the resultant regression coefficients for relative diameter (d_i/d) squared as a function of relative height (h_i/h) are unitless. An estimate of upper stem diameter (d_i) is obtained as:

$$d_i = d\sqrt{\mathrm{b}_0 + \mathrm{b}_1\,(h_i/h) + \mathrm{b}_2\left({h_i}^2/h^2\right)} \tag{2.3}$$

with units of measure identical to d. This simple quadratic equation fits the mid-section of tree boles reasonably well but it tends to be biased in the upper (more conic) portion and it fails to account for the butt flare near ground level. Because fitted equations can be biased in the upper stem portions, the equation is sometimes conditioned so that when h_i equals h estimated diameter is zero. This condition can be ensured by imposing the constraint $\mathrm{b}_0 = -\mathrm{b}_1 - \mathrm{b}_2$. Substitution of this condition into Eq. 2.2 and simplifying gives

$$d_i^2/d^2 = \mathrm{b}_1\left((h_i/h) - 1\right) + \mathrm{b}_2\left(\left(h_i^2/h^2\right) - 1\right) \tag{2.4}$$

which is a two-parameter regression equation without a constant term.

By applying the quadratic formula to the parabolic taper Eq. 2.2, it can be inverted to obtain an expression for height h_i to any specified diameter d_i for trees of given *dbh* and total height values:

$$h_i = \frac{-\mathrm{b}_1 h - \sqrt{(\mathrm{b}_1 h)^2 - 4\mathrm{b}_2(\mathrm{b}_0 h^2 - d_i^2 h^2/d^2)}}{2\mathrm{b}_2} \tag{2.5}$$

Using the quadratic taper model (2.2) and substituting the expression for d_i^2 into the integral for volume results in

$$v_{h_1-h_2} = kd^2 \int_{h_1}^{h_2} \left[\mathrm{b}_0 + \mathrm{b}_1(h_i/h) + \mathrm{b}_2\left(h_i^2/h^2\right)\right] dh_i$$

Solving the definite integral results in

$$v_{h_1-h_2} = \mathrm{k}d^2\left[\mathrm{b}_0 h_i + (\mathrm{b}_1/2)\left(h_i^2/h\right) + (\mathrm{b}_2/3)\left(h_i^3/h^2\right)\right]\Big|_{h_1}^{h_2} \tag{2.6}$$

If the taper equation is integrated over the total bole length (i.e., from 0 to h, where h denotes total tree height), an expression for total tree volume will result. Following through with the same parabolic taper function, we have:

$$\begin{aligned} v &= kd^2\left[\mathrm{b}_0 h_i + (\mathrm{b}_1/2)\left(h_i^2/h\right) + (\mathrm{b}_2/3)\left(h_i^3/h^2\right)\right]\Big|_0^h \\ &= \left\{kd^2\left[\mathrm{b}_0 h + (\mathrm{b}_1/2)\left(h^2/h\right) + (\mathrm{b}_2/3)\left(h^3/h^2\right)\right]\right\} - 0 \end{aligned}$$

Noting that the above expression can be written as

$$\mathrm{k}\left[\mathrm{b}_0 + (\mathrm{b}_1/2) + (\mathrm{b}_2/3)\right] d^2 h$$

and setting $k\,[b_0 + (b_1/2) + (b_2/3)]$ equal to b, the implied volume equation is

$$v = bd^2h$$

which is a logical model for total tree volume v (stem volume equations are discussed in detail in Chap. 3).

In addition to the quadratic equation in relative height, a number of studies have applied higher order polynomials, sometimes with a combination of lower and higher order terms, to describe tree stem profiles (Bruce et al. 1968; Hilt 1980 and others). These studies have used different definitions of relative height. In the following sections of this chapter we will use the notation x for relative height h_i/h and z for all other definitions of relative height; each time z appears, its particular definition will be given. The notation y is used for the dependent variable and, due to the different definitions that have been used by different authors, y will be defined in each particular context.

Hilt (1980) used a modification of the taper model proposed by Bruce et al. (1968) when describing taper for upland oaks:

$$y = z^{\frac{3}{2}} + b_1\left(z^{\frac{3}{2}} - z^3\right)h + b_2\left(z^{\frac{3}{2}} - z^3\right)dh + b_3\left(z^{\frac{3}{2}} - z^{30}\right)d + b_4\left(z^{\frac{3}{2}} - z^{30}\right)dh$$

where

$y = d_i^2/d^2$
$z = (h - h_i)/(h - h_d)$
h_i = height at measurement point
h_d = height to *dbh*

Another functional form for taper equations was proposed by Laasasenaho (1982). The function involves a polynomial in accordance with a Fibonacci series of exponents:

$$y = b_1x + b_2x^2 + b_3x^3 + b_4x^5 + b_5x^8 + b_6x^{13} + b_7x^{21} + b_8x^{34}$$

Where $y = d_i/d0.2h$ and $x = (1 - h_i/h)$. The diameter at 20% of total tree height (d0.2h) was used as the reference diameter in this work, rather than *dbh*.

Other examples of simple taper functions include that of Ormerod (1973):

$$d_i = d\left(\frac{h - h_i}{h - h_d}\right)^b \tag{2.7}$$

which involves only one parameter b. In Ormerod's function h_d is height to *dbh*.

Sharma and Oderwald (2001), in deriving dimensionally compatible volume and taper functions, noted that the following mathematical form can accommodate the overall shape of a tree bole:

$$d_i^{\,2} = ad^{b_1}\left(1 - \frac{h_i}{h}\right)h_i^{\,b_2} \tag{2.8}$$

where b_1 and b_2 are dimensionless parameters to be estimated. The dimensionless term $(1-h_i/h)$ ensures that $d_i=0$ when $h_i=h$. The dimensions of d_i^2, d, and h_i are L^2, L, and L, respectively. Hence, the dimensional form of Eq. 2.8 (where M denotes mass, L denotes length, and T denotes time) can be written as

$$M^0L^2T^0 = M^0L^{b_1+b_2}T^0 \tag{2.9}$$

Therefore, this equation is balanced if and only if $b_1+b_2=2$. The specific values of b_1+b_2 are indeterminate, but their sum must equal two. Consequently, the parameter b_2 can be eliminated from Eq. 2.8 to obtain the taper equation

$$d_i{}^2 = ad^{b_1}\left(1-\frac{h_i}{h}\right)h_i{}^{2-b_1} \tag{2.10}$$

By applying the constraint that $h_i=h_d$ when $d_i=d$ to Eq. 2.10 the constant a equals

$$a = \frac{(d/h_d)^{2-b_1}}{(1-h_d/h)}$$

Substituting this value of a into Eq. 2.10 and rearranging the terms results in the final taper equation of Sharma and Oderwald (2001):

$$d_i{}^2 = d^2\left(\frac{h}{h_d}\right)^{2-b_1}\left(\frac{h-h_i}{h-h_d}\right) \tag{2.11}$$

Thomas and Parresol (1991) used trigonometric based functions to model taper of thinned and unthinned slash pine, oak and sweet gum trees:

$$d_i{}^2/d^2 = b_1(x-1) + b_2\sin(b_3\pi x) + b_4\cot(\pi x/2)$$

where the symbols are as before and b_1–b_4 are parameters to be estimated. The use of trigonometric functions guarantee that the fitted function has an appropriate shape: the first term represents a linear relationship between squared relative diameter and relative height, the second term reflects an increase of diameter in the middle part of the tree and the third gives the neiloid shape to the bottom portion. Comparison of the trigonometric model with the generally used Max and Burkhart (1976) segmented model, presented in the next section, indicated that the equations based on trigonometric functions have a similar performance, with advantages in terms of parsimony and simplicity of integration for volume computation. It is important to note that the trigonometric equation does not predict $d_i/d=0$ when $h_i=h$ $(x=0)$.

Biging (1984) presented a taper equation derived from the Chapman-Richards function to describe stem form for second-growth mixed-conifers in northern California:

$$d_i = d\left[b_1 + b_2\ln\left(1-\left(1-e^{-b_1/b_2}\right)(h_i/h)^{1/3}\right)\right]$$

where symbols are as before. The function is parsimonious concerning the number of parameters and can be integrated to directly yield volumes to any height. Fitting of this equation to six conifer species (ponderosa pine, Douglas-fir, white fir, red fir, sugar pine, and incense cedar) showed performance similar to that of the Max and Burkhart (1976) segmented function fitted to the same data set.

Taper functions provide maximum flexibility for computing volumes of any specified portions of tree stems. Tree boles are highly variable, however, with multiple inflection points and it is difficult to describe their shapes over entire lengths without rather complex functions. Two widely-adopted approaches (segmented taper functions and variable exponent models) aimed at developing more accurate descriptions of tree stems from ground to tip than what simple functions provide will be covered in detail.

2.2.2 Segmented Functions

Although tree boles cannot be completely described in mathematical terms, it is common and convenient to assume that segments of tree stems approximate various geometric solids. The lower bole portion is generally assumed to be a neiloid frustum, the middle portion a paraboloid frustum, and the upper portion a cone (Fig. 2.2). This suggests that three functions are needed to describe expected tree taper, one each for the lower, middle, and upper segments of the bole. These three functions can be joined to form a single model which can be fitted using segmented regression techniques.

Max and Burkhart (1976), using the quadratic function of Kozak et al. (1969) to describe taper in three bole segments, fitted segmented polynomial regression models to describe stem taper for loblolly pine trees. Segmented polynomial models consist of a sequence of grafted or joined sub-models. In the case of one independent variable, the domain is partitioned and a different polynomial sub-model is defined on each section of the partition. These sub-models are then grafted together to form the segmented polynomial model. In general terms, a segmented polynomial model can be written as

$$y = f(x) + e$$

where

$$\begin{aligned} f(x) &= f_1(x, \mathbf{b_1}), \quad && \mathrm{a}_0 \le x \le \mathrm{a}_1 \\ &= f_2(x, \mathbf{b_2}), \quad && \mathrm{a}_1 < x \le \mathrm{a}_2 \\ &= \ldots && \ldots \\ &= f_r(x, \mathbf{b_r}), \quad && \mathrm{a}_{r-1} < x \le \mathrm{a}_r \end{aligned}$$

where the $\mathbf{b_i}$ are vectors of parameters to be estimated. These functions are then grafted together at the join points $a_1, a_2, \ldots, a_r$, by imposing restrictions in such a manner that f is continuous and has continuous first or higher order derivatives.

If the join points are known fixed constants, then f is linear in the unknown parameters $\mathbf{b_i}$'s, and the $\mathbf{b_i}$'s can be estimated using multiple linear regression methods. If the join points must be estimated and $f_1, f_2, \ldots, f_r$ are polynomials then f can be rewritten with the restrictions that f be continuous and have a continuous first partial derivative at each join point. The reparameterized model is linear in the $\mathbf{b_i}$'s but nonlinear in the a_i's, thus non-linear least squares procedures must be used.

Using the notation of Max and Burkhart (1976) for the dependent and independent variables, the quadratic taper model of Kozak et al. (1969), with no restrictions, can be written as:

$$y = b_0 + b_1 x + b_2 x^2 \quad (2.12)$$

where

$$y = d_i^2/d^2$$

$$x = h_i/h$$

Dividing the tree stem into three segments, each to be described by fitting Eq. 2.12, results in

$$y = b_{11} + b_{12}x + b_{13}x^2 \text{ (conoid)} \quad (2.13)$$

$$y = b_{21} + b_{22}x + b_{23}x^2 \text{ (paraboloid)} \quad (2.14)$$

$$y = b_{31} + b_{32}x + b_{33}x^2 \text{ (neiloid)} \quad (2.15)$$

Let the height from ground at join point between (2.13) and (2.14) be a_1 and that of the join point between (2.14) and (2.15) be a_2.

The three submodels should be continuous at the each of the join points, which implies (1) that the two segments are equal at the join points and (2) that the first derivatives at the join points are also equal. A third restriction (3) that should be imposed is that $d_i = 0 (y = 0)$ when $h_i = h (x = 1)$. Based on these restrictions, the equations for the top, middle and bottom segments can be found.

Top segment: If restriction (3) is applied to Eq. 2.13, then

$$b_{11} + b_{12} + b_{13} = 0 \Rightarrow b_{11} = -b_{12} - b_{13}$$

and the equation for the top segment is

$$y = b_{12}(x - 1) + b_{13}(x^2 - 1) \quad (2.16)$$

Middle segment: Applying restrictions (1) and (2) to Eq. 2.14 to guarantee continuity at a_1 implies that

$$b_{11} + b_{12}a_1 + b_{13}{a_1}^2 = b_{21} + b_{22}a_1 + b_{23}{a_1}^2 \tag{2.17a}$$

$$b_{12} + 2b_{13}a_1 = b_{22} + 2b_{23}a_1 \tag{2.17b}$$

Manipulation of Eqs. 2.17a and 2.17b leads to the following expressions for b_{21} and b_{22}:

$$b_{21} = b_{11} - b_{13}{a_1}^2 + b_{23}{a_1}^2 = -b_{12} - b_{13} - b_{13}{a_1}^2 + b_{23}{a_1}^2 \tag{2.18a}$$

$$b_{22} = b_{12} + 2b_{13}a_1 - 2b_{23}a_1 \tag{2.18b}$$

Substituting (2.18a) and (2.18b) in Eq. 2.14 originates the equation for the middle segment:

$$y = b_{12}\,(x - 1) + b_{13}\left(x^2 - 1\right) + (b_{23} - b_{13})\,(a_1 - x)^2 \tag{2.19}$$

Bottom segment: The equation for the bottom segment can be obtained by applying restrictions (1) and (2) to Eq. 2.15 to guarantee continuity at a_2 which implies that:

$$b_{21} + b_{22}a_2 + b_{23}{a_2}^2 = b_{31} + b_{32}a_2 + b_{33}{a_2}^2 \tag{2.20a}$$

$$b_{22} + 2b_{23}a_2 = b_{32} + 2b_{33}a_2 \tag{2.20b}$$

Manipulation of Eqs. 2.20a and 2.20b leads to the following expressions for b_{31} and b_{32}:

$$b_{31} = b_{21} - b_{23}{a_2}^2 + b_{33}{a_2}^2 \tag{2.21a}$$

$$b_{32} = b_{22} + 2b_{23}a_2 - 2b_{33}a_2 \tag{2.21b}$$

Substituting (2.21a) and (2.21b) in Eq. 2.15 yields the following equation:

$$y = b_{21} + b_{22}x + b_{23}x^2 + (b_{33} - b_{23})\,(a_2 - x)^2 \tag{2.22}$$

Noting that the three first terms of Eq. 2.22 correspond to the equation for the middle segment the equation for the bottom segment is:

$$y = b_{12}(x - 1) + b_{13}(x^2 - 1) + (b_{23} - b_{13})(a_1 - x)^2 + (b_{33} - b_{23})(a_2 - x)^2 \tag{2.23}$$

By using two dummy variables I_1 and I_2 and renaming the parameters Eqs. 2.16, 2.19 and 2.23 can be combined into a single model:

$$y = b_1(x-1) + b_2(x^2-1) + b_3(a_1 - x)^2 I_1 + b_4(a_2 - x)^2 I_2 \tag{2.24}$$

where

$I_1 = 1$ if $x \leq a_1, = 0$ otherwise
$I_2 = 1$ if $x \leq a_2, = 0$ otherwise

which is the six-parameter quadratic-quadratic-quadratic form fitted by Max and Burkhart (1976).

In summary a segmented polynomial taper function fitted to three submodels with three parameters each and two join points would require estimation of 11 parameters. By imposing the restriction that diameter at the tip of the tree is zero, the number of parameters is reduced by one. Two further constraints that the grafted functions are continuous and smooth (i.e. first partial derivatives are equal) at the join points reduce the number of parameters by four leaving a total of six parameters to be estimated.

Sharma and Burkhart (2003) further considered the restriction that forces Eq. 2.24 to have continuous second partial derivatives with respect to x at points a_1 and a_2, then:

$$b_{13} = b_{23} = b_{33}$$

Applying this restriction and rearranging the terms, Eq. 2.24 collapses to the two-parameter quadratic taper model of Kozak et al. (1969), namely:

$$y = b_1(x-1) + b_2(x^2-1)$$

Typically, each constraint or restriction results in a decrease of one parameter in the model to be fitted, but that is not always the case as demonstrated here.

Sharma and Burkhart (2003) searched for an optimal specification, that is the minimum number of parameters required, to estimate diameters along the tree stem accurately for the segmented polynomial taper equation of Max and Burkhart (1976). Results from fitting with a large set of measurements of loblolly pine stems showed that an eight-parameter model with minimum constraints (diameter at the tip of the tree equals zero and the adjacent functions are continuous at the join points) did not perform better than the six-parameter model (Eq. 2.24) with an additional smoothness constraint in terms of fit and predictive ability. A four-parameter model with the join points assumed to be known as 11% and 75% of total tree height was slightly superior to the six- and eight-parameter models in estimating tree diameters.

A number of additional studies involving segmented taper functions have been published. For example, Demaerschalk and Kozak (1977) considered two submodels sufficient to describe the tree profile for Douglas-fir and broad leaf maple and found that the inflection (or join) point fell between 20% and 25% of total tree height. In another instance, Fang et al. (2000) described tree profiles by combining three segments and using a segmented-stem taper model based on a variable-form differential equation. They estimated the upper inflection points for slash and loblolly pine trees as 54% and 57% of total tree height, respectively; the lower inflexion point was estimated at 7% of total tree height for both species. Petersson (1999) developed a segmented stem profile model for Scots pine in Scandinavia. A separate function was specified for segments with relative tree heights of 20% and 80% being selected as fixed join points for the sub-functions.

It should be noted that segmented taper functions are a form of spline. The utility of cubic smoothing splines for describing tree stem taper and for estimating volume of segments of stems has been explored by a number of researchers including Goulding (1979), Lahtinen and Laasasenaho (1979), Liu (1980), Figueiredo-Filho et al. (1996a), and Koskela et al. (2006).

2.2.3 Variable-Exponent Functions

Variable-exponent or variable form taper functions are simple continuous functions that describe the shape of tree boles with a varying exponent from ground to top to account for neiloid, paraboloid and conic forms. Taper functions of this type were first developed by Newnham (1988) and Kozak (1988). Since their introduction, a variety of approaches to the variable-exponent taper function form have been developed (Perez et al. 1990; Newnham 1992; Muhairwe et al. 1994; Kozak 1997, 1998; Bi 2000 and others). The model of Kozak (1988), and variations and extensions from that initial publication, will be used to illustrate the variable-exponent approach.

The variable-exponent approach is rooted in the idea that, if the values of z and y are selected appropriately, a simple power function

$$y = z^c \tag{2.25}$$

where c is a "variable-form" exponent that varies along the entire length of the stem profile in order to accommodate the tree stem variation in shape. In developing the variable-form taper function Kozak (1988) imposed the following definitions:

$$y = d_i / dI$$

$$z = \frac{1 - \sqrt{h_i/h}}{1 - \sqrt{p}}$$

where

d_i = diameter at h_i
h_i = height from ground, $0 \leq h_i \leq h$

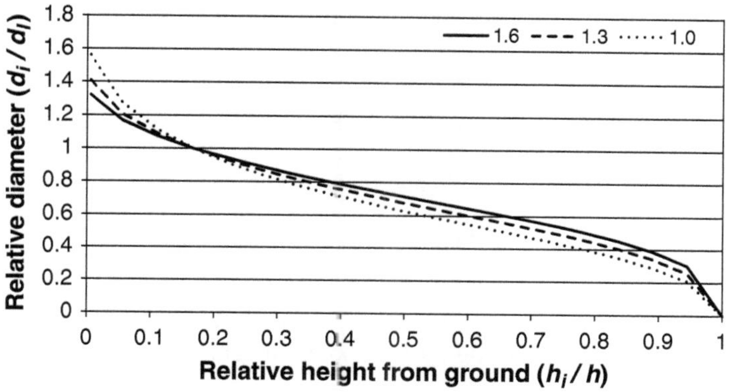

Fig. 2.3 Taper curves obtained with the Eq. 2.25 with different values of *k* (Adapted from Kozak 1988)

h = total height of the tree
hI = height of the inflection point from ground
$p = (hI / h)$
dI = diameter at the inflection point

Kozak's variable-exponent function was developed for diameter inside bark, in which case dI has to be also measured under bark, but the model has been applied either for inside or outside bark diameter prediction.

Demaerschalk and Kozak (1977) indicated that the relationship between d_i/d and h_i/h changes from neiloid to paraboloid at a fixed proportion (p) of the total height of the tree, and they called it the inflection point. They found that the inflection point ranged from 20% to 25% of total height from the ground for their sample of commercial tree species in British Colombia, Canada. Further, the relative height of the inflection point was fairly constant within a species, regardless of tree size. By expressing the exponent c as a function of h_i/h and of a parameter k

$$c = (h_i / h + \mathrm{k})^{-1} \tag{2.26}$$

and by varying k one can obtain a family of curves (Fig. 2.3) that is useful for describing the shape of tree boles. However, to obtain a good fit between d_i/dI and h_i/h, Kozak found the form of the exponent to be much more complicated than indicated in Eq. 2.26. A detailed study of the various forms of the exponent indicated that it can be expressed as a multiple curvilinear regression:

$$c = \mathrm{b}_0 + \mathrm{b}_1 x + \mathrm{b}_2 x^2 + \mathrm{b}_3/x + \mathrm{b}_4 \ln(x + 0.0001) + \mathrm{b}_5\sqrt{x} + \mathrm{b}_6 e^x + \mathrm{b}_7(d/h) \tag{2.27}$$

where $x = h_i / h$

Analysis of Eq. 2.27 for 33 species groups indicated that the best subset of variables is

$$c = \mathrm{b}_2 x^2 + \mathrm{b}_4 \ln(x + 0.0001) + \mathrm{b}_5 \sqrt{x} + \mathrm{b}_6 e^x + \mathrm{b}_7 (d/h) \tag{2.28}$$

Substituting Eq. 2.28 in Eq. 2.25 and renumbering the coefficients, we obtain:

$$d_i/dI = z^{\mathrm{b}_1 x^2 + \mathrm{b}_2 \ln(x+0.001) + \mathrm{b}_3 \sqrt{x} + \mathrm{b}_4 e^x + \mathrm{b}_5 (d/h)} \tag{2.29}$$

In Eq. 2.29 the diameter at the inflection point (*dI*) is not known but can be estimated from diameter outside bark at breast height (*d*). The function used to calculate *dI*, was

$$dI = \mathrm{a}_0 d^{a_1} \mathrm{a}_2{}^d \tag{2.30}$$

Substituting Eq. 2.30 into Eq. 2.29 and rearranging the terms, we get:

$$d_i = \mathrm{a}_0 d^{\mathrm{a}_1} \mathrm{a}_2{}^d z^{\mathrm{b}_1 x^2 + \mathrm{b}_2 \ln(x+0.001) + \mathrm{b}_3 \sqrt{x} + \mathrm{b}_4 e^x + \mathrm{b}_5 (d/h)} = \mathrm{a}_0 d^{\mathrm{a}_1} \mathrm{a}_2{}^d z^c \tag{2.31}$$

Although Eq. 2.31 can be fitted directly with nonlinear least squares, a logarithmic transformation is often employed. Using logarithmic transformation, Eq. 2.31 can be linearized as

$$\begin{aligned} \ln(d_i) &= \ln(\mathrm{a}_0) + \mathrm{a}_1 \ln(d) + \ln(\mathrm{a}_2)\, d + \mathrm{b}_1 \ln(z) x^2 + \mathrm{b}_2 \ln(z) \ln(x + 0.001) \\ &\quad + \mathrm{b}_3 \ln(z) \sqrt{x} + \mathrm{b}_4 \ln(z) e^x + \mathrm{b}_5 \ln(z)(d/h) \\ &= \ln(\mathrm{a}_0) + \mathrm{a}_1 \ln(d) + \ln(\mathrm{a}_2)\, d + c \ln(z) \end{aligned} \tag{2.32}$$

Coefficients in (2.32) can be obtained using multiple linear regression methods. The properties of this model are: (*i*) $d_i = 0$ when $h_i/h = 1$ (top of the tree); *(ii)* $d_i = dI$ (estimated) when $h_i/h = p$; *(iii)* the function changes direction when $h_i/h = p(z = 1)$.

The variable-exponent taper function has been very effective for estimating diameters throughout the length of the bole but it has the drawbacks that (1) it cannot be analytically integrated to compute stem volumes, (2) the expression for the exponent of x has multiple terms in x implies that the model is likely to exhibit strong multicollinearity (see Sect. 2.6.2), and (3) iterative methods must be used to estimate heights at specified stem diameters. However, it should be noted that numerical techniques for obtaining volume and height estimates are now reasonably easy to implement.

After the pioneering works of Newnham (1988, 1992) and Kozak (1988, 1997) to derive variable-exponent taper functions, other authors followed the same line of thinking but using different base taper equations and/or exponents. Sharma and Zhang (2004) developed a variable-exponent taper equation for jack pine, black spruce, and balsam fir trees grown in eastern Canada based on the dimensionally compatible taper equation of Sharma and Oderwald (2001) presented in Eq. 2.11 by expressing the parameter b_1 of the exponent as a quadratic function of relative height $x = h_i / h$:

$$d_i^2 = d^2 \left(\frac{h_i}{h_d} \right)^{2-(b_1+b_2 x+b_3 x^2)} \left(\frac{h - h_i}{h - h_d} \right) \tag{2.33}$$

In an effort to overcome the weakness of unstable specification in the variable-form taper models introduced by Newnham (1988, 1992) and Kozak (1988, 1997), Bi (2000) constructed the base function from trigonometric equations. Bi's specification for the exponent includes variables for detecting changes in stem form along the bole and variables for taking into account differences in stem form among trees of different sizes. When fitted to data from 25 species of Australian eucalyptus, the model was found to be stable in specification, flexible in fitting data for trees of varying species and with varying stem forms, and accurate in predictions of taper and merchantable height.

Valentine and Gregoire (2001) developed a variable-exponent taper function using numerical switches to achieve the variable exponent. Numerical switches are functions widely used in process models of plant growth (e.g. Thornley and Johnson 1990) to switch on or switch off some process. In taper modeling, these functions can be used to switch from a neiloid to a paraboloid and then to a cone. The authors considered each bole segment modeled with a modification of Ormerod's (1973) function (Eq. 2.7) that uses tree basal area as the dependent variable instead of diameter, with the needed restrictions to guarantee continuity at the join points (*dbh* and height to the base of the crown):

$g_i = g\left(\frac{h-h_i}{h-h_d}\right)^{b_1}$	Middle segment (paraboloid if $b_1 \sim 1$, $h_d \leq h_i \leq h_{cb}$)
$g_i = g\left(\frac{h-h_i}{h-h_d}\right)^{b_1}\left(\frac{h-h_i}{h-h_{cb}}\right)^{b_2}$	Top segment (cone if $b_1 + b_2 \sim 2$, $h_i > h_{cb}$)
$g_i = g\left(\frac{h-h_i}{h-h_d}\right)^{b_{00}+b_1}$	Basal segment (neiloid if $b_{00} + b_1 > 2$, $h_i < h_d$)

where:

g_i = cross-sectional area at height above ground h_i
g = breast height basal area
h_{cb} = height to the base of the crown

$$b_{00} = b_0 h$$

and other symbols have the definitions given before.

The switching model for the whole bole was:

$$g_i = g \left(\frac{h - h_i}{h - h_d} \right)^{b_1 + S_1(h_i)} \left(\frac{h - h_i}{h - h_{cb}} \right)^{b_2 S_2(h_i)} \tag{2.34}$$

where

$$S_1(h_i) = \frac{\mathrm{a}_0}{1 + [h_i/(\mathrm{a}_1 h)]^{\mathrm{a}_2}}, \quad 0 < \mathrm{a}_1 < h_{cb}/h$$

$$S_2(h_i) = \frac{(h_i/h_{cb})^{\mathrm{a}_3}}{1 + (h_i/h_{cb})^{\mathrm{a}_3}}$$

Switching off function $S_1(h_i)$ decreases in value from a_0 towards 0 as h_i increases in value from 0 to h, while switching on function $S_2(h_i)$ increases from 0 to 1 as h_i increases, being = 1/2 when $h_i = h_{cb}$. If $h_i = 0$ the equation reduces in form to the neiloid. A transition from neiloid to paraboloid occurs with the decrease in value of $S_1(h_i)$ with increasing height. $S_1(h_i) = \mathrm{a}_0/2$ when $h_i = \mathrm{a}_1 h$. Midway between $h_i = \mathrm{a}_1 h$ and $h_i = h_{cb}$ (where $S_1(h_i)$ approaches 0 and $S_2(h_i) \sim 1$), the bole takes the form of a paraboloid. The second numerical switch is centered at h_{cb}. When $h_i > h_{cb} S_1(h_i)$ will be close to 0 and $S_2(h_i)$ will approach 1, implying that Eq. 2.34 reduces in form to the cone. This taper model proved to be reasonably precise for ponderosa pine, slash pine, sweet gum and yellow poplar, with easily interpreted parameters.

Westfall and Scott (2010) used a modification of the switching model of Valentine and Gregoire (2001) by incorporating estimated join points and modifying one of the switching functions. Fitting the model to data from 19 species groups collected across the northeast United States indicated that the model generally performed well and that the modifications made to the original model allowed for significant improvement in describing the observed data.

2.3 Inclusion of Additional Predictor Variables

In developing tree stem taper and volume equations researchers have strived for a system that is simple, accurate, and flexible while requiring only a few easily measured tree characteristics such as diameter at breast height and total height as predictors. However, even for a given species, there is considerable variation in form among individual trees.

Management practices, such as thinning, pruning, and applying fertilizers, also affect tree form and stem taper. Hence there has been considerable interest in including additional predictor variables to better account for tree-to-tree variation. These additional or auxiliary variables have generally taken one of three forms: (1) crown dimensions, (2) stand and site variables, or (3) upper stem diameter measurements.

2.3.1 Crown Dimensions

Crown dimensions have been considered as auxiliary variables for describing tree profiles because of the relationship between crown and stem form development (Larson 1963). Because variations in bole form can be attributed to changes in the size of the live crown, its distribution along the stem, and the length of the branch-free bole (Larson 1963), a number of attempts have been made to include crown dimensions, along with d and h, in taper functions.

Burkhart and Walton (1985) fitted a taper model in *dbh* and total height after dividing the data into crown ratio (c_r) classes and examined the relationship between mean crown ratio of the classes and the coefficients of the taper model. Trends of coefficients over discrete crown ratio can be useful for specifying an appropriate model for inclusion of crown ratio as a continuous variable. After dividing the data into three groups in a way that the number of trees in each category would be approximately equal, the relatively simple model of Kozak et al. (1969) presented in Eq. 2.12 was fitted to the data in each crown ratio class. As crown ratio (c_r) increased, parameter estimates increased in absolute value, indicating that the parameters are related to crown ratio. The variables c_r, xc_r and x/c_r were added to the second degree polynomial in relative height (x) and used jointly with x and x^2 in a stepwise regression procedure to find the best subset of variables to predict y. All three variables containing crown ratio entered the equation after the relative height terms (0.10 probability level), but the percentage of variance explained increased only from 87.8 to 90.1 by adding c_r, xc_r and x/c_r after x and x^2. From these results, it was concluded that either (1) the functions of crown ratio tested in the stepwise procedure were not appropriate for incorporating the effect of crown ratio, or (2) the effect of crown ratio on parameters in this particular model is not marked.

In a second attempt to incorporate crown ratio into a taper model, Burkhart and Walton (1985) fitted the segmented taper function of Max and Burkhart (1976) – Eq. 2.24 – to loblolly pine trees after dividing the data into crown ratio classes. There were no discernable trends in the submodel coefficients (the b_i values), but the upper join point showed a trend with crown ratio. In general, the upper join point was estimated to occur lower on the stem for trees with proportionally larger crowns, a logical trend from a biological standpoint as the point of maximum annual ring increment occurs lower on the stems of large-crowned trees (Farrar 1961). One would expect the location of the estimated upper join point to be related to this point of maximum ring width and, hence, lower on the stems of trees with large crowns. The lower join point was less affected by crown ratio than the upper one. Linear and nonlinear functions relating the join points to crown ratio were specified and the following alternatives were examined: neither join point written as a function of crown ratio; either the upper join point or the lower join point written as a function of crown ratio; both join points written as functions of crown ratio. The reduction in the error sum of squares due to adding crown ratio after *dbh* and total height was slight and it was concluded that the inclusion of crown ratio as a predictor variable would not be warranted for most applications.

In another study involving the segmented taper function, Valenti and Cao (1986) evaluated various ways to express the coefficients (b_i) and the join points (a_i) of the segmented model of Max and Burkhart (1976) as functions of crown ratio. The best combination resulted from specifying b_i and a_i as functions of crown ratio. While the resulting three-segment taper equation with crown ratio as an additional independent variable was more flexible and provided more accurate predictions of upper stem diameters, the gains in fit statistics were modest (percent of variation explained increased from 96.6 for the base model to 96.9 for the model with two parameters written as functions of crown ratio). The best model with crown ratio included required estimation of eight parameters as opposed to six for the base form involving *dbh* and total height only as predictors. A larger number of parameters to be estimated by nonlinear least squares, associated with high correlation among the predictor variables, can lead to difficulties in obtaining stable solutions because of multicollinearity.

The use of tree age, height to the base of the crown, crown length, as well as combinations of these variables with tree height and diameter and a distance-independent competition index were introduced in the model of Petersson (1999) as auxiliary variables. The best model with auxiliary variables included height to the base of the crown, tree age and the slenderness coefficient (*d/h*). Even if the coefficients of the auxiliary variables were significantly different from zero, their introduction in the model led to a slightly higher residual mean square error.

Leites and Robinson (2004) also used the Max and Burkhart (1976) segmented taper equation to explore relationships between taper parameters and crown variables. They used mixed-effects techniques to fit the model and tested the regression of the random parameters against crown variables (crown length and crown ratio). The best combination of mixed effects was that in which the random-effects parameters affected both linear and quadratic terms of the equation submodels. Supplementing parameters b_1 and b_2 of Eq. 2.24 by crown variables improved the equation fit but the additional improvement from adding random effects indicated that additional variation might be explained by some other variables. Validation with an independent data set showed a slight improvement on bias and precision for the stem middle section but for the upper stem section bias was slightly increased by adding these variables.

As a general conclusion of this section it can be said that the combination of *dbh* and total height provides a great deal of information about crown size for various species of trees. Consequently, including crown variables as predictors is generally not warranted due to the practical difficulties and costs of obtaining crown measurements coupled with a low level of increased precision in the statistical fit to data.

2.3.2 Site and Stand Variables

Crown development is affected by site and stand conditions, which suggests inclusion of site and stand variables in stem profile equations might prove effective.

Certain site and stand measures could potentially be available or could be computed as part of an inventory summary, thus obviating the need for taking additional standing-tree measurements in addition to *dbh* and total height. Muhairwe et al. (1994) examined the effect of including crown class, site class, and age in the exponent part of Kozak's (1988) variable-exponent taper equation for three species: Douglas-fir, western red cedar, and aspen. Effects on prediction of diameter along the stem, total stem volume, and three merchantable heights were assessed. The additional variables resulted in only marginal improvements to the published version of Kozak's taper function, and the authors concluded that including these variables would not be justified in practical implementation. Muhairwe et al. (1994) noted that the exponent of Kozak's variable-exponent function includes the ratio d/h which is highly correlated with crown dimensions, site and stand variables. Models that include d/h may not be improved by addition of variables with which the ratio is strongly related.

Site index and stand variables were also tested as auxiliary variables in the study of Petersson (1999), described previously, but none of these variables was included in the selected model, as tree variables (age, height to the base of the crown and the d/h ratio) performed better.

Sharma and Zhang (2004) modified Eq. 2.33 to accommodate stand density effects. Since the exponent of the height solely determines the tree form, the stand density effect can be incorporated by adding a stand density function to the exponent, that is:

$$d_i^2 = d^2\left(\frac{h_i}{h_d}\right)^{2-(b_1+b_2x+b_3x^2+b_4g(SD))}\left(\frac{h-h_i}{h-h_d}\right) \tag{2.35}$$

where $g(SD)$ is a function of stand density and b_4 is a parameter to be estimated. The best function that described the stand density effect in this case was

$$g(SD) = 1/N$$

where N is number of trees per ha at the time of plot establishment.

The stand density variable was not effective for estimating stem taper in jack pine or balsam fir, but the taper of black spruce trees was affected by stand density or thinning (Fig. 2.4).

In a following study, Sharma and Parton (2009) used a similar methodology, i.e. a dimensional analysis approach, to model taper in jack pine and black spruce plantations growing at varying stand densities. Different ways of expressing stand density were examined and the model obtained was:

$$\frac{d_i}{d} = b_0\left(\frac{h_i}{h_d}\right)^{b_1+b_2x+b_3x^2+b_4\sqrt{G}/d}\left(\frac{h-h_i}{h-h_d}\right)$$

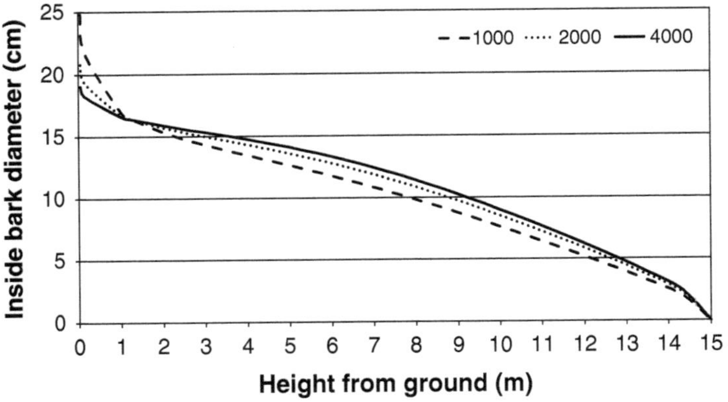

Fig. 2.4 Tree profiles generated from Eq. 2.35 using $d = 17$ cm and $h = 15$ m at different stand densities (1,000, 2,000, and 4,000 trees ha^{-1}) for black spruce (Adapted from Sharma and Zhang 2004)

where G is stand basal area and other symbols are defined as before. The density effect on taper was more pronounced for jack pine than for black spruce trees, but significant for both.

2.3.3 *Upper-Stem Diameters*

After *dbh* and total height the most informative additional variable for prediction of stem taper is an upper-stem diameter measurement (that is a stem diameter in the section between *dbh* and the top of the tree). Czaplewski and McClure (1988) constrained the segmented taper function of Max and Burkhart (1976) to pass through *dbh* and an upper stem diameter measured outside bark at 5.3 m above ground. Variance of residuals was reduced, but bias was approximately the same for both the conditioned and unconditioned models. The authors concluded that the reduction in variance may be of practical importance, depending on the particular prediction objectives and the relative cost of constraining an upper stem measurement compared to the cost of measuring more trees.

Kozak (1998) examined the impact of an additional upper-stem diameter outside bark measurement on predictive ability of his variable-exponent taper equation (Kozak 1988). His analysis indicated that improvements were small and were mainly restricted to increasing the precision of the estimates. It was also demonstrated that additional diameter measurements should be taken between 40% and 50% of the height above breast height for greatest improvement. Measurement errors in upper stem diameters and in their heights affected both the precision and bias of predictions.

Other work on using upper-stem measurements in stem taper modeling includes that of Rustagi and Loveless (1991) who developed compatible variable-form volume and stem-profile equations for Douglas-fir. Flewelling and Raynes (1993) and Flewelling (1993) proposed a multipoint stem profile system that conditioned the models to pass through measured upper stem diameters.

Cao (2009) pointed out that with the recent advances in laser technology, accurate and affordable upper-stem diameters can be obtained, making its use more appealing in taper functions. Therefore he developed and evaluated two methods for calibrating predictions from the Max and Burkhart (1976) segmented model, one for *dbh* and another for both *dbh* and an upper-stem diameter. The calibration was obtained by imposing restrictions to some of the parameters (arbitrarily chosen) so that the predictions at certain values of relative height equal the measured diameters. Results were promising for outside-bark diameter but varied depending on where the diameter was measured, with optimum gains when the upper-stem diameter was measured at the midpoint between breast height and the tree tip. He also tested a calibration for inside-bark diameters but in this case calibration with only *dbh* produced inferior predictions, whereas the calibration based on both *dbh* and an upper-stem diameter offered only modest improvements over the unadjusted predictions.

Taper equations have been fitted using mixed-models techniques in order to take into account systematic and random variation associated with the jth observation (diameter measurement) along the bole for the ith individual (tree). One of the advantages of the use of mixed-models is the possibility to calibrate the parameters for each future prediction by estimating the random components of the parameters from a subsample of a particular tree. This allows the use of upper-stem diameters to improve the quality of the predictions, without considering them explicitly in the fitted model. Examples of calibration of parameters for future predictions with the measurement of a subsample of upper-stem diameters are given in Trincado and Burkhart (2006) and Sharma and Parton (2009).

Cao and Wang (2011) evaluated two approaches for incorporating the midpoint upper-stem diameter to improve the accuracy of diameter predictions along the bole of loblolly pine trees: (1) calibrating a segmented taper equation by constraining a parameter, and (2) localizing the taper model by predicting random effects for each tree. The calibration technique is simpler and produced less-biased prediction of diameters. Results from calibration were similar for both fixed- and mixed-effects taper models; however, a slight gain in accuracy and precision was attained with the mixed-effects model.

2.4 Compatible Prediction of Inside and Outside Bark Diameters

Taper functions may be used to predict diameters inside bark or outside bark. When both diameters are of interest (such as in the computation of bark volume) it is important to guarantee the compatibility between the two estimates (i.e., the relationship between the estimates of diameter over and under bark at the same height above ground must be logical).

In the development of a system for prediction of total and merchantable volumes allowing for different definitions of tree volume, Nunes et al. (2010) proposed a set of compatible models for over and underbark diameters (d_i and du_i, respectively) along the bole by assuming that an upper stem diameter over bark can be effectively obtained by adding the bark thickness to the respective stem diameter under bark, both modeled by the equation of Demaerschalk (1973):

$$du_i = \mathrm{a}_1 d^{\mathrm{a}_2} \frac{(h - h_i)^{\mathrm{a}_3}}{h^{\mathrm{a}_4}}$$

$$d_i = du_i + \mathrm{a}_5 d^{\mathrm{a}_6} \frac{(h - h_i)^{\mathrm{a}_7}}{h^{\mathrm{a}_8}}$$

In their model Nunes et al. used heights above stump but the same methodology can be applied with heights above ground and other taper models can be used for estimating diameter under or over bark.

2.5 Taper-Volume Compatible Functions

Demaerschalk (1972, 1973) defined a taper and volume system as compatible when the integration of the taper equation yields the same total volume as that given by the volume equation. Taper and volume data are often considered independently when, in fact, they can be analyzed as mathematically dependent quantities. An important benefit of compatible taper-volume models is that numerically consistent results are obtained. Another advantage of compatible systems is that appropriate taper models may be suggested through knowledge of volume models and vice versa. Here we introduce the concept of taper-volume compatibility for the case where integration of the taper function over the entire bole length gives an estimate of total stem volume that is the same as that given by a stem volume equation. More complex systems of equations involving total and merchantable volume as well as stem taper are addressed in Chap. 3.

Given a taper function that expresses diameter (d_i) at any given height above ground (h_i) total stem volume (v) can be obtained as

$$v = \mathrm{k} \int_0^h d_i^{\,2} dh$$

where h is total tree height and k is a constant that includes $\pi/4$ multiplied by a factor to obtain the desired units of volume. One can fit a taper equation and then integrate for volume but estimates for volume are generally biased and not fully efficient. Alternatively, a volume equation can be fitted and an implied taper function can often be derived. In this case the taper equation will not be "best" in a statistical sense. Hence there has been much interest in treating taper-volume relationships as a system of interrelated equations.

Cao et al. (1980) modified Max and Burkhart's (1976) segmented taper function to be approximately compatible with total volume by fitting

$$(kd_i^2h/v - 2z) = b_1(3z^2 - 2z) + b_2(z - a_1)^2I_1 + b_3(z - a_2)^2I_2 \tag{2.36}$$

where $z = (h - h_i)/h$, relative height from the tip to top diameter d_i, and other symbols remain as previously defined.

Integrating model (2.36) over total tree height gives an estimate of total volume

$$\begin{aligned} v &= \hat{v}\left[1 + b_2(1 - a_3)^3/3 + b_3(1 - a_2)^3/3\right] \\ &= \hat{v}c \end{aligned} \tag{2.37}$$

where $\hat{v}$ is tree total volume estimated from $v = b_1' + b_2'd^2h$ and

$$c = 1 + b_2(1 - a_1)^3/3 + b_3(1 - a_2)^3/3$$

The values for c ranged from 0.9896 to 0.9959 for the four combinations of sample and measurement data fitted; thus the model was labeled "essentially" a compatible taper equation.

McClure and Czaplewski (1986) noted that by imposing an additional constraint, which decreased the number of parameters to be estimated by one, exact compatibility in Cao et al.'s formulation can be ensured with the following:

$$0 = (1 - a_1)^3b_2 + (1 - a_2)^3b_3$$

When ranking candidate taper models for different prediction objectives, Cao et al. (1980) found that the segmented taper function of Max and Burkhart (1976) performed best for estimating stem diameters but the segmented taper-volume model was better for estimating volumes to various heights and various top diameters. The choice of model and of which parameters are fitted and which are derived when compatibility is desired depends on the objectives for use of the results. It is common to have multiple prediction objectives, hence fitting the models as a system of equations with no individual component being "optimized" but rather attempting to minimize prediction error for the system is often an attractive alternative. Approaches to fitting taper-volume relationships as a system of equations are explored more completely in Chap. 3.

2.6 Statistical Considerations

2.6.1 *Model Assumptions*

As when fitting any linear or non-linear regression models, the regression assumptions in relation to the model error must be checked when fitting taper models.

Data required to fit taper equations imply the measurement of several diameters along the bole of each individual of a set of sample trees. These characteristics of the data imply that the observations from the same individual (tree) are spatially correlated, violating the assumption of independent errors. Correlation between residuals from the same individual (within-individual correlation) in taper equations has been reported by several authors (e.g. Valentine and Gregoire 2001; Garber and Maguire 2003; Trincado and Burkhart 2006; Rojo et al. 2005). One way to overcome this problem is to model the error structure with some type of autocorrelation structure. As the intervals between measurements within the tree are often not regular, a continuous autoregressive structure (CAR(p)) has been often used (e.g. Rojo et al. 2005). Some authors found that fitting the model with mixed-effects techniques (see Sect. 2.6.4) would take care of the within-individual correlation (Valentine and Gregoire 2001) while others found that the model was improved by incorporating, even when using mixed-modeling techniques, an autoregressive continuous autocorrelation structure into the model (Garber and Maguire 2003; Trincado and Burkhart 2006).

Heteroscedasticity has also been reported when fitting taper curves. It is usual to fit taper curves using a unitless variable such as d_i/d or $(d_i/d)^2$ as the dependent variable and this option has been found to reduce or overcome heteroscedasticity. Although it is usually assumed that variance of d_i/d or $(d_i/d)^2$ is homogeneous, that is not generally strictly true; variance is generally related with changing levels of relative height h_i/h. Additionally, the use of a transformed dependent variable has unwanted implications due to the problem of re-transformation bias (Sect. 2.6.3).

More recently, taper curves have been fitted with d_i or d_i^2 as the dependent variable leading to the need to use weighted non-linear regression to homogenize variance of the model errors. When using mixed-modeling techniques both within- and between-tree variance must be considered when selecting the variance functions to be used as weights in the fitting procedure (e.g. Trincado and Burkhart 2006).

2.6.2 *Multicollinearity*

Multicollinearity in multiple linear regression occurs when there are strong linear dependencies among the independent variables. Regression coefficients for models exhibiting strong multicollinearity may not be precisely estimated, thus resulting in areas of the regressor space where prediction could be poor. Taper equations with a large number of independent variables that might be correlated have been used in several studies (e.g. Bruce et al. 1968; Laasasenaho 1982; Kozak 1988). Potential problems of multicollinearity should be considered when selecting a taper curve for a particular data set.

For example Kozak's variable-exponent (Eq. 2.32) requires estimation of eight parameters and, given that the expression for the exponent of x has multiple terms in x, is likely to exhibit strong multicollinearity. One means to combat multicollinearity is to eliminate variables but not to the point that the quality of fit is severely compromised. With this goal in mind, Perez et al. (1990) explored possibilities for specifying a more parsimonious version of Kozak's model. Perez et al. (1990) fitted alternative variable-form taper models using multiple linear regression following logarithmic transformation and assuming that the inflection point occurred at 25% of total height. Setting the location of the inflection point at 15%, 20%, 30%, and 35% of total height had little effect on the predictive properties of any model. From the potential models with comparable fitting quality, preference was given to the one with the least number of parameters. The "best" or most parsimonious model that did not show much loss in predictive ability as compared to the "full" model of Kozak (1988) had the following form:

$$d_i = b_0 d^{b_1} z^{b_3 x^2} + b_4 \ln(x + 0.001) + b_5 (d/h) \tag{2.38}$$

where symbols are as defined before.

When fitting Eq. 2.38, after logarithmic transformation, to data from *Pinus oocarpa* in Honduras, the regressor variables did not show strong linear dependence and, as a result, their coefficient estimates are expected to be more stable because of less inflated variances. Fit statistics for models (2.32) and (2.38) were basically equivalent, with $R^2 = 0.959$ for both models. When tested with data that were withheld and not used in the estimation of parameters the mean bias and standard deviation of the two models were very similar.

2.6.3 Retransformation Bias

When fitting some of the taper models presented in the previous sections, for instance the variable-exponent Eq. 2.32, a commonly applied transformation involves taking the logarithm of both sides of the prediction equation. The logarithmic transformation allows the use of linear regression and, in many cases, reduces heterogeneity of the variance. Predictions are, however, desired in arithmetic units and retransformation of the fitted equation results in bias. The use of a logarithm transformation results in under estimates of the y variable and therefore some type of correction is often used.

Using plant biomass prediction as a specific example, Baskerville (1972) pointed out that the use of a logarithm transformation results in under estimates of biomass. Baskerville recommended the retransformed estimator

$$\hat{y} = e^{\hat{\mu}} e^{\hat{\sigma}^2/2}$$

where $\hat{y}$ is the estimated mean in arithmetic units, $\hat{\mu}$ is the estimated mean of y for a given level of x after logarithmic transformation, and $\hat{\sigma}^2$ is the estimated variance about the fitted regression after logarithmic transformation.

Flewelling and Pienaar (1981) reviewed several estimators for the correction of bias associated with the fitting of models in logarithmic form. Logarithmic transformation has often been applied in biomass estimation and the problem of the transformation bias is generally addressed; however, the same issues apply to other prediction objectives such as fitting tree profile equations. When a model is fitted in logarithmic form, assuming a normal error for the logarithmic transformed model implies that the error associated with the multiplicative model will be lognormal. Flewelling and Pienaar analysed the properties of different estimators and related them to the intended use of the regression equations and characteristics of the sample data.

Snowdon (1991) noted that the estimator recommended by Baskerville (1972) is consistent but itself biased with a tendency to overestimate the true proportional bias. He suggested that the bias in logarithm regressions could be estimated from the ratio of the arithmetic mean of the back-transformed predicted values from the regression. Under the assumption of a lognormal distribution of errors, the conditions of application of this ratio estimator are optimal. A simulated sampling study has shown that this method gives more reliable results than the method recommended by Baskerville.

In addition to logarithmic transformation, the dependent variance in stem profile equations is often expressed as a dimensionless variable, for example as d_i/d or $(d_i/d)^2$. While these transformations tend to reduce heterogeneity of variance, predictions of these transformed dependent variables must be subsequently re-transformed to produce estimates of d_i and d_i^2, which are needed to predict stem diameters and volumes. In the case of relative diameter and relative basal area, retransformations of model predictions are also generally biased.

Czaplewski and Bruce (1990) pointed out that if d is a known value an unbiased estimate of $(d_i/d)^2$ can be directly transformed to an unbiased estimate of cross-sectional area and therefore an unbiased estimate of volume. However, the squared retransformation of an unbiased estimate of d_i/d will produce a biased estimate of $(d_i/d)^2$ and hence a biased estimate of volume.

When illustrating the possible extent of retransformation bias, Czaplewski and Bruce (1990) fitted the segmented regression model of Max and Burkhart (1976) for two dependent variables, d_i/d and $(d_i/d)^2$ (Table 2.1). The segmented regression taper equation with d_i/d as the dependent variable produced volume estimates that were consistently higher than those for the same model that used $(d_i/d)^2$ as the dependent variable. The mean difference in estimated log volume varied from 0.5% to 1.0% for butt and midbole logs and from 1.0% to 3.5% for the top logs.

The fitted segmented regression models were also used to estimate upper-stem diameters (d_i). The square-root retransformation was applied to the predictions that involve $(d_i/d)^2$. The fitted profile model with $(d_i/d)^2$ as the dependent variable produced retransformed diameter estimates that were smaller than those from the same model fitted to d_i/d (differences averaged 0.08–0.30 cm (0.2–0.7%) for butt and midbole log end diameters and 0.12–0.56 cm (0.7–2.1%) for the merchantable top diameter). The authors proposed equations for the reduction of the magnitude of the above mentioned biases but pointed out the need to develop accurate variance models.

Table 2.1 Estimated coefficients for the Max and Burkhart (1976) stem profile equation using data from grand fir (From Czaplewski and Bruce 1990)

Regression response variable	b_1	b_2	b_3	b_4	a_1	a_2	MSE[a]
d_i^2/d^2	−2.010	0.957	−0.858	118.5	0.722	0.062	0.01113
d_i/d	−2.162	0.942	−0.858	121.1	0.699	0.061	0.00415

[a]Residual mean squared error (*MSE*) has dimensionless units of $(d_i/d)^4$ for the $(d_i/d)^2$ model, and $(d_i/d)^2$ for the d_i/d model

Gregoire et al. (2000) investigated the statistical properties of volume predictions from an integrated taper model. They derived the expected volume and its prediction variance when the underlying taper model of stem cross-sectional area was linear in its parameters and had a Gaussian error distribution. The integration of the taper model for this special case does not insinuate a bias into the derived model of stem volume. When the underlying Gaussian taper model for cross-sectional area is nonlinear in its parameters, there is the usual bias which accrues from the nonlinearity, but not from the integration of the taper model itself.

When the underlying model portrays the taper of stem diameter rather than cross-sectional area, the integral of the squared taper model to predict volume is considerably more complicated. As Gregoire et al. (2000) commented, the prediction variance of volume in this situation involves the first four moments of the error distribution of the taper model, making an analytic determination of the statistical properties of volume predictions rather intractable.

2.6.4 Mixed-Effects Approach

Requisite data to fit taper models are clustered in the sense that multiple measurements of diameter, d_i, at successively greater heights, h_i, form a cluster of data from each tree. Taper models customarily are fitted to clusters of data from multiple trees, so that even if observations from different trees are statistically independent, there is the strong possibility that data within a tree are correlated. If the within-tree co-variance structure is not modeled appropriately, a bias is introduced into the estimates of standard errors of the fixed-effect parameter estimates and the related hypothesis tests and confidence interval estimations (either for model parameters or for prediction of stem taper) are invalid. Whereas ignoring the co-variance structure does not introduce bias into the estimates of the fixed-effect parameters or predictions from the fitted model, these statistics are likely to be less precise than they would be under an appropriately specified covariance matrix.

An alternative to modeling the so-called marginal covariance structure when dealing with clustered data is to account for this correlation structure by introducing subject-specific random effects into the model. This technique is facilitated by considering stem profile data as repeated measurements done along the stem of individuals. Mixed-effects models account directly for within- and between-tree variation in stem form. An important characteristic in contrast to traditional regression is that mixed-effects modeling allows for both *population-specific* and *subject-specific* models. A *population-specific* model considers both fixed and random effects parameters. If prior information for the response variable is available for a new individual k, then random-effects parameters can be predicted and an adjusted response (*subject-specific*) rather than a mean response (*population-specific*) can be obtained. A population-specific response can be obtained by assuming that the vector of random effects $\mathbf{b}_k$ for a new individual k has expected value $E(\mathbf{b_k}) = 0$.

A very readable introduction to both linear and nonlinear mixed-effects models appears in Schabenberger and Pierce (2002; Chapters 7 and 8).

An initial application of mixed-effects modeling for stem form was presented by Lappi (1986). In this novel application Lappi specified a mixed linear model to analyze and predict variation in the taper of Scots pines by decomposing the variation in diameters along the bole into one component due to random stand effects and another component due to individual tree effects. That is, he was able to distinguish between-stand effects from within-stand effects. Since that time, a number of additional researchers have used mixed linear and nonlinear models for stem taper.

The mixed-effects model has a number of appealing features, and it has been the subject of many recent treatises both within the biometrical forestry literature (Tasissa and Burkhart 1998; Valentine and Gregoire 2001; Trincado and Burkhart 2006; Garber and Maguire 2003; Leites and Robinson 2004; Eerikäinen 2001; Yang et al. 2009) and the wider statistical literature.

In the following paragraphs, we summarize for illustrative purposes the nonlinear mixed-effects model for upper-stem diameter presented by Trincado and Burkhart (2006). Letting y_{ij} symbolize the square of d_i/d at the jth height, h_{ij}, on the ith tree, a segmented model for y_{ij} was put forth as

$$\begin{aligned} y_{ij} = {} & (\beta_1 + \mathrm{b}_{1i})(x_{ij} - 1) + (\beta_2 + \mathrm{b}_{2i})({x_{ij}}^2 - 1) + (\beta_3 + \mathrm{b}_{3i})(\alpha_1 - x_{ij})^2 \mathrm{I}_1 \\ & + (\beta_4 + \mathrm{b}_{4i})(\alpha_2 - x_{ij})^2 \mathrm{I}_2 + \varepsilon_{ij} \end{aligned} \tag{2.39}$$

where $x_{ij} = h_{ij}/h_i$ is the relative height of the measurement and other variables are as before. In (2.39), the $\beta_1, \ldots, \beta_4$ and α_1, α_2 are the fixed-effects parameters and the $\mathrm{b}_{1i}, \ldots, \mathrm{b}_{4i}$ are random effects specific to the ith tree.

Letting, $\mathbf{b_i} = (\mathrm{b}_{1i}, \ldots, \mathrm{b}_{4i})'$ they specified that

$$\mathbf{b_i} \sim N(\mathbf{0}, \mathbf{D})$$

with $E(\mathbf{b_i}) = \mathbf{0}$ and where $\mathbf{D}$ is unstructured covariance matrix of the random effects. Given the random effects, the errors ε_{ij} were assumed to be mutually uncorrelated. The β's and α's are fixed effect parameters to be estimated, in addition to σ^2 and whatever parameters are stipulated in $\mathbf{D}$.

The inclusion of a vector of random effects permits the model of tree taper to be individualized to each tree, while also permitting a pooling of all the data when

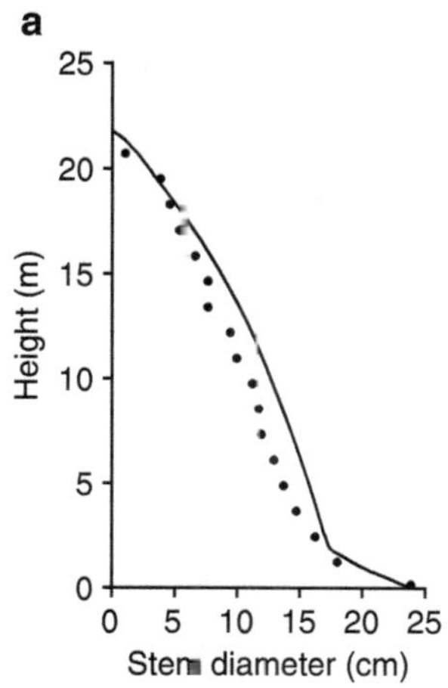

Fig. 2.5 Stem profile curves generated using (**a**) mean and (**b**) calibrated response based on two upper-stem diameters at 3.6 and 6.1 m for one sample tree ($d = 17.8$ cm and $h = 21.8$ m). The stem diameters used for calibration are represented by *open circles* (Adapted from Trincado and Burkhart 2006)

fitting the model. As noted by Gregoire et al. (1995), the specification of random effects does not presume equally spaced measurements along the stem, nor that the number of measurements on each stem be identical. The flexibility permitted by a mixed model of this sort comes at the price of added complexity, both statistical and computational. Importantly, the random individual effects induce an intra-individual correlation structure that accounts for the lack of independence among measurements on the same stem. It is this feature that has great appeal for modelers of correlated data, even though the induced correlation function may not be easily discernible (Schabenberger and Gregoire 1996). In the applications of mixed-effects models in the references given earlier in this section, such models were superior to purely fixed-effects models when using likelihood or information criteria to quantify the quality of the model to data.

It is possible to predict the value of $\mathbf{b}_i$, as shown in Trincado and Burkhart (2006). In a linear mixed model setting, this empirical best linear unbiased predictor, say $\widehat{\mathrm{b}}_i$, or EBLUP, may be used when generating predictions from the fitted model. (The EBLUP is also known as an empirical Bayes estimator, because it is the mean of the posterior distribution of $\widehat{\mathrm{b}}_i$ in an empirical Bayes framework). The EBLUPs may be used to construct the ith tree's taper profile. While a mixed-effects model is necessarily subject-specific, predictions that incorporate the EBLUPs may be subject-specific, or not.

A subject-specific response profile requires knowledge of a sample of upper stem diameter(s) so that the EBLUP may be computed. Trincado and Burkhart investigated two different measurement scenarios in selecting upper stem diameter(s) for calibration: (1) only one upper stem diameter at absolute heights of 2.4, 4.9, or 6.1 m (8, 16, or 20 ft); and (2) two upper stem diameters measured at 3.7 and 6.1 m (12 and 20 ft). This last scenario gave the best results (Fig. 2.5) when testing the tree functions fitted using mixed-effects modeling with an independent data set. Mixed-model predictions of random effects improved the predictive capability of the segmented taper equation mainly in the lower portion of the bole. The method was effective for localizing stem curves for trees growing under different site and management conditions. Mixed-effects models provide a general framework that can be applied to various taper equation forms, increasing their flexibility and efficiency in predicting for local conditions.

References

Amidon EL (1984) A general taper functional form to predict bole volume for five mixed-conifer species in California. For Sci 30:166–171

Avery TE, Burkhart HE (2002) Forest measurements. 5th edn. McCraw-Hill, New York

Baskerville GL (1972) Use of logarithmic regression in the estimation of plant biomass. Can J For Res 2:49–53

Bi H (2000) Trigonometric variable-form taper equations for Australian eucalypts. For Sci 46: 397–409

Biging GS (1984) Taper equations for second-growth mixed conifers of northern California. For Sci 30:1103–1117

Brooks JR, Jiang L, Ozçelik R (2008) Compatible stem volume and taper equations for Brutian pine, Cedar of Lebanon, and Cilicica fir in Turkey. For Ecol Manage 256:147–151

Bruce D, Curtis RO, Vancoevering C (1968) Development of a system of taper and volume tables for red alder. For Sci 14:339–350

Burkhart HE, Walton SB (1985) Incorporating crown ratio into taper equations for loblolly pine trees. For Sci 31:478–484

Cao QV (2009) Calibrating a segmented taper equation with two diameter measurements. South J Appl For 33:58–61

Cao QV, Wang J (2011) Calibrating fixed- and mixed-effects taper equations. For Ecol Manage 262:671–673

Cao QV, Burkhart HE, Max TA (1980) Evaluation of two methods for cubic-volume prediction of loblolly pine to any merchantable limit. For Sci 26:71–80

Clark A III, Souter RA, Schlaegel BE (1991) Stem profile equations for southern tree species. USDA Forest Service, Southeastern Forest Experiment Station, Asheville, Research Paper SE-282

Czaplewski RL, Bruce D (1990) Retransformation bias in a stem profile model. Can J For Res 20:1623–1630

Czaplewski RL, McClure JP (1988) Conditioning a segmented stem profile model for two diameter measurements. For Sci 34:512–522

Demaerschalk JP (1972) Converting volume equations to compatible taper equations. For Sci 18:241–245

Demaerschalk JP (1973) Integrated systems for estimation of tree taper and volume. Can J For Res 3:90–94

Demaerschalk JP, Kozak A (1977) The whole-bole system: a conditioned dual-equation system for precise prediction of tree profiles. Can J For Res 7:488–497

Diéguez-Aranda U, Castedo-Dorado F, Álvarez-González JG, Rojo A (2006) Compatible taper function for Scots pine plantations northwestern Spain. Can J For Res 36:1190–1205

Eerikäinen K (2001) Stem volume models with random coefficients for *Pinus kesiya* in Tanzania, Zambia, and Zimbabwe. Can J For Res 31:879–888

Fang Z, Borders BE, Bailey RL (2000) Compatible volume-taper models for loblolly and slash pine based on a system with segmented-stem form factors. For Sci 46:1–12

Farrar JL (1961) Longitudinal variation in the thickness of the annual ring. For Chron 37:323–330

Figueiredo-Filho A, Borders BE, Hitch KL (1996a) Number of diameters required to represent stem profiles using interpolated cubic splines. Can J For Res 26:1113–1121

Figueiredo-Filho A, Borders BE, Hitch KL (1996b) Taper equations for *Pinus taeda* plantations in Southern Brazil. For Ecol Manage 83:39–46

Flewelling JW (1993) Variable-shape stem-profile predictions for western hemlock. Part II. Predictions from DBH, total height, and upper stem measurements. Can J For Res 23: 537–544

Flewelling JW, Pienaar LV (1981) Multiplicative regression with lognormal errors. For Sci 27: 281–289

Flewelling JW, Raynes LM (1993) Variable-shape stem-profile predictions for western hemlock. Part I. Predictions from DBH and total height. Can J For Res 23:520–536

Fonweban J, Gardiner B, Macdonald E, Auty D (2011) Taper functions for Scots pine (*Pinus sylvestris* L.) and Sitka spruce (*Picea sitchensis* (Bong.) Carr.) in Northern Britain. Forestry 84:49–60

Fowler JH, Rennie JC (1988) Merchantable height in lieu of total height in stem profile equations. For Sci 34:505–511

Garber SM, Maguire DA (2003) Modeling stem taper of three central Oregon species using nonlinear mixed effects models and autoregressive error structures. For Ecol Manage 179: 507–522

Gomat HY, Deleporte P, Moukini R, Mialounguila G, Ognouabi N, Saya AR, Vigneron P, Saint-Andre L (2011) What factors influence the stem taper of *Eucalyptus*: growth, environmental conditions, or genetics? Ann For Sci 68:109–120

Goulding A (1979) Cubic spline curves and calculation of volume of sectionally measured trees (Pines). N Z J For Sci 9:89–99

Gregoire TG, Schabenberger O, Barrett JP (1995) Linear modelling of irregularly spaced, unbalanced, longitudinal data from permanent-plot measurements. Can J For Res 25:137–156

Gregoire TG, Schabenberger O, Kong FZ (2000) Prediction from an integrated regression equation: a forestry application. Biometrics 56:414–419

Hilt DE (1980) Taper-based system for estimating stem volumes of upland oaks. USDA Forest Service, Northeastern Forest Experiment Station, Broomall, Research Paper NE-458

Klos RJ, Wang GG, Dang Q-L, East EW (2007) Taper equations for five major commercial tree species in Manitoba. Can West J Appl For 22:163–170

Knoebel BR, Burkhart HE, Beck DE (1984) Stem volume and taper functions for yellow-poplar in the Southern Appalachians. South J Appl For 8:185–188

Koskela L, Nummi T, Wenzel S, Kivinen V-P (2006) On the analysis of cubic smoothing spline-based stem curve prediction for forest harvesters. Can J For Res 36:2909–2919

Kozak A (1988) A variable-exponent taper equation. Can J For Res 18:1363–1368

Kozak A (1997) Effects of multicollinearity and autocorrelation on the variable-exponent taper functions. Can J For Res 27:619–629

Kozak A (1998) Effects of upper stem measurements on the predictive ability of a variable-exponent taper equation. Can J For Res 28:1078–1083

Kozak A (2004) My last words on taper equations. For Chron 80:507–515

Kozak A, Munro DD, Smith HG (1969) Taper functions and their application in forest inventory. For Chron 45:278–283

Laasasenaho J (1982) Taper curve and volume functions for pine, spruce and birch (*Pinus sylvestris, Picea abies, Betula pendula, Betula pubescens*). Commun Inst For Fenn 108:74

Laasasenaho J, Melkas T, Aldén S (2005) Modelling bark thickness of *Picea abies* with taper curves. For Ecol Manage 206:35–47

Lahtinen A, Laasasenaho J (1979) On the construction of taper curves by using spline functions. Commun Inst For Fenn 95(8):1–63

Lappi J (1986) Mixed linear models for analyzing and predicting stem form variation of Scots pine. Commun Inst For Fenn 134:69

Larson PR (1963) Stem form development of forest trees. For Sci Monogr 5:42pp

Lee W-K, Seo J-H, Son Y-M, Lee K-H, Gadow Kv (2003) Modeling stem profiles for Pinus densiflora in Korea. For Ecol Manage 172:69–77
Leites LP, Robinson AP (2004) Improving taper equations of loblolly pine with crown dimensions in a mixed-effects modeling framework. For Sci 50:204–212
Liu CJ (1980) Log volume estimation with spline approximation. For Sci 26:361–369
Max TA, Burkhart HE (1976) Segmented polynomial regression applied to taper equations. For Sci 22:283–289
McClure JP, Czaplewski RL (1986) Compatible taper equation for loblolly pine. Can J For Res 16:1272–1277
Muhairwe CK (1999) Taper equations for *Eucalyptus pilularis* and *Eucalyptus grandis* for the north coast in New South Wales, Australia. For Ecol Manage 113:251–269
Muhairwe CK, LeMay VM, Kozak A (1994) Effects of adding tree, stand, and site variables to Kozak's variable-exponent taper equation. Can J For Res 24:252–259
Newberry JD, Burkhart HE (1986) Variable-form stem profile models for loblolly pine. Can J For Res 16:109–114
Newnham D (1988) A variable-form taper function. Forestry Canada, Petawawa National Forestry Institute Information Report, PI-X-83
Newnham RM (1992) Variable-form taper functions for four Alberta tree species. Can J For Res 22:210–223
Nunes L, Tomé J, Tomé M (2010) A system for compatible prediction of total and merchantable volumes allowing for different definitions of tree volume. Can J For Res 40:747–760
Ormerod DW (1973) A simple bole model. For Chron 49:136–138
Özçelik R, Brooks JR (2011) Modeling stem profile of Lebanon cedar, Brutian pine, and Cilicica fir in Southern Turkey using nonlinear mixed-effects models. Eur J For Res 130:613–621
Perez DN, Burkhart HE, Stiff CT (1990) A variable-form taper function for *Pinus oocarpa* Schiede in Central Honduras. For Sci 36:186–191
Petersson H (1999) A segmented stem profile model for *Pinus sylvestris*. For Ecol Manage 124: 13–26
Rojo A, Perales X, Sánchez-Rodríguez F, Álvarez-González J, Gadow Kv (2005) Stem taper functions for maritime pine (*Pinus pinaster* Ait.) in Galicia (Northwestern Spain). Eur J For Res 124:177–186
Rustagi KP, Loveless RS Jr (1991) Compatible variable-form volume and stem-profile equations for Douglas-fir. Can J For Res 21:143–151
Schabenberger O, Gregoire TG (1996) Population-averaged and subject-specific approaches for clustered categorical data. J Stat Comput Simul 54:231–253
Schabenberger O, Pierce FJ (2002) Contemporary statistical models for the plant and soil sciences. CRC Press LLC, London
Sharma M, Burkhart HE (2003) Selecting a level of conditioning for the segmented polynomial taper equation. For Sci 49:324–330
Sharma M, Oderwald RG (2001) Dimensionally compatible volume and taper equations. Can J For Res 31:797–803
Sharma M, Parton J (2009) Modeling stand density effects on taper for jack pine and black spruce plantations using dimensional analysis. For Sci 55:268–282
Sharma M, Zhang SY (2004) Variable-exponent taper equations for jack pine, black spruce, and balsam fir in eastern Canada. For Ecol Manage 198:39–53
Snowdon P (1991) A ratio estimator for bias correction in logarithmic regressions. Can J For Res 21:720–724
Tasissa G, Burkhart HE (1998) An application of mixed effects analysis to modeling thinning effects on stem profile of loblolly pine. For Ecol Manage 103:87–101
Thomas CE, Parresol BR (1991) Simple, flexible, trigonometric taper equations. Can J For Res 21:1132–1137
Thornley JHM, Johnson IR (1990) Plant and crop models. Clarendon, Oxford
Trincado G, Burkhart HE (2006) A generalized approach for modeling and localizing stem profile curves. For Sci 52:670–682

Valenti MA, Cao QV (1986) Use of crown ratio to improve loblolly pine taper equations. Can J For Res 16:1141–1145

Valentine HT, Gregoire TG (2001) A switching model of bole taper. Can J For Res 31:1400–1409

Westfall JA, Scott CT (2010) Taper models for commercial tree species in the northeastern United States. For Sci 56:515–528

Yang Y, Huang S, Trincado G, Meng S (2009) Nonlinear mixed-effects modeling of variable-exponent taper equations for lodgepole pine in Alberta. Can Eur J For Res 128:415–429

Younger NL, Temesgen H, Garber SM (2008) Taper and volume responses of Douglas-fir to sulfur treatments for control of Swiss needle cast in the Coast Range of Oregon. West J Appl For 23:142–148

Zakrzewski WT (1999) A mathematically tractable stem profile model for jack pine in Ontario. North J Appl For 16:138–143

Chapter 3
Tree-Stem Volume Equations

3.1 Developing Volume Equations

Volume equations are used to predict the content of stems of standing trees as a function of easily measured tree attributes such as diameter at breast height and tree height. The predictor variables required in order to achieve acceptable accuracy vary by tree form. Excurrent crown form results when terminal growth exceeds branch lateral growth; this form is typical of many conifers and a few hardwood species such as yellow-poplar. Decurrent (also called deliquescent) crown forms result when lateral branches grow as fast or faster than the terminal leader; decurrent crowns are typical of many hardwood species including elms, oaks and maples. Shrubs are woody, perennial plants that generally lack a well-defined main stem.[1] For excurrent forms, the usual predictors for stem volume are *dbh* and total tree height (Fig. 3.1a). Total tree height is generally not highly correlated with the volume of the main stem of interest for decurrent tree forms, and a measure of merchantable height may be employed instead (Fig. 3.1b). For shrub forms estimation of volume in multiple stems requires additional independent variables, as well as use of diameter at root collar in lieu of diameter at breast height. In order to achieve acceptable precision for estimates of volume in the multiple stems of shrubs, predictors needed often include diameter at root collar, total height, and number of stems and perhaps crown width (Fig. 3.1c).

Much of the research on estimating stem volume of trees has been directed towards excurrent forms and involves *dbh* and total tree height as predictors. Varying units for the dependent variable have been employed but cubic units are most commonly used; general conclusions reached for estimating cubic volume of stems apply if other measures of volume are used. Hence, subsequent discussion in this

[1]For species that can be regenerated by coppice (such as eucalyptus), multiple stems from each stool may be kept until the next harvest. In such cases, each stem is usually treated as an individual excurrent tree but with height to *dbh* being measured from the top of the stool.

H.E. Burkhart and M. Tomé, *Modeling Forest Trees and Stands*,
DOI 10.1007/978-90-481-3170-9_3,

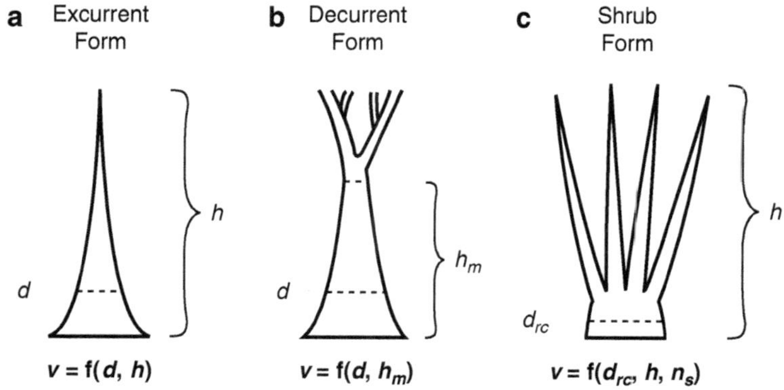

Fig. 3.1 Predictor variables needed for estimating stem volume varies for excurrent (**a**), decurrent (**b**), and shrub (**c**) forms. Where $d = dbh$, $h =$ total height, h_m = merchantable height, d_{rc} = diameter at root collar, and n_s = number of stems

chapter of tree-stem volume estimation will assume the objective is to predict cubic volume for single-stemmed trees using *dbh* and total height as predictors unless otherwise noted.

In some instances stem volume is related to *dbh* only. Such single-variable volume equations are often termed "local volume equations" in American forestry parlance. Alternatively, and what is more common practice, a sample of heights and diameters may be obtained as part of a timber inventory. The data are used to fit an equation to estimate height from *dbh*. This fitted relationship can then be used to convert a standard volume equation (that is an equation that involves *dbh* and total height as predictors) to a form to estimate volume from *dbh* only. Because so called standard volume equations are often applied in conjunction with a height-diameter curve when estimates based on *dbh* only are desired, functions involving *dbh* only will not be further considered in this chapter.

When constructing volume equations, precise information is needed on stem volume and the predictors for a sample of trees. This information has generally been gathered by felling sample trees and acquiring the needed measurements, but measurements of standing trees have, on occasion, been used for obtaining the data required. The true volume of a tree stem can most accurately be determined through water displacement methods (Martin 1984; Figueiredo Filho et al. 2000), but this approach is generally not practical. Rather stems of felled trees are measured for diameter and length over relatively small sections, the volume of each section is then computed (commonly using Smalian's formula which involves the average of the cross-sectional area of the large and small ends times the length to acquire cubic volume). Volume of the sections is summed to acquire the portion of stem volume of interest. The top section of excurrent form trees is typically treated as a cone when total stem volume is computed (Avery and Burkhart 2002 provide additional detail on tree measurements).

3.2 Equations for Total Stem Volume

Equations can be fitted to estimate contents of any specified portion of the stem of interest. In illustrating general approaches and methods for predicting tree volume from measurements of *dbh* and total tree height, we will assume that the dependent variable is total stem (stump to tip) cubic volume.

3.2.1 Combined Variable Equations

Numerous tree-stem volume models have been proposed but one of the most effective is the "combined variable" equation which combines *dbh* and height into a single predictor as:

$$v = \mathrm{b}_0 + \mathrm{b}_1 d^2 h \tag{3.1}$$

where v = stem volume, d = diameter at breast height, h = total tree height, and b_0, b_1 are estimated using linear regression techniques. Various additional terms of d and h have been added after d^2h, but invariably these additional terms contribute little to the fit statistics. When total stem volume is the dependent variable of interest the intercept term is sometimes omitted and a single parameter is estimated in the "constant form factor" equation

$$v = \mathrm{b}_1 d^2 h \tag{3.2}$$

The intercept term in the combined variable Eq. 3.1 is expected to be a small positive number when the dependent variable is total-stem volume because trees with total height less than height to *dbh* will have positive stem volume but a d^2h value of zero. Equations 3.1 and 3.2 will generally produce similar predicted values of total stem volume. However, with the possible exception of fitting a sample that does not include an adequate representation of small trees and when predictions for small trees are desired, there is no advantage to be gained by forcing the total stem volume equation through the origin. Thus, the combined-variable function is generally preferred for total stem volume prediction. If the dependent variable is a specified merchantable portion, rather than total volume, of the stem the constant term is expected to be negative because there is zero merchantable volume for trees with positive d^2h values until sufficient tree size for the specified merchantable portion is reached (Fig. 3.2).

The combined-variable equation is a particular form of a simple linear regression $y = \mathrm{b}_0 + \mathrm{b}_1 x$ where y = tree stem volume and $x = d^2h$. When fitting simple linear regression equations with ordinary least squares (OLS), the following assumptions are commonly invoked:

1. x and y are linearly related
2. x is measured without error

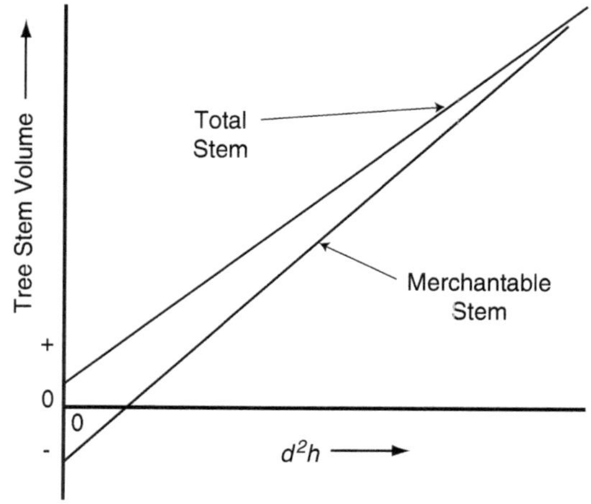

Fig. 3.2 Expected relationship between total and merchantable tree stem volume for the combined-variable function: $v = b_0 + b_1 d^2 h$; where v = cubic volume, d = diameter at breast height, and h = total tree height

3. the variance of y is constant at all levels of x
4. the observations are independent (i.e. the residuals (e_i) are not correlated), and
5. when testing hypotheses and computing confidence intervals, one further assumes that the y values are normally distributed at any given level of x.

When tree volumes are plotted versus d^2h values, a straight-line relationship results but the variance of volume (y) increases with increasing levels of d^2h (x). This phenomenon of increasing variance of y with increasing values of x is seen in many biological relationships, but it has received special attention in the case of fitting tree stem volume functions. When a non homogeneous variance is exhibited, the resultant estimates of the regression parameters are still unbiased but they are not the minimum variance estimators. To acquire the best estimators (unbiased, minimum variance), a weighted regression analysis is required.

When the assumptions for ordinary least squares are fully satisfied the distribution of the dependent variable can be written:

$$y_i \sim N(\beta_0 + \beta_1 x_i, \sigma^2_{y.x})$$

where $\sim N$ indicates distributed normally, $\beta_0 + \beta_1 x_i$, represents the mean of y_i at any specified value of x_i, and $\sigma^2_{y.x}$ symbolizes the constant variance at all levels of x_i. In the case of non homogeneous variance, the distribution becomes

$$y_i \sim N(\beta_0 + \beta_1 x_i, \sigma^2_{y,x_i})$$

where $\sigma^2_{y.x_i}$ varies according to the level of x_i. One can apply weights $1/\sigma^2_{y,x_i}$ to obtain unbiased, minimum variance estimators or one can transform the independent variable in order to stabilize the variance. If the variance of y_i is proportional to some function of x_i, that is $\sigma^2_{y,x_i} = \mathrm{k} f(x_i)$ where k is a constant of proportionality, then the distribution of the y_i observations can be written

$$y_i \sim N(\beta_0 + \beta_1 x_i, \mathrm{k} f(x_i))$$

Recalling that a constant, c, times a normally distributed random variable distributes normally with mean $c\mu$ and variance $c^2\sigma^2$, the expression for the distribution of y_i with variance $kf(x_i)$ can be multiplied by $1/\sqrt{f(x_i)}$ to obtain an expression for a transformed dependent variable with constant variance k, that is:

$$y_i/\sqrt{f(x_i)} \sim N((\beta_0 + \beta_1 x_i)/\sqrt{f(x_i)}, \text{k})$$

If the combined-variable formula is assumed and the variance of tree volume is taken to be proportional to $(d_i^2 h_i)^2$, then

$$v_i \sim N(\beta_0 + \beta_1 d_i^2 h_i, k(d_i^2 h_i)^2)$$

and

$$v_i/d_i^2 h_i \sim N((\beta_0 + \beta_1 d_i^2 h_i)/d_i^2 h_i, \text{k})$$

The simple linear regression

$$v/d^2 h = \text{b}_0(1/d^2 h) + \text{b}_1$$

is fitted and then algebraically rearranged as

$$v = \text{b}_0 + \text{b}_1 d^2 h$$

for stem volume prediction.

A general expression for the weighing function can be written $(1/d_i^2 h_i)^k$ for the combined-variable equation. Although a value of 2 for k has often been assumed since it was recommended by Cunia (1964), a number of investigations into the most appropriate weighting function have been conducted. McClure et al. (1983) suggested using k = 1.5 for white oak and loblolly pine. Gregoire and Dyer (1989) fitted the weight function for loblolly pine and red pine using a number of techniques. They obtained k = 1.70–1.84 for loblolly pine and k = 1.01–2.07 for red pine. Williams et al. (1992) evaluated a number of weight functions for two loblolly pine samples and one white oak data set. All weight functions tested produced similar results in terms of goodness of fit. The most robust weight function studied was $(1/d_i^2 h_i)^{\text{k}}$ where k ranged from 1.80 to 2.07 for the three data sets. In a follow-up analysis, Williams and Gregoire (1993) recommended a weight function $(1/d_i^{2.3} h_i^{0.7})^{\text{k}_1}$ for all three data sets previously studied for which k_1 values ranged from 1.80 to 2.07.

While the primary impetus for using weights when estimating coefficients in the combined-variable tree volume equation has been to meet the assumption of homogeneous variance, the principal impact has been to reduce the influence of high-leverage data points in the larger tree sizes. For typical samples of tree stem volume, trees in larger size classes are fewer in number but larger in variance

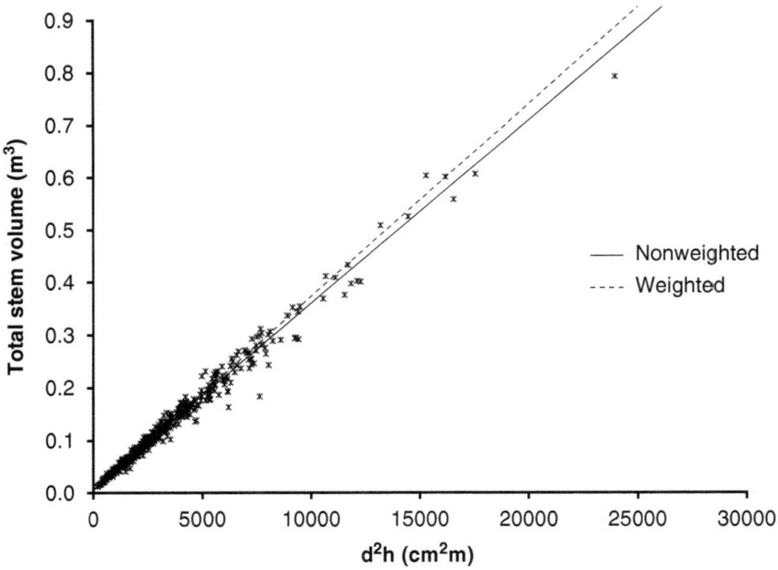

Fig. 3.3 Plot of data for 427 loblolly pine trees versus d^2h and regression of $v = \mathrm{b}_0 + \mathrm{b}_1 d^2 h$ using ordinary least squares and fitting with weights under the assumption that the variance of tree-stem volume is proportional to $(d^2h)^2$

than trees in smaller size classes. This condition generally leads to a few highly influential points with regard to the slope of the straight-line regression. When weighted regression is applied, the weights are inversely proportional to variance and more weight is given to the smaller trees than the larger ones which results in a decreased influence of the high-leverage data point(s). An illustration of this phenomenon is shown in Fig. 3.3 where the raw data along with the unweighted ordinary least squares of fit of

$$v = \mathrm{b}_0 + \mathrm{b}_1 d^2 h$$

as well as the coefficients for a weighted regression where variance was assumed to be proportional to $(d^2h)^2$. The estimated coefficients for the unweighted fit for these data from Burkhart (1977) are $\mathrm{b}_0 = 0.01076$, $\mathrm{b}_1 = 0.00003487$ and for the weighted regression $\mathrm{b}_0 = 0.00626$, $\mathrm{b}_1 = 0.00003666$. The influence of the single data point at d^2h of around 25,000 is mitigated when weights are applied.

3.2.2 Logarithmic Volume Equations

A commonly used function for estimating tree stem volume is the logarithmic, or Schumacher and Hall (1933), equation:

$$v = \mathrm{b}_1 d^{\mathrm{b}_2} h^{\mathrm{b}_3} \tag{3.3}$$

Equation 3.3, which goes through the (0,0) point, can be fitted directly using nonlinear regression techniques, but, in past practice, the functional form has generally been converted to a form that is linear in the parameters through logarithmic transformation and linear regression techniques have been applied, giving:

$$\ln v = b_0 + b_1 \ln d + b_2 \ln h \tag{3.4}$$

The logarithmic transformation has the advantage of stabilizing the variance of v across the range of d and h. However, the equivalent coefficients (b_1 and e^{b_0}, b_2 and b_1, b_3 and b_2) for Eqs. 3.3 and 3.4 will differ considerably for the two methods of parameter estimation largely because of differences in assumptions about the error term. Furthermore, taking logarithms leads to a transformation bias. Instead of passing through the arithmetic mean of tree volume Eq. 3.4 goes through the geometric mean, which must be less than the arithmetic mean. Consequently, a correction factor is sometimes applied to account for this bias. The most commonly applied correction involves adding one half on the estimated variance about the fitted regression before exponentiating (Baskerville 1972), namely

$$\hat{y}_i = \exp(\hat{\mu}_i + \hat{\sigma}^2/2) \tag{3.5}$$

where $\hat{y}_i$ is the "corrected" estimate of the dependent variable, $\hat{\mu}_i$ represents the expression for the mean as estimated by regression analysis, and $\hat{\sigma}^2/2$ is half the estimated variance (on a logarithmic scale) about the fitted regression. A number of different factors for correcting logarithmic transformation bias have been proposed (Flewelling and Pienaar 1981; Snowdon 1991), but the correction factor is often ignored when applying logarithmic tree stem volume functions.

As noted, the so-called logarithmic equation is conditioned through the origin, which is reasonable for total stem volume but not tenable for estimating a merchantable portion of the bole. Schumacher and Hall (1933) recommended transferring the coordinates of the origin of the logarithmic volume equation from (0,0) to an appropriate location when merchantable volume is the dependent variable. Alternatively, a constant term can be added leading to an intrinsically non linear equation that is sometimes designated the "generalized logarithmic equation" (Table 3.1):

$$v = b_0 + b_1 d^{b_2} h^{b_3} \tag{3.6}$$

With the constant term added, Eq. 3.6 cannot be transformed to a linear function through taking logarithms. Equation 3.6 can be thought of as a generalization of Eq. 3.3 or as a generalization of the combined-variable Eq. 3.1 where the exponents of d and h are estimated rather than being specified as 2 and 1, respectively.

Table 3.1 Equation forms commonly used for estimation of individual tree stem volumes

Name	Equation form
1. Combined variable	$v = b_0 + b_1 d^2 h$
2. Constant form factor	$v = b_1 d^2 h$
3. Logarithmic	$v = b_1 d^{b_2} h^{b_3}$
4. Generalized logarithmic	$v = b_0 + b_1 d^{b_2} h^{b_3}$
5. Honer transformed variable	$v = d^2/(b_0 + b_1 h^{-1})$
	$v =$ *stem volume*
	$d = dbh$
	$h =$ *a measure of tree height*
	$b_0, b_1, b_2, b_3 =$ *constants*

3.2.3 Honer Volume Equation

Honer (1965) noted that the relationship between d^2/v and h^{-1} is linear with homogeneous variance for total stem volume. Thus, he proposed fitting the simple linear regression

$$d^2/v = \mathrm{b}_0 + \mathrm{b}_1 h^{-1} \tag{3.7}$$

which is then algebraically rearranged as

$$v = d^2/(\mathrm{b}_0 + \mathrm{b}_1 h^{-1}) \tag{3.8}$$

for estimating total stem volume (v). In lieu of transforming Eq. 3.8 in order to estimate the parameters using linear regression methods (3.7), nonlinear regression can be applied to equation form (3.8). The empirical estimates of b_0 and b_1 will differ considerably depending on whether direct estimation is applied to form (3.8) or linear regression is used with the transformed version (3.7).

Although Honer (1965) found Eq. 3.7 to provide accurate estimates of total stem volume, other comparisons with alternative estimating equations (e.g. Burkhart 1977) have shown that the combined-variable form is preferred.

3.3 Estimating Merchantable Stem Volume

Standard volume equations are used to estimate volume of a specified portion of the bole from stump height to a fixed top diameter of standing trees using measurements of *dbh* and total or merchantable height. Because estimates of stem volume for a variety of products are needed, each with different size requirements, and changing merchantability standards, it is desirable to obtain volume estimates for various top diameter limits. Separate regression equations can be fitted to stem volume for a range of top diameters but unconstrained, independent equations to several top limits may have the undesirable characteristic of crossing within the range of observed data, because the slope of the straight-line regression for merchantable volume is generally greater than that for total stem volume (Fig. 3.2). These considerations have led to the concept of predicting volume ratios. By predicting ratios of merchantable volume to total volume, and multiplying the ratio times total stem volume, one can obtain volumes to any specified top diameter limit or the volume between specified diameters on the stem by subtraction. Imposing appropriate constraints on the volume-ratio prediction equation insures that estimates of volumes for various parts of the bole will be logically related.

3.3.1 *Volume Ratio Equations*

Equations to predict the ratio (R_{di}) of merchantable stem volume (v_{di}) to total stem volume (v) for varying top diameter limits have been developed by a number of researchers (Honer 1964; Burkhart 1977; Van Deusen et al. 1981; and others). Burkhart's (1977) model

$$R_{di} = 1 + \mathrm{b}_1 \left(d_i^{\mathrm{b}_2} d^{\mathrm{b}_3} \right) \tag{3.9}$$

has three parameters and is conditioned such that when top diameter (d_i) is zero the ratio equals one. Parameters in Eq. 3.9 can be estimated directly using nonlinear least squares or by applying linear regression methods following transformation. Typically nonlinear estimation has been used. Using data from a sample of loblolly pine trees, the combined-variable equation was fitted to estimate total stem volume and Eq. 3.9 was fitted using nonlinear regression in order to estimate volume to any desired top diameter limit (Burkhart 1977). Plotting the resultant volume predictions showed that for small top diameters the volume lines were nearly parallel to the total stem volume line, while the slopes became increasingly steeper, but without illogical crossing, as top diameter became larger. Thus, the approach gave logical and consistent results when converting total stem volume to merchantable volume for any desired top limits.

Van Deusen et al. (1981) noted that Eq. 3.9 is constrained only at the tree tip (when $d_i = 0$, $R_{di} = 1$). Because empirically b_1 must be negative, the ratio Eq. 3.9 can assume negative values as the diameter ratio approaches one (i.e. as top diameter d_i approaches *dbh*). This behavior generally presents no practical difficulty, because interest is usually in stem volumes for top diameters well above *dbh*. However, Van Deusen et al. (1981) proposed an exponential model that results in ratio values between one and zero, namely:

$$R_{di} = \mathrm{e}^{\mathrm{b}_1 (d_i d^{-1})^{\mathrm{b}_2}} \tag{3.10}$$

For this two-parameter model, estimates of b_1 will be negative and the resultant curve will equal one where d_i is zero and will approach zero as d increases without bound. Hence, a desirable double conditioning is insured. A comparison of the

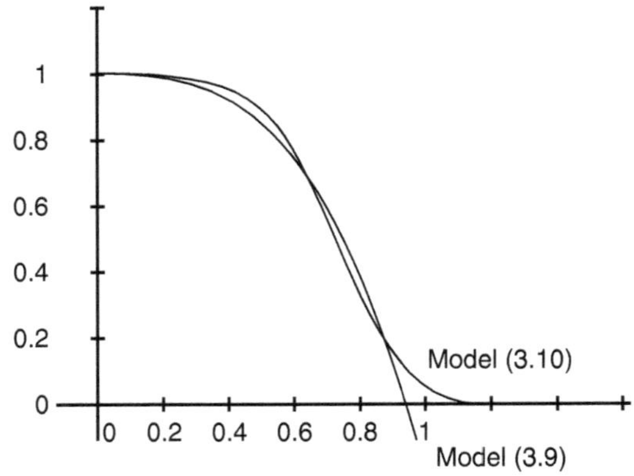

Fig. 3.4 Comparison of volume ratio model forms (3.9) and (3.10). The axes are merchantable volume/total volume vs. merchantable height/total height (From Van Deusen et al. 1981)

behavior of Eqs. 3.9 and 3.10 is shown in Fig. 3.4. While the exponential expression (3.10) has the desirable property of non–negativity as d_i approaches *dbh* it is relatively inflexible when only two parameters are estimated, which results in bias over parts of the range of top diameters with especially pronounced bias for large top diameters (i.e. for the area where d_i approaches *dbh*). Experience with fitting Eq. 3.10 to tree stem volume and weight data has shown that the more flexible three-parameter version

$$R_{di} = e^{\mathrm{b}_1 (d_i{}^{\mathrm{b}_2} d^{\mathrm{b}_3})} \tag{3.11}$$

results in less bias and superior predictive performance.

Cao and Burkhart (1980) noted that products are sometimes defined in fixed length multiples. Consequently equations for estimating the volume in any specified length of the bole are required. They formulated the volume ratio model:

$$R_{hi} = 1 + a_1((h - h_i)^{a_2} h^{a_3}) \tag{3.12}$$

where

$(h - h_i)$ = distance from the tree tip to the limit of utilization
h = total tree height
$R_{hi} = v_{hi}/v$
a_1, a_2, a_3 = regression coefficients to be estimated

Equation 3.12 is conditioned so that R_{hi} is 1 when $h_i = h$ (i.e. when the height desired is at the tree tip and the volume estimated is for the complete stem).

A pair of volume ratio equations such as (3.9) and (3.12) provides information needed for computing volumes to specified top diameter limits and for specified lengths of tree boles, but product specifications often include small end diameter minimums for fixed length multiples, thus necessitating the need for a taper function. By equating the two volume ratio equations

$$1 + \mathrm{b}_1(d_i{}^{\mathrm{b}_2} d^{\mathrm{b}_3}) = 1 + a_1((h - h_i)^{a_2} h^{a_3}) \tag{3.13}$$

and solving for d_i one obtains the implied taper function

$$d_i = \left[\frac{a_1((h - h_i)^{a_2} h^{a_3}) d^{b_3}}{b_1}\right]^{\frac{1}{b_2}} \tag{3.14}$$

which gives estimated stem diameter d_i as a function of tree *dbh* (d), total height (h) and height above ground (h_i).

Equation 3.14 can be rearranged to obtain an expression for height at any given top diameter for a tree with specified *dbh* and total height as

$$h_i = h - \left[\frac{b_1 (d_i{}^{b_2} d^{b_3}) h^{a_3}}{a_1}\right]^{\frac{1}{a_2}} \tag{3.15}$$

Implied, or derived, taper functions are reasonably accurate and they provide the additional capability of estimating diameters for specified lengths of bole (or length of bole to a specified top diameter limit). The derived relationships are not, however, "best" in a statistical sense because the sum of squared error minimized to obtain estimates of b_1, b_2, b_3 and a_1, a_2, a_3 was for the ratio of merchantable to total stem volume. Several studies aimed at estimating merchantable stem volume and deriving the implied taper relationships (Knoebel et al. 1984; Amateis and Burkhart 1987; Tassisa et al. 1997; Bullock and Burkhart 2003) have concluded that fitting the pair of equations

$$v_{di} = (v)(R_{di}) \tag{3.16}$$

$$\text{and } v_{hi} = (v)(R_{hi})$$

results in better performance of the system for estimating merchantable volume and taper relationships. The observed total stem volume (v) for each sample tree, along with appropriate variables in the ratio expression (R_{di} or R_{hi}) and the associated merchantable volume (v_{di} or v_{hi}), is used when fitting equations in form (3.16). To implement the system, an expression for v, with estimated coefficients, is required.

In most applications, a common form of the ratio expression (R) has been used in fitting that is

$$v_{di} = v(1 + b_1 d_i{}^{b_2} d^{b_3}) \tag{3.17a}$$

$$v_{hi} = v(1 + b_1 (h - h_i)^{b_2} h^{b_3}) \tag{3.17b}$$

or

$$v_{di} = v \ \exp(b_1 d_i{}^{b_2} d^{b_3}) \tag{3.18a}$$

$$v_{hi} = v \ \exp(a_1 (h - h_i)^{a_2} h^{a_3}) \tag{3.18b}$$

Empirical evaluation has shown that the exponential form (3.18a) is generally preferred over (3.17a) when top diameters are used but (3.17b) is superior to (3.18b) when height is employed. This led Bullock and Burkhart (2003) to fit the "mixed" set

$$v_{di} = v\ \exp(\mathrm{b}_1 {d_i}^{\mathrm{b}_2} d^{\mathrm{b}_3}) \tag{3.19}$$

$$v_{hi} = v(1 + a_1(h - h_i)^{a_2} h^{a_3}) \tag{3.20}$$

to data for predicting tree weight to varying merchantable limits. Equating (3.19) and (3.20) and solving for the implied taper relationships gives:

$$d_i = \left\{ (d^{\mathrm{b}_3}/\mathrm{b}_1)^{\frac{1}{\mathrm{b}_2}} \left[\log\left(1 + a_1(h - h_i)^{a_2} h^{a_3}\right)\right]^{\frac{1}{\mathrm{b}_2}} \right\} \tag{3.21}$$

$$h_i = h - \left\{ \left(h^{a_3}/a_1\right)^{\frac{1}{a_2}} \left[\left(\exp\left(\mathrm{b}_1 {d_i}^{\mathrm{b}_2} d^{\mathrm{b}_3}\right)\right) - 1\right]^{\frac{1}{a_2}} \right\} \tag{3.22}$$

When fitting data using least squares one assumes that the random errors are independent. Use of multiple observations from each sample tree when fitting merchantable volume relationships violates the assumption of independence of observations, which leads to biased estimates of the variances of estimated parameters and unreliable tests of significance and confidence intervals. Estimation methods which account for the correlation among observations (Gregoire et al. 1995) have been applied to the problem of predicting merchantable stem volume (Gregoire and Schabenberger 1996). Although correlated observations lead to biased estimates of standard errors, the parameter estimates are unbiased. Consequently, when the primary interest is prediction rather than hypothesis testing, ordinary least squares is often applied.

3.3.2 Deriving Taper Functions from Volume Equations

Demaerschalk (1972) showed that an existing total stem volume equation can be used in conjunction with taper data to develop a taper function that is compatible with the volume equation (compatible in the sense that integration of the taper function over the limits zero to total tree height produces the volume equation). Clutter (1980) reversed the volume-taper compatibility process by noting that any variable-top merchantable stem volume equation implicitly defines an associated taper function. Assuming that $v = \mathrm{b}_1 d^{\mathrm{b}_2} h^{\mathrm{b}_3}$ and that merchantable volume can be estimated as

$$v_{di} = v\left[1 + \mathrm{b}_4 {d_i}^{\mathrm{b}_5} d^{\mathrm{b}_6}\right] \tag{3.23}$$

and applying methods of differential calculus yields the taper function

$$d_i = b_7 d^{b_8} h^{b_9} (h - h_i)^{b_{10}} \tag{3.24}$$

with the inverse equation

$$h_i = h - b_{11} d_i^{b_{12}} d^{b_{13}} h^{b_{14}} \tag{3.25}$$

While the numeric values of the coefficients will vary depending on which equations are fitted and which coefficients are subsequently derived, the form of the implied taper function and the associated inverse function is the same when ratio Eqs. 3.9 and 3.12 are equated (as in (3.13)) to derive (3.14) and (3.15) or when (3.23) is used as the starting point to derive (3.24) and (3.25). Empirical evaluations with several different data sets have shown, however, that fitting two equations as indicated in (3.16) and computing implied coefficient values results in better predictions of tree taper than deriving the coefficients from the estimation of (3.23).

3.3.3 Compatible Stem Volume and Taper Functions

The concept of defining a taper and volume system that is compatible, in that integration of the taper equation yields the same total volume as that estimated by the volume equation, was introduced in Sect. 2.5. This basic concept has been extended by a number of researchers to include compatibility of merchantable volume, total volume, and taper relationships (Reed and Green 1984; Byrne and Reed 1986; McTague and Bailey 1987; Jordan et al. 2005 and others). The equation system developed by Byrne and Reed (1986) will be used as an example for demonstrating concepts and methods involved in such volume-taper systems. Byrne and Reed (1986) described five equations systems that can be applied to estimate upper stem diameter, total stem volume, and merchantable volumes to any merchantability limit (expressed in terms of diameter or height). They calibrated their system of equations using stem analysis data from red and loblolly pine. The equation system that provided the best overall fit to validation data for both species was based on a segmented taper equation. The best performing system of equations from the alternatives evaluated by Byrne and Reed (1986) consisted of a volume estimation system derived from the segmented taper equation of Cao et al. (1980).

The notation of Byrne and Reed (1986), after some minor changes, will be used in the presentation of their compatible taper, total and merchantable stem volume equations.

a_i, b_i, c = regression coefficients estimated from sample data, where, $i = 1, 2, 3, \ldots,$

d = diameter at breast height,

d_i = top diameter at height h_i,
h = total tree height,
h_i = height above the ground to top diameter d_i,
K = constant to convert diameter squared to cross-sectional area in square units on the scale of interest,
v = total cubic volume above the ground,
v_m = cubic volume from the ground to some top diameter or height limit,
$p = h - h_i$,
$z = (h - h_i)/h$, relative tree height from the tip to top diameter d,
R = ratio of merchantable volume to total stem volume,
R_{hi} = volume ratio for v_m prediction to an upper height limit (h_i), and
R_{di} = volume ratio for v_m prediction to an upper diameter limit (d_i).

A volume estimation system was derived from the segmented taper equation given by Cao et al. (1980)

$$(d^2\mathrm{K}h/v) - 2z = \mathrm{b}_1(3z^2 - 2z) + \mathrm{b}_2(z - a_1)^2\mathrm{I}_1 + \mathrm{b}_3(z - a_2)^2\mathrm{I}_2 \tag{3.26}$$

where

$$I_i = \begin{cases} 1 \; if \; z \geq a_i \\ 0 \; if \; z < a_i, \; i = 1, 2. \end{cases}$$

By considering $v = cd^2h$ (the constant form factor total volume equation), then

$$d^2 = v/ch \tag{3.27}$$

and the taper equation can be rewritten as

$$d_i{}^2 = d^2(c/\mathrm{K})\left[2z + \mathrm{b}_1(3z^2 - 2z) + \mathrm{b}_2(z - a_1)^2 I_1 + b_3(z - a_2)^2 I_2\right] \tag{3.28}$$

where I_1 and I_2 are as previously defined. A total stem volume equation can be derived by summing the following three integrals

$$v = \int_0^{h_1} Kd_i^2 dh_i + \int_{h_1}^{h_2} Kd_i^2 dh_i + \int_{h_2}^{h} Kd_i^2 dh_i \tag{3.29}$$

where h_1 and h_2 are variables representing the heights at the two join points of the model (i.e. a_1 and a_2, respectively). Integrating gives

$$v = \mathrm{c}\left[1 + (\mathrm{b}_2/3)(1 - a_1)^3 + (\mathrm{b}_3/3)(1 - a_2)^3\right] d^2h \tag{3.30}$$

This Eq. 3.30 is in the form of a constant form factor total volume equation. An equation for merchantable volume to a height limit can be found in a similar way as the total stem volume equation by integrating to height h_i instead of h. The resulting equation is

$$v_m = \mathrm{c}\Big[1 + (\mathrm{b}_1 - 1)z^2 - \mathrm{b}_1 z^3 - (\mathrm{b}_2/3)\Big\{(z - a_1)^3 I_1 - (1 - a_1)^3\Big\} - (\mathrm{b}_3/3)\Big\{(z - a_2)^3 I_2 - (1 - a_2)^3\Big\}\Big] d^2 h \tag{3.31}$$

A volume ratio equation to a height limit (R_{hi}) is found by dividing the above equation by the total stem volume equation

$$R_{hi} = (1/\gamma)\Big[1 + (\mathrm{b}_1 - 1)z^2 - \mathrm{b}_1 z^3 - (\mathrm{b}_2/3)\Big\{(z - a_1)^3 I_1 - (1 - a_1)^3\Big\} - (\mathrm{b}_3/3)\Big\{(z - a_2)^3 I_2 - (1 - a_2)^3\Big\}\Big] \tag{3.32}$$

where

$$\gamma = 1 + (\mathrm{b}_2/3)(1 - a_1)^3 + (\mathrm{b}_3/3)(1 - a_2)^3$$

To obtain a volume ratio equation to a diameter limit (R_{di}), the taper equation is algebraically redefined in terms of d_i, d, and h:

$$h_i = h\Big[1 - \Big\{\Big(-B \pm (B^2 - 4AC)^{1/2}\Big)/2A\Big\}\Big] \tag{3.33}$$

where

$$A = (\mathrm{c}/\mathrm{k})(3\mathrm{b}_1 + \mathrm{b}_2 J_1 + \mathrm{b}_3 J_2)$$
$$B = (2\mathrm{c}/\mathrm{K})(1 - \mathrm{b}_1 - \mathrm{a}_1\mathrm{b}_2 J_1 - \mathrm{a}_2\mathrm{b}_3 J_2)$$
$$C = (\mathrm{c}/\mathrm{K})(\mathrm{a}_1{}^2\mathrm{b}_2 J_2 + \mathrm{a}_2{}^2\mathrm{b}_3 J_2) - d^2/D^2)$$

$$J_i = \begin{cases} 1 \textit{ if } d_i \geq M_i \\ 0 \textit{ if } d_i < M_i,\ i = 1, 2. \end{cases}$$

$$M_i = \text{estimated diameter at } h_i$$
$$= d\Big\{(\mathrm{c}/\mathrm{K})\Big[2a_i + \mathrm{b}_1(3c_i{}^2 - 2a_i) + \mathrm{b}_3(a_i - a_2)^2\Big]\Big\}^{1/2}$$

Substitution of Eq. 3.33 into Eq. 3.32 results in the following R_{di} equation:

$$R_{di} = (1/\gamma)\Big[1 + (\mathrm{b}_1 - 1)w^2 - \mathrm{b}_1 w^3 - (\mathrm{b}_2/3)\Big\{(w - a_1)^3 I_1 - (1 - a_1)^3\Big\} - (\mathrm{b}_3/3)\Big\{(w - a_2)^3 I_2 - (1 - a_2)^3\Big\}\Big] \tag{3.34}$$

where

$$w = \left[-B \pm \left(B^2 - 4AC\right)^{1/2}\right] \Big/ 2A$$

and A, B, C and γ are as previously defined.

Byrne and Reed (1986) used two methods when fitting the volume estimation system to data:

1. least squares techniques to estimate parameters in the taper equation and algebraically solve for the coefficients of the other equations based on the fitted taper equation coefficients, and
2. simultaneously fit all equations using a numerical procedure to minimize the total system sum of squared error.

In the simultaneous fitting procedure all four equations in the volume estimation system (taper, v, R_{hi}, R_{di}) were fit to the data at the same time. This procedure minimizes the total system squared error (*TSSE*) for each model. *TSSE* is defined as the summation of the squared observed minus predicted values for each of the equations in a system.

$$TSSE = \sum_{i=1}^{N} \frac{(y_i - \hat{y}_i)^2}{\hat{\sigma}_y^2} + \sum_{i=1}^{n} \frac{(v_i - \hat{v}_i)^2}{\hat{\sigma}_v^2} + \sum_{i=1}^{N} \frac{\left(R_{hi} - \hat{R}_{hi}\right)^2}{\hat{\sigma}_{Rh}^2} + \sum_{i=1}^{N} \frac{\left(R_{di} - \hat{R}_{di}\right)^2}{\hat{\sigma}_{Rd}^2}$$

Where

y_i, $\hat{y}_i$ = observed and predicted diameters for the taper function, respectively,
v_i, $\hat{v}_i$ = observed and predicted total cubic volume,
R_{hi}, $\hat{R}_{hi}$ = observed and predicted volume ratios for v_m prediction to an upper height limit,
R_{di}, $\hat{R}_{di}$ = observed and predicted volume ratios for v_m prediction to an upper diameter limit,
$\hat{\sigma}_y^2$ = mean square error from the least squares fit of the taper equation,
$\hat{\sigma}_v^2$ = mean square error from the least squares fit of the total volume equation,
$\hat{\sigma}_{Rh}^2$ = mean square error form the least squares fit of the volume ratio to a height limit equation,
$\hat{\sigma}_{Rd}^2$ = mean square error from the least squares fit of the volume ratio to a diameter limit equation,
N = number of height/diameter observations for fitting the equation, and
n = number of trees for fitting the equation.

The fitted equations were evaluated for estimation of taper, total stem volume, and volume ratios from each height-diameter observation in a validation data set using four criteria:

1. average residual or bias,
2. standard deviation of the residual or precision,

3. average of the absolute values of residuals, and
4. percent variation explained (pseudo R^2).

For red pine the fitting approaches were similar in predictive ability, but for loblolly pine the simultaneous fitting resulted in marked improvements. Additional testing and ranking of the model forms and fitting procedures considered led to the final conclusion that the segmented taper equation of Cao et al. (1980) utilized in a simultaneous fitting procedure is the most accurate and precise predictor of taper, total volume, and volume ratios for both red and loblolly pine.

3.4 Inclusion of Variables in Addition to *dbh* and Total Height

Predictor variables in addition to *dbh* and total height have been used to increase the generality and the precision of tree stem volume equations. The most commonly used additional variable is a stem diameter above *dbh*. Although specifying the upper-stem diameter at some fixed percentage of stem length from *dbh* to tree tip (e.g. 50% or at the mid point) would be desirable from the standpoint of being biologically meaningful and statistically effective, it is more practical to measure an upper-stem diameter at a specified height for all tree sizes. The most common form quotient that has been used in the United States is Girard Form Class defined as the inside bark diameter at 17.3 ft (5.27 m) above ground divided by *dbh* (outside bark) (Avery and Burkhart 2002). Volume equations that incorporate form measurements, such as form quotients based on the ratio of upper-stem diameters to *dbh*, can be easily fitted with stem analysis data but their application in field situations is problematic. Advances in the technology of tree measuring devices have made measurements of outside bark upper-stem diameters more accurate and feasible in instances where the stem is visible at the point of measurement, but ability to sight the point of measurement on trees growing in stand conditions is sometimes limited.

The combined–variable function has sometimes been generalized to predict stem volume as a function of *dbh*, total height, and a measure of form (F)

$$v = b_0 + b_1(d^2h)(F) \tag{3.35}$$

A wide variety of different equations forms and definitions of tree form (F) have been used.

Due to difficulties with measuring an upper-stem diameter, and because tree stem form is related to crown size, measures of crown (typically crown ratio, defined as the length of live crown divided by total tree height) have been incorporated into tree volume equations. However, in many cases (Laasasenaho 1982, and others), the reduction in residual variance due to adding crown ratio after tree *dbh* and height has been marginal. An exception is the results of Hann et al. (1987) which showed significant improvement due to incorporating crown ratio into Douglas–fir total stem volume equations.

Because the inclusion of an expression for tree form can be difficult in many field situations and the inclusion of crown ratio has generally resulted in only marginal improvement in prediction of stem volume, Burkhart (1977) examined stand characteristics (age, site index, number of trees and basal area per unit area) which might serve as surrogates for form in predicting stem volume of natural and plantation-grown loblolly pine; adding stand variables after d^2h did not improve predictive ability for an independent validation data set.

3.5 Volume Prediction for Irregular Stems

Hann and Bare (1978) developed a comprehensive set of tree volume equations for unforked and forked trees of the major species in Arizona and New Mexico, USA. Their system consists of estimating the total stem cubic volume for unforked trees using the combined-variable function with d^2h equal to dbh squared times total tree height. Total cubic volume of forked trees is then estimated as:

$$v_f = v \times R_{t,f}$$

where

v_f = predicted total stem gross cubic volume of a forked tree
v = predicted total stem gross cubic volume of an unforked tree
$R_{t,f}$ = predicted ratio of actual total stem gross cubic volume in a forked tree divided by predicted total stem gross cubic volume in an unforked tree of the same dimensions.

Equations for estimating merchantable volumes of stems, as well as total cubic volume, were also developed to provide predictions of volume for both regular (unforked) and irregular (forked) stems.

Merchantable height (h_m), rather than total height, is sometimes employed for decurrent tree forms (Fig. 3.1b). The combined variable and other equation forms that are useful for estimating tree volume when using total tree height are generally satisfactory when substituting merchantable tree height (h_m), that is

$$v = b_0 + b_1 d^2 h_m \tag{3.36}$$

When estimating volume for shrub forms (Fig. 3.1c) additional predictor variables are needed. Because shrubs do not normally have a useful dbh value, diameter at the root collar (d_{rc}) is employed. Some studies have found the relatively simple equation

$$v = b_0 + b_1 d_{rc}^2 h \tag{3.37}$$

to be sufficient for shrubs, but in other instances a count of the number of stems involved has been included in addition to d_{rc} and shrub height h. Chojnacky (1985) developed equations to predict gross cubic volume for trees in the pinyon-juniper type in the western United States using measurements of diameter at the root collar d_{rc}, total height (h) and number of basal stems (n_s). The variables d_{rc} squared times total height and a dummy variable to distinguish single- from multiple-stem trees ($n_s = 1$ if single-stem; 0 if multiple-stem) were used in a combined-variable-like volume estimating equation.

3.6 Stem Quality Assessment and Prediction

In addition to predicting the total volume and the volume in given portions of stems, an assessment of wood quality may also be needed. Quality is often determined by grading which entails placing logs, veneer bolts, or other product categories into quality classes. The primary variables associated with grades of round wood are diameter, sweep, and indicators of internal defects.

As an alternative to assigning portions of boles into grade or quality classes, there have been studies aimed at developing quality classification systems based on empirical measurements of individual trees. Product specifications and utilization standards change over time, resulting in obsolence for classification systems that assign grades based on current products. However, if appropriate tree characteristics, in addition to *dbh* and total height, are measured these characteristics can be related to quality for varying product categories regardless of changes in standards over time.

Sonderman and Brisbin (1978) evaluated a number of tree characteristics that could affect potential product quality and quantity of hardwood trees, namely *dbh*, crown class, crown ratio, total height, fork height, sweep and crook, rot and seams, and limb count and limb-related defects. Using these measurements they constructed an index system that included crown class, stem sweep, and number of limbs for ranking sample trees.

In a study of stem quality of planted loblolly pine trees in plots that were measured at 3-year intervals, Choi et al. (2008) classified individual trees as (i) single stem or forked, (ii) normal top or broken top, (iii) straight, bole sweep, butt sweep, or short crook, and (iv) no disease or disease. This hierarchical classification scheme results in 32 possible categories for each individual tree at each measurement occasion. Tree quality was then estimated using tree and stand characteristics as predictors, starting with the general form:

$$Quality = f(d, h, cc, h_r, t, S, N) \tag{3.38}$$

where d is diameter at breast height, h is total height, cc is crown class (dominant, codominant, intermediate, or suppressed), h_r is relative height ($h/\bar{h}$), t is stand age (years), S is site index, and N is trees per hectare.

Significant variables were selected using a backward elimination procedure. The dependent variable "Quality" is a discrete nominal response variable with *J* possible categories; hence a multinominal distribution results. Stem characteristics over time were modeled by fitting a system of multicategorical logit models.

There are inherent difficulties with quantifying wood quality as opposed to quantity, but statistical methodology for categorical data can be effectively applied to this problem. While quantitative studies of tree stem quality are relatively rare, previous results with varying objectives provide a beginning point for future work. Strub et al. (1986) modeled the proportion of saw timber trees in unthinned loblolly pine plantations. Burkhart and Bredenkamp (1989) and Amateis and Burkhart (2005) used similar methods to model the proportion of pulpwood, sawtimber, and peeler trees in thinned and unthinned loblolly pine plantations. Prestemon and Buongiorno (2000) developed an ordered-probit model to predict tree grade from tree and stand-level variables in natural uneven-aged southern pine stands. Zhang et al. (2005) evaluated the relationship of tree growth and stem quality characteristics with initial spacing in black spruce.

References

Alegria C, Tomé M (2011) A set of models for individual tree merchantable volume prediction for *Pinus pinaster* Aiton in central inland of Portugal. Eur J For Res 130:871–879

Amateis RL, Burkhart HE (1987) Cubic-foot volume equations for loblolly pine trees in cutover, site- prepared plantations. South J Appl For 11:190–192

Amateis RL, Burkhart HE (2005) The influence of thinning on the proportion of peeler, sawtimber, and pulpwood trees in loblolly pine plantations. South J Appl For 29:158–162

Avery TE, Burkhart HE (2002) Forest measurements, 5th edn. McGraw-Hill, New York

Baskerville GL (1972) Use of logarithmic regression in the estimation of plant biomass. Can J For Res 2:49–53

Brooks JR, Wiant HV Jr (2008) Ecoregion-based local volume equations for Appalachian hardwoods. North J Appl For 25:87–92

Bullock BP, Burkhart HE (2003) Equations for predicting green weight of loblolly pine trees in the South. South J Appl For 27:153–159

Burkhart HE (1977) Cubic-foot volume of loblolly pine to any merchantable top limit. South J Appl For 1(2):7–9

Burkhart HE, Bredenkamp BV (1989) Product-class proportions for thinned and unthinned loblolly pine plantations. South J Appl For 13:192–195

Byrne JC, Reed DD (1986) Complex compatible taper and volume estimation systems for red and loblolly pine. For Sci 32:423–443

Cao QV, Burkhart HE (1980) Cubic-foot volume of loblolly pine to any height limit. South J Appl For 4:166–168

Cao QV, Burkhart HE, Max TA (1980) Evaluation of two methods for cubic-volume prediction of loblolly pine to any merchantable limit. For Sci 26:71–80

Choi J, Burkhart HE, Amateis RL (2008) Modeling trends in stem quality characteristics of loblolly pine trees in unthinned plantations. Can J For Res 38:1446–1457

Chojnacky DC (1985) Pinyon-juniper volume equations for the Central Rocky Mountain States. USDA Forest Service, Intermountain Forest and Range Experiment Station, Ogden, Research Paper INT-339

Clutter JL (1980) Development of taper functions from variable-top merchantable volume equations. For Sci 26:117–120

Cunia T (1964) Weighted least squares method and construction of volume tables. For Sci 10: 180–191

Demaerschalk JP (1972) Converting volume equations to compatible taper equations. For Sci 18:241–245

Eerikäinen K (2001) Stem volume models with random coefficients for *Pinus kesiya* in Tanzania, Zambia, and Zimbabwe. Can J For Res 31:879–888

Fang Z, Baily RL (1999) Compatible volume and taper models with coefficients for tropical species on Hainan Island in southern China. For Sci 45:85–100

Figueiredo Filho A, Amaral Machado S, Araujo Carneiro MR (2000) Testing accuracy of log volume calculation procedures against water displacement techniques (xylometer). Can J For Res 30:990–997

Flewelling JW, Pienaar LV (1981) Multiplicative regression with lognormal errors. For Sci 27: 281–289

Gregoire TG, Dyer ME (1989) Model fitting under patterned heterogeneity of variance. For Sci 35:105–125

Gregoire TG, Schabenberger O (1996) A non-linear mixed-effects model to predict cumulative bole volume of standing trees. J Appl Stat 23:257–272

Gregoire TG, Schabenberger O, Barrett JP (1995) Linear modeling of irregularly spaced, unbalanced, longitudinal data from permanent-plot measurements. Can J For Res 25:137–156

Hann DW, Bare BB (1978) Comprehensive tree volume equations for major species of New Mexico and Arizona. USDA Forest Service, Intermountain Forest and Range Experiment Station, Ogden, Research Paper INT-209

Hann DW, Walters DK, Scrivani JA (1987) Incorporating crown ratio prediction equations for Douglas-fir stem volume. Can J For Res 17:17–22

Hibbs D, Bluhm A, Garber S (2007) Stem taper and volume of managed red alder. West J Appl For 22:61–66

Honer TG (1964) The use of height and squared diameter ratios for the estimation of merchantable cubic foot volume. For Chron 40:324–331

Honer TG (1965) A new total cubic foot volume function. For Chron 41:476–493

Johansson T (2005) Stem volume equations and basic density for grey alder and common alder in Sweden. Forestry 78:249–262

Jordan L, Berenhaut K, Souter R, Daniels RF (2005) Parsimonious and completely compatible taper, total, and merchantable volume models. For Sci 51:578–584

Knoebel BR, Burkhart HE, Beck DE (1984) Stem volume and taper functions for yellow-poplar in the southern Appalachians. South J Appl For 8:185–188

Laasasenaho J (1982) Taper curve and volume functions for pine, spruce and birch (*Pinus sylvestris, Picea abies, Betula pendula, Betula pubescens*). Commun Inst For Fenn 108:74

MacDonald GB, Forslund RR (1986) Application of a geometrical volume equation to species with different bole forms. Can J For Res 16:311–314

Martin AJ (1984) Testing volume equation accuracy with water displacement techniques. For Sci 30:41–50

McClure JP, Schreuder HT, Wilson RL (1983) A comparison of several volume table equations for loblolly pine and white oak. USDA Forest Service, Southeastern Forest Experiment Station, Asheville, Research Paper SE-240

McTague JP, Bailey RL (1987) Simultaneous total and merchantable volume equations and a compatible taper function for loblolly pine. Can J For Res 17:87–92

Newberry JD, Burkhart HE, Amateis RL (1989) Individual tree merchantable volume to total volume ratios based on geometric solids. Can J For Res 19:679–683

Nunes L, Tomé J, Tomé M (2010) A system for compatible prediction of total and merchantable volumes allowing for different definitions of tree volume. Can J For Res 40:747–760

Özçelik R (2008) Comparison of formulae for estimating tree bole volumes of *Pinus sylvestris*. Scand J For Res 23:412–418

Prestemon JP, Buongiorno J (2000) Determinants of tree quality and lumber value in natural uneven-aged southern pine stands. Can J For Res 30:211–219

Reed DD, Green EJ (1984) Compatible stem taper and volume ratio equations. For Sci 30:977–990

Rustagi KP, Loveless RS Jr (1991) Improved cubic volume prediction using a new measure of form-factor. For Ecol Manage 40:1–11

Schumacher FX, Hall FS (1933) Logarithmic expression of timber-tree volume. J Agric Res 47:719–734

Sherrill JR, Bullock BP, Mullin TJ, McKeand SE, Purnell RC (2011) Total and merchantable stem volume equations for midrotation loblolly pine (*Pinus taeda* L.). South J Appl For 35:105–108

Snowdon P (1991) A ratio estimator for bias correction in logarithmic regressions. Can J For Res 21:720–724

Sonderman DL, Brisbin RL (1978) A quality classification system for young hardwood trees- the first step in predicting future products. USDA Forest Service, Northeastern Forest Experiment Station, Broomall, Research Paper NE-419

Strub MR, Green EJ, Burkhart HE, Pirie WR (1986) Merchantability of loblolly pine – an application for nonlinear regression with a discrete dependent variable. For Sci 32:254–261

Tassisa G, Burkhart HE, Amateis RL (1997) Volume and taper equations for thinned and unthinned loblolly pine trees in cutover, site-prepared plantations. South J Appl For 21:146–152

Teshome T (2005) A ratio method for predicting stem merchantable volume and associated taper equations for *Cupressus lusitanica*, Ethiopia. For Ecol Manage 204:171–179

Van Deusen PC, Sullivan AD, Matney TG (1981) A prediction system for cubic foot volume of loblolly pine applicable through much of its range. South J Appl For 5:186–189

Van Deusen PC, Matney TG, Sullivan AD (1982) A compatible system for predicting the volume and diameter of sweetgum trees to any height. South J Appl For 6:159–163

Williams MS, Gregoire TG (1993) Estimating weights when fitting linear regression models for tree volume. Can J For Res 23:1725–1731

Williams MS, Schreuder HT, Gregoire TG, Bechtold WA (1992) Estimating variance functions for weighted linear regression. In: Proceedings of the conference on applied statistics in agriculture, Kansas State University, pp 153–160

Zhang Y, Borders BE, Bailey RL (2002) Derivation, fitting and implication of a compatible stem taper-volume-weight system for intensively managed, fast growing loblolly pine. For Sci 48:595–607

Zhang SY, Lei YC, Bowling C (2005) Quantifying stem quality characteristics in relation to initial spacing and modeling their relationship with tree characteristics in black spruce (*Picea mariana*). North J Appl For 22:85–93

Chapter 4
Tree Weight and Biomass Estimation

Weight scaling—that is, determining the weight of tree boles that have been cut and delivered to a converting facility—has become common practice because it is direct, efficient, and it encourages delivery of freshly cut material. As a result, much effort has been devoted to developing equations to estimate the weight, rather than the volume, of the commercial portion of tree stems. Furthermore, weight estimates are more easily obtained and more practical to apply than volume units when estimates for parts of trees such as branches and foliage in addition to stems are also desired.

Increasing interest in predicting fuel loads and potential fire behavior, quantifying carbon cycles and storage, and sustainable forest management has heightened the need for reliable allometric equations to estimate tree and shrub biomass. Accordingly, modelers have developed improved methods for estimating biomass of component parts as well as total biomass for a variety of tree and shrub species.

4.1 Estimating Green Weight of Stems

Green weight (i.e. weight prior to any seasoning or drying) per unit volume of tree stems varies according to wood density and moisture content. While there is some variation in green weight per unit volume throughout the length of tree boles, this variation is typically not large. Thus model forms applied and research results obtained from work with estimating tree-stem cubic volume apply equally well when weight units are used as the dependent variable. The tree-stem equation forms discussed in Chap. 3 and summarized in Table 3.1 for estimating cubic volume have been used with equivalent results for estimating weight. Fitting a prediction equation requires accurate determination of tree weight, along with measurements of the predictor variables, on a sample of felled trees.

Baldwin (1987) used a modified form of Schumacher and Hall's (1933) volume model to predict green and dry weight of the total bole of loblolly pine trees namely:

$$w = b_1 d^{b_2} h^{b_3} p^{b_4} \tag{4.1}$$

H.E. Burkhart and M. Tomé, *Modeling Forest Trees and Stands*,
DOI 10.1007/978-90-481-3170-9_4,

where

w = predicted weight
d = diameter at breast height
h = total height
p = any additional predictor variable specified

The final form of the total bole weight prediction equation with $p = \exp(t^2)$ was

$$\ln w = b_1 + b_2 \ln d + b_3 \ln h + b_4 t^2 \tag{4.2}$$

where

t = tree age

Bole weights from the stump to any top diameter were predicted from total bole weight and a weight ratio equation. The ratio equation, to predict the proportion of total stem weight below a given upper-stem diameter limit, employed was

$$R_{di} = \exp(b_1 d_i^{b_2} d^{b_3}) \tag{4.3}$$

where

R_{di} = estimated ratio of partial to total bole weight
d_i = upper-stem diameter limit
d = diameter at breast height
b_1, b_2, b_3 = coefficients estimated from data.

In a study involving data on green weight of 970 loblolly pine trees from across the species commercial range in the southeastern USA, Bullock and Burkhart (2003) used the combined-variable equation to estimate total bole weight:

$$w = b_0 + b_1 d^2 h \tag{4.4}$$

Where w = total stem green weight and other symbols remain as previously defined. Two forms of a weight ratio equation for varying top-diameter limits were evaluated:

$$R_{di} = 1 + b_1 (d_i^{b_2} d^{b_3}) \tag{4.5}$$

and

$$R_{di} = \exp(b_4 d_i^{b_5} d^{b_6}) \tag{4.6}$$

as well as the analogous two forms for weight ratios to varying upper-height limits

$$R_{hi} = 1 + a_1 ((h - h_i)^{a_2} h^{a_3}) \tag{4.7}$$

$$R_{hi} = \exp(a_4 (h - h_i)^{a_5} h^{a_6}) \tag{4.8}$$

In comparing the two forms of ratio equations, Eq. 4.6 was best for predicting green weight to any upper stem diameter limit, and Eq. 4.7 was recommended for predictions to any upper height limit. These results with stem weight as the variable of interest are consistent with results reported when cubic volume was the unit of measure used (e.g. Tasissa et al. 1997).

As a further means of evaluating the relative performance of the two ratio model forms, Bullock and Burkhart (2003) compared the implicit taper functions for Eqs. 4.5 and 4.7 and for 4.6 and 4.8. Mixed taper forms were also derived by using combinations of the two ratio model forms. The combination of ratio models (4.6) for top diameter limits and (4.7) for upper height limits (the best form for predicting, respectively, weight to a given top diameter and to a given height limit) resulted in the most accurate implicit taper relationships for predicting diameter at specified heights and for predicting height at a specified diameter. (The implicit taper function for this "mixed forms" combination is given in Sect. 3.3.1.)

4.2 Estimating Dry Weight of Stems

Although wood is commonly bought and sold on a green weight basis, estimates of dry weight may be desired when the wood will be utilized for pulped products or conversion to energy. Dry weight is more highly correlated with the yield of pulped products than is green weight, which is influenced by variations in moisture content. Dry weight may also be the desired metric in studies of biomass productivity. For some species wood density is not closely related to tree age and dry weight of stems can be estimated with the predictors *dbh* and total tree height. An application of the combined-variable equation

$$w = b_0 + b_1 d^2 h$$

to estimate the dry weight of wood and bark (w) for plantation-grown sycamore trees resulted in an r^2 value of 0.99 (Belanger 1973).

Wood density (specific gravity) for some species, however, varies by age to an extent that age, in addition to tree diameter and height, should be included when predicting the dry weight of stems. Tree age can be directly incorporated by specifying

$$\text{Dry weight} = f\,(d, h, t)$$

or a separate equation to estimate specific gravity $= f\,(t)$ can be fitted and dry weight can be estimated as

$$\text{Dry weight} = (\text{cubic volume})\,(\text{specific gravity})\,(\text{K}) \tag{4.9}$$

where cubic volume can be m^3 or ft^3, specific gravity is the dry weight of wood divided by the weight of an equal volume of water, and K is the weight of a cubic unit of water (1 g/cm^3 or 1,000 kg/m^3 or 62.4 lbs/ft^3).

Specific gravity for the whole stem or for a specific merchantable part of the stem has been estimated (Burkhart and Beckwith 1970, and others) from tree age (t) with the following model

$$SG = b_0 + b_1 t^{-1} \tag{4.10}$$

If a cubic volume equation is available or fitted for estimating contents of the same portion of the stem as the specific gravity function, dry weight values for trees of varying species and ages can be obtained. Assuming the combined-variable function for cubic volume

$$v = b_0 + b_1 d^2 h \tag{4.11}$$

multiplying Eqs. 4.10 and 4.11 and rewriting as

$$w = c_0 + c_1 t^{-1} + c_2 d^2 h + c_3 (d^2 h)(t^{-1}) \tag{4.12}$$

results in an implied dry weight prediction equation of the combined-variable form with intercept and slope varying by tree age.

Equations such as (4.10) are useful for estimating specific gravity of a specified portion of tree stems as a function of the tree's age. Wood density varies from stump to tip and from pith to bark for many tree species, which has resulted in efforts to model specific gravity on an individual ring and relative stem height basis (Tasissa and Burkhart 1998, and others). A taper function can be used in conjunction with tree diameter and height increment prediction equations to establish an estimated sheath of wood year by year. Wood density of annual rings at specified heights can then be predicted and dry weight of given portions of stems can be determined.

Parresol and Thomas (1989) developed a generalized density-integral model for estimating tree-stem biomass. The mass (M) for a lamina with a continuous density function $\rho(x, y)$ is commonly defined as:

$$M = \iint_R \rho(x, y) dA$$

The 3-dimensional structure of a tree can be represented as a 2-dimensional lamina by taking cross-sectional area (y) as one dimension and height (h_i) as the other dimension (Fig. 4.1). Parresol and Thomas (1989) noted that moving from lower limit h_1 to upper limit h_u in Fig. 4.1, the variable y is seen to vary from 0 to upper limit $f(h_i)$, which establishes the following model for stem biomass:

$$w = \int_{h_1}^{h_u} \int_0^{f(h_i)} \rho(h_i, y) dy dh_i$$

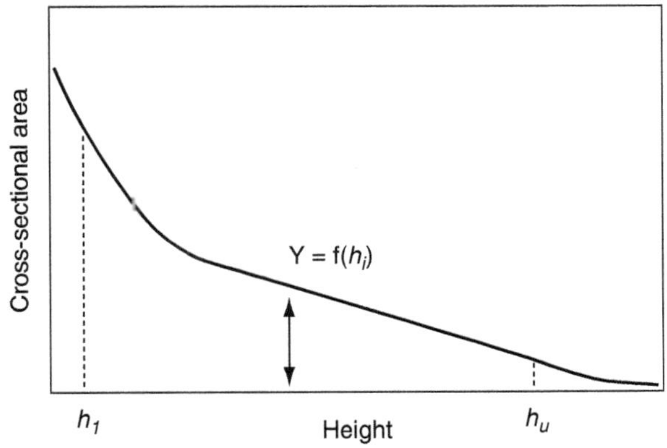

Fig. 4.1 Excurrent tree form represented as a 2-dimensional lamina (From Parresol and Thomas 1989)

where $f(h_i)$ is an equation expressing taper in cross-sectional area as a function of height, and w is bole dry weight of wood between limits h_1 and h_u. Typically stem profile (cross-sectional area) is modeled using relative height ($x = h_i / h$) instead of actual height. Performing a change of variable from h to x results in the following generalized stem biomass model:

$$w = h \int_{x_1}^{x_u} \int_{0}^{f(x)} \rho(x, y) dy dx$$

For a specific biomass model one needs to define ρ and f. As a demonstration of the density integral approach to tree bole biomass estimation, Parresol and Thomas (1989) derived specific equations for slash pine in the West Gulf region of the United States.

A number of prediction variables were examined but only relative height and age were important for estimating bole specific gravity:

$$SG = \mathrm{b}_0 + \mathrm{b}_1 x + \mathrm{b}_2 t \tag{4.13}$$

where SG is specific gravity, x is relative height, t is plantation age (years) and b_0, b_1, b_2 are coefficients estimated from data. For this initial investigation of the density-integral approach, Parresol and Thomas (1989) used the simple taper function of Kozak et al. (1969)

$$d_i^2/d^2 = \mathrm{c}_1(x - 1) + \mathrm{c}_2(x^2 - 1)$$

where d_i is upper stem diameter, d is dbh, and c_1, c_2 are coefficients estimated from data. The specific integral weight equation for slash pine stem dry weight was then derived as follows:

$$\rho(x, y) = (\mathrm{b}_0 + \mathrm{b}_1 x + \mathrm{b}_2 t)\mathrm{k}_1$$

and

$$f(x) = d^2\left[c_1(x-1) + c_2(x^2-1)\right]k_2$$

where $k_1 = 1{,}000$ (the factor for converting *SG* to density in kg/m^3) and $k_2 = \pi/40{,}000$ (the factor converting d_i^2 to cross-sectional area), the following specific model results:

$$w = h\int_{x_1}^{x_u}\int_0^{d^2\left[c_1(x-1)+c_2(x^2-1)\right]k_2} k_1(b_0 + b_1x + b_2t)dydx$$

Performing the integration gives:

$$w = \frac{\pi}{40}d^2h\left\{\begin{array}{l}[b_0 + b_2t]\left[-(c_1+c_2)(x_u-x_1) + \dfrac{c_1(x_u^2-x_1^2)}{2} + \dfrac{c_2(x_u^3-x_1^3)}{3}\right]\\ +b_1\left[\dfrac{-(c_1+c_2)}{2}(x_u^2-x_1^2) + \dfrac{c_1(x_u^3-x_1^3)}{3} + \dfrac{c_2(x_u^4-x_1^4)}{4}\right]\end{array}\right\}$$

where w is dry weight (g), and $\pi/40$ is k_1k_2.

Parresol and Thomas (1989) compared the integrated model and total stem and weight ratio equations for predicting sectional stem weight and found that the density-integral model performed well. They conjectured that employing a more complex taper function in the integrated modeling approach would lead to more accurate estimates of sectional bole weights. In two follow-up studies (Thomas et al. 1995; Parresol and Thomas 1996) a trigonometric taper function (Thomas and Parresol 1991) was used to describe stem profiles in the density-integral system:

$$d_i{}^2/d^2 = b_1(x-1) + b_2\sin(b_3\pi x) + b_4\cot(\pi x/2) \qquad (4.14)$$

where x is relative height (h_i/h). Bole specific gravity (*SG*) was predicted from relative height and tree age (t) with Eq. 4.13.

The integral weight equation was then derived using Eqs. 4.13 and 4.14 forming a system of three linear statistical models with non-linear cross-equation constraints (Parresol and Thomas 1996). When contemporaneous correlations are present in systems of equations, seemingly unrelated regression (SUR) estimation yields more efficient parameter estimates than ordinary least squares. The null hypothesis that the contemporaneous covariances are zero was rejected; therefore the system of equations was fitted using SUR as well as OLS. Parameter estimates for fitting with OLS and with SUR to data for slash pine and for willow oak were similar but the standard errors of the parameter estimates were smaller for SUR in all cases. For slash pine, standard errors were reduced by 11–29% and for willow oak by 5–20%.

4.3 Biomass Estimation

Estimates of forest biomass are needed for forest management and for scientific study purposes. Woody biomass may be converted to various end uses including firewood, biofuels, and pulp products. Studies of ecosystem productivity, energy and nutrient flows, and carbon cycles require the ability to assess forest biomass. In many instances the estimates of above ground biomass (stem, branches, foliage) suffice, but in some cases below ground components (roots) are also included. Tree biomass (components and total) is commonly estimated using fitted regression relationships. A sample of trees is chosen through an appropriate selection procedure for destructive sampling, and the dry weights of the components and total weight of each tree are determined and related by regression analysis to easily measured dimensions of standing trees.

4.3.1 Models for Biomass Estimation

Researchers have used a variety of regression models for estimating total-tree and tree-component biomass; prediction equations typically are of three forms (Parresol 1999):

$$\text{Linear (additive error)}: \ y = \beta_0 + \beta_1 x_1 + \cdots \beta_j x_j + \varepsilon \tag{4.15}$$

$$\text{Nonlinear (additive error)}: \ y = \beta_0 x_1^{\beta_1} x_2^{\beta_2} \ldots x_j^{\beta_j} + \varepsilon \tag{4.16}$$

$$\text{Nonlinear (multiplicative)}: \ y = \beta_0 x_1^{\beta_1} x_2^{\beta_2} \ldots x_j^{\beta_j} \varepsilon \tag{4.17}$$

where $y =$ total or component biomass, $x_j =$ tree dimension variable, $\beta_j =$ model parameter, and $\varepsilon =$ error term. Some commonly used tree dimension variables are diameter at breast height (*d*), d^2, total height (*h*), d^2h, age (*t*) and live crown length (c_l). Diameter at the base of the live crown has proved to be one of the best predictor variables for crown weight (Clark 1982). Biomass data often exhibit heteroscedasticity; that is the error variance is not constant at all levels of x. If models (4.15) and (4.16) are fitted to such data, weighted analysis is necessary to achieve minimum variance parameter estimates (assuming all other regression assumptions are met). Nonlinear regression equations such as (4.17) are often transformed into linear (additive error) regression equations by taking the logarithm of both sides of the equation. In this form, the equation parameters can easily be estimated by least squares procedures. Typically, the variance of y is not uniform across the domain of one or more of the x_j's; however, when transformed to logarithms, model (4.17) generally has homoscedastic (constant) variance. The logarithmic equation form is

$$\ln y = \ln b_0 + b_1 \ln x_1 + \cdots + b_j \ln x_j + \ln \varepsilon \tag{4.18}$$

where ln is natural logarithm. Goodness-of-fit statistics related to the transformed equation and are not directly comparable with the same statistics produced through use of either model (4.15) or (4.16). Although logarithmic transformation may be used, it is usually desirable to express estimated values of y in arithmetic (i.e., untransformed) units. The antilogarithm of $\ln y$ is not an unbiased estimate of the arithmetic mean of y. If $\hat{\mu} = \hat{y}$ and $\hat{\sigma}^2$ = sample variance of the logarithmic equation, then $\hat{y}$ is often estimated as

$$\hat{y} = \exp(\hat{\mu} + \hat{\sigma}^2/2)$$

However, some researchers (e.g. Madgwick and Satoo 1975; Hepp and Brister 1982) have indicated that the logarithmic bias correction factors commonly employed tend to overestimate and hence to overcorrect, the true bias. As an alternative to logarithmic transformation for stabilizing error variance, if the error variance is a function of a small number of unknown parameters and if these parameters can be consistently estimated, generalized least squares may be used to obtain asymptotically efficient estimates of the model parameters.

4.3.2 Additivity of Linear Biomass Equations

A desirable feature of tree biomass component regression equations is that the predictions for the components sum to the prediction for biomass of the total tree. Kozak (1970), Chiyenda and Kozak (1984), and Cunia and Briggs (1984, 1985a) have discussed the problem of ensuring additivity for a set of tree biomass functions. In a review paper on assessing tree and stand biomass, Parresol (1999) summarized methods for forcing additivity by identifying three procedures.

In procedure 1, the total biomass regression function is defined as the sum of the individually calculated best regression functions of the k biomass components:

$$\hat{y}_1 = f_1(\mathbf{x'}_1)$$

$$\hat{y}_2 = f_2(\mathbf{x'}_2)$$

$$\vdots$$

$$\hat{y}_k = f_k(\mathbf{x'}_k)$$

$$\hat{y}_{total} = \hat{y}_1 + \hat{y}_2 + \cdots + \hat{y}_k$$

In procedure 2, the additivity of the components is ensured by using the same independent variables and the same weight function in the weighted least squares linear regressions of the biomass of each component and that of the total. Under this method, the regression coefficients of the total equation are computed by summing the regression coefficients of the (assumed independent) component equations (the $\mathbf{b_i}$ vectors), that is:

$$\hat{y}_1 = \mathbf{x}'\mathbf{b_1}$$

$$\hat{y}_2 = \mathbf{x}'\mathbf{b_2}$$

$$\hat{y}_k = \mathbf{x}'\mathbf{b_k}$$

$$\hat{y}_{total} = \mathbf{x}'\left[\mathbf{b_1} + \mathbf{b_2} + \cdots + \mathbf{b_k}\right]$$

This result holds only under the restrictive assumption that the k components $y_i\,(i = 1, \ldots, \mathrm{k})$ are independent, which implies that the $\varepsilon_i\,(i = 1, \ldots, \mathrm{k})$ are uncorrelated.

In procedure 3, which is general and flexible, statistical dependencies among sample data are accounted for using generalized least squares regression with dummy variables techniques to calculate a set of regression functions such that: (1) each component regression contains its own independent variables, and the total-tree regression is a function of all independent variables used; (2) each regression can use its own weight function; and (3) additivity is ensured by setting constraints (i.e., linear restrictions) on the regression coefficients. In the seemingly unrelated regression (SUR) formulation for a set of contemporaneously correlated linear statistical models with cross-equation constraints, the structural equations for a system of models of biomass additivity can be specified as

$$\mathrm{y_1} = f_1(\mathbf{X_1}) + \varepsilon_1$$

$$\mathrm{y_2} = f_2(\mathbf{X_2}) + \varepsilon_2$$

$$\vdots$$

$$\mathrm{y_k} = f_k(\mathbf{X_k}) + \varepsilon_\mathrm{k}$$

$$\mathrm{y}_{total} = f_{total}(\mathbf{X_1}, \mathbf{X_2}, \ldots, \mathbf{X_k}) + \varepsilon_{total}$$

and redundant columns in f_{total} are eliminated. When the stochastic properties of the error vectors are specified, along with the linear restrictions, the structural equations form a statistical model for efficient parameter estimates and reliable prediction intervals.

Parresol (1999) presented an example of fitting biomass data according to procedures 1, 2, and 3 using data from 39 willow oak trees. The individual "best" biomass component equations were:

$$\hat{y}_{wood} = \mathrm{b}_0 + \mathrm{b}_1 d^2 h$$

$$\hat{y}_{bark} = \mathrm{b}_0 + \mathrm{b}_1 d^2$$

$$\hat{y}_{crown} = \mathrm{b}_0 + \mathrm{b}_1 (d^2 h)(c_l) + \mathrm{b}_2 h$$

For total tree biomass, the best individual equation was:

$$\hat{y}_{total} = \mathrm{b}_0 + \mathrm{b}_1 d^2 h$$

Scatter plots of residuals indicated increasing error variance, and error models were specified to account for the heteroscedasticity. By way of example, for procedure 1 for additivity, total tree biomass is simply the sum of the components. In an example for a tree with $d = 30$ cm, $h = 18$ m, and $c_l = 10$ m the total biomass is estimated as 603.4 kg with an approximate 95% confidence interval of ± 111.7 kg. By contrast, OLS fit for total tree biomass gives a mean and approximate 95% prediction interval of 557.6 kg $\pm$ 102.8 kg. Ensuring additivity using procedure 1 resulted in the prediction interval increasing from ± 102.8 to ± 111.7, indicating a loss in efficiency.

Continuing with the example using data from 39 willow oak trees, a set of linear models with statistical dependence among components and total tree biomass was estimated using SUR methods. The following equations and variance functions were employed for the willow oak sample data:

$$\hat{y}_{wood} = \mathrm{b}_{10} + \mathrm{b}_{11} d^2 h;$$

$$\hat{\sigma}^2 = (d^2 h)^{1.95}$$

$$\hat{y}_{bark} = \mathrm{b}_{20} + \mathrm{b}_{21} d^2 h;$$

$$\hat{\sigma}^2 = (d^2 h)^{1.745}$$

$$\hat{y}_{crown} = \mathrm{b}_{30} + \mathrm{b}_{31} (d^2 h)(c_l) + \mathrm{b}_{32} h;$$

$$\hat{\sigma}^2 = \left[\frac{(d^2 h)(c_l)}{10{,}000}\right]^{1.646} \times \exp\left[-0.00406 h^2\right]$$

$$\hat{y}_{total} = \mathrm{b}_{40} + \mathrm{b}_{41} d^2 h + \mathrm{b}_{42} (d^2 h)(c_l) + \mathrm{b}_{43} h;$$

$$\hat{\sigma}^2 = (d^2 h)^{1.844}$$

where $b_{40} = b_{10} + b_{20} + b_{30}$, $b_{41} = b_{11} + b_{21}$, $b_{42} = b_{31}$, and $b_{43} = b_{32}$. For system parsimony, the $\hat{y}_{bark}$ equation was modified by using d^2h rather than d^2 (the predictor in the "best" individual equation).

The estimated total biomass for the example willow oak tree was 583.8 kg when using parameter estimates from the SUR fitting, with prediction limits of ±93.8 kg. This prediction interval on $\hat{y}_{total}$ is narrower than the least squares prediction interval on $\hat{y}_{total}$. One might expect the individually best regression on $\hat{y}_{total}$ to have the smallest variance, because it is the best estimator that is a linear unbiased function of $\hat{y}_{total}$. However, when contemporaneous correlations are present, it is possible to obtain a better linear unbiased estimator that is a function of $\hat{y}_{wood}$, $\hat{y}_{bark}$, $\hat{y}_{crown}$ and $\hat{y}_{total}$. Thus, even under the constraint of additivity, the SUR estimator can achieve lower variance and be a more efficient estimator than OLS.

Russell et al. (2009) developed a system of linear equations to predict the component and total tree masses of young loblolly pine trees grown in a high-density spacing trial. Their system of equations, fitted using seemingly unrelated regression (SUR) methods which ensured that the sum of the component masses equaled the tree mass, was:

$$\hat{y}_{stem} = b_{10} + b_{11}{d_{gl}}^2h + b_{12}t$$

$$\hat{y}_{foliage} = b_{20} + b_{21}{d_{gl}}^2c_l$$

$$\hat{y}_{branch} = b_{30} + b_{31}{d_{gl}}^2h$$

$$\hat{y}_{root} = b_{40} + b_{41}{d_{gl}}^2h$$

$$\hat{y}_{total} = b_{50} + b_{51}{d_{gl}}^2h + b_{52}{d_{gl}}^2c_l + b_{53}t$$

where $\hat{y}_c$ ($c =$ stem, foliage, branch, root, total) is the observed component (or total tree) mass (g), d_{gl} is the groundline diameter (mm), h is the total tree height (cm), t is the tree age (years), c_l is the live crown length (cm), and b_{ij} are the coefficients to be estimated for the jth parameter in the ith equation.

4.3.3 Additivity of Nonlinear Biomass Equations

Three procedures for insuring additivity for a set of linear tree biomass equations were presented by Parresol (1999). In a subsequent paper, Parresol (2001) illustrated two procedures for forcing additivity for nonlinear models. The procedures differ, depending on how the separate components are aggregated. The following discussion of these methods is based on the presentation of Parresol (2001).

In procedure 1, subscripts refer to tree biomass components (stem, branches, etc.), and the total biomass regression function is defined as the sum of the separately calculated best regression functions for the components of biomass:

$$\hat{y}_1 = f_1(\mathbf{x}_1, \mathbf{b}_1)$$
$$\hat{y}_2 = f_2(\mathbf{x}_2, \mathbf{b}_2)$$
$$\vdots$$
$$\hat{y}_c = f_c(\mathbf{x}_c, \mathbf{b}_c)$$
$$\hat{y}_{total} = \hat{y}_1 + \hat{y}_2 + \cdots + \hat{y}_c$$

Reliability (i.e., confidence intervals) of the total biomass prediction can be determined from variance properties of linear combinations:

$$\operatorname{var}(\hat{y}_{total}) = \sum_{i=1}^{c} \operatorname{var}(\hat{y}_i) + 2\sum\sum_{i<j} \operatorname{cov}(\hat{y}_i, \hat{y}_j)$$

where
$\operatorname{cov}(\hat{y}_i, \hat{y}_j) = \hat{\rho}_{\hat{y}_i,\hat{y}_j} \sqrt{\operatorname{var}(\hat{y}_i)\operatorname{var}(\hat{y}_j)}$
$\hat{\rho}_{y_i,y_j}$ = estimated correlation between $\hat{y}_i$ and $\hat{y}_j$.

In procedure 2, which is more general and flexible than procedure 1, statistical dependencies (i.e., contemporaneous correlations) among sample data are accounted for using nonlinear joint-generalized least squares regression, also known as nonlinear seemingly unrelated regressions (NSUR). A set of nonlinear regression functions is specified such that (i) each component regression constrains its own independent variables, and the total-tree regression equation is a function of all independent variables used; (ii) each regression equation can employ an appropriate weight function; and (iii) additivity is ensured by setting constraints (restrictions) on the regression coefficients. The structural equations for the system of nonlinear models of biomass prediction with additivity ensured can be specified as:

$$y_1 = f_1(X_1, \beta_1) + \varepsilon_1$$
$$\mathrm{y}_2 = f_2(\mathrm{X}_2, \beta_2) + \varepsilon_2$$
$$\vdots$$
$$\mathrm{y}_c = f_c(\mathrm{X}_c, \beta_c) + \varepsilon_c$$
$$\mathrm{y}_{total} = f_{total}(\mathrm{X}_1, \mathrm{X}_2, \ldots, \mathrm{X}_c, \beta_1, \beta_2, \ldots, \beta_c) + \varepsilon_{total}$$

When the stochastic properties of the error of the vectors are specified along with the coefficient restrictions, the structural equations become a statistical model that allows for efficient parameter estimation and reliable prediction intervals.

The procedure 2 or NSUR method is generally preferable to procedure 1. If disturbances or errors in the different equations are correlated (the usual situation for biomass models), then procedure 2 is superior to procedure 1, because NSUR takes into account the contemporaneous correlations and results in lower variance (Parresol 2001).

Procedures 1 and 2 were illustrated by Parresol (2001) with a numeric example using data from a sample of 40 slash pine trees. The structural equations for procedure 1 were

$$\hat{y}_{wood} = \mathrm{b}_1 (d^2 h)^{\mathrm{b}_2}$$

$$\hat{y}_{bark} = \mathrm{b}_1 d^{\mathrm{b}_2}$$

$$\hat{y}_{crown} = \mathrm{b}_1 d^{\mathrm{b}_2} h^{\mathrm{b}_3}$$

$$\hat{y}_{total} = \text{sum of components}$$

where d is diameter at breast height and h is total tree height. Appropriate weighting functions were applied to account for heteroscedasticity.

Procedure 2 involves a set of nonlinear models with allowance for statistical dependence among components and the total tree biomass. Parameters are estimated using nonlinear seemingly unrelated regression (NSUR) with parameter restrictions that ensure additive predictions. The model for total biomass must consist of a combination of the component biomass models to be additive, thus using the same component equations as those used in procedure 1, the NSUR formation is:

$$\hat{y}_{wood} = \mathrm{b}_{11} (d^2 h)^{\mathrm{b}_{12}}$$

$$\hat{y}_{bark} = \mathrm{b}_{21} d^{\mathrm{b}_{22}}$$

$$\hat{y}_{crown} = \mathrm{b}_{31} d^{\mathrm{b}_{32}} h^{\mathrm{b}_{33}}$$

$$\hat{y}_{total} = \mathrm{b}_{11} (d^2 h)^{\mathrm{b}_{12}} + \mathrm{b}_{21} d^{\mathrm{b}_{22}} + \mathrm{b}_{31} d^{\mathrm{b}_{32}} h^{\mathrm{b}_{33}}$$

where $\hat{y}_{total}$ is restricted to have the same independent variables and coefficients as the component equations. The same error or weight functions that were used for the three component models in procedure 1 were applied and an error function for $\hat{y}_{total}$ was determined.

For a tree with $d = 20$ cm and $h = 17$ m, Parresol (2001) computed the estimated total tree mass for the equations fitted to slash pine data using procedure 1 to be 268.1 kg with an approximate 95% confidence interval of ±9.3 kg. The comparable figures for procedure 2 were 264.2 ± 6.3 kg. The NSUR confidence interval was 32% narrower (±6.3 versus ±9.3 kg) than that computed using procedure 1.

In the example just presented the tree above ground was divided into three components: bole wood, bole bark, and crown. If the crown component were separated into two subcomponents, foliage and branches, one might wish to apply restrictions on the subcomponent equations to give predictions that sum to the prediction from the crown regression equation, while still maintaining the overall additivity to the total. To maintain the property of additivity under this scenario while employing NSUR estimation procedures requires that the crown component model be a combination of the subcomponent models with appropriate parameter restrictions. Parresol (2001) noted that the system of biomass equations could be expanded by adding equations for foliage and branches as:

$$\hat{y}_{wood} = \mathrm{b}_{11}(d^2h)^{\mathrm{b}_{12}}$$

$$\hat{y}_{bark} = \mathrm{b}_{21}d^{\mathrm{b}_{22}}$$

$$\hat{y}_{foliage} = \mathrm{b}_{31}d^{\mathrm{b}_{32}}h^{\mathrm{b}_{33}}$$

$$\hat{y}_{branches} = \mathrm{b}_{41}(d\,c_l)^{\mathrm{b}_{42}}$$

$$\hat{y}_{crown} = \mathrm{b}_{31}d^{\mathrm{b}_{32}}h^{\mathrm{b}_{33}} + \mathrm{b}_{41}(d\,c_l)^{\mathrm{b}_{42}}$$

$$\hat{y}_{total} = \mathrm{b}_{11}(d^2h)^{\mathrm{b}_{12}} + \mathrm{b}_{21}d^{\mathrm{b}_{22}} + \mathrm{b}_{31}d^{\mathrm{b}_{32}}h^{\mathrm{b}_{33}} + \mathrm{b}_{41}(d\,c_l)^{\mathrm{b}_{42}}$$

The expressed restrictions in this system guarantee that the foliage and branch predictions will sum to equal the crown prediction, plus the total prediction will still equal the sum of the component predictions. With appropriate restrictions many layers can be added to a system while maintaining the property of additivity.

Nonlinear seemingly unrelated regression methods have been applied in a number of studies aimed at producing a consistent set of tree biomass prediction equations. Carvalho and Parresol (2003) applied NSUR for estimating mass of the stem, crown and tree for Pyrenean oak in Portugal. A system of nonlinear additive biomass equations was developed for 15 native species of eucalyptus in Australia by Bi et al. (2004). Data from 441 trees sampled on several sites representative of eucalyptus plantings in Portugal were used by António et al. (2007) to simultaneously fit, using seemingly unrelated regression, biomass of stem wood, stem bark, foliage, and branches. Sabatia et al. (2008) utilized NSUR to fit biomass component equations for shortleaf pine; the component equations were constrained in the estimation process to sum to total tree biomass.

4.3.4 Inclusion of Additional Predictor Variables

In developing tree biomass equations researchers have evaluated the use of various tree variables, as well as stand and site variables, as predictors. Diameter at breast height is a universally used predictor because it shows a good relationship with tree biomass components and it is easy to obtain.

Several studies have concluded that height, as an additional predictor, only adds marginally to the predictive ability of diameter-based biomass regressions (e.g. Ter-Mikaelian and Korzukhin 1997; Johansson 1999; Verwijst and Telenius 1999; Porté et al. 2002; Jenkins et al. 2003). In the models fitted by Parresol (1999), height was a useful predictor for stem wood but not for stem bark. Lambert et al. (2005) found that tree height improved the accuracy of the stem equations but not of the crown equations. Bi et al. (2004) reported similar conclusions. The combined variable of diameter and tree height (d^2h) performed better as a predictor for stem and bark components than diameter alone, but not for branch and leaf components. Pitt and Bell (2004) and Ter-Mikaelin and Parker (2000) also found large increases in predictive ability of stem biomass, but found needle and branch biomass models to be invariant to the addition of height. It has been argued that height can serve to adjust predictions for effects such as stand age, density and site quality (e.g. Ter-Mikaelin and Parker 2000).

In the equations developed by António et al. (2007) the use of height implied decreases in the sums of squares for press residuals (*SSrp*) by 72%, 8%, 12% and 10%, respectively, for stem wood, stem bark, leaves and branches. Furthermore, the use of crown length instead of height in the models for the crown components led to greater decreases in *SSrp* of 29% and 19%, respectively, for leaves and branches. António et al. (2007) detected no impact of stand density on tree allometry. The addition of tree height compensated for effects of differences in stand density, as found by Pitt and Bell (2004) for thinning treatments.

Porté et al. (2002) reported that stand age greatly improved the predictive ability of the branch and stem models they developed for maritime pine. A change in the allometric relationships with stand development was also noted by Bond-Lamberty et al. (2002), Saint-André et al. (2005) and Shaiek et al. (2011) who introduced age as a supplementary variable in biomass prediction equations.

Harrington and Fownes (1993) found that the allometry of woody biomass and leaf area did not differ between planted and coppice stands in fast-growing tropical tree species. For António et al. (2007) the introduction of a dummy variable representing coppice produced a significant reduction in the residual sum of squares but the gain in predictive ability was not considered large enough to justify this additional model complexity.

Crow (1978) and Ketterings et al. (2001) found an influence of site factors on tree allometry but this result has not been reported by other authors (e.g. António et al. 2007 or Shaiek et al. 2011). Most studies suggest that it is possible to use the same model for biomass prediction across different sites and regions provided that height is included and that the stage of development of the stand is taken into account.

References

Alemdag IS, Hort KW (1981) Single-tree equations for estimating biomass of trembling aspen, largetooth aspen and white birch in Ontario. For Chron 57:169–173

Amateis RL, Burkhart HE, Dunham PH (1992) Estimating dry weight of dormant-season foliage of loblolly pine. Biomass Bioenergy 3:319–322

António N, Tomé M, Tomé J, Soares P, Fontes L (2007) Effect of tree, stand, and site variables on the allometry of *Eucalyptus globulus* tree biomass. Can J For Res 37:895–906
Attiwill PM, Ovington JD (1968) Determination of forest biomass. For Sci 14:13–15
Baldwin VC Jr (1987) Green and dry-weight equations for above-ground components of planted loblolly pine trees in the West Gulf region. South J Appl For 11:212–218
Barney RJ, VanCleve K, Schlentner R (1978) Biomass distribution and crown characteristics in two Alaskan *Picea mariana* ecosystems. Can J For Res 8:36–41
Bartelink HH (1997) Allometric relationships for biomass and leaf area of beech (*Fagus sylvatica* L.). Ann For Sci 54:39–50
Belanger RP (1973) Volume and weight tables for plantation-grown sycamore. USDA Forest Service, Southeastern Forest Experiment Station, Asheville, Research Paper SE-107
Bi H, Birk E, Turner J, Lambert M, Jurskis V (2001) Converting stem volume to biomass with additivity, bias correction, and confidence bands for two Australian tree species. N Z J For Sci 31:298–319
Bi H, Turner J, Lambert MJ (2004) Additive biomass equations for native eucalypt forest trees of temperate Australia. Trees 18:467–479
Bond-Lamberty B, Wang C, Gower ST (2002) Aboveground and belowground biomass and sapwood area allometric equations for six boreal tree species in northern Manitoba. Can J For Res 32:1441–1450
Brown JK (1976) Estimating shrub biomass from basal stem diameters. Can J For Res 6:153–158
Brown S, Gillespie AJR, Lugo AE (1989) Biomass estimation methods for tropical forests with applications to forest inventory data. For Sci 35:881–902
Brown S, Gillespie AJR, Lugo AE (1991) Biomass of tropical forests of south and southeast Asia. Can J For Res 21:111–117
Bullock BP, Burkhart HE (2003) Equations for predicting green weight of loblolly pine trees in the South. South J Appl For 27:153–159
Burkhart HE, Beckwith JR III (1970) Specific gravity prediction and dry-weight yield estimation. Tappi 53:603–604
Carvalho JP, Parresol BR (2003) Additivity in tree biomass components of Pyrenean oak (*Quercus pyrenaica* Willd). For Ecol Manage 179:269–276
Chiyenda SS, Kozak A (1984) Additivity of component biomass regression equations when the underlying model is linear. Can J For Res 14:441–446
Clark A III (1978) Total tree and its utilization in the Southern United States. For Prod J 28:47–52
Clark A III (1982) Predicting biomass production in the South. In: Hotvedt JE, Jackson BD (eds) Predicting growth and yield in the Mid-South. 31st annual forestry symposium. Louisiana State University, Baton Rouge, pp 119–139
Clark A III, Phillips DR, Schroeder JG (1980) Predicted weights and volumes of red oak trees in Western North Carolina. USDA Forest Service, Southeastern Forest Experiment Station, Asheville, Research Paper SE-209
Crow TR (1978) Common regressions to estimate tree biomass in tropical stands. For Sci 24: 110–114
Crow TR, Laidly PR (1980) Alternative models for estimating woody plant biomass. Can J For Res 10:367–370
Crow TR, Schlaegel BE (1988) A guide to using regression equations for estimating tree biomass. North J Appl For 5:15–22
Cunia T, Briggs RD (1984) Forcing additivity of biomass tables: some empirical results. Can J For Res 14:376–384
Cunia T, Briggs RD (1985a) Forcing additivity of biomass tables: use of the generalized least squares method. Can J For Res 15:23–28
Cunia T, Briggs RD (1985b) Harmonizing biomass tables by generalized least squares. Can J For Res 15:331–340
Deans JD, Moran J, Grace J (1996) Biomass relationships for tree species in regenerating semi-deciduous tropical moist forest in Cameroon. For Ecol Manage 88:215–225

Fatemi FR, Yanai RD, Hamburg SP, Vadeboncoeur MA, Arthur MA, Briggs RD, Levine CR (2011) Allometric equations for young northern hardwoods: the importance of age-specific equations for estimating aboveground biomass. Can J For Res 41:881–891

Francis JK (1984) Biomass accumulation by single- and multiple-stemmed young sycamore. For Sci 30:372–374

Geudens G, Staelens J, Kint V, Goris R, Lust N (2004) Allometric biomass equations for Scots pine (*Pinus syvestris* L.) seedlings during the first years of establishment in dense natural regeneration. Ann For Sci 61:653–659

Harrington RA, Fownes JH (1993) Allometry and growth of planted versus coppice stands of four fast-growing tropical tree species. For Ecol Manage 56:315–327

Hepp TD, Brister GH (1982) Estimating crown biomass in loblolly pine plantations in the Carolina flatwoods. For Sci 28:115–127

Jacobs MW, Cunia T (1980) Use of dummy variables to harmonize tree biomass tables. Can J For Res 10:483–490

Jenkins JC, Chojnacky DC, Heath LS, Birdsey RA (2003) National-scale biomass estimators for United States tree species. For Sci 49:12–35

Johansson T (1999) Biomass equations for determining fractions of pendula and pubescent birches growing on abandoned farmland and some practical implications. Biomass Bioenergy 16:223–238

Ketterings QM, Coe R, van Noordwijk M, Ambagau Y, Palm CA (2001) Reducing uncertainty in the use of allometric biomass equations for predicting above-ground tree biomass in mixed secondary forests. For Ecol Manage 146:199–209

King JS, Giardina CP, Pregitzer KS, Friend AL (2007) Biomass partitioning in red pine (*Pinus resinosa*) along a chronosequence in the Upper Peninsula of Michigan. Can J For Res 37:93–102

Kozak A (1970) Methods for ensuring additivity of biomass components by regression analysis. For Chron 46:402–404

Kozak A, Munro DD, Smith HG (1969) Taper functions and their application in forest inventory. For Chron 45:278–283

Lambert M-C, Ung C-H, Raulier F (2005) Canadian national tree aboveground biomass equations. Can J For Res 35:1996–2018

Lenhart JD, Hyink DM (1973) Direct estimation of the ovendry weight of plantation-grown loblolly pine trees in the interior West Gulf Coastal Plain. For Prod J 23:49–50

Levy PE, Hale SE, Nicoll BC (2004) Biomass expansion factors and root: shoot ratios for coniferous tree species in Great Britain. Forestry 77:421–430

Madgwick HAI (1983) Above-ground weight of forest plots – comparison of seven methods of estimation. N Z J For Sci 13:100–107

Madgwick HAI, Satoo T (1975) On estimating the aboveground weights of tree stands. Ecology 56:1446–1450

McGinnis TW, Shook CD, Keeley JE (2010) Estimating aboveground biomass for broadleaf woody plants and young conifers in Sierra Nevada, California, forests. West J Appl For 24:203–209

Meeuwig RO, Cooper SV (1981) Stand estimates of biomass and growth in pinyon-juniper woodlands in Nevada. USDA Forest Service, Intermountain Forest and Range Experiment Station, Ogden, Research Note INT-311

Miller EL, Meeuwig RO, Budy JD (1981) Biomass of singleleaf pinyon and Utah juniper. USDA Forest Service, Intermountain Forest and Range Experiment Station, Ogden, Research Paper INT-273

Monserud RA, Onuchin AA, Tchebakova NM (1996) Needle, crown, stem, and root phytomass of *Pinus sylvestris* stands in Russia. For Ecol Manage 82:59–67

Moore JR (2010) Allometric equations to predict the total above-ground biomass of radiata pine trees. Ann For Sci 67:806p1–806p11

Oswald BP, Botting RR, Coble DW, Farrish KW (2010) Aboveground biomass estimation for three common woody species in the post oak savannah of Texas. South J Appl For 34:91–94

Parresol BR (1999) Assessing tree and stand biomass: a review with examples and critical comparisons. For Sci 45:573–593
Parresol BR (2001) Additivity of nonlinear biomass equations. Can J For Res 31:865–878
Parresol BR, Thomas CE (1989) A density-integral approach to estimating stem biomass. For Ecol Manage 26:285–297
Parresol BR, Thomas CE (1996) A simultaneous density-integral system for estimating stem profile and biomass: slash pine and willow oak. Can J For Res 26:773–781
Peichl M, Arain MA (2007) Allometry and partitioning of above-and belowground tree biomass in an age-sequence of white pine forests. For Ecol Manage 253:68–80
Pitt DG, Bell FW (2004) Effects of stand tending on the estimation of aboveground biomass of planted juvenile white spruce. Can J For Res 34:649–658
Porté A, Trichet P, Bert D, Loustau D (2002) Allometric relationships for branch and tree woody biomass of Maritime pine (*Pinus pinaster* Ait.). For Ecol Manage 158:71–83
Reed DD, Liechty HO, Jones EA, Zhang Y (1996) Above- and belowground dry matter accumulation pattern derived from dimensional biomass relationships. For Sci 42:236–241
Russell MB, Burkhart HE, Amateis RL (2009) Biomass partitioning in a miniature-scale loblolly pine spacing trial. Can J For Res 39:320–329
Sabatia CO, Lynch TB, Will RB (2008) Tree biomass equations for naturally regenerated shortleaf pine. South J Appl For 32:163–167
Saint-André L, M'Bou AT, Mabiala A, Mouvondy W, Jourdan C, Roupsard O, Deleporte P, Hamel O, Nouvellon Y (2005) Age-related equations for above- and below-ground biomass of *Eucalyptus* hybrid in Congo. For Ecol Manage 205:199–214
Satoo T (1982) Forest biomass. Martinus Nijhoff, The Hague
Schumacher FX, Hall F (1933) Logarithmic expression of timber tree volume. J Agric Res 47:719–734
Shaiek O, Loustau D, Trichet P, Meredieu C, Bachtobji B, Garchi S, El Aouni MH (2011) Generalized biomass equations for the main aboveground biomass components of maritime pine across contrasting environments. Ann For Sci 68:443–452
Smith WB, Brand GJ (1983) Allometric biomass equations for 98 species of herbs, shrubs, and small trees. USDA Forest Service, North Central Forest Experiment Station, St. Paul, Research Paper NC-299
Snell JAK, Brown JK (1978) Comparison of tree biomass estimators – DBH and sapwood area. For Sci 24:455–457
Tasissa G, Burkhart HE (1998) Modeling thinning effects on ring specific gravity of loblolly pine (*Pinus taeda* L.). For Sci 44:212–223
Tasissa G, Burkhart HE, Amateis RL (1997) Volume and taper equations for thinned and unthinned loblolly pine trees in cutover, site-prepared plantations. South J Appl For 21:146–152
Ter-Mikaelian MT, Korzukhin MD (1997) Biomass equations for sixty-five North American tree species. For Ecol Manage 97:1–24
Ter-Mikaelin MT, Parker WC (2000) Estimating biomass of white spruce seedlings with vertical photo imagery. New For 20:145–162
Thomas CE, Parresol BR (1991) Simple, flexible, trigonometric taper equations. Can J For Res 21:1132–1137
Thomas CE, Parresol BR, Lê KHN, Lohrey RE (1995) Biomass and taper for trees in thinned and unthinned longleaf pine plantations. South J Appl For 19:29–35
Tinker D, Stakes GK, Arcano RM (2010) Allometric equation development, biomass, and aboveground productivity in ponderosa pine forests, Black Hills, Wyoming. West J Appl For 25:112–119
Van Deusen PC, Baldwin VC Jr (1993) Sampling and predicting tree dry weight. Can J For Res 23:1826–1829
Verwijst T, Telenius B (1999) Biomass estimation in short rotation forestry. For Ecol Manage 121:137–146
Zhang Y, Borders BE, Will RE, De Los Santos Posadas H (2004) A model for foliage and branch biomass prediction for intensively managed fast growing loblolly pine. For Sci 50:65–80

Zianis D, Mencuccini M (2004) On simplifying allometric analyses of forest biomass. For Ecol Manage 187:311–332

Zianis D, Xanthopoulos G, Kalabokidis K, Kazakis G, Ghosn D, Roussou O (2011) Allometric equations for aboveground biomass estimation by size class for *Pinus Brutia* Ten. trees growing in North and South Aegean Islands, Greece. Eur J For Res 130:145–160

Chapter 5
Quantifying Tree Crowns

5.1 Approximating Tree Crowns with Geometric Shapes

Larger crown sizes generally produce higher rates of growth for trees of a given species and age. Crown characteristics have also been found effective for predicting responses to silvicultural treatments, such as thinning and fertilizing applications. Consequently, crown measures are often incorporated in growth and yield models to improve predictions of stand development and response to management practices.

Crown size, which is a surrogate for the amount of photosynthetically active foliage, is typically quantified using measures of crown width and length. Crown profiles are irregular and the branches of neighboring trees are often interlocked, making the measurement of crown width difficult. As with tree boles, the area of a tree's crown projection is typically determined by calculating an average diameter and then assuming a circular shape. The length of green crown may be defined as the vertical distance from the tip (highest growing point) to the lowest live foliage. While the upper limit can be objectively defined, the base of the tree crowns is often very difficult to ascertain (additional detail on measuring tree crowns can be found in Avery and Burkhart 2002).

With measurements of crown length and width, crown surface area and crown volume can be approximated by assuming the crown is some regular geometric solid. It is common to assume that tree crowns are cones or paraboloids. If one assumes that a cone is a reasonable approximation for the crown, the surface area and volume can be computed as

$$\text{Surface area} \quad (\text{m}^2) = \frac{\pi c_w}{2}\sqrt{c_l^2 + \left(\frac{c_w}{2}\right)^2}$$

$$\text{Volume} \quad (\text{m}^3) = \frac{\pi c_w c_l}{12}$$

H.E. Burkhart and M. Tomé, *Modeling Forest Trees and Stands*,
DOI 10.1007/978-90-481-3170-9_5,

where

c_w = diameter at the crown base (m)
c_l = crown length (m)

While invoking the assumption that tree crowns assume a regular geometric shape may provide acceptable approximations of surface area and volume for some purposes, more exact values for given species have been derived through empirical modeling.

5.2 Modeling Crown Profiles

Biging and Wensel (1990) defined crown volume as the geometric space occupied by the crown and developed models to describe the crown volume and related measures for six conifer species in northern California. Crown sectional volumes were computed from felled tree measurements and total crown volume was determined by summing the sectional volumes. Regression equations of the following form were fitted to estimate crown volume (c_v) from the easily measured variables of *dbh* (d), total tree height (h) and live-crown ratio (c_r):

$$c_v = ad^b h^c c_r^d$$

where c_r is length of live crown divided by total tree height and a, b, c and d are coefficients estimated by regression techniques. Estimates of cumulative crown volume at any height in the crown, crown cross-sectional area, and crown surface were derived from the total crown volume relationship.

Hann (1999) and Marshall et al. (2003) modeled crown profiles by dividing the crown into segments: the portion of the crown above the point of largest crown width (c_{wmax}) and the portion below that point. Hann's equation for the upper portion of Douglas-fir crowns predicted a crown profile that ranged in shape from nearly conic to parabolic, depending on position within the crown and the social status of the tree as indicated by the ratio of total height divided by *dbh* for the tree. The equation for the lower portion predicted a crown profile with a cylindrical shape.

Crown profile for the light crown (that portion above c_{wmax}) was expressed by the equation:

$$c_{wah} = (c_{wmax}) R_{pah}^{\left[a_0 + a_1 RPA_h^{1/2} + a_2(h/d)\right]} \tag{5.1}$$

where

c_{wah} = crown width at height h above ground for the portion of the crown above the height where largest crown width occurs
c_{wmax} = largest crown width of a stand-grown tree

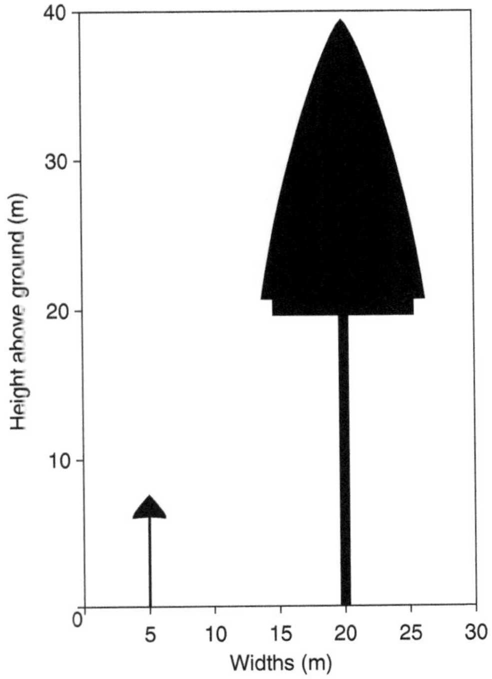

Fig. 5.1 Predicted Douglas-fir crown profiles for a dominant tree ($dbh = 81.3$ cm, total height = 39.6 m, and crown ratio = 0.5), and a suppressed tree ($dbh = 5.1$ cm, total height = 7.6 m, and crown ratio = 0.2) (From Hann 1999)

R_{pah} = relative position of the crown width within the length of the crown occurring above the portion of the largest crown width
h = total tree height
d = tree diameter at breast height

Crown width for the shade portion of the crown was predicted by

$$c_{wbh} = \mathrm{b}_1(c_{wmax}) \tag{5.2}$$

where c_{wbh} = crown width at height h above the ground for the portion of the crown below the height where c_{wmax} occurs.

Figure 5.1 illustrates the application of this segmented approach for predicting crown profiles for dominant and suppressed Douglas-fir trees.

Baldwin and Peterson (1997) proposed a function of the following form to model the crown shape of loblolly pine trees growing in stands with closed canopies:

$$O_{cr} = \mathrm{b}_1\left(\frac{R_{ch} - 1}{R_{ch} + 1}\right) + \mathrm{b}_2(R_{ch} - 1) \tag{5.3}$$

where O_{cr} is outer crown radius (m), R_{ch} is relative crown height (height within the crown/crown length), and b_1 and b_2 are coefficients to be determined from individual-tree data. Note that when relative crown height is equal to one (top of

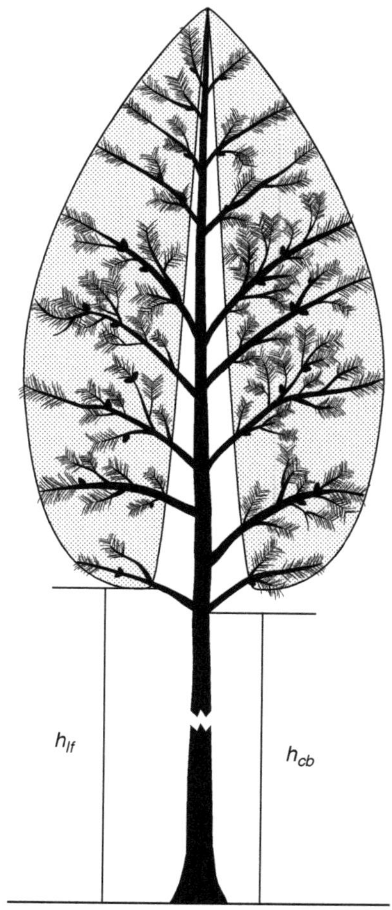

Fig. 5.2 Diagram of the modeled inner and outer shape of a loblolly pine crown showing the height to the base of the full live crown (h_{cb}) and height to the base of the live foliage (h_{lf}) (From Baldwin and Peterson 1997)

crown), crown radius is equal to zero. From data it was observed that maximum crown width occurs above the crown base, except for cone-shaped crowns, and varies with stand and tree characteristics.

The area next to the tree bole from the crown base to almost the tree tip is essentially defoliated in loblolly pine, as well as in many other conifers. It is roughly conical (not consistently concave or convex) in shape, with the apex near the crown tip. Therefore, this inner area was modeled with a straight line. The inner shape begins at a predicted radius at the crown base and extends upwards to an intersection with the bole below the crown tip (Fig. 5.2). This intersection is the point at which all inner crown radius measurements above it are zero. Typically, the point is just below 1–2 years height growth, where foliage is growing immediately off the bole.

The model for the inner crown shape is

$$I_{cr} = \mathrm{b}_3 + \mathrm{b}_4(R_{ch}) \tag{5.4}$$

where I_{cr} is inner crown radius (m) and b_3 and b_4 are coefficients to be determined from individual-tree data. The maximum crown radius is found directly as the maximum of the outer shape function. The intercept for the inner function is the maximum radius of inner defoliation occurring at the base of the live crown.

In trees sampled by Baldwin and Peterson (1997), foliage typically did not occur until 0.3–0.9 m above the measured height to base of live crown (h_{cb}). Consequently an equation was fitted to predict height to the base of live foliage (h_{lf}) from the outer crown measurements and used to redefine crown base (Fig. 5.2), thus redefining R_{ch} in (5.3) and (5.4). The sample data were fitted to model (5.3) using the new definition of R_{ch}. Foliated crown length (c_{lf}) was then defined as total tree height (h) minus h_{lf}, and crown ratio (c_r) as c_{lf} divided by h. The height of the live foliage, in meters, is estimated to occur at

$$h_{lf} = 0.9326h_{cb} - 0.0267d + 0.1006h$$

where

h_{cb} is height to the base of the full live crown (m)
d is bole diameter 1.37 m above ground line (cm)
h is total tree height (m)

Crown shape changes due to tree growth, competition, and age. Thus the parameters in Eqs. 5.3 and 5.4 were related to individual tree characteristics to capture the dynamic nature of crown shape. The coefficients b_1 and b_2 in Eq. 5.3 were found to be a function of tree *dbh* and crown ratio while b_3 in (5.4) was a function of *dbh* and foliated crown length and b_4 was a function of *dbh*. The best fitted equation that included individual tree characteristics to predict outer crown shape was

$$\begin{aligned} O_{cr} =& (-4.5121 + 0.5176d + 4.3529c_r) \times \left(\frac{R_{ch} - 1}{R_{ch} + 1}\right) \\ &+ (4.4749 - 0.0175t - 0.4985d - 6.0414c_r)(R_{ch} - 1) \end{aligned} \quad (5.5)$$

where t is age (years) and c_r is crown ratio $= (h - h_{cb})/h$

The prediction equation for the inner crown shape model that includes tree characteristics was

$$I_{cr} = (0.0168d + 0.0155c_{lf}) + (-0.0233d)(R_{ch}) \quad (5.6)$$

where c_{lf} is foliated crown length (m).

Outer crown radius and area of inner defoliation were predicted for two representative sample trees using Eqs. 5.5 and 5.6 and plotted in Fig. 5.3 to illustrate the models' flexibility in predicting height to maximum radius and crown form for trees of different sizes and ages. Baldwin and Peterson (1997) demonstrated how their equation system can be applied to determine key crown measures such

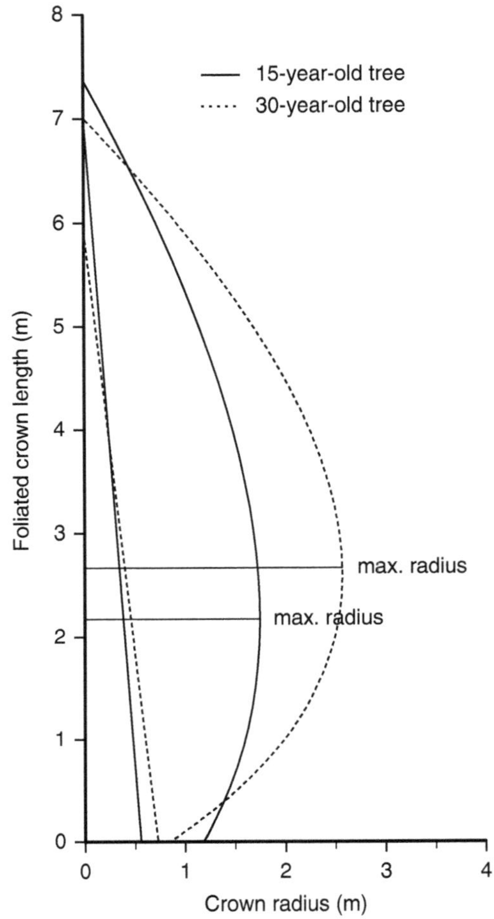

Fig. 5.3 Predicted inner and outer crown shape of a 15-year-old and a 30-year-old loblolly pine sample tree (From Baldwin and Peterson 1997)

as maximum crown radius and its height, crown volume, and crown surface area. Additional examples of approaches for describing crown profiles, with emphasis on models developed in central Europe, can be found in Pretzsch (2009).

5.2.1 Incorporating Stochastic Variation

While relatively simple functions may be suitable for describing tree crown profiles for many purposes, individual trees will deviate significantly from the mean profile for a given species. Hence it may be desirable to apply stochastic models to represent the high degree of unexplained variability in tree crown form. Biging and Gill (1997) performed a feasibility study of using stochastic models to describe the profile of tree crowns for five conifer species from the mixed conifer region of northern

California, USA. Their pilot approach was later extended with an enhanced data set for conifers (Gill and Biging 2002a) and for two hardwood species with highly variable crown forms (Gill and Biging 2002b).

Biging and Gill (1997) viewed the crown radii to be composed of three components: (1) the trend line, (2) the autoregressive moving average time-series model (ARMA), and (3) the random error. The trend line can be thought of as the basic morphological term, the ARMA term incorporates spatial correlation of branch length, and the error or white noise term is unaccounted for variation. Instead of having a function that changes over time, as in most time-series analyses, Biging and Gill (1997) considered the radius of tree crowns as functions of the height increment (that is, height increment was considered analogous to the time variable in standard time-series analyses).

Because equidistant measurements help facilitate the analysis of a time-series, a cubic spline was fitted to the data and then crown measurements were interpolated to 0.3–0.6 m height increments depending on the length of the tree crown.

When analyzing time-series data it is desirable to have a series that has a constant mean (i.e., no trend) and homogeneous variance. Such a series is referred to as a stationary series. From plots of the data, Biging and Gill observed that data of most of the tree crown series contained a quadratic trend. Figure 5.4 shows the interpolated data and the quadratic trend for tree number 24, a representative tree. After removal of the trend, residuals were plotted and analyzed to ensure that stationarity was achieved.

Biging and Gill fitted ARMA models to each tree. An ARMA (p, q) has autoregressive (AR) factors up to order p and moving average (MA) factors of order q. An AR process expresses the series in terms of past observations and the current disturbance (random error) whereas a MA process expresses the series in terms of current and past disturbances (series of random errors). The spatial autocorrelation approach can improve predictive ability because nearby branches experienced the same microenvironment, biological interactions, and history. For these crown models, one might interpret an AR process as the crown radius at a given height expressed as a weighted linear combination of, and thus influenced by, the previous (lower) crown radii. A MA process is used to model a variable that is in equilibrium but is disturbed from the equilibrium by outside forces. Thus, an interpretation of a MA crown profile model is that the profile follows the deterministic trend, but is moved from this trend by outside forces such as competition and climatic variables.

The form of an ARMA (p, q) (Biging and Gill 1997) is:

$$z_j = \phi_1 z_{j1} + \phi_2 z_{j-2} + \cdots + \phi_p z_{j-p} + \delta + u_j - \theta_1 u_{j-1} - \cdots - \theta_q u_{j-q}$$

where

z_j = crown radius at height j minus the quadratic or cubic crown trend at height j $(j = 1, \ldots, J)$

ϕ_i = parameters of the autoregressive factors $(i = 1, \ldots, p)$

Fig. 5.4 Plot of interpolated crown profile data with estimated quadratic trend superimposed (From Biging and Gill 1997)

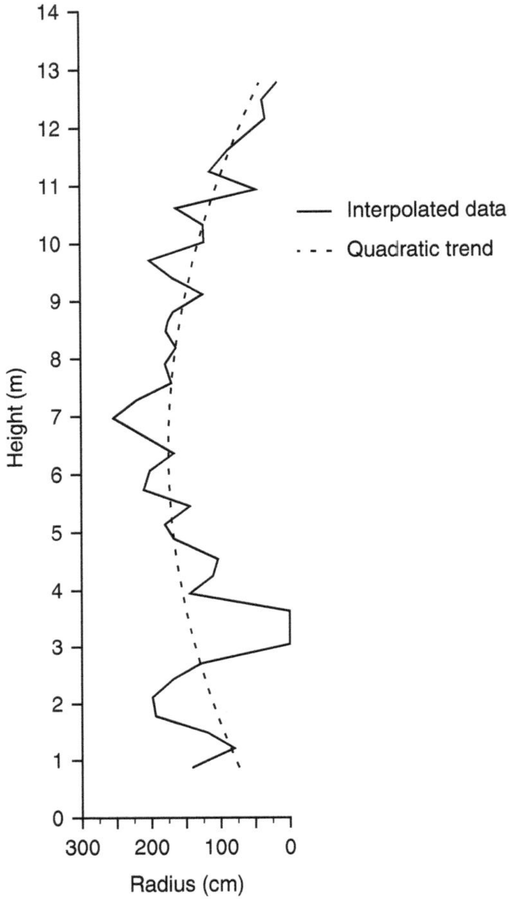

θ_k = parameters of the moving average factors ($k = 1, \ldots, q$)
δ = constant
u_j = white noise (a sequence of identically and independently distributed random disturbances with mean zero and variance σ^2).

Biging and Gill (1997) found that for 70% of the sample trees the crown profile could be modeled as a quadratic or cubic trend in conjunction with a simple autoregressive moving average model (ARMA). In the remaining cases they used a quadratic or cubic trend in conjunction with white noise. The stochastic ARMA models were judged to be visually and statistically an improvement over using Euclidean geometric crown profile models.

5.2.2 *Additional Techniques for Describing Tree Crowns*

Additional methods that have been applied to describe tree crowns include stochastic frontier models (Nepal et al. 1996), non-parametric models (Doruska and Mays 1998), and use of fractal geometry (Zeide and Gresham 1991; Zeide and Pfeifer 1991; Zeide 1998). Nepal et al. (1996) demonstrated the stochastic frontier estimation technique by fitting a nonlinear function to the shape of loblolly pine tree crowns. Their crown shape model related the maximum observed crown radius to the relative crown length. Doruska and Mays (1998) modeled crown profile of loblolly pine by nonparametric regression analysis. Nonparametric regression involves selection of polynomial order and bandwidth which determine curvature and degree of smoothing. Noting that tree crowns are difficult to quantify in terms of classical geometry, Zeide and Pfeifer (1991) proposed that fractal geometry may prove a fruitful approach. Unlike Euclidean dimensions, which are integers (one for lines, two for areas, three for volumes) common to all objects, fractal dimensions are fractional numbers that are unique for each object. Zeide and Gresham (1991) calculated fractal dimension of loblolly pine from the regression of foliage area (or mass) on the area (or volume) of the convex hull that envelopes the crown.

5.3 Modeling Crown Morphology

As an alternative to assuming a crown shape or predicting the crown profile, crown morphology (branch diameter, location, angle and length) can be modeled. A number of models of branch size, inclination and location have been developed for the primary purpose of characterizing wood quality. For instance, Colin and Houllier (1992) predicted the maximum and average diameter of branches within a whorl, distance from the tip of the tree to the whorl, and branch angle for Norway spruce trees. Standard inventory tree measurements were used as predictor variables of branch characteristics. Maguire et al. (1994) predicted the number, diameter and distribution of primary branches along the bole of young Douglas-fir trees. Crown structure for 9- to 30-year old loblolly pine trees was quantified via analysis of branch diameters and location, both along and around the bole, by Doruska and Burkhart (1994). Three equations were developed to describe diameter distribution of branches, and circular statistics were used to examine branching patterns around the bole. The total number of branches within a crown was predicted and three equations were employed to describe the mean and range of diameters within a whorl. Trincado and Burkhart (2009) developed a stochastic model to simulate the processes of initiation, diameter growth, death, and self pruning of branches in loblolly pine trees. Three components – whorls, branches and knots – were modeled and hierarchically connected. The branch model was linked to an individual tree model to simulate the dynamics of first-order branches. Information on the trend of branch diameters along and around the stem, volume of knots, and spatial location of knots is provided by the modeling system.

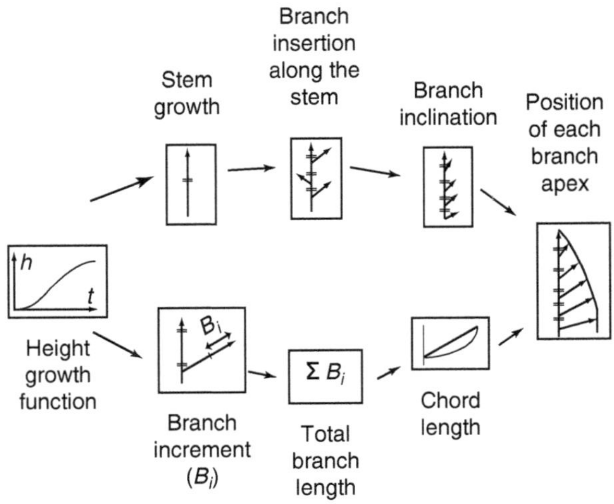

Fig. 5.5 Diagram of a model of crown shape implemented from the height growth function (From Deleuze et al. 1996)

Because of the role of crown volume and shape for determining individual tree stem growth and form, which in turn impacts stand dynamics, growth and yield, a great deal of effort has been devoted to quantifying crown morphology. Crown variables of branch length and foliage distribution, in addition to branch diameters and locations along the stem, are of interest. It is often desirable to predict crown variables from the input variables commonly used to implement forest growth and yield models.

In an effort aimed at modeling crown shape, Deleuze et al. (1996) developed prediction equations for branch length increment and inclination for Norway spruce. The model of branch extension is based on height growth and on the year of elongation of the branch. A second model describes the change in branch inclination. Branch spread increased with increased between-tree spacing, whereas branch inclination was found to be affected by crown contact. The result of the study was a relatively simple model of crown shape development that can be used as an input in growth predictions; in addition the relationship between branch diameter and branch length can be used in assessment of wood quality (knottiness of the stem). The overall structure of the height-growth driven system for modeling crown shape is displayed in Fig. 5.5.

When developing crown profile models based on branch attributes of Douglas-fir trees, Roeh and Maguire (1997) observed that despite use of primary branching structure as a modeling basis, a practical prediction system for crown profile must be able to accept easily measured whole tree characteristics as the sole or primary input. Thus, in their study, branch diameter was modeled as a function of tree *dbh*, height, and crown length. Although branch angle and branch length each were expressed as functions of branch diameter in addition to whole-tree variables, branch diameter was regarded as an endogenous variable or, in other words, was predicted within the system; hence, both branch angle and branch length at an arbitrary depth into the crown were ultimately functions of only whole-tree variables and crown depth of interest. The developed system therefore preserves the capacity to estimate crown profile from standard tree measurements while allowing parameters to be estimated from branch data collected on standing or felled sample trees.

Roeh and Maguire's (1997) models for estimating crown profile in Douglas-fir were developed from attributes of individual whorl branches, including basal diameter, total length, angle of origin, and height from ground. The prediction system consisted of four equations to predict whorl branch basal diameter, branch length, branch angle of origin, and corresponding crown radius. The system is entered with three standard individual-tree measurements: diameter at breast height, total height, and height to crown base. Four approaches were developed for modeling crown profile with this system of equations, distinguished by parameter estimation method, modeling data subset, and the specific form by which whorl branch diameter was represented (mean or maximum diameter). Modified three-stage least squares was applied to account for the correlation of error terms across the equations, and this procedure was compared with ordinary and nonlinear least squares methods.

Roeh and Maguire found the following segmented polynomial model to be the best model for estimating mean whorl branch diameter for a given position in the crown:

$$br_{dmean} = a_1(h-hi) + a_2(h-hi)^2 + a_3 d(h-hi) + a_4 d(h-hi)^2 + a_5 h(h-hi) + a_6 c_l(h-hi)^2 + a_7 I\left[h(K-(h-hi))^2\right] \quad (5.7)$$

where br_{dmean} = mean whorl branch diameter (mm); $(h-hi)$ = depth into crown (m); d = diameter at breast height (cm); h = total tree height (m); c_l = crown length (m); $I = 1$ if $(h-hi) > K$, otherwise $I = 0$; $K = 2/3\ c_l$.

A similar range of model forms as evaluated for mean branch diameter was explored for maximum branch diameter. The best model was close in form to the selected mean branch diameter model, Eq. 5.7, but did not require a second quadratic segment near the base of the crown:

$$br_{d\max} = b_0 + b_1(h-hi) + b_2(h-hi)^2 + b_3 d(h-hi) + b_4 d(h-hi)^2 + b_5 h(h-hi) + b_6 h(h-hi)^2 + b_7 c_l(h-hi)^2 \quad (5.8)$$

where $br_{d\max}$ = maximum whorl branch diameter (mm). $(h-hi)$, d, h, and c_l remain as defined before.

Angle of origin was expressed as the angle subtended by the bole and base of the branch (that is angle from vertical). This angle became progressively larger (branches more horizontal) with increasing depth into crown. However, the smoothed graph of branch angle on $(h-hi)$ revealed that the average trend in branch angle down the stem was curvilinear with an asymptote of approximately 90°. After exploring various model forms for branch angle, Roeh and Maguire selected:

$$br_{vang} = [c_0 + c_1 h]\left(1 - e^{c_2(h-hi)+c_3 S}\right)^{c_4 br_d} \quad (5.9)$$

where br_{vang} = angle of origin, from vertical (°); h = total tree height (m); S = site index (m); br_d = branch basal diameter (mm). Graphical analysis of the Douglas-fir data suggested that, like branch diameter, branch length peaked within the crown. Analysis of numerous models indicated that the most appropriate form for predicting branch length was:

$$br_l = e^{d_1}(h - hi)^{d_2} e^{d_3(h-hi)} br_d^{d_4} h^{d_5} \tag{5.10}$$

where br_l = branch length (cm). $(h - hi)$, br_d, and h are as above.

Equations 5.7, 5.9, and 5.10 were combined with the trigonometric formula for crown radius (c_{rad}) to produce the following system of equations:

$$\begin{aligned} br_{dmean} &= s_1(h-hi) + s_2(h-hi)^2 + s_3 d(h-hi) + s_4 d(h-hi)^2 \\ &\quad + s_5 h(h-hi) + s_6 c_l (h-hi)^2 + s_7 I\left[h(K-(h-hi))^2\right] \end{aligned} \tag{5.11}$$

$$\begin{aligned} br_{vang} &= (s_8 + s_9 h)\left(1 - e^{s_{10}(h-hi)+s_{11}S}\right)^{s_{12} br_{d_i}} \\ br_l &= e^{s_{13}}(h-hi)^{s_{14}} e^{s_{15}(h-hi)} br_{d_i}^{s_{16}} h^{s_{17}} \\ c_{rad} &= br_{li}\left[\pi \sin(br_{vang_i}/180)\right] \end{aligned}$$

where br_{di}, br_{vang_i}, and br_{li} were the instrumental variables for br_{dmean}, br_{vang} and br_l, respectively. The three stage least squares procedure for parameter estimation produced seven nonsignificant coefficients s_3, s_5, s_6, s_7, s_9, s_{11}, and s_{17}. Although the system of equations approach accounted for across-equation error correlation, it would be expected to change the estimated mean squared error (MSE) for each component equation relative to the individual equation MSE estimates, and, hence, it would also be expected to change the standard errors for parameter estimates. Some otherwise significant independent variables therefore could be expected to become nonsignificant in the system parameter estimation. Independent variables were dropped successively from the Eq. 5.11 system based on parameter estimate P-values until only significant parameter estimates remained ($\alpha = 0.05$). The final system of equations reduced to the following form:

$$br_{dmean} = s_{18}(h-hi) + s_{19}(h-hi)^2 + s_{20} d(h-hi)^2 \tag{5.12}$$

$$\begin{aligned} br_{vang} &= s_{21}\left(1 - e^{s_{22}(h-hi)}\right)^{s_{23} br_d} \\ br_l &= e^{s_{24}}(h-hi)^{s_{25}} e^{s_{26}(h-hi)} br_d^{s_{27}} \\ c_{rad} &= br_l\left[\pi \sin(br_{vang}/180)\right] \end{aligned}$$

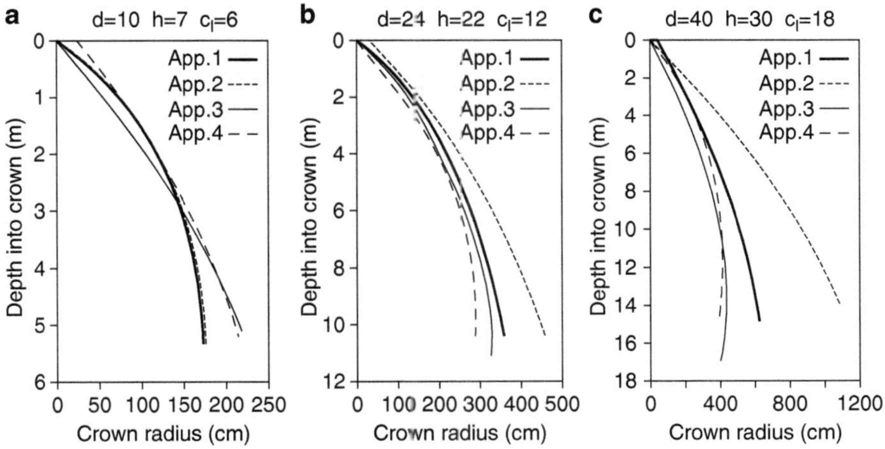

Fig. 5.6 Predicted crown profiles for a (**a**) small, (**b**) medium, and (**c**) large Douglas-fir tree. Site index for Approach 4 was held constant at 38 m (From Roeh and Maguire 1997)

In their study Roeh and Maguire (1997) evaluated four modeling approaches consisting of combinations of use of mean branch diameter or maximum branch diameter as the driving variable and different parameter estimation techniques. Figure 5.6 illustrates variability in Douglas-fir crown profiles predicted for small, medium, and large trees by the four approaches (approach 2, consisting of OLS/NLS estimation techniques and use of mean branch diameter, was judged best based on overall performance for representing the branch characteristics and crown profile of the validation trees).

While much attention has been devoted to modeling excurrent form conifer crowns, Cluzeau et al. (1994) provides an example of quantifying the crown morphology of a decurrent branching species, common ash. The relationship between stem and branch growth was studied from cumulated annual increments of stems and branches of unforked ash trees. An allometric relationship between branch length and cumulated stem growth above the branch base was modified to account for an additional tree height effect and fitted to data. Crown dimensions of trees are then derived from the branch growth equation and an average mean branching angle by solving a crown radius equation. Thus crown shape results from branch spread and the distance from the stem apex to the branch tip projection (Fig. 5.7).

For practical purposes many crown profile models are based on easily measured and readily available tree variables and branch variables are related to tree stem variables rather than vice versa. Mitchell (1975) conducted a detailed examination of the growth of stems and branches of Douglas-fir trees and developed equations and numeric methods to simulate dynamics and yields that can be depicted as follows: age, site quality and tree vigor → height growth → branch extension → size and shape of the crown → foliar volume → quantity and distribution of annual

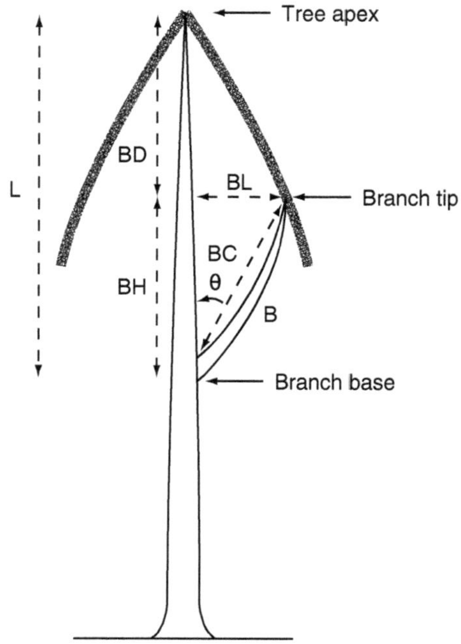

Fig. 5.7 Branch and stem basic characteristics. *BD* distance from stem apex to branch tip projection, *BL* branch spread, *L* distance from stem apex of the base of the branch, *BH* branch reach, *BC* branch chord, *Θ* branching angle, *B* branch length. *BD*, *BC*, and Θ are derived from the other measured characteristics (From Cluzeau et al. 1994)

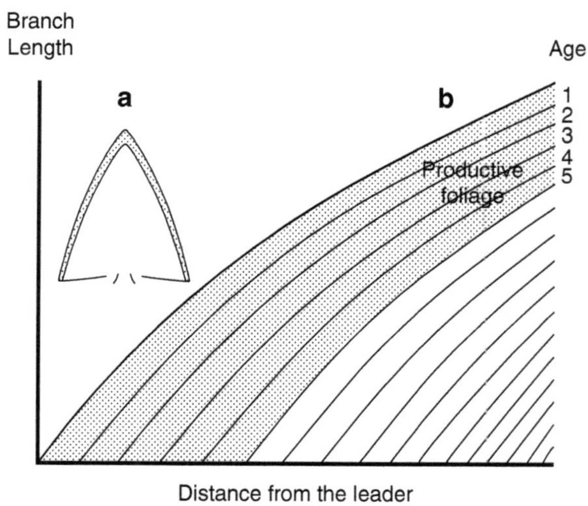

Fig. 5.8 (**a**) Change in shoot length over the surface of the crown (**b**) The present (*bold curve*) and past crown profiles that relate length of branches to distance from leader (From Mitchell 1975)

bole increment, and subsequent height growth. These and other relations were incorporated into a dynamic mathematical model that allows the simulated crowns of individual trees to expand and contract asymmetrically in a three-dimensional growing space in response to internal growth processes and the physical restrictions imposed by competitors. The crowns add a shell of foliage each year that benefits the tree in diminishing amounts for 5 years (Fig. 5.8). The volume increment produced

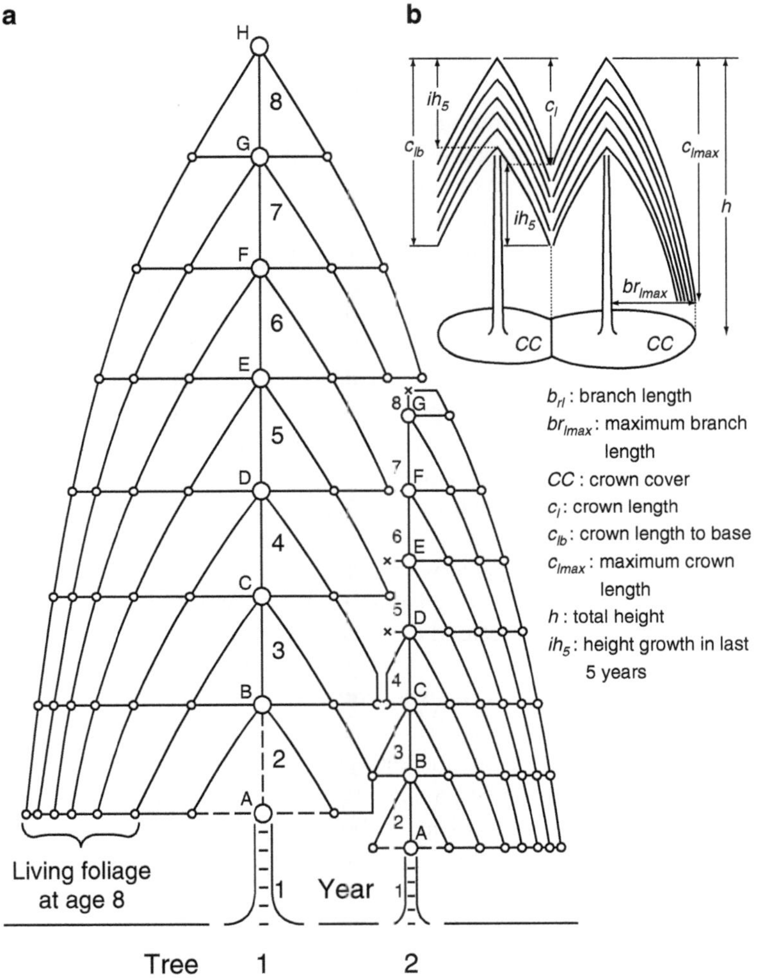

Fig. 5.9 Inter-tree competition showing (**a**) the method by which a relatively vigorous tree overtops a slow growing competitor, and (**b**) the variables that characterize the crown (From Mitchell 1975)

by the 1- to 5-year-old foliage is distributed over the bole annually and accumulated to provide tree and stand statistics. Other trees in a stand are assumed to be relevant to the model of a single tree only if branch extension is obstructed by neighbors. The technique of constraining growing space in relation to location and size of competing crowns is depicted in Fig. 5.9.

5.4 Tree Crowns and Growth

A number of studies have shown close relationships between tree crown size and/or morphology and individual tree and stand growth. Sprinz and Burkhart (1987) investigated empirical and theoretical relationships between tree crown, stem, and stand characteristics for unthinned stands of planted loblolly pine. Readily measured crown variables representing the amount of photosynthetic area or distance of the translocation process were identified. Various functions of these variables were defined and evaluated with regard to efficacy in predicting stem and stand attributes. The stem attributes modeled included basal area, basal area growth, diameter at breast height, and diameter growth. Crown diameter and crown projection area were particularly important in contributing to model fit and prediction of individual stem characteristics, while sum of crown projection areas was found especially important in stand level equations. As these crown measures developed over time so did corresponding stem and stand attributes.

Larocque and Marshall (1994a) related changes over time of crown relative growth measures of crown width/crown length (crown shape ratio), crown surface/crown volume, and foliage/crown volume to changes in relative growth rates (RGR) for red pine plantations originating from different initial spacings. Crown shape ratio decreased with increased *dbh* in the absence of severe competition and increased with *dbh* under severe competitive stress. The other two crown relative growth measures were always negatively correlated with *dbh*, indicating that large trees use their aerial growing space less efficiently than small trees at all stages of stand development. Crown shape ratio changed in the same manner as RGR.

Raulier et al. (1996) developed a model of bole volume increment based on crown dimensions and tree social status to predict bole volume increment independently of stand structure. Data were collected in two boreal black spruce stands in Quebec. A varying parameter approach was taken to show that the crown profile of black spruce depends on competition. Formal expressions for crown surface area and volume were developed from the crown profile and were used to derive a potential growth function for bole volume. Three social status indices were considered to characterize competition experienced by a subject tree. These indices were combined with the potential growth function to successfully model bole volume increment.

Because of the relationship between tree crown measures and tree and stand development, crown variables are often included in growth models. The most practical measure to include is crown length or crown ratio (length of green crown divided by total tree height) because height to the base of the live crown can be measured at the same time total height is assessed. While the tip of the tree crown is generally straight forward to determine, the base of the crown is often somewhat ambiguous. The crown base has been variously defined; examples include the lowest branch with green foliage, the lowest whorl which contains at least three live branches, and the point half way between the first whorl with one or more live branches and the whorl with at least four live branches. The United States Forest Service generally defines crown ratio in terms of compacted or uncompacted measurements (Toney and Reeves 2009). Measurement of compacted crown ratio (CC_r) involves envisioning the transfer of lower branches of trees with asymmetric crowns to fill holes in the upper portion of the crown. Uncompacted crown ratio (UNC_r) is measured without adjustment for holes in the crown and may be a more appropriate measurement when interest is on height to the first live branches in the crown. Toney and Reeves developed equations to convert CC_r to UNC_r for tree species in the western USA. UNC_r was modeled as a logistic function of CC_r and tree diameter; species-specific equations were fitted by nonlinear regression.

5.4.1 Modeling Crown Ratio

Due to its effectiveness as a predictor variable in many growth and yield relationships, a great deal of effort has been aimed at modeling crown ratio. Crown ratio (c_r) has been related to various tree and stand variables including *dbh*, height, stand density, and age. Holdaway (1986) developed the following model for predicting the crown ratio from tree and stand variables:

$$c_r = \mathrm{b}_1\left(\frac{1}{1+\mathrm{b}_2 G}\right) + \mathrm{b}_3(1 - e^{-\mathrm{b}_4 d})$$

where G = basal area per hectare and d = diameter at breast height of the subject tree.

Dyer and Burkhart (1987) started with a function to constrain crown ratio predictions between 0 and 1, namely:

$$c_r = 1 - \exp(-\Phi(x)) \tag{5.13}$$

where c_r is crown ratio and $\Phi(x)$ is some function of tree and stand attributes. When $\Phi(x)$ is positive c_r will be within its logical range. The function $\Phi(x)$ was determined by first linearizing the basic model structure. All possible regressions of no more than four independent variables were constructed for various functions of tree and stand attributes. Several functions for $\Phi(x)$ were chosen based on Mallow's C_p statistic, mean square error, percent variation explained, and analysis of residuals. The final function for $\Phi(x)$ was based on analysis of non-linear fit of the model. This procedure resulted in the following crown ratio model:

$$c_r = 1 - \exp\left[-(\mathrm{b}_0 + \mathrm{b}_1 t^{-1})d/h\right] \tag{5.14}$$

where

t = stand age
d = tree *dbh*
h = tree total height
b_0 and b_1 are positive coefficients

Dyer and Burkhart's final model form (5.14) has several desirable properties for modeling crown ratio. The variable d/h ensures that trees with more taper have higher crown ratios than those trees with less taper. With age entering as its reciprocal, crown ratio will decrease with age and the rate of this decrease will level off with time (provided that diameter and height are modeled with asymptotic functions). The inclusion of various stand density measures did not significantly reduce the residual sum of squares over Eq. 5.14, indicating that the effect of density on crown ratio is apparently largely accounted for in the d/h variable.

Hynynen (1995) also used the general model form (5.13) when developing crown ratio prediction equations for unthinned and thinned Scots pine stands in Finland. The function $\Phi(x)$ was specified in terms of tree and stand variables. The analysis of the relationships between crown ratio and tree and stand characteristics resulted in the following crown ratio model form for unthinned stands:

$$c_r = 1 - \exp\left\{-\left[a_0(\exp(-a_1 G)) + a_2 h_{dom}^{-1}\right]\left(\frac{d}{h}\right)^{a_3}\right\} \tag{5.15}$$

where

G is stand basal area (m^2/ha)
h_{dom} is dominant height (m)
d is tree diameter at breast height (cm)
h is total tree height (m)
a_0, a_1, a_2, a_3 are parameters

Thinning changes stand density, which has a strong effect on the development of tree crown ratio. At the time of thinning, crown ratio of a tree will be equal to the crown ratio of a similar tree in an unthinned stand. After thinning, crown recession for most trees in thinned stands is temporarily arrested because of the increased growing space, and tree crown ratios start to build back up by height increment. In developing a model for thinning response, Hynynen assumed that in thinned stands, trees crown ratio approaches the crown ratio of a tree growing in an unthinned stand with initial basal area equal to the basal area of the thinned stand after thinning.

Hynynen further assumed that the effect of thinning on crown ratio is affected by thinning intensity and time elapsed after thinning. The effect of thinning intensity was described by the difference between stand basal area before thinning (G_b), and stand basal area after thinning (G_a), which modifies the thinning response function. The difference between current dominant height (h_{dom}) and dominant height at the time of thinning (h_{domt}) was applied to describe the effect of time after thinning. The thinning response function was expressed as:

$$THIN = (G_b - G_a)\exp\left[-\left(\frac{h_{dom} - h_{domt}}{a_4}\right)^{a_5}\right] \tag{5.16}$$

where

G_b is stand basal area before thinning (m^2/ha)
G_a is stand basal area after thinning (m^2/ha)

h_{dom} is dominant height (m)
h_{domt} is dominant height at the time of thinning (m)
a_4, a_5 are parameters

Function (5.16) was incorporated in model (5.15) in connection with stand basal area, resulting in the crown ratio model for thinned stands:

$$c_r = 1 - \exp\left\{-\left[a_0 \exp(-a_1(G + THIN)) + a_2 {h_{dom}}^{-1}\right]\left(\frac{d}{h}\right)^{a_3}\right\} \quad (5.17)$$

where *THIN* is given by Eq. 5.16.

Liu et al. (1995) introduced a thinning response variable (*T*) into the crown ratio model of Dyer and Burkhart (1987) (Eq. 5.14) as follows:

$$c_r = 1 - T \exp\left[-(\mathrm{b}_0 + \mathrm{b}_1 t_s^{-1})d/h\right] \quad (5.18)$$

Equation 5.18 is bounded between 0 and 1 given that T is between 0 and 1. In the initial specification of (5.18) T was defined as I^{t_t/t_s} where I is the ratio of after thinning basal area to before thinning basal area, t_s is stand age and t_t is the stand age at time of thinning (from Short and Burkhart 1992).

The thinning effect function, *T*, found in Eq. 5.18 has certain desirable properties. First, when no thinning has occurred, the before to after thinning ratio, *I*, is 1 which means T has no effect on the prediction of crown ratio. Second, as the ratio t_t/t_s becomes smaller through time the effect of T becomes smaller. This suggests that over time the effect of thinning diminishes, and crown development approaches that of an unthinned stand condition. This specification of *T*, however, ensures a monotonically decreasing response to thinning over time, which implies that the maximum response of crown size to thinning occurs at the time of thinning. Biologically, however, there should be no immediate response at the time of thinning. Instead, response to thinning should begin at zero and increase to some maximum as the crowns of the residual trees respond to extra growing space and additional sunlight. Then, as the stand again closes, the response should diminish and approach an unthinned condition. With these considerations in mind, Liu et al. (1995) derived a thinning response function:

$$T = I^{\frac{-(t_s - t_t)^2 + k(t_s - t_t)}{t_s^2}} \quad (5.19)$$

where

T = thinning response
t_s = stand age
t_t = age of stand at time of thinning
I = ratio of after thinning basal area to before thinning basal area
k = duration parameter.

The duration of thinning response (in years) is determined by the value of the duration parameter, k. The first derivative of the exponential part of Eq. 5.19 with respect to $t_s - t_t$, the time elapsed since thinning, indicates that the maximum thinning response will occur at

$$\frac{kt_t}{k + 2t_t}$$

years after thinning. Thus, age of maximum response depends on age of the stand at time of thinning and k. Using Eq. 5.19, a new allometric crown ratio model was specified:

$$c_r = 1 - I^{\frac{r\left[-(t_s - t_t)^2 + k(t_s - t_t)\right]}{t_s^2}} \exp\left[-(\mathrm{b}_0 + \mathrm{b}_1/t_s)d/h\right] \tag{5.20}$$

where all variables are as previously defined. The rate parameter, r, is dimensionless, and along with I, t_s, and t_t define the shape of the response function.

In a study aimed at developing crown ratio equations for Austrian forests, Hasenauer and Monserud (1996) employed a logistic function:

$$c_r = \frac{1}{1 + e^{-\varphi x}} \tag{5.21}$$

where ϕx is a linear function of input variables and estimated coefficients. To select a set of variables for the linear function ϕx, Hasenauer and Monserud defined the variable groups of size characteristics (*SIZE*), competition measures (*COMP*), and site factors (*SITE*), thus:

$$c_r = \frac{1}{1 + e^{-[\mathrm{b}_0 + \mathrm{b}_1 SIZE + \mathrm{b}_2 COMP + \mathrm{b}_3 SITE]}} \tag{5.22}$$

Data from the Austrian National Forest Inventory consisting of more than 42,000 trees were used to fit Eq. 5.22 with the total variation explained by the model varying from 49% for larch to 17% for "other broadleafed species".

Soares and Tomé (2001) used data from spacing trials and permanent plots of *Eucalyptus globulus* to evaluate several nonlinear equations (exponential, logistic, Richards, and Weibull functions) restricted to the interval [0,1] for crown ratio prediction. Based on measurement of fit and prediction ability, they recommended the following Richards function for tree crown ratio prediction in Eucalyptus stands in Portugal:

$$c_r = \frac{1}{\left[1 + e^{-(-b_0 + b_1 t^{-1} + b_2 N + b_3 h_{dom} + b_4 d)}\right]^{1/6}} \tag{5.23}$$

where t, age (years); N, number of trees per ha; h_{dom}, dominant height (m); d, diameter at breast height (cm).

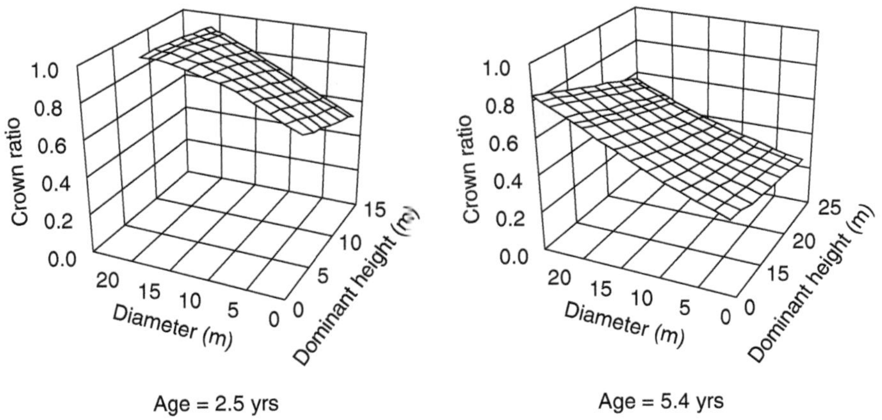

Fig. 5.10 Relationship between crown ratio and tree diameter and stand dominant height for two age classes of eucalyptus plantations (From Soares and Tomé 2001)

Function (5.23) is age and density dependent, reflecting the importance of competition; age was expressed by its inverse and the number of live trees per hectare was the best expression of density; an initial tree dimension (diameter) and a measure of stand productivity (dominant height) were also required as explanatory variables. In function (5.23) greater values for age, number of trees or dominant height resulted in smaller crown ratio values; in the same stand, at a specific age, an increase in diameter resulted in higher crown ratio values (Fig. 5.10).

Temesgen et al. (2005) noted that crown ratio (c_r) is often an important predictor variable for tree-level growth equations in multi-species and multi-layered stands. Accordingly, they developed models to predict c_r from size, competition and site variables for several coniferous and one hardwood tree species growing in complex stands of southeastern British Columbia, Canada. The data indicated c_r decreasing with increasing height, and with increasing competition. A logistic model form was used to constrain predicted c_r values to the interval [0,1]. Predictors were divided into tree size, stand competition, and site measures, and the contribution of each set of contributors was examined. For all models, height was an important predictor. The stand competition measure, basal area of larger trees, contributed significantly to predicting c_r given that crown competition factor was also included as a measure of competition. Site variables slightly improved predictions for some species, but much of the variability in c_r was not accounted for, indicating that other variables are important for explaining c_r changes in these complex stands.

Leites et al. (2009) evaluated crown ratio estimation accuracy for the c_r equations in two variants of the Forest Vegetation Simulator. They also assessed the effects of using measured crown ratio for estimating 10-year diameter growth and 30-year basal area increment. Differences between measured and predicted crown ratio were positively correlated with differences in diameter growth predictions; differences in the 30-year basal area increment predictions were not large when either predicted or measured crown ratio values were used.

Although crown ratio is often the dependent variable used for estimating changes in tree crown size, direct modeling of crown rise (crown recession) is sometimes employed. Maguire and Hann (1990a) estimated crown recession rates for Douglas-fir trees via branch mortality dating. The logarithmic model recommended by the authors predicts crown recession from current crown ratio, total length, breast height age, height growth, and crown competition factor. Short and Burkhart (1992) developed individual tree crown-height increment equations to predict annual crown height increment in thinned and unthinned loblolly pine plantations. The selected model contained tree height, crown ratio, age, and a measure of competition as predictors. Liu et al. (1995) found little difference in predictive ability between using a crown height increment equation or an allometric model, which derives crown ratio from tree diameter and height information.

5.4.2 Crown Relationships for Open-Grown Trees

In addition to models for estimating crown variables for stand-grown trees, relationships between crown size and tree dimensions of open-grown trees have been found useful in forest modeling. Open-grown trees represent the empirical maximum for certain tree dimensions. These maximum values have been found valuable for modeling competition and crown closure. Honer (1971) developed crown radius equations for open-grown balsam fir and black spruce, Strub et al. (1975) regressed *dbh* on crown width for open-grown loblolly pine trees, Leech (1984) estimated crown width from *dbh* for open-grown radiata pine trees in South Australia, and Farr et al. (1989) presented height diameter and crown-width equations for western hemlock and Sitka spruce. Smith et al. (1992) collected data on open-grown loblolly, longleaf, and shortleaf pines to provide predictive equations of crown width and maximum potential basal area growth. Hasenauer (1997) investigated crown width, *dbh*, height to live crown base, and taper rates for open-grown trees in Austria.

References

Avery TE, Burkhart HE (2002) Forest measurements, 5th edn. McGraw-Hill, New York

Baldwin VC Jr, Peterson KD (1997) Predicting the crown shape of loblolly pine trees. Can J For Res 27:102–107

Baldwin VC Jr, Peterson KD, Burkhart HE, Amateis RL, Dougherty PM (1997) Equations for estimating loblolly pine branch and foliage weight and surface area distributions. Can J For Res 27:918–927

Biging GS, Gill SJ (1997) Stochastic models for conifer tree crown profiles. For Sci 43:25–34

Biging GS, Wensel LC (1990) Estimation of crown form for six conifer species of northern California. Can J For Res 20:1137–1142

Brown JK (1978) Weight and density of crowns of Rocky Mountain conifers. USDA Forest Service, Intermountain Forest and Range Experiment Station, Ogden, Research Paper INT-197

Cluzeau C, LeGoff N, Ottorini J-M (1994) Development of primary branches and crown profile of *Fraxinus excelsior*. Can J For Res 24:2315–2323

Colin F, Houllier F (1992) Branchiness of Norway spruce in northeastern France: predicting the main crown characteristics from usual tree measurements. Ann Sci For 49:511–538

Courbet F, Sabatier S, Guédon Y (2007) Predicting the vertical location of branches along Atlas cedar stem (*Cedrus atlantica Manetti*) in relation to annual shoot length. Ann For Sci 64: 707–718

Crecente-Campo F, Marshall P, LeMay V, Diéguez-Aranda U (2009) A crown profile model for *Pinus radiata* D. Don in northwestern Spain. For Ecol Manage 257:2370–2379

Curtis RO, Reukema DL (1970) Crown development and site estimates in a Douglas-fir plantation spacing test. For Sci 16:287–301

Deleuze C, Hervé J-C, Colin F, Ribeyrolles L (1996) Modelling crown shape of *Picea abies* spacing effects. Can J For Res 26:1957–1966

Doruska PF, Burkhart HE (1994) Modeling the diameter and locational distribution of branches within the crowns of loblolly pine trees in unthinned plantations. Can J For Res 24:2362–2376

Doruska PF, Mays JE (1998) Crown profile modeling of loblolly pine by nonparametric regression analysis. For Sci 44:445–453

Dyer ME, Burkhart HE (1987) Compatible crown ratio and crown height models. Can J For Res 17:572–574

Farr WA, DeMars DJ, Dealy JE (1989) Height and crown width related to diameter for open-grown western hemlock and Sitka spruce. Can J For Res 19:1203–1207

Ford R, Ford ED (1990) Structure and basic equations of simulator for branch growth in the Pinaceae. J Theor Biol 146:1–13

Garber SM, Maguire DA (2005) The response of vertical foliage distribution to spacing and species composition in mixed conifer stands in central Oregon. For Ecol Manage 211:341–355

Gary HL (1976) Crown structure and distribution of biomass in a lodgepole pine stand. USDA Forest Service, Rocky Mountain Forest and Range Experiment Station, Fort Collins, Research Paper RM-165

Gill SJ, Biging GS (2002a) Autoregressive moving average models of crown profiles for two California hardwood species. Ecol Mod 152:213–226

Gill SJ, Biging GS (2002b) Autoregressive moving average models of conifer crown profiles. J Agric Biol Environ Stat 7:558–573

Hann DW (1999) An adjustable predictor of crown profile for stand-grown Douglas-fir trees. For Sci 45:217–225

Hans P (1997) Functions for predicting crown height of *Pinus sylvestris* and *Picea abies* in Sweden. Scand J For Res 12:179–188

Hasenauer H (1997) Dimensional relationships of open-grown trees in Austria. For Ecol Manage 96:197–206

Hasenauer H, Monserud RA (1996) A crown ratio model for Austrian forests. For Ecol Manage 84:49–60

Holdaway MR (1986) Modeling tree crown ratio. For Chron 62:451–455

Honer TG (1971) Crown shape in open-and forest-grown balsam fir and black spruce. Can J For Res 1:203–207

Hynynen J (1995) Predicting tree crown ratio for unthinned and thinned Scots pine stands. Can J For Res 25:57–62

Larocque GR, Marshall PL (1994a) Crown development in pine stands. I. Absolute and relative growth measures. Can J For Res 24:762–774

Larocque GR, Marshall PL (1994b) Crown development in red pine stands. II. Relationships with stem growth. Can J For Res 24:775–784

Ledermann T (2011) A non-linear model to predict crown recession of Norway spruce (*Picea abies* [L.] Karst.) in Austria. Eur J For Res 130:521–531

Leech JW (1984) Estimating crown width from diameter at breast height for open-grown radiata pine trees in south Australia. Aust For Res 14:333–337

Leites LP, Robinson AP, Crookston NL (2009) Accuracy and equivalence testing of crown ratio models and assessment of their impact on diameter growth and basal area increment predictions of two variants of the forest vegetation simulator. Can J For Res 39:655–665

Liu J, Burkhart HE, Amateis RL (1995) Projecting crown measures for loblolly pine trees using a generalized thinning response function. For Sci 41:43–53

Madgwick HAI, Jackson DS (1974) Estimating crown weights of *Pinus radiata* from branch variables. N Z J For Sci 4:520–528

Maguire DA, Hann DW (1990a) Constructing models for direct prediction of 5-year crown recession in southwestern Oregon Douglas-fir. Can J For Res 20:1044–1052

Maguire DA, Hann DW (1990b) A sampling strategy for estimating past crown recession on temporary growth plots. For Sci 36:549–563

Maguire DA, Moeur M, Bennett WS (1994) Models for describing basal diameter and vertical distribution of primary branches in young Douglas-fir. For Ecol Manage 63:23–55

Mäkinen H, Colin F (1998) Predicting branch angle and branch diameter of Scots pine from usual tree measurements and stand structural information. Can J For Res 28:1686–1696

Mäkinen H, Colin F (1999) Predicting the number, death, and self-pruning of branches in Scots pine. Can J For Res 29:1225–1236

Mäkinen H, Ojansuu R, Sairanen P, Yli-Kojola H (2003) Predicting branch characteristics of Norway spruce (*Picea abies* (L.) Karst.) from simple stand and tree measurements. Forestry 76:525–546

Marshall DD, Johnson GP, Hann DW (2003) Crown profile equations for stand-grown western hemlock trees in northwestern Oregon. Can J For Res 33:2059–2066

Mitchell KJ (1975) Dynamics and simulated yield of Douglas-fir. For Sci Monogr 17:1–39

Moeur M (1981) Crown width and foliage weight of northern Rocky Mountain conifers. USDA Forest Service, Intermountain Forest and Range Experiment Station, Ogden, Research Paper INT-283

Nepal SK, Somers GL, Caudill SB (1996) A stochastic frontier model for fitting tree crown shape in loblolly pine (*Pinus taeda* L.). J Agric Biol Environ Stat 1:336–353

Pretzsch H (2009) Forest dynamics, growth and yield. Springer, Berlin

Raulier F, Ung C-H, Ouellet D (1996) Influence of social status on crown geometry and volume increment in regular and irregular black spruce stands. Can J For Res 26:1742–1753

Remphrey WR, Powell GR (1984) Crown architecture of *Larix laricina* saplings: quantitative analysis and modeling of (nonsylleptic) order 1 branching in relation to development of the main stem. Can J Bot 62:1904–1915

Roeh RL, Maguire DA (1997) Crown profile models based on branch attributes in coastal Douglas-fir. For Ecol Manage 96:77–100

Short EA III, Burkhart HE (1992) Predicting crown-height increment for thinned loblolly pine plantations. For Sci 38:594–610

Siemon GR, Wood GB, Forrest WG (1976) Effects of thinning on crown structure in radiata pine. N Z J For Sci 6:57–66

Smith WR, Farrar RM Jr, Murphy PA, Yeiser JL, Meldahl RS, Kush JS (1992) Crown and basal area relationships of open-grown southern pines for modeling competition and growth. Can J For Res 22:341–347

Soares P, Tomé M (2001) A tree crown ratio prediction equation for eucalyptus plantations. Ann For Sci 58:193–202

Sprinz PT, Burkhart HE (1987) Relationships between tree crown, stem, and stand characteristics in unthinned loblolly pine plantations. Can J For Res 17:534–538

Strub MR, Vasey RB, Burkhart HE (1975) Comparison of diameter growth and crown competition factor in loblolly pine plantations. For Sci 21:427–431

Temesgen H, LeMay V, Mitchell SJ (2005) Tree crown ratio models for multi-species and multi-layered stands of southeastern British Columbia. For Chron 81:133–141

Toney C, Reeves MC (2009) Equations to convert compacted crown ratio to uncompacted crown ratio for trees in the Interior West. West J Appl For 24:76–82

Trincado G, Burkhart HE (2009) A framework for modeling the dynamics of first-order branches and spatial distribution of knots in loblolly pine trees. Can J For Res 39:566–579

Tucker GF, Lassoie JP, Fahey TJ (1993) Crown architecture of stand-grown sugar maple (*Acer saccharum* Marsh.) in the Adirondack Mountains. Tree Physiol 13:297–310

Zeide B (1991) Fractal geometry in forestry applications. For Ecol Manage 46:179–188

Zeide B (1998) Fractal analysis of foliage distribution in loblolly pine crowns. Can J For Res 28:106–114

Zeide B, Gresham CA (1991) Fractal dimensions of tree crowns in three loblolly pine plantations of coastal South Carolina. Can J For Res 21:1208–1212

Zeide B, Pfeifer P (1991) A method for estimation of fractal dimension of tree crowns. For Sci 37:1253–1265

Chapter 6
Growth Functions

6.1 Introduction

Growth functions describe the change in size of an individual or population with time. The selection of appropriate growth functions for tree and stand modeling is an important aspect in the development of growth and yield models. Here we present information on the forms and characteristics of the more commonly-used growth functions for modeling forest development. When fitted to data, a number of these functions will give essentially equivalent results within the range of the observations used for estimating the equation's coefficients. However, their behavior when extrapolated may be quite different depending on the underlying mathematical properties involved. Hence, understanding these properties is helpful to modelers to determine which candidate functions to consider for specific applications.

Unless the data available for modeling cover a very small range of time, there are certain properties that a growth function should exhibit to be consistent with the principles of biological growth (Fig. 6.1):

(i) The curve is often limited by the value zero at a specific beginning ($t = 0$ or $t = t_0$), depending if the variable that is being modeled starts at $t = 0$, as is the case for the great majority of the tree and stand variables, or later on, as happens with tree diameter at breast height or stand basal area;
(ii) The curve generally should exhibit a maximum value usually achieved at an older age (existence of an asymptote);
(iii) The slope of the curve should increase with increasing growth rate in the initial phase and decrease in the final stages (show an inflection point).

At this point it is important to understand the concepts of growth and yield. Growth is the increase in size of an individual or population per unit of time (for instance volume growth in m^3ha^{-1} year^{-1}) while yield is the size of the tree or population at a certain point in time (for instance total volume at age 50

H.E. Burkhart and M. Tomé, *Modeling Forest Trees and Stands*,
DOI 10.1007/978-90-481-3170-9_6,

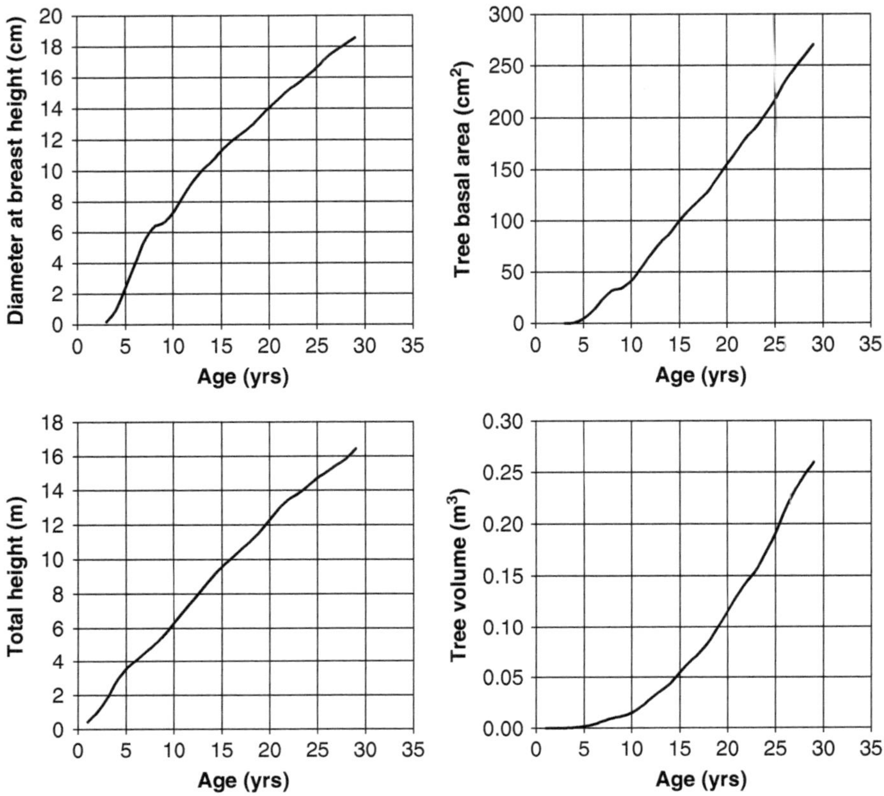

Fig. 6.1 Development of a single tree over time (Data from stem analysis of a maritime pine tree at age 29 years)

in m^3ha^{-1}). In the case of continuous functions, a yield equation can be obtained by integrating the growth equation and, conversely, a growth equation can be obtained by differentiating a specified yield equation.

6.2 Empirical Versus Mechanistic or Theoretical Growth Functions

Functions used to model growth have been classified into one of two broad divisions (Thornley 1976; Hunt 1982; Vanclay 1994), empirical and theoretical or mechanistic models. The last are conceived in terms of the mechanism of the system (Thornley 1976), usually having an underlying hypothesis associated with the cause or function of the phenomenon described by the response variable (Vanclay 1994), while empirical models describe the behavior of the response variable without trying to identify the causes and explaining the phenomenon. The distinction between the two is not sharp and most modeling applications contain both empiricism and mechanism in varying mixtures. According to Thornley (1976), the mechanistic modeler will tend to construct models before doing the experiments or analyzing the data, thinking of possible mechanisms and deducting their consequences by means of a model while the empirical modeler will describe the behavior of the response variable based on the data. If a mathematical model has a theoretical/biological basis, the parameter estimates, even if obtained from empirical data, can provide insight into the phenomenon that is being modeled.

Many equations have been used to model tree and stand growth, usually under the integral form. Table 6.1 summarizes some non-sigmoid functions that are commonly used in forest modeling (Grosenbaugh 1965; Prodan 1968). A longer list can be seen in Kiviste (1988) or Kiviste (2002).[1] More than one differential form may exist for a given function; the differential form shown in Table 6.1 is the one most often used in forestry applications. None of these functions exhibit all the desirable properties listed previously; thus, when using them for modeling, one needs to be cautious with extrapolations outside the range of data used to fit the models and with the signs that the parameters can take so that they exhibit a shape compatible with biological growth.

The functions known as Freese, Hossfeld I and Korsun show extremes (maximum and/or minimum) that, depending on the signs of the coefficients, may follow within the range of time values relevant for forest growth modeling. However they follow approximately the shape of a growth curve for a limited range of ages and have been used in several growth modeling applications, especially the Hossfeld I function.

Theoretical growth functions have commonly been developed in their growth form – either absolute or relative growth – and the respective yield form has been obtained by integration. Generally this approach allows interpretation of the function parameters and helps to impose restrictions on the values that the parameters can take to be biologically consistent.

Several authors (e.g. Grosenbaugh 1965; Pienaar and Turnbull 1973; Causton and Venus 1981; Hunt 1982; Zeide 1993; Kiviste et al. 2002) have analysed growth function properties. Table 6.2 summarizes the properties of the sigmoid growth functions that are more commonly used in forest modeling. In this table, the asymptote is designated by A, the k parameter is related to the slope of the curve (growth rate), the m parameter is a shape parameter, and, usually, the c parameter relates to the initial condition used. An analysis of the assumptions from which the functions in Table 6.2 were derived, as well as the interpretation of the parameters, is given in the following sections. In these sections the functions are grouped according to their functional form, with the groups being designated by the most general or best known function of the group: (1) Lundqvist-Korf; (2) Richards; (3) Hossfeld IV/McDill-Amateis; (4) other growth functions.

---

[1]The analysis of growth functions published by Kiviste (1988) in Russian was later partially translated into Spanish (Kiviste et al. 2002). We are not aware of an English version of this work.

Table 6.1 Some non-sigmoid functions commonly used in forest growth modeling

Author or designation	Mathematical expression: Integral form (yield)	Differential form (growth)	Restrictions in the parameters	Properties: Value at the origin	Inflection point	Asymptote $t \to \infty$
Hyperbola	$Y = a_0 + a_1 \frac{1}{t}$	$\frac{dY}{dt} = -a_1 \frac{1}{t^2}$	$a_1 < 0$	$t \to 0; Y \to -\infty$	None	$Y = a_0$
–	$Y = a_0 + a_1 \frac{1}{t} + a_2 t$	$\frac{dY}{dt} = a_2 - a_1 \frac{1}{t^2}$	$a_1 < 0; a_2 > 0$	$t \to 0; Y \to -\infty$	None	$Y \to \infty$
Linear-logarithmic	$Y = a_0 + a_1 \log t$	$\frac{dY}{dt} = a_1 - \frac{1}{t}$	$a_1 > 0$	$t \to 0; Y \to -\infty$	None	$Y \to \infty$
–	$Y = a_0 + a_1 t^{a_2}$	$\frac{dY}{dt} = \frac{a_2}{t}(Y - a_0)$	$a_1, a_2 > 0; a_2 < 1$	$t = 0; Y = a_0$	None	$Y \to \infty$
			$a_1, a_2 < 0$	$t \to 0; Y \to -\infty$	None	$Y \to a_0$
–	$Y = (a_0 + a_1 t)^{a_2}$	$\frac{dY}{dt} = \frac{a_0 a_2 Y}{a_0 + a_1 t}$	$a_1, a_2 > 0; a_2 < 1$	$t = 0; Y = a_0^{a2}$	None	$Y \to \infty$
Exponential	$Y = a_0 + a_1 e^{a_2 t}$	$\frac{dY}{dt} = a_2 (a_0 - Y)$	$a_1, a_2 < 0$	$t = 0; Y = a_0 + a_1$	None	$Y \to a_0$
Freese	$Y = a_0 t^{a_1} + a_2^{t}$	$\frac{dY}{dt} = Y\left(\log a_2 + \frac{a_1}{t}\right)$	$a_0, a_1 > 0; \ln a_2 < 0$	$t = 0; Y = 1$	$t = \frac{-a_1 \pm \sqrt{a_1}}{\ln a_2}$	$Y \to -\infty$
			$a_0, a_1 > 0; \ln a_2 > 0$			
			$a_0, a_1 > 0; \ln a_2 = 0$		None	
Hossfeld I	$Y = \frac{t^2}{a_0 + a_1 + a_2 t^2}$	$\frac{dY}{dt} = Y^2\left(\frac{2a_0 + a_1 t}{t^3}\right)$	$a_0 > 0; a_1 < 0$	$t = 0; Y = 0$	$a_1 a_2 t^3 + 3a_0 a_2 t^2$	$Y \to \frac{1}{a_2}$
			$a_0 > 0; a_1 > 0$	$t = 0; Y = 0$	$-a_0^2 = 0$	
Korsun (logarithmic parabola)	$Y = a_0 t^{a_1 + a_2 \log t}$ $a_0 > 0$	$\frac{dY}{dt} = \frac{Y}{t}(a_1 + 2a_2 \log t)$	$a_1 > 0; a_2 < 0$	$t \to 0; Y \to 0$	$z^2 - z + 2a_2 = 0$ $z = \beta_1 + 2a_2 \ln t$	$Y \to 0$

Table 6.2 Theoretical/sigmoid growth functions commonly used in forest growth modeling

Author or designation	Mathematical expression: Integral form (yield)	Differential form (growth)	Restrictions in the parameters	Properties: Value at the origin	Inflection point	Asymptote $t \to \infty$
Schumacher	$Y = Ae^{-\frac{k}{t}}$	$\frac{dY}{dt} = Y\frac{k}{t^2}$	$k > 0$	$t \to 0; Y \to 0$	$t = \frac{k}{2}; \quad Y = \frac{A}{e^2}$	$Y \to A$
Johnson-Schumacher	$Y = Ae^{-\frac{k}{t+a}}$	$\frac{dY}{dt} = Y\frac{k}{(t+a)^2}$	$k > 0$	$t \to 0; Y \to Ae^{\frac{k}{a}}$	$t = \frac{k}{2} - a;$ $Y = \frac{A}{e^2}$	$Y \to A$
Lundqvist-Korf	$Y = Ae^{-\frac{k}{t^m}}$	$\frac{dY}{dt} = mY\frac{k}{t^{m+1}}$	$k > 0$	$t \to 0; Y \to 0$	$t = \left(\frac{mk}{m+1}\right)^{\frac{1}{m}};$ $Y = Ae^{-\frac{m+1}{m}}$	$Y \to A$
			$n > 0$			
Monomolecular	$Y = A\left(1 - ce^{-kt}\right)$	$\frac{dY}{dt} = k\,(A - Y)$	$k > 0$	$t = 0; Y = A\,(1 - c)$	None	$Y \to A$
Logistic	$Y = \frac{A}{(1+ce^{-kt})}$	$\frac{dY}{dt} = \frac{k}{A}\left(AY - Y^2\right)$	$k > 0$	$t = 0; Y = \frac{A}{1+c}$	$t = \frac{1}{k}\log c;$ $Y = \frac{A}{2}$	$Y \to A$
				$t \to -\infty; Y \to 0$		
Pearl-Reed[a]	$Y = \frac{A}{1+ce^{-f(\mathbf{X},t)}}$	$\frac{dY}{dt} = \frac{AY-Y^2}{A}f(\mathbf{X}, t)$	$c > 0$	$t = 0; Y = \frac{A}{1+c}$	Inflection depends on *f*(***X***,*t*)	$Y \to A$
				$t \to -\infty; Y \to 0$		
Gompertz	$Y = Ae^{-ce^{-kt}}$	$\frac{dY}{dt} = kY\ln\frac{A}{Y}$	$k > 0$	$t = 0; Y = Ae^{-c}$	$t = \frac{\log c}{k};$ $Y = \frac{A}{e}$	$Y \to A$
			$c > 0$	$t \to -\infty; Y = 0$		
Richards	$Y = A\left(1 - ce^{-kt}\right)^{-\frac{1}{1-m}}$	$\frac{dY}{dt} = \frac{kY}{1-m}\left[\left(\frac{A}{Y}\right)^{1-m} - 1\right]$	$k > 0$	$t = 0; Y = A(1 - c)^{\frac{1}{1-m}}$	$t = \frac{\log\left(\frac{c}{1-m}\right)}{k};$ $Y = Am^{\frac{1}{1-m}}$	$Y \to A$
McDill-Amateis/ Hossfeld/ log-logistic	$Y = A\frac{t^k}{c+t^k}$	$\frac{dY}{dt} = k\frac{Y}{t}\left(1 - \frac{Y}{A}\right)$	$k > 1$	$t \to 0; Y \to 0$	$t = \left[\frac{c(k-1)}{k+1}\right]^{1/k};$ $Y = \frac{A}{2}\left(1 - \frac{1}{k}\right)$	$Y \to A$

[a]Depending on the function f(X,t) may not be a sigmoidal function

6.3 Growth Functions of the Lundqvist-Korf Type

6.3.1 Schumacher Function

Schumacher's function (Schumacher 1939) represents an early attempt in forestry to develop a growth function from biologically sound assumptions. The model proposed by this author for "generalized use" relies on the hypothesis that the relative growth rate increases linearly with the squared inverse of time (which means that it decreases nonlinearly with time):

$$\frac{1}{Y}\frac{dY}{dt} = k\frac{1}{t^2} \tag{6.1}$$

The yield function is

$$Y = A\,e^{-k\frac{1}{t}} \tag{6.2}$$

where $A = Y_0 e^{k/t_0}$ is the asymptote and (t_0, Y_0) are the initial values.

The k parameter expresses the rate of decrease of the relative growth rate and is therefore inversely related to the growth rate. The location of the inflection point depends on the value of k; the value of Y at the time at which the inflection occurs depends on A and k.

The Johnson-Schumacher function (Grosenbaugh 1965) is a generalization of Schumacher's function that includes cases when the initial value for $t = 0$ is not zero, that is:

$$Y = A\,e^{-k\frac{1}{t+b}}$$

The additional parameter (b) implies that, for $t = 0$, the tree/stand has already attained the dimension $Ae^{-k/b}$. The addition of an extra parameter is of limited value except in the cases where the trees are planted and one wants to consider the initial size of the seedlings.

6.3.2 Lundqvist-Korf Function

Another generalization is the Lundqvist-Korf function (Korf 1939; Lundqvist 1957):

$$\frac{1}{Y}\frac{dY}{dt} = k\frac{m}{t^{(m+1)}} \tag{6.3}$$

The corresponding yield function is:

$$Y = A\,e^{-k\frac{1}{t^m}} \tag{6.4}$$

where $A = Y_0 e^{k/t_0^m}$ is the asymptote and $(t_0,\ Y_0)$ are the initial values.

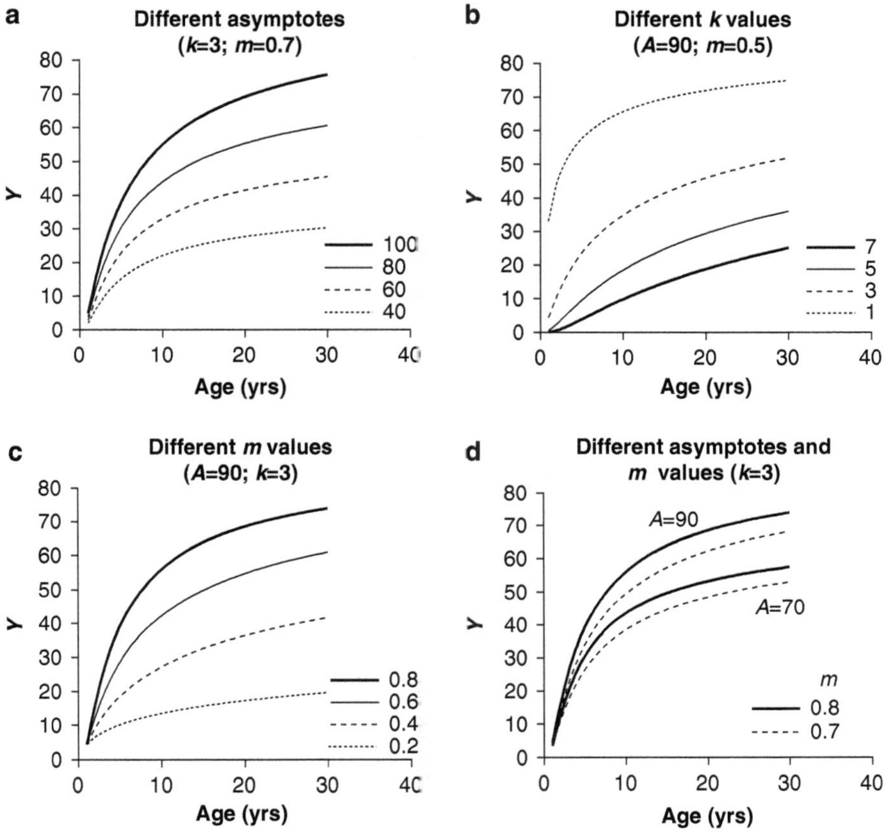

Fig. 6.2 Flexibility of the Lundqvist-Korf growth function exhibited through specifying varying values for the parameters

The *k* parameter is inversely related to the growth rate but this is also influenced by the *m* parameter, adding flexibility to the curve.

Figure 6.2 illustrates the flexibility of the Lundqvist function. By changing the parameters (asymptote, *k* and *m*) it is possible to cover a large range of shapes. When fixing the other parameters, the *k* parameter has an inverse relationship with the growth rate, while the reverse is true for the *m* parameter. It is important to stress that the three parameters interact as is shown on Fig. 6.2; it is possible to obtain higher growth with a smaller value of *m*, if the latter is combined with a higher asymptote. The location of the inflection point does not depend on the value of the asymptote, but on the combined values of *k* and *m*, increasing with *k*. The effect of *m* is highly dependent on the *k* value. Figure 6.3 shows the combined effect of the two shape parameters on the age at which the inflection occurs. The respective *Y* value, however, does not depend on the *k* parameter, but on the asymptote and *m* parameter values (Fig. 6.4), occurring at higher *Y* values, the higher the asymptote and the *m* value.

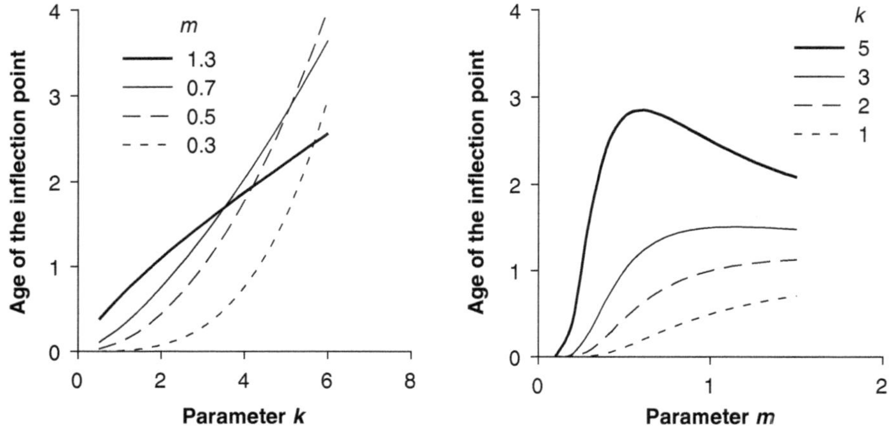

Fig. 6.3 Effect of the k and m parameters of the Lundqvist-Korf's function on the location of the inflection point

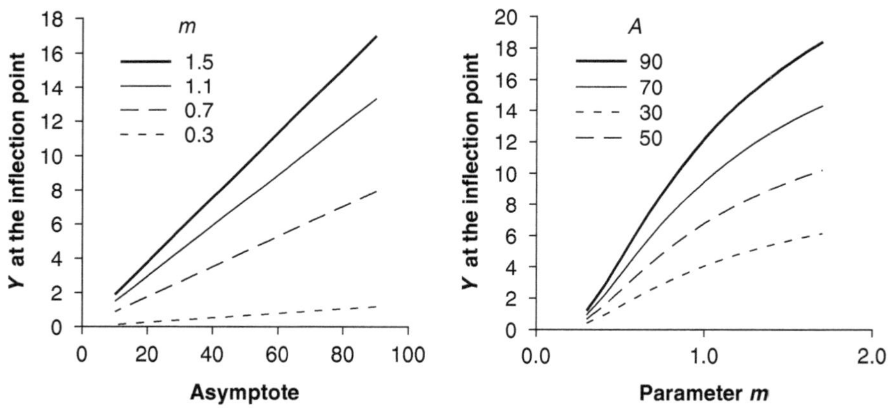

Fig. 6.4 Effect of the asymptote (A) and m parameters of the Lundqvist-Korf function on the inflection point of Y

6.4 Growth Functions of the Richards Type

6.4.1 Monomolecular Function

The monomolecular function, sometimes referred in agriculture and economics as the Mitscherlich function or law of diminishing returns (Zeide 1993), can be obtained under the assumption that the absolute growth rate is proportional to the difference between the maximum value (asymptote) and the present dimension:

$$\frac{dY}{dt} = k\,(A - Y) = kA - kY \tag{6.5}$$

The absolute growth rate decreases linearly with the size of the individual or population (*Y*).

The yield form is

$$Y = A\left(1 - ce^{-kt}\right), \tag{6.6}$$

with $c = e^{kt_0}\left(1 - \frac{Y_0}{A}\right)$. Using the initial condition $Y(0) = 0$ leads to $c = 1$.

The function has an upper asymptote A, but no inflection point. The k parameter expresses the rate of decrease of the absolute growth rate.

6.4.2 *Logistic and Generalized Logistic Functions*

The logistic function, first developed for population growth, is one of the best known sigmoid functions. It was applied to predicting yield of loblolly pine in 1937 (MacKinney et al. 1937) and is based on the assumption that the relative growth rate is equal to a biotic potential k, reduced according to the size/dimension of the population (in the present case tree/stand):

$$\frac{1}{Y}\frac{dY}{dt} = (k - mY) \tag{6.7}$$

The relative growth rate is therefore a declining linear function of the dimension.

The yield function is

$$Y = \frac{A}{\left(1 + ce^{-kt}\right)} \tag{6.8}$$

with $c = \frac{Y_0/m}{k - mY_0}e^{kY_0}$ and $A = \frac{k}{m}$

The inflection point of the logistic function occurs at $t = \log(c)/k$ and $Y = A/2$ and the curve is symmetric around the inflection point. The parameter k is the maximum relative growth rate and corresponds to the initial stage of growth.

A generalization of the logistic function (Pearl and Reed 1923), formulated to overcome the symmetry of the logistic curve, usually has the integral form:

$$Y = \frac{A}{\left(1 + c\,e^{-\left(a_1 t + a_2 t^2 + a_3 t^3\right)}\right)}$$

where a_1, a_2 and a_3 are parameters that define the shape of the curve. The inflection point is variable and almost incalculable. Note that the function may have more than one inflection point. A detailed study of this function as well as a reparameterization of it was published by Grosenbaugh (1965).

Monserud (1984) proposed the following function, which has been referred to as a generalization of the logistic growth function:

$$Y = \frac{A}{\left(1 + ce^{-f(\mathbf{X},t)}\right)} \tag{6.9}$$

where f (***X***,t) is a function of age (t) and of several independent variables (***X***). A is the asymptote and c is the half-saturation parameter that defines the value of $e^{-f(\mathbf{X},t)}$ at which $Y(t) = A/2$ (Cieszewski 2002).

6.4.3 Gompertz Function

The Gompertz equation (Gompertz 1825) was designed to describe age distribution in human populations and later on it was applied as a growth model (Winsor 1932). This function can be directly obtained from the following differential equation:

$$\frac{1}{Y}\frac{dY}{dt} = k\,(\log A - \log Y) = -k \log\left(\frac{Y}{A}\right) \tag{6.10}$$

The function assumes that the relative growth rate is inversely related to the logarithm of the ratio between the present dimension of Y and the respective asymptotic value.

The relative growth function can also be defined as a decreasing exponential function of time:

$$\frac{1}{Y}\frac{dY}{dt} = ke^{-ce^{-kt}}$$

Integration of Eq. 6.10 leads to the yield function:

$$Y = A\,e^{-ce^{-kt}} \tag{6.11}$$

with $c = (\log A - \log Y_0)e^{kt_0} = \log\left(\frac{A}{Y_0}\right)e^{kt_0}$

6.4.4 Richards Function

Richards (1959) generalized the function presented by von Bertalanffy (1938) for animal growth (Pienaar and Turnbull 1973). This function describes the absolute growth rate as the difference between an anabolic rate (constructive metabolism), which in plants is proportional to the photosynthetically active area, and a catabolic rate (destructive metabolism) that is proportional to biomass. If the photosynthetically active area is expressed as an allometric relationship with biomass, these relationships can be expressed as:

Anabolic rate	$c_1 S = c_1(c_0 Y^m) = c_2 Y^m$
Catabolic rate	$c_3 Y$
Potential growth rate	$c_2 Y^m - c_3 Y$
Growth rate	$c_4(c_2 Y^m - c_3 Y)$,

where S is the photosynthetically active area; Y is the biomass (or other tree/stand variable); m is the allometric constant of the relationship between S and Y; c_0, c_1, c_2, c_3 are proportionality coefficients; and c_4 is an efficacy coefficient.

The following differential form of the Richards function is then obtained:

$$\frac{dY}{dt} = aY^m - bY \tag{6.12}$$

By integration (Bernoulli differential equation), the corresponding yield function is obtained:

$$Y = A\left(1 - ce^{-kt}\right)^{\frac{1}{1-m}}, \tag{6.13}$$

where the parameters c, k and A are:

$$c = e^{-(1-m)bt_0} = e^{-kt_0}$$

$$k = (1-m)\,b$$

$$A = \left(\frac{a}{b}\right)^{\frac{1}{1-m}} \quad \text{(asymptote)}$$

The m exponent is often taken to equal 2/3. It is important to note that the monomolecular, logistic and Gompertz functions are particular cases of the Richards function when the parameter m takes, respectively, the values 0, 2, or tends to 1.

Figure 6.5 shows the flexibility of the Richards function as well as the effect of the three parameters on the respective shape. Higher values of k produce higher growth rates while, on the contrary, smaller values of m result in higher growth rates. As expected, the asymptote is also positively related with higher yields.

Figures 6.6 and 6.7 show effects of changing parameter values on the location of the inflection point and the corresponding Y value. Figure 6.6 shows that higher values of the k parameter result in earlier inflection points, while the opposite relationship can be observed with the m parameter. The value of Y at the time when the inflection point occurs is higher for higher asymptote values, but it is inversely related with the value of the m parameter.

Causton and Venus (1981) present a detailed study on the application of the Richards function to plant growth modeling.

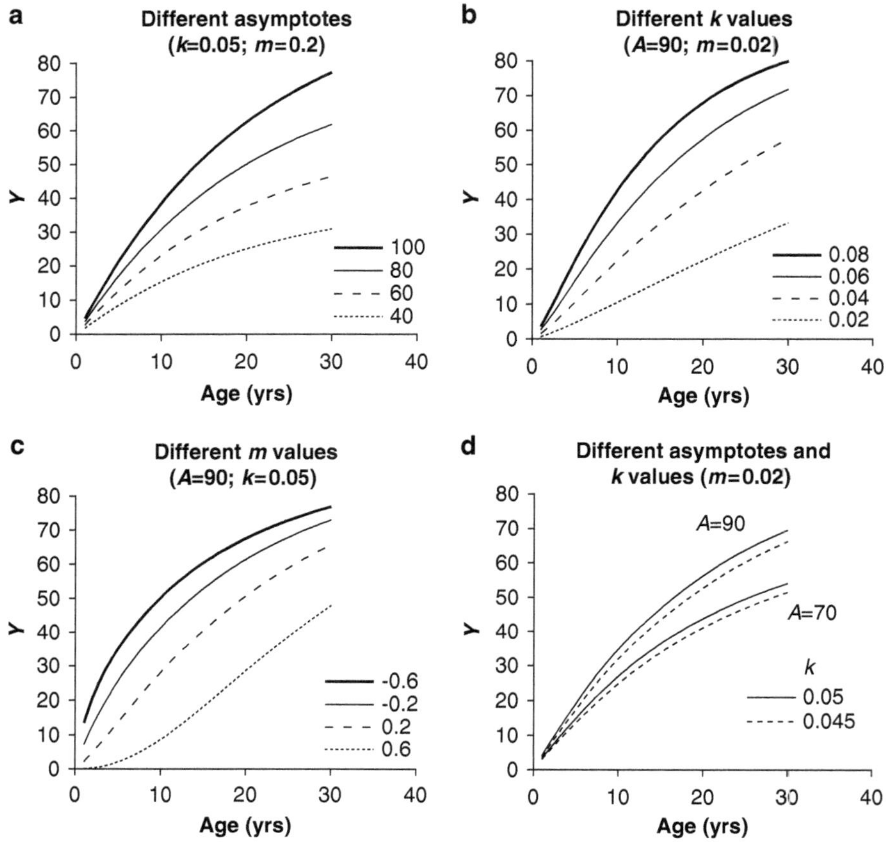

Fig. 6.5 Flexibility of the Richards growth functions with changing values of the parameters

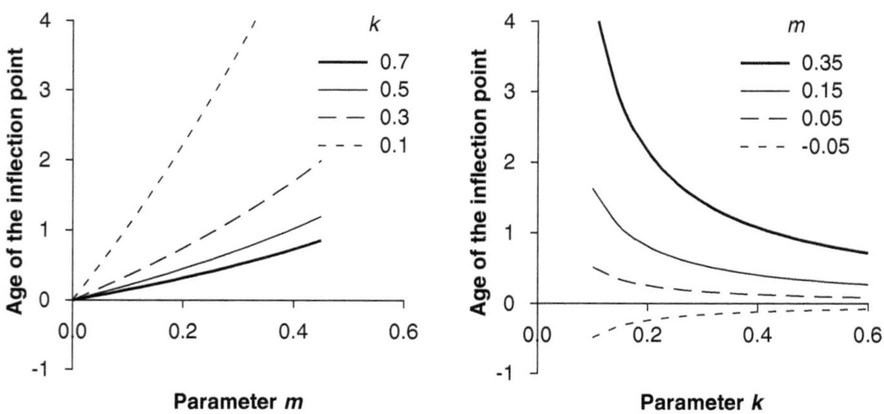

Fig. 6.6 Effect of the k and m parameters of the Richards function on the location of the inflection point

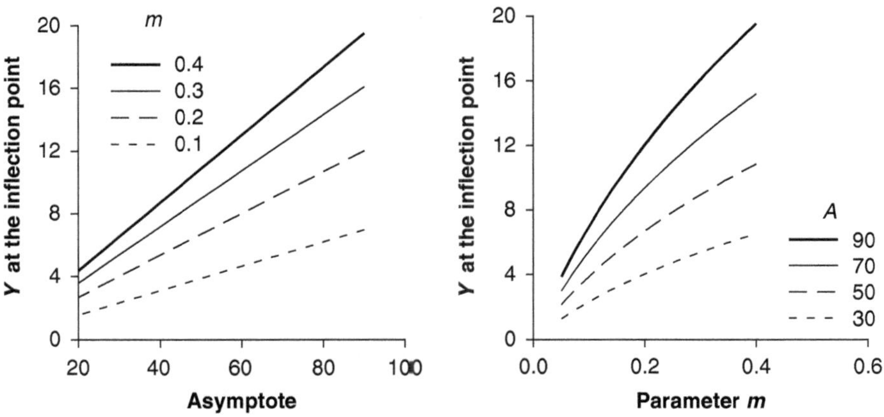

Fig. 6.7 Effect of the asymptote (A) and m parameters of the Richards's function on the inflection point of Y

Since its introduction for forestry applications by Pienaar and Turnbull (1973), this equation has been used extensively in studies of tree and stand growth (Zeide 1993). The Richards function is flexible (Fig. 6.5) and typically fits growth data well. However, some authors question the usefulness of the Richards function due to its intrinsic properties. Ratkowsky (1983, pp. 83–84) showed that this equation is "the only model that has an unacceptable intrinsic nonlinearity as the solution locus departs significantly from an hyperplane". These properties may lead, in practice, to instability in the parameter estimates. One way to overcome this problem is to use expert judgement for an estimate for the asymptote and apply nonlinear regression to estimate the remaining parameters from data fitting.

6.5 Functions of the Hossfeld IV Type

6.5.1 The Hossfeld IV Function

The Hossfeld IV function is a sigmoid function, originally proposed in 1822 (Zeide 1993), for the description of tree growth:

$$Y = \frac{t^k}{c + t^k/A} = A\frac{t^k}{Ac + t^k} = A\frac{t^k}{c_1 + t^k} \tag{6.14}$$

where A is the asymptote, t is the age and c and k are parameters.

The function can also be obtained from the generalized logistic in Eq. 6.9 by using $f(\mathrm{X}, t) = -k\ln(t)$. Consequently some authors (e.g. Cieszewski 2000; Dieguéz-Aranda et al. 2006) designate it as the log-logistic growth function.

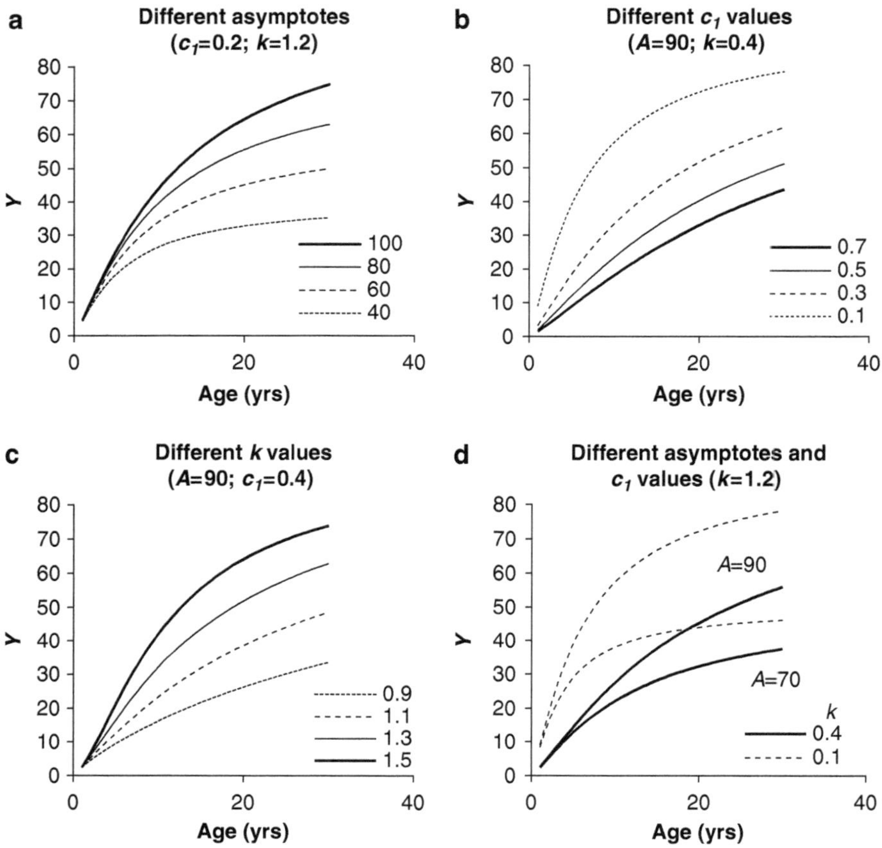

Fig. 6.8 Flexibility of the Hossfeld IV growth functions when the parameters take different values

To our knowledge the equation is not based on any specific biological rationale, but it generally performs well. According to Kiviste (1988), it is the third most accurate of 31 three-parameter equations when the three main stand variables (total tree height, stem diameter and volume) are considered together. Kiviste further found it to be the best equation for volume growth.

The Hossfeld IV function is able to take several shapes (Fig. 6.8) and to produce inflection points located earlier or later in the life of the tree or stand (Figs. 6.9 and 6.10).

6.5.2 *McDill-Amateis/Hossfeld IV Function*

McDill and Amateis (1992) proposed the use of a growth function, written in differential form, whose integral form is equivalent to the Hossfeld IV function. The McDill-Amateis function was developed in order to guarantee compatibility

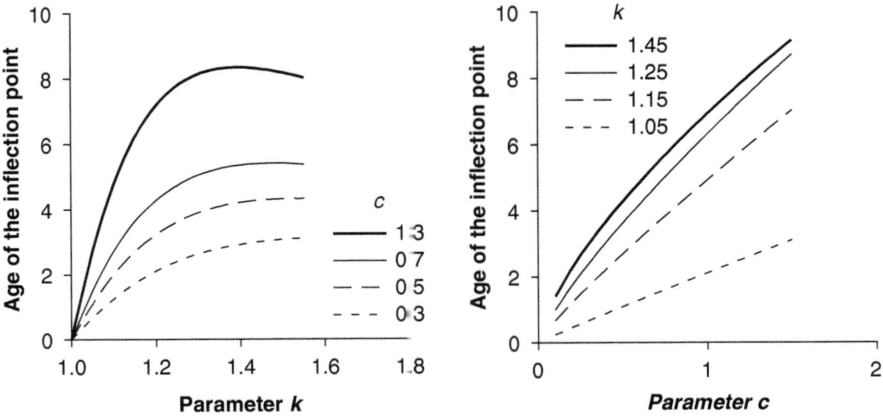

Fig. 6.9 Effect of the *k* and *c* parameters of the Hossfeld IV function on the location of the inflection point

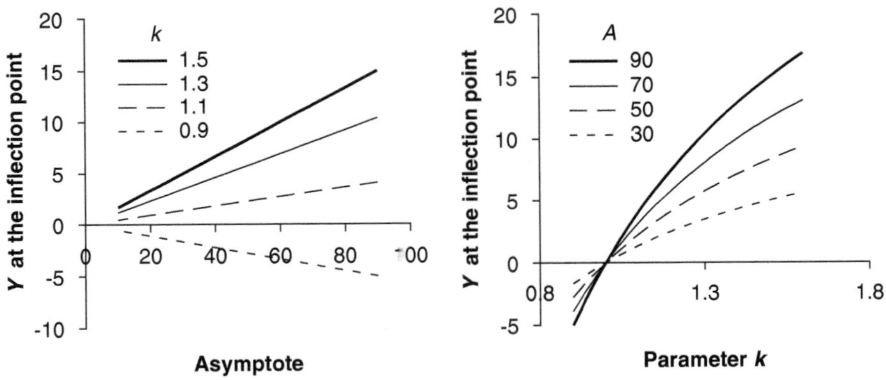

Fig. 6.10 Effect of the asymptote (*A*) and *k* parameters of the Hossfeld IV function in the *Y* value at the time when the inflection point occurs

of dimensions and also to take into account the biological properties expected from growth functions. The variables considered for the growth function and the respective dimensions were:

Variable	dY/dt	t	Y	A
Dimension	LT^{-1}	T	L	L

where L indicates length, T is time and A is the asymptote for variable T.

Applying dimensional analysis to these variables (McDill and Amateis 1992) and taking into account, at the same time, that the growth rate tends to zero when Y tends to the asymptote, the following differential form is obtained:

$$\frac{dY}{dt} = k\frac{Y}{t}\left(1 - \frac{Y}{A}\right)$$

In this equation, k is a parameter related to the growth rate. The function has one less parameter than the Richards function which can be an advantage when fitting the function to empirical data.

The solution for the differential equation leads to the following yield function, known in forestry literature as the McDill-Amateis function:

$$Y = \frac{A}{1 - \left(1 - \frac{A}{Y_0}\right)\left(\frac{t_0}{t}\right)^k} \tag{6.15}$$

where (t_0, Y_0) is the initial condition.

By making $c = \left(\frac{1}{Y_0} - \frac{1}{A}\right)t_0^k$ the integral form of the McDill-Amateis function coincides with the Hossfeld IV function (Eq. 6.14). McDill and Amateis's formulation allows a better explanation for the parameters in Eq. 6.15; the k parameter expresses the growth rate and c is related to the initial conditions. The inflection point occurs when

$$Y = \frac{A}{2}\left(1 - \frac{1}{k}\right)$$

6.5.3 Generalizations of the Hossfeld IV Function

In Kiviste's study the most accurate equations with three (Levakovic III equation) and more (Levakovic I and Yoshida I equations) parameters are modifications of the Hossfeld IV equation, namely:

Levakovic I function:

$$Y = A\left(\frac{t^k}{c_1 + t^k}\right)^{c_2}$$

Levakovic III function:

$$Y = A\left(\frac{t^2}{c_1 + t^2}\right)^{c_2}$$

Yoshida I function:

$$Y = A\frac{t^k}{c_1 + t^k} + c_2$$

6.6 Other Growth Functions

Two growth functions that do not fall within any of the previous categories but merit mention, as they have been used in forest modeling with success (Zeide 1993; Kiviste et al. 2002), are the Weibull and the Sloboda functions.

The Weibull function multiplied by a parameter A has been used successfully in forest modeling (Yang et al. 1978; Payandeh and Wang 1995):

$$Y = A\left(1 - e^{-kt^b}\right)$$

The Sloboda equation is a generalization of the Gompertz equation by adding a parameter (b):

$$Y = Ae^{-ce^{-kt^b}}$$

6.7 Zeide Decomposition of Growth Functions

When analyzing a large set of growth functions in differential form Zeide (1993) found that all the investigated equations could be decomposed into two components: growth expansion and growth decline. The expansion component represents the innate tendency towards exponential multiplication and is associated with biotic potential, photosynthetic activity, absorption of nutrients, constructive metabolism, anabolism, and the like. The decline component represents the constraints imposed by external (competition, limited resources, respiration, and stress) and internal (self-regulatory mechanisms and aging) factors. Those factors that adversely affect growth have been referred to as environmental resistance, destructive metabolism, catabolism, respiration, and so on.

The decomposition can be achieved either by a subtraction or a division (subtraction of logarithms) of the two effects. As was shown before, the Richards function differential form was defined using these concepts. Analysing the decomposition of the growth functions by division, and rewriting the equations in order to simplify the notation of the constant parameters, Zeide (1993) found that all the equations analyzed, except Weibull's, are particular cases of the two following forms:

$$\ln\left(y'\right) = k + p\ln(y) + q\ln(t) \quad \leftrightarrow \quad y' = k_1 y^p t^q$$

$$\ln\left(y'\right) = k + p\ln(y) + qt \quad \leftrightarrow \quad y' = k_1 y^p e^{qt}$$

where $p > 0$, $q < 0$ and $k_1 = e^k$.

In both forms the expansion component is proportional to ln(y) or, in the antilog form, is a power function of size. The forms differ in the decline component: in the first form, designated *LTD*, the decline component is proportional to the logarithm of age while in the second form, *TD*, it is either a power function or an exponential function of age. Depending on the values of p and q, several distinct integral equations can be obtained from the same equation form. For the functions listed in Table 6.2, the *LTD* form includes the Lundqvist-Korf and the McDill-Amateis (Hossfeld IV) functions; the *TD* form includes the Richards function and all its particular cases.

The transformed equations reveal quite different and simple relationships between the growth functions. Despite the difference of their integral forms, Hossfeld IV and Korf equations are varieties of the same basic form. On the other hand, differentiation shows that the outward similarity between the Chapman-Richards and Weibull equations is misleading.

Zeide (1993) proposed a third form in which the declining component is expressed as a function of size instead of age, the *YD* form:

$$\ln\left(y'\right) = k + p\ln(y) + qy \quad \leftrightarrow \quad y' = k_1 y^p e^{qy}$$

The three forms above are very useful for the direct modeling of tree and/or stand growth. These forms provide some assurance that the resulting model will display appropriate behavior from a biological stand point.

6.8 Formulating Growth Functions Without Age Explicit

Age is commonly employed as a variable in growth and yield modeling of even-aged stands. However, age is not always readily available. Increment cores (to the pith) can be used for age determination in tree species with well-defined annual rings, but coring may not be possible for certain species or in tropical zones. In uneven-aged stands, composed of trees that differ markedly in age, stand age cannot be used as a predictor of growth and yield. Hence, an alternative to the typical applications of growth functions for age determinate trees and stands is sometimes needed.

Tomé et al. (2006) showed that it is possible to obtain formulations of the growth functions as difference equations in which age is not explicit. Age independent difference forms can be obtained by solving the equation for age (t) and substituting it in the expression of the growth function for age equal to $t + a$. To illustrate the procedure, assume the Lundqvist function and solve for age t:

$$Y_t = A\, e^{-k\frac{1}{t^m}} \Rightarrow t = \left[\frac{-k}{\ln\left(Y_t/A\right)}\right]^{\frac{1}{m}}$$

The expression for t can be substituted in the growth function written for age $t + a$, where a is the projection length:

$$Y_{t+a} = Ae^{-k\frac{1}{\left(\left[\frac{-k}{\ln(Y_t/A)}\right]^{\frac{1}{m}}+a\right)^m}} \quad \text{(Lundqvist function without } t \text{ explicit)}$$

which results in the formulation of the Lundqvist function as an age independent equation. The expression seems complex, but parameter estimates converge quickly when fitted to data.

Using a similar procedure, the following formulations of the Richards and Hossfeld IV functions as age independent equations can be obtained:

$$Y_{t+a} = A\left(1 - e^{-ka}\left(1 - \left(\frac{Y_t}{A}\right)^{1-m}\right)\right)^{\frac{1}{1-m}} \quad \text{(Richards function without } t \text{ explicit)}$$

$$Y_{t+a} = A\frac{\left(\left(\frac{c\,Y_t}{A-Y_t}\right)^{\frac{1}{k}} + a\right)^k}{c + \left(\left(\frac{c\,Y_t}{A-Y_t}\right)^{\frac{1}{k}} + a\right)^k} \quad \text{(Hossfeld IV function without } t \text{ explicit)}$$

In order to model growth with these equations, at least one of the parameters has to be expressed as a function of site variables (e.g. soils and climate) and stand characteristics. It is important to stress that the age-independent difference equations are invariant for projection length only if the parameters are expressed solely as a function of other variables that are invariant with time. Otherwise, if the parameters are expressed as a function of variables that vary with time, the projections will depend on the projection interval used.

References

Causton DR, Venus JC (1981) The biometry of plant growth. Edward Arnold, London

Cieszewski CJ (2000) Analytical site index solution for the generalized log-logistic height equation. For Sci 46:291–296

Cieszewski CJ (2002) Comparing fixed- and variable-base-age site equations having single versus multiple asymptotes. For Sci 48:7–23

Diéguez-Aranda U, Grandas-Arias JA, Álvaraz-González JG, Gadow Kv (2006) Site quality curves for birch stands in north-western Spain. Silva Fennica 40:631–644

García O (2008) Visualization of a general family of growth functions and probability distributions – the growth-curve explorer. Environ Model Softw 23:1474–1475

Gompertz B (1825) On the nature of the function expressive of the law of human mortality, and on a new mode of determining the value of life contingencies. Philos Trans R Soc Lond 115: 513–583

Grosenbaugh LR (1965) Generalization and reparameterization of some sigmoid and other nonlinear functions. Biometrics 21:708–714

Hunt R (1982) Plant growth curves. The functional approach to plant growth analysis. Edward Arnold Ltd, London

Kiviste A (1988) Mathematical functions of forest growth [in Russian]. Estonian Agricultural Academy, Tartu

Kiviste A, Álvarez-González JG, Rojo-Alboreca A, Ruíz-González AD (2002) Funciones de crecimiento de aplicación en el ámbito forestal. Instituto nacional de investigación y tecnologia agraria y alimentaria, Madrid

Korf V (1939) A mathematical definition of stand volume growth law [in Czech: Příspěvek k matematické definici vzrůstového zákona lesních porostů]. Lesnická práce 18:339–379

Lundqvist B (1957) On the height growth in cultivated stands of pine and spruce in Northern Sweden [In Swedish: Om Höjdutveck-lingrn i kulturbestand av tall och gran i Norrland]. Medd Skogsforskn Inst 47:64

MacKinney AL, Schumacher FX, Chaiken LF (1937) Construction of yield tables for nonnormal loblolly pine stands. J Agric Res 54:531–545

McDill ME, Amateis RL (1992) Measuring forest site quality using the parameters of a dimensionally compatible height growth function. For Sci 38:409–429

Monserud RA (1984) Height growth and site index curves for inland Douglas-fir based on stem analysis data and forest habitat type. For Sci 30:943–965

Payandeh B, Wang Y (1995) Comparison of the modified Weibull and Richards growth function for developing site index equations. New For 9:147–155

Pearl R, Reed LJ (1923) On the mathematical theory of population growth. Metron 3:6–19

Pienaar LV, Turnbull KJ (1973) The Chapman-Richards generalization of von Bertalanffy's growth model for basal area growth and yield in even-aged stands. For Sci 19:2–22

Prodan M (1968) Forest biometrics (English translation from German by Gardiner SH). Pergamon Press, Oxford

Ratkowsky DA (1983) Nonlinear regression modeling: a unified approach. In: Statistics: textbooks and monographs, vol 48. Marcel Dekker, New York

Richards FJ (1959) A flexible growth function for empirical use. J Exp Bot 10:290–301

Schumacher FX (1939) A new growth curve and its application to timber yield studies. J For 37:819–820

Thornley JHM (1976) Mathematical models in plant physiology: a quantitative approach to problems in plant and crop physiology. Academic, London/New York

Tomé J, Tomé M, Barreiro S, Paulo JA (2006) Age-independent difference equations for modeling tree and stand growth. Can J For Res 36:1621–1630

Vanclay JL (1994) Modelling forest growth and yield: applications to mixed tropical forests. CAB International, Wallingford

von Bertalanffy L (1938) A quantitative theory of organic growth (Inquiries on growth laws (2)). Human Biol 10:181–213

Winsor CP (1932) The Gompertz curve as a growth curve. Proc Natl Acad Sci USA 18:1–8

Yang RC, Kozak A, Smith JHG (1978) The potential of Weibull-type functions as flexible growth curves. Can J For Res 8:424–431

Zeide B (1993) Analysis of growth equations. For Sci 39:594–616

Chapter 7
Evaluating Site Quality

7.1 Need to Quantify Site Quality

Assessment of site quality is essential for identifying the productive potential of land and for providing a frame of reference for silvicultural diagnosis and prescription. To be most useful for modeling and prediction, a measure of site quality must be quantitative – that is expressed by a number. In the context of practical forest management, a quantitative assessment of site quality should be objective, easily determined, and, when using measurements of trees in the evaluation, free from the influence of stand density. Although site quality evaluation for volume production is desired, it is seldom feasible to use direct measures of volume productivity when quantifying wood-growing potential. Typically there are limited historical records of yields from forested sites. Furthermore, volume productivity is highly dependent on stand density (Chap. 8) and product definitions, merchantability limits, and tree volume estimating functions used. Thus an indirect measure – dominant stand height development – has become the most widely used means of evaluating site quality, especially for even-aged monocultures.

Site index – that is, the average height of the dominant portion of the stand at an arbitrarily chosen age – is the most commonly used indicator of site quality. Height growth is highly correlated with volume productivity, and dominant height is not greatly affected by stand density and thinning treatments (assuming thinning from below). A number of different definitions of dominant or top height are applied around the world (Sect. 7.2), but all involve measurement of the trees in the upper part of the canopy.

Knowing dominant height provides little information about site quality unless it is standardized at a particular age. If a stand is measured at the index age, the dominant height can be taken as the site index. However, it is rare that measurements coincide with the index age, but rather they usually occur before or after. Consequently a means for projecting the height of the measured stand forward or backward to the index age is required. These height-age relationships are traditionally termed site index curves or equations.

H.E. Burkhart and M. Tomé, *Modeling Forest Trees and Stands*,
DOI 10.1007/978-90-481-3170-9_7,

Several reviews of methods for evaluating forest site productivity have been published, including those by Carmean (1975), Hägglund (1981), Vanclay (1992), and Skovsgaard and Vanclay (2008). Because of the importance and prevalent use of the site index concept in empirical modeling of growth and yield for even-aged stands, this chapter explores in detail the development of height-age-site index relationships.

7.2 Computing Top Height

The site index concept is based on development of the dominant portion of stands over time. Unfortunately there is no generally agreed upon definition of which trees to include and how many to measure as site trees. One commonly used definition is average height of trees in the dominant and codominant crown classes. There is no standard definition of how many trees to measure or what should be the mix between dominant and codominant trees in the sample.

In an effort to avoid subjective judgment, such as assigning trees to crown classes, dominant or top height is often defined in terms of tree size. A specified number of the tallest trees per hectare is one measure that has been used, but the number per unit area is not standardized. Alternatively trees of the largest diameter might be measured as site trees. Selecting the largest diameter trees is more easily implemented in the field than selecting those of largest height.

Sharma et al. (2002a) evaluated seven definitions of top height with data collected over a 15-year period in thinned and unthinned loblolly pine plantations. With the exception of a few cases at certain measurements, all seven definitions of top height were significantly different from each other. Predictions of site index with the various definitions of top height indicated that using trees that have always been dominant or codominant over the life of the stand is more precise than estimates using other definitions.

Determining which trees on sample plots to measure as site trees is done subjectively, not randomly. Consequently, estimates of top height, unlike average height, are a function of the size of the sampling unit. Two commonly-used sampling practices involve measuring a fixed number of site trees per unit area and selecting a fixed proportion of trees per unit area. Zeide and Zakrzewski (1993) pointed out that these two procedures (constant number or constant proportion) are both biased and their bias is opposite in sign. The number or proportion can be varied over time, or, as recommended by Zeide and Zakrzewski, a weighted average of estimates from the two methods can be computed. When applied to permanent plots established in jack pine in Ontario, Canada, the authors found that the weights needed to provide an unbiased estimate were nearly equal.

The impact of sample plot size on estimates of top height has been the subject of a number of studies, including those of Garcia (1998), Magnussen (1999), and Garcia and Batho (2005).

Modeling principles involved in projecting top stand height development over time remain essentially the same regardless of the particular definition of top height employed. Thus, for purposes of presenting height/age/site index modeling, we will refer to top height or dominant height without necessarily providing an exact definition of how the terms were applied in various analyses. For simplicity, throughout this chapter, h_{dom} will be used to designate a measure of dominant or top height unless further clarification is needed in the context of the discussion. Age in site index estimation may be defined as total tree age, age at breast height, or, in the case of plantations, as years since planting. Regardless of details of how age was defined in specific instances, unless necessary to the presentation, we will simply refer to "age" without further specificity and designate it as t. The base age for site index (S) will be denoted by t_b. Before applying site index functions it behooves users to check carefully definitions used for stand height and age determination and to collect field data appropriate to the sampling and measurement protocols used in developing the prediction equations.

7.3 Data Sources for Developing Site Index Curves

7.3.1 Temporary Plots

Site index equations have often been developed by measuring height-age pairs on temporary plots in stands of varying site qualities and ages and then fitting a "guide curve" with the data. (Methods for fitting guide curves are given in Sect. 7.4.) The guide curve represents the height development for the average site index in the data. Heights at all ages for all other site classes are typically assumed to be proportional to that of the guide curve.[1] Thus if the guide curve passed through 20 m at index age, the heights for all ages for the site index curve 22 m would be assumed to be equal to 1.1 times the guide curve, those for the 18 m site index curve would be 0.9 times the guide curve values, etc. This procedure produces a family of anamorphic site index curves.

7.3.2 Permanent Plots

Although the guide curve method has the advantage of being relatively inexpensive in time and money resources to apply, site index curves from longitudinal or remeasurement data are generally preferred. The remeasurement data may come

[1]The so-called "coefficient of variation method" (e.g. Brickell 1968) can be applied to circumvent the proportionality assumption, but it is difficult to develop satisfactory site index curves using this technique and it is seldom applied in practice.

from repeated observations on permanent sample plots or from a sample of stem analysis trees. Obtaining height growth data from remeasured plots is time consuming and expensive, but it provides highly useful data. Selecting a sample of trees for stem analyses to establish height-age pairs over time is a means of obtaining growth data more quickly and less expensively than is the case with permanent growth plots.

Periodic measurements of dominant height and age on permanent plots provide the best source of data for fitting site index functions. Data on stand height development over time allow for fitting of anamorphic (one shape) or polymorphic (many shapes) site index equations. However, permanent plot measurement data are relatively expensive to obtain and results are not available until a number of years after the plots are initially established.

When remeasurement data are available, height can be predicted directly or height growth can be modeled. If measurement error is small and the measurements are close together (e.g. annual measurements) finite differences in height and age $(\frac{h_{dom2}-h_{dom1}}{t_2-t_1})$ can be used to approximate the instantaneous rate of height growth dh_{dom}/dt. The differential form $dh_{dom}/dt = f(t)$ or $dh_{dom}/dt = f(t, h_{dom})$ can be fitted directly and height-age curves can be produced via integration. The resultant site index curves may be anamorphic or polymorphic for growth predicted as a function of age or of age and height depending on the form of the differential equation used.

In cases where measurement error is large and/or the interval between measurements is long, an accurate approximation of instantaneous growth is not possible and direct fitting of height is preferred.

7.3.3 *Stem Analysis*

Stem analysis may be performed by making internodal measurements for tree species that have a determinate growth pattern. Alternatively, the main stem may be split along the pith so that heights at various ages can be measured directly. As a third alternative, cross-sectional cuts may be made at given heights and rings counted to determine tree age at that height.

If internodal measurements are feasible to obtain, or if it is possible to split the stem along the pith, measurement error for the height-age pairs is minimal. If, however, cross-sectional cuts are made along the stem and ring counts are recorded, a bias in height determination for given ages results because the cut will seldom occur at the tip of a terminal leader (Fig. 7.1).

7.3.3.1 Height Correction Methods

Various methods for estimating the height corresponding to a given age with sectioned-tree stem analysis data have been proposed (Carmean 1972; Lenhart

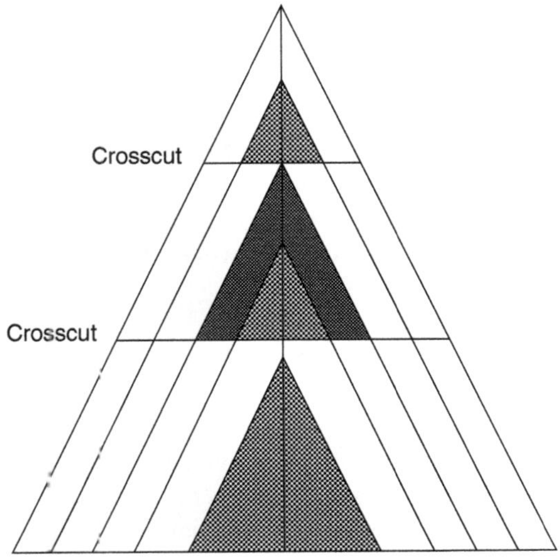

Fig. 7.1 Diagrammatic representation of the stem growth development of a 5-year-old tree. *Shaded* portions represent "hidden tips" in stem sections (From Dyer and Bailey 1987)

1972; Newberry 1991; Fabbio et al. 1994; Kariuki 2002). Dyer and Bailey (1987) compared six height correction methods for tree-section data using actual heights at known ages for 28 loblolly pine trees. In their comparison, Carmean's (1972) method, which assumes that height growth is constant for each year contained within a bolt and that each crosscut occurs at the middle of a year's growth, was most accurate. Lenhart's (1972) correction method, although somewhat less accurate than the method of Carmean (1972), produced data that when used to fit the Chapman-Richards function gave very similar results.

Procedures commonly used for estimating heights at different ages of stem-analyzed trees consist of ring counts at crosscuts made at predetermined heights. The accuracy of estimates of the length of the hidden tips in each bolt is dependent on the bolt length and the correction algorithm employed. Fabbio et al. (1994) proposed a procedure called *Issa* where the distance between the sampled section and the apex of the successive hidden tip varies according to the number of tips within each bolt. Data from 27 black pine trees were used to compare precision of the *Issa* method with that of Carmean's (1972) and Lenhart's (1972) methods and with a branch whorl method. When applying the alternative correction methods to crosscutting frequencies of 50, 100, and 200 cm, the *Issa* procedure was the most precise at a cutting frequency of 50 cm and the Carmean method was the most precise for a frequency of 200 cm.

Kariuki (2002) developed an approach (called TARG) that uses annual ring width to estimate where annual height growth ceased within a bolt section. The performance of the TARG method was assessed against other documented procedures using data taken at sampling intervals of 1.5 and 3 m. While Carmean's (1972) and Lenhart's (1972) procedures gave acceptable estimates for the 1.5 m sampling interval, both methods significantly overestimated the tree height when

3 m sampling intervals were used. The method of Fabbio et al. (1994) significantly overestimated the true tree heights for both sampling intervals, while the TARG method did not show a significant difference for either sampling interval. For height-age curves fitted using Richards (1959) equation, the data corrected using the TARG method gave more accurate height estimates than those provided by data adjusted by other methods at both sampling intervals. Application of the TARG approach does, however, require the additional time and expense of acquiring highly accurate measurements of annual growth ring widths.

7.3.3.2 Selection of Stem Analysis Trees

The most serious drawback to constructing site index curves from stem analysis data is that height growth of individual trees does not necessarily represent dominant height growth of stands. Shifts in relative position of individual trees in the height distribution occur during stand development. For site index curves to be applicable, the degree of dominance of predicted heights should remain constant over time. Stem analysis trees are typically selected at older ages in order to provide height-age pairs over a span approximating rotation lengths of interest. Individual trees that represent dominant or top height in more mature stands may not have occupied a similar position in the canopy at earlier ages. Thus a bias can, and often does, occur.

The average height of stem analysis trees at younger ages is generally less than that of the trees in the dominant canopy at those ages. Thus, when equations are fitted to data derived from a sample of stem analysis trees, the resultant curves are too steep and their application results in over estimates of site index.

The problem of selecting stem analysis trees to represent height development of the dominant portion of stands where relative dominance of individuals changes over time has received considerable attention (Dahms 1963; Curtis 1964; Zeide and Zakrzewski 1993; Magnussen and Penner 1996; Raulier et al. 2003; Feng et al. 2006). Dahms (1963) showed that using data only from the single tallest sectioned tree per plot substantially reduced bias when developing site index curves for lodgepole pine. In addition to changes in height growth patterns of individual trees over time, mortality among the tallest trees is another factor that can cause height development of stands and individual trees to differ. In the case of the lodgepole pine data, mortality did not appear to have exerted a large influence on the development of site trees.

Magnussen and Penner (1996) presented an algorithm to recover an estimate of dominant height from stem analysis data. Their procedure consists of estimating dominant height as a function of the population mean, the standardized height difference, and the population variance of tree height. By relating the reconstructed dominant height from the stem analysis trees to their theoretical expectations, one can calculate the expected bias. The recovery algorithm minimizes the expected bias in dominant height when the only growth history of the population is the one derived from the trees selected for dominance at or near rotation age. Results confirmed that the recovery procedure very effectively reduced the relative bias. Bias was reduced by over 40% for the first half of a rotation; small but insignificant increases in bias were incurred for the second half of a rotation.

When modeling height growth of spruce-dominated, even-aged stands in British Columbia, Canada, with a stochastic differential equation formulation of the Bertalanffy-Richards function, Hu and Garcia (2010) used both stem analysis (SA) and permanent sample plot (PSP) data. A comparison of SA data with PSP observations showed a statistically significant bias was present. This bias was, however, smaller than that indicated in theoretical calculations of Magnussen and Penner (1996) and Feng et al. (2006). There was also a significant difference in error structure between the two data sources. Hu and Garcia were able to combine observations from SA and PSP sources using methods that reduced bias and increased the precision of the final model.

7.4 Fitting Site Index Guide Curves

Numerous height-age models have been successfully applied for fitting site index guide curves. One of the most widely applied model forms is the logarithm of height-reciprocal of age model, namely

$$\ln h_{dom} = \mathrm{b}_0 + \mathrm{b}_1 t^{-1} \tag{7.1}$$

where $\ln h_{dom}$ is the logarithm of dominant height and t is the stand age (monospecific, even-aged stands are assumed). The logarithm-reciprocal transformations result in a sigmoid shaped growth function, stabilize the variance of the dependent variable, provide for an upper asymptote in height as age increases without bound, and allow fitting with linear regression methods. Assuming parameter estimates for b_0 and b_1 in (7.1), we note that if age is equal to index age (that is $t = t_b$), by definition dominant height equals site index (i.e., $h_{dom} = S$):

$$\ln S = \mathrm{b}_0 + \mathrm{b}_1 t_\mathrm{b}^{-1}$$

which implies that

$$\mathrm{b}_0 = \ln S - \mathrm{b}_1 t_\mathrm{b}^{-1}$$

Substituting the definition of b_0 into (7.1) and simplifying gives

$$\ln h_{dom} = \ln S + \mathrm{b}_1 (t^{-1} - t_\mathrm{b}^{-1}) \tag{7.2}$$

which can be used to generate height at given ages for specified values of S and t_b. Equation 7.2 can also be rearranged to estimate site index (S) for specified values of height (h_{dom}) and age (t).

Another commonly used model for fitting guide curves for site index equation purposes is the so-called Chapman-Richards model

$$h_{dom} = \mathrm{b}_0(1 - \mathrm{e}^{\mathrm{b}_1 t})^{\mathrm{b}_2} \tag{7.3}$$

which must be fitted using nonlinear regression techniques. Writing the guide curve Eq. 7.3 in the equivalent form of Eq. 7.2 gives

$$h_{dom} = S\left[\frac{1 - \mathrm{e}^{\mathrm{b}_1 t}}{1 - \mathrm{e}^{\mathrm{b}_1 t_\mathrm{b}}}\right]^{\mathrm{b}_2} \tag{7.4}$$

where all symbols remain as previously defined.

Although widely used, the guide curve method of site index curve construction, which results in what is known as "anamorphic" site index curves, has a number of shortcomings. When the same shape is assumed for all site classes, the inflection point (age of maximum height growth) is the same regardless of site quality. This is not a biologically reasonable assumption, as one would expect an earlier culmination of height growth on better quality sites.

A second problem often encountered is bias in the fitted guide curve. In order to have an unbiased trend of dominant height over time when fitting temporary plot data with ordinary least squares methods, all site qualities must be approximately equally represented at all ages. Often a negative correlation exists between site quality and age for a given population. This negative correlation stems from the fact that better quality sites tend to be managed on shorter rotations, resulting in a disproportionate number of the older sample observations being in stands on poorer quality sites. A negative correlation between stand age and site classes will cause the guide curve to flatten excessively.

While the assumption of a common shape for all site classes is often cited as the culprit in biased estimates of site index when using guide-curve-based anamorphic equations, the primary problem often rests with bias in the guide curve itself. The assumption of a common shape is not in serious error for site classes reasonably close to the mean, but if the guide curve is biased the entire family of site curves will be biased.

A test for the validity of the common-shape (anamorphic) assumption can be conducted by dividing the data into age classes and computing the coefficient of variation (standard deviation divided by the mean) of heights for each class. If the coefficient of variation is not related to age, the assumption of proportionality of heights at all ages is reasonable.

When site and hence the error term are correlated with the predictor (age), biased and inconsistent estimates will result if OLS is applied. Walters et al. (1989) compared instrumental variable (IV) estimation to ordinary least squares when using temporary plot data to fit a site index guide curve. If an appropriate instrument that is uncorrelated with site index but correlated with age is available, bias can be reduced and consistent coefficient estimates can be obtained. Measures of stand density

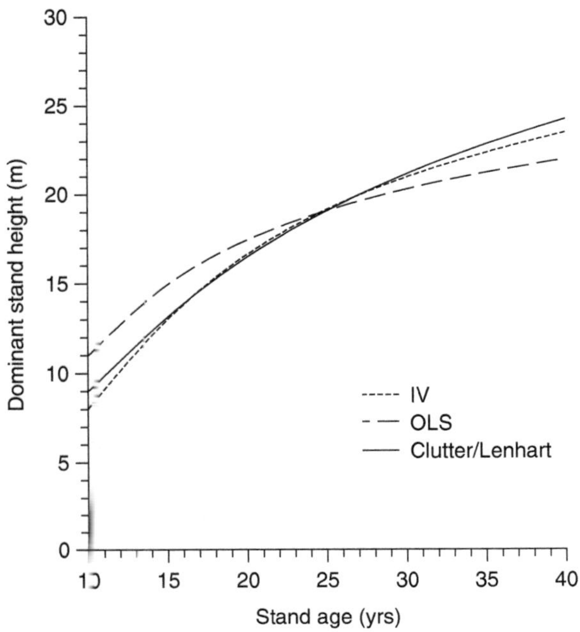

Fig. 7.2 Comparison of instrumental variable (*IV*), ordinary least squares (*OLS*), and height-age curves of Clutter and Lenhart (1968), which are assumed to present the correct trend for loblolly pine plantations. The figure is based on the average site index of 19.5 m (From Walters et al. 1989)

(Chap. 8) were considered as instruments. However, for five data sets examined, the IV techniques proved effective in only one case (Fig. 7.2). Nevertheless, these results suggest that the use of instrumental variables may, in some instances, prove effective for reducing bias in guide curves fitted to temporary plot data.

7.4.1 Comparisons of Stem-Analysis and Guide-Curve Based Site Index Equations

In a study involving Douglas-fir in the Pacific Northwest region of the United States, Monserud (1985) found that the type of data and resulting methodology used to develop the site index equations were strongly related to the similarity of the resulting curves. Site index curves derived from stem analysis studies were quite similar to each other, but differed substantially from those developed by guide-curve methods. The magnitude of possible differences due solely to different methods of site index curve construction (stem analysis vs. guide curve) was demonstrated to be quite large when both methods were applied to the same data (Fig. 7.3). Other comparisons (e.g. Thrower and Goudie 1992, for Douglas-fir in British Columbia, Canada; Milner 1992, for ponderosa pine, western larch, lodgepole pine, and Douglas-fir in western Montana, USA) have shown similar trends to those observed by Monserud.

Fig. 7.3 Comparison of stem-analysis model to a proportional guide-curve model fitted to the same sample of Douglas-fir trees (From Monserud 1985)

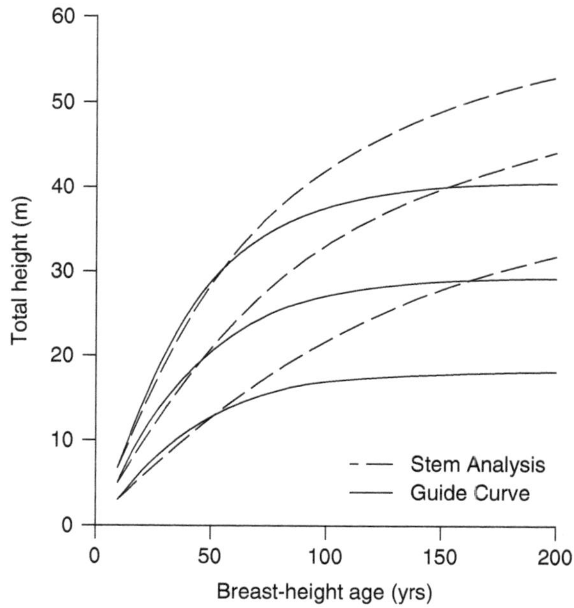

7.5 Site Index Equations Using Age and Height at Index Age

As noted previously, differential and difference equations can produce both anamorphic and polymorphic curve forms. Another approach to developing polymorphic site index equations that has been employed in a number of instances (King 1966; Beck 1971; Graney and Burkhart 1973; Trousdell et al. 1974; Burkhart and Tennent 1977a, b; Carmean and Lenthall 1989), (and is sometimes referred to as the "parameter prediction method"), involves predicting height from age and site index, that is $h_{dom} = f(t, S)$.

The parameters of a number of different height functions have been related to site index to develop polymorphic site curves, but the Chapman-Richards equation has been most commonly employed. Site index equations for radiata pine in New Zealand (Burkhart and Tennent 1977a) provide an example of this approach. Relating constants in the Chapman-Richards function to site index (observed or interpolated height at index age) gave

$$h_{dom} = \mathrm{b}_1 S^{\mathrm{b}_2}(1 - \exp(\mathrm{b}_3 S \cdot t))^{\mathrm{b}_4} \tag{7.5}$$

Equation 7.5 was conditioned to insure that predicted height equals site index when age equals index age (t_{b}) by writing b_1 in terms of the other coefficients. Conditioning led to the following expression for height as a function of age and site index:

$$h_{dom} = \frac{S(1-\exp(b_3S \cdot t))^{b_4}}{(1-\exp(b_3S \cdot t_b))^{b_4}} \tag{7.6}$$

The two-parameter function (7.6) was then fitted to data.

Expressing height development as a function of age and an arbitrarily designated height-age pair (site index) has produced satisfactory results from an applications standpoint. However, the procedure has a number of drawbacks: (1) height measurements at or near the index age are required, (2) the resultant site index curves are specific to the base age used to determine site index in fitting, and (3) for many functions (including the Chapman-Richards), the fitted equation cannot be solved explicitly for S, given h_{dom} and t.

7.6 Segmented Models for Site Index Curves

In reviewing various site index models, Devan and Burkhart (1982) found that some had very good characteristics for young tree ages while others fit better at older ages. This observation suggested use of a segmented modeling approach as a possible means to obtaining good predictive ability across the full range of ages of interest. Hence a segmented polynomial regression approach was taken to develop a system of equations where different functions would apply over a given domain.

Height increment submodels were fitted to data from loblolly pine for various combinations of the independent variables of height, age, and age squared for two segments with a join point being a function of age. Height increment, dh/dt, was approximated as the finite height increment calculated from adjacent bolts of stem analysis trees after transforming height to logarithms and ages to reciprocals. The differential equation model used was a generalization of that of Clutter and Lenhart (1968). The resultant curves were polymorphic, did not require specification of a base age for fitting, provided an estimate of height equal to site index at the base age, and could be analytically solved for both height and site index. A disadvantage of the differential equation form used to describe height growth in the two segments was that it was not conditioned through the origin (i.e., when t was equal zero, height was not equal zero).

Borders et al. (1984) noted that no single published equation form conformed well to height-age data for slash pine plantations. Anamorphic curve forms appeared to fit best at young plantation ages (less than 15 years) whereas polymorphic forms seemed best at older ages. This dichotomy indicated that a spline of these two model forms might provide a good overall description of slash pine height development. An algebraic difference formulation (Sect. 7.8) was applied because it could be easily fitted with nonlinear least squares and it avoided the necessity of estimating instantaneous growth rates. A difference form of the logarithm of height-reciprocal of age anamorphic height-age model was fitted for young ages and joined to a difference formulation based on Clutter and Jones (1980) polymorphic height increment model. The site index curve from this spline reproduced observed trends in the data and gave estimates with the desirable properties of (1) height is zero when age is zero, (2) height at base age equals site index, (3) each site index has a separate asymptote, and (4) the curves are invariant with respect to choice of base age.

Jones and Reed (1991) sought to improve predictions at younger ages of red pine plantations by extending existing site index equations through segmented regression techniques. The segmented models join a polynomial equation with site index curves published by Lundgren and Dolid (1970) that are widely used in the Lake States region of the USA. Two segmented models were fitted, one joining a polynomial function in age to Lundgren and Dolid's site index equation based on the simple monomolecular function and the other to their exponential-monomolecular function. In both cases the join point was set at 20 years, because the data used to fit the original equations was for ages 20 and above. The joined models produced identical predictions to the published equations of Lundgren and Dolid (1970) for ages older that the join point. The two segments were conditioned to insure that the predicted values for height and the slopes for the two segments were the same at the join point. Evaluation showed that the segmented model involving the simple monomolecular function was considerably improved over the original for predicting height at young ages. For the exponential-monomolecular function, the segmented model produced results for young ages similar to those of the original function.

7.7 Differential Equation Approach

As a simple illustration of the differential equation approach, assume that adequate data are available for modeling height growth and that height growth divided by total height is proportional to the reciprocal of age squared. These assumptions suggest the differential equation:

$$\left(h_{dom}^{-1}\right) dh_{dom}/dt = -\mathrm{b}t^{-2} \tag{7.7}$$

Separating variables and integrating (7.7)

$$\int h_{dom}^{-1} dh_{dom} = -\mathrm{b} \int t^{-2} dt$$

results in

$$\ln h_{dom} = \mathrm{b}t^{-1} + \mathrm{c}$$

then letting $c = \ln S - bt_b^{-1}$ and collecting terms and simplifying one obtains

$$\ln h_{dom} = \ln S + b\left(t^{-1} - t_b^{-1}\right) \quad (7.8)$$

which is the familiar and often used Schumacher-type function approach to site index curve construction. Equation 7.8 could be fitted directly to the height-age data to estimate the slope coefficient b which determines a family of anamorphic site curves, but the least squares estimate will not be identical to that obtained by fitting (7.7). Use of differenced data is often preferred as a means of mitigating effects of age-site correlation, and, hence bias, in guide-curve fitting.

Differential equations used to describe height growth are typically more complex than (7.7) and they involve terms in both age and height on the right-hand side. Having an analytical solution to the differential equation is an advantage, but numerical solutions can be employed if necessary. Amateis and Burkhart (1985) fitted the differential equation

$$d \ln h_{dom}/dt^{-1} = b_1 \ln h_{dom}/t^{-1} + b_2 \ln h_{dom} \quad (7.9)$$

to loblolly pine height-age measurements. Equation 7.9 is separable with a closed form solution; it expresses height growth as a function of both height and age and produces polymorphic site index curves. The integrated form of Eq. 7.9 yields the total height model:

$$\ln h_{dom} = \ln S (t_b/t)^{b_1} e^{b_2(t^{-1}-t_b^{-1})} \quad (7.10)$$

Height–age functions have generally been treated as deterministic and have been fitted using linear or nonlinear regression techniques. In a departure from usual procedure, Garcia (1983) modeled height growth by a stochastic differential equation. Stochastic differential equations consist of deterministic and random components. The form employed by Garcia can be written:

$$dh_{dom}^c/dt = b\left(a^c - h_{dom}^c\right) + \sqrt{b}\,\sigma \dot{w}(t)$$

with $\dot{w}(t)$ being "white noise," the formal derivative of a Wiener stochastic process (García 1983). The deterministic part integrates to the Richards (1959) growth curve

$$h_{dom} = a\left[\left(1 - h_{dom0}^c/a^c\right) e^{-b(t-t_0)}\right]^{1/c}$$

The serially independent white noise perturbations cause a particular stand to deviate from the most likely trajectory given by Richards growth function. Stochastic differential equations have been used to model height growth for a number of species; examples include radiata pine in New Zealand (García 1999), Maritime pine in Spain (García 2005), southern beeches (*Nothofagus*) in Chile (Salas and García 2006), and spruce in Canada (Hu and García 2010).

7.8 Difference Equation Approach

The difference equation approach requires data from plots or individual trees. It can be applied to any height-age equation to produce families of anamorphic or polymorphic curves. To develop a difference form of a height-age equation, height at a future time must be expressed as a function of future age, current age and current height, that is:

$$h_{dom} = f(t_2, t_1, h_{dom1})$$

As an example, the difference equation form of the Schumacher height-age model can be written

$$\ln h_{dom2} = \ln h_{dom1} + \mathrm{b}\left(t_2^{-1} - t_1^{-1}\right) \tag{7.11}$$

After computing an estimate of β using regression analysis, a site index equation is obtained by letting t_2 equal t_b so that h_{dom2} is by definition S and rewriting Eq. 7.11 as

$$\ln h_{dom} = \ln S + \mathrm{b}\left(t^{-1} - t_{\mathrm{b}}^{-1}\right) \tag{7.12}$$

The difference equation formation may be somewhat difficult to develop with more complex height-age models, but a usable form can generally be obtained. An example application of this approach is the site index equation developed by Clutter and Jones (1980) for slash pine plantations in the southeastern coastal plain of the USA. The difference equation model

$$\ln h_{dom2} = \mathrm{b}_2 t_2^{-1} - \mathrm{b}_3 + \left(\ln h_{dom1} - \mathrm{b}_2 t_1^{-1} + \mathrm{b}_3\right) \mathrm{e}^{\mathrm{b}_1\left(t_1^{-1} - t_2^{-1}\right)} \tag{7.13}$$

was fitted to plot remeasurement data to produce a polymorphic site index equation of the form

$$\ln S = \mathrm{b}_0 + \mathrm{b}_1 e^{\mathrm{b}_2 t^{-1}}\left[\ln h_{dom} + \mathrm{b}_3 t^{-1} + \mathrm{b}_4\right] \tag{7.14}$$

where

S = site index (index age of 25 years)
t = plantation age
h_{dom} = average height of dominant and codominant trees

Cao (1993) evaluated three different methods for estimating coefficients in selected site index models. Method one involved obtaining coefficients from either a height-age model or a differential equation model. The second method utilized a height growth, or difference equation, of the form $\ln h_{dom2} = f(t_2, t_1, h_{dom1})$. Method three was also a difference equation approach but the dependent variable was h_{dom} rather than $\ln h_{dom}$, that is $h_{dom2} = f(t_2, t_1, h_{dom})$. The height-age model forms considered were those of Schumacher (1939), Richards (1959), Bailey and Clutter (1974), Clutter and Lenhart (1968), and Amateis and Burkhart (1985). Remeasurement data from loblolly pine plantations were used to estimate coefficients of the height-age functions using the three different forms. Evaluation statistics consisting of the means of the differences between observed and predicted stand heights, mean of the absolute value of the differences, and square root of mean squared error indicated that in the majority of cases coefficients of the site index models considered should be obtained using the height growth (difference) equation formulation.

7.8.1 Algebraic Difference Approach

Bailey and Clutter (1974) developed base-age invariant polymorphic site index curves using a technique that has come to be known as the algebraic difference approach. Anamorphic (proportional) site index curves are commonly based on the assumption that $\ln h_{dom}$ and $(1/t)^{\mathrm{c}}$ are linearly related. Given height-age data from m sites (plots), an anamorphic system results from fitting:

$$\ln h_{dom} = \mathrm{a}_i + \mathrm{b}(1/t)^{\mathrm{c}} \qquad (i = 1, 2, \ldots m) \tag{7.15}$$

where

a_i = a parameter specific to the ith site
b = a common regression slope parameter

Equation 7.15 can be fitted by nonlinear least squares or c can be specified (commonly taken to be +1) and linear regression can be applied.

Estimates of the a_i's are not required in the final form of the model. Setting $t = t_b$ implies that $h_{dom} = S$, that is:

$$\begin{aligned} \ln S_i &= \mathrm{a}_i + \mathrm{b}(1/t_\mathrm{b})^{\mathrm{c}} \\ \mathrm{a}_i &= \ln S - \mathrm{b}(1/t_\mathrm{b})^{\mathrm{c}} \end{aligned} \tag{7.16}$$

Substituting (7.16) for a_i in (7.15) gives:

$$\begin{aligned} \ln h_{dom} &= \ln S_i + \mathrm{b}(t^{-1} - t_\mathrm{b}{}^{-\mathrm{c}}) \\ \text{or} \quad h_{dom} &= S_i\,[\exp(\mathrm{b}(t^{-\mathrm{c}} - t_\mathrm{b}{}^{-\mathrm{c}}))] \end{aligned} \tag{7.17}$$

an equation for h_{dom} in terms of t, t_b, and S_i.

If $b < 0$, height over age curves produced by (7.17) start at the origin and tend to different asymptotes as age increases without bound. The asymptotes for a curve through the point (t_b, S) will be

$$S e^{-b t_b^{-c}}$$

Models based on (7.15) generate proportional site index curves with a constant relative growth rate, $(dh_{dom}/dt)/h_{dom}$, across all sites at a given age. If a polymorphic (non proportional) system is desired (7.15) can be written:

$$\ln h_{dom} = a + b_i (1/t)^c \qquad (7.18)$$
$$(i = 1, 2 \ldots m)$$

In (7.18) the b_i's are site-specific parameters and $(dh_{dom}/dt)/h_{dom}$ contains them; thus relative growth rate is a function of both age and site. Expressing b_i as a function of site index, as was done with a_i in (7.15) results in

$$\ln h_{dom} = a + (\ln S_i - a)(t_b/t)^c \qquad (7.19)$$
$$\text{or} \quad h_{dom} = e^a (S_i/e^a)^{[t_b/t]^c}$$

If $S < e^a$ and $c > 0$, the limits for h_{dom} are zero as t approaches zero and e^a as t tends towards infinity. The first limit is desirable, but the second indicates that the asymptote for height is the same for all sites. Thus, curves from (7.19) for given values of S_i at index age t_b will start at the origin, have rates of increase dependent on values of the S_i's, and tend to a common upper asymptote.

Bailey and Clutter (1974) fitted Eq. 7.19 to data from radiata pine plantations in New Zealand. They went on to note that the method used to obtain polymorphic curves with an Eq. 7.18 model can be applied to other equation forms by identifying a parameter in the equation responsible for curve shape and allowing that parameter be site-specific. An estimation technique not requiring specification of site index in the data before fitting must then be derived.

Since the seminal paper by Bailey and Clutter (1974), a number of studies have used an algebraic difference type of approach for the development of site index equations (Goelz and Burk 1992; McDill and Amateis 1992; Amaro et al. 1998; Palahí et al. 2004; Bravo-Oviedo et al. 2004; Diéguez-Aranda et al. 2005b). Various functions have been applied (Chapman Richards, Hossfeld, Lundquist-Korf, Schumacher, Sloboda) and different constraints on the model parameters have been evaluated.

McDill and Amateis (1992) developed a growth function that is dimensionally compatible (Amateis and McDill 1989) with two parameters: a rate parameter and a maximum size parameter. They proposed a parameter estimation procedure that specifies some parameters as being the same for all stands (global parameters) and others that vary from stand to stand (stand-level) parameters.

The dimensionally compatible differential equation formulated by McDill and Amateis for height growth as a function of age and height was:

$$dh_{dom}/dt = \mathrm{a}(h_{dom}/t)(1 - h_{dom}/h_{\max}) \tag{7.20}$$

The rate parameter in (7.20) is denoted by a and the asymptotic maximum height parameter by $h_{\max}$. The equation was constrained so that dh_{dom}/dt goes to zero as h_{dom} approaches $h_{\max}$. Separating variables and integrating Eq. 7.20 gives a difference equation form:

$$h_{dom2} = \frac{h_{\max}}{1 - (1 - h_{\max}/h_{dom1})\,(t_1/t_2)^{\mathrm{a}}} \tag{7.21}$$

where h_{dom1} and t_1 are initial conditions.

Amaro et al. (1998) used the algebraic difference equation approach for modeling dominant height of eucalyptus plantations in Portugal. Difference forms of the Chapman-Richards and the Lundquist-Korf functions were derived and compared with the performance of the McDill-Amateis difference model. Three difference equations were derived for the Chapman-Richards and the Lundquist-Korf functions by successively specifying one of the three parameters in functions as the "free" parameter, leaving two parameters to be estimated statistically. The model form and free parameter combination that best described the data was selected. In the application with data from eucalyptus plantations, the Chapman-Richards equation with the shape parameter as the free or varying parameter (the asymptote and rate parameters were estimated) was selected. Thus the difference equation approach provided a basis for selecting both which model form and which combination of global and stand-specific parameters to use.

Similar approaches have been taken in other studies but the model as well as the free parameter that produced the best results has varied. Palahí et al. (2004) evaluated the McDill-Amateis difference model and difference forms of the Chapman-Richards, Hossfeld, Lundquist-Korf, and Schumacher models for estimating dominant height growth of Scots pine in north-east Spain. They concluded that the polymorphic difference equation derived from the Hossfeld function resulted in the best compromise between biological and statistical aspects, producing the most adequate site index curves. Bravo-Oviedo et al. (2004) allowed the rate and asymptote parameters in the Richards and Schumacher equation forms to be free parameters when evaluating candidate site index models for Mediterranean maritime pine in Spain. In this application, the polymorphic form of the Schumacher model with a common asymptote was judged most appropriate. Difference equations based on McDill-Amateis, Richards, Korf and Hossfeld growth functions were formulated for developing site index equations for Scots pine plantations in the Galicia region of Spain by Diéguez-Aranda et al. (2005b); the function proposed by McDill and Amateis (1992) produced the most adequate site curves.

7.8.2 Generalized Algebraic Difference Approach for Dynamic Site Equations

Desirable characteristics of site index equations include passing through the origin, exhibiting polymorphism, having multiple asymptote values, and predicting height at base age to be equal to site index. Bailey and Clutter (1974) introduced the concept of base-age invariance in which predictions from dynamic equations are unaffected by arbitrary changes in base-age. Their algebraic difference approach (ADA) was expanded by Cieszewski and Bailey (2000) to the generalized algebraic approach (GADA). The GADA methodology allows derivation of flexible dynamic equations that are base-age invariant, polymorphic with varying asymptotes, and with predicted height equal to site index at base age. Base-age invariant equations also have the path invariance property, which means that one-step predictions or multiple iterations will result in the same predicted value at a given final age.

To facilitate the generalized algebraic difference approach, Cieszewski and Bailey (2000) identified a theoretical variable called a growth intensity factor x and defined it to be a quantification of growth dynamics associated with a site. x can be a variable or a function of any number of variables such as climate, soils, and genetic components. Because it is not practically obtainable, x is replaced with initial conditions that are measurable so that the equation is operationally useful. This replacement, however, does not occur until after the equation is explicitly formulated in a way that it exhibits desirable properties such as polymorphism and varying asymptotes.

The first step in the GADA process is to select a base equation and to identify in it the site-specific parameters. Then, one defines explicitly how the site-specific parameters change across different sites by replacing them with explicit functions of x and new parameters. Thus, the initially selected two-dimensional base equation is expanded into an explicit three-dimensional base equation describing both independent variables t and x. In the final step, a solution for x replaces all x's with implicit definitions using the equation's initial conditions t_0 and Y_0 (Cieszewski and Bailey 2000).

Proceeding with a more formal definition of GADA, Cieszewski and Bailey (2000) noted that symbolically the base equation is the same as in the ADA:

$$Y(t) = f(t, \rho_1 \dots \rho_{n-1}, \rho_n) \tag{7.22}$$

where $\rho_1 \dots \rho_n$ are the equation parameters. If in the base equation, a given site-specific parameter ρ_i is defined as a function g_i of x and any number of j new parameters, viz., $\rho_i \equiv g_i(x, \rho_{il} \dots \rho_{ij})$, the base Eq. 7.22 with multiple site-specific parameters is changed to the explicit three-dimensional site equation with two independent variables t and x:

$$Y(t, x) = f(t, \rho_1 \dots \rho_{m-1}, g_m(x, \rho_{m1} \dots \rho_{mk}) \dots g_n(x, \rho_{n1} \dots \rho_{ni})) \tag{7.23}$$

where $Y(t, x)$ is a function of t, x and $m + \mathrm{k} + 1 - 1$ parameters.

If Eq. 7.23 can be solved for x, the right hand side of this solution, with initial condition values for t and Y, that is:

$$x = u(t, Y, \rho_1 \ldots \rho_{n_i}) = u(t_0, Y_0, \rho_1 \ldots \rho_{n_i}) \tag{7.24}$$

can be substituted in Eq. 7.23 in place of x so the dynamic equation

$$Y(t, t_0, Y_0) = f(t, \rho_1 \ldots \rho_m, u(t_0, Y_0, \rho_1 \ldots \rho_{n_1}))$$

after reformulation and elimination of redundant parameters, becomes the dynamic equation with an implicitly defined initial condition:

$$Y(t, t_0, Y_0) = f(t, t_0, Y_0, \rho_1 \ldots \rho_w) \tag{7.25}$$

where

$$n - 1 \leq w \leq m + k + \cdots + l - 1 \tag{7.26}$$

The result in Eq. 7.26 means that Eq. 7.25 has a *smaller* or *equal* number of parameters than Eq. 7.23.

Practical applications of the GADA involve different levels of complexity and difficulty in equation derivations. Cieszewski and Bailey (2000) classified the equations as simple or complex, depending on whether the solutions involved are based on just a reformation of an equation (simple) or on finding its roots (complex).

Nord-Larsen (2006) summarized the steps in formulating GADA equations as follows:

1. Select the basic equation form.
2. Identify the site-specific parameter(s) in the basic equation.
3. Generalize the site-specific parameters in the equation as functions of X, where X is an unknown measure of site quality.
4. Solve for X and substitute for the independent variables, t and h_{dom}, the initial conditions, t_0 and h_{dom0}. The X in the equation developed in step 3 is finally replaced by this solution for X.

7.8.2.1 GADA Applied to Simple Equations

As presented in Sect. 7.8.1 ADA procedures can be readily applied to produce anamorphic site curves with varying asymptotes or polymorphic curves with a common asymptote. Continuing with using the Schumacher equation as an example, that is

$$\ln Y(t) = \mathrm{a} - \mathrm{b}/t \tag{7.27}$$

where Y is height and t represents age, the GADA approach can be applied when more than one simultaneous site-specific parameter is required to adequately describe changes in curve shapes across different sites. For example, if one assumes concurrent polymorphism and varying asymptotes, both a and b in (7.27) could be dependent on x while x could define the limiting size, that is

$$\ln Y(t, x) = x + \mathrm{b}x/t \tag{7.28}$$

The solution for x would then be

$$x = \frac{\ln Y}{1 - \mathrm{b}/t} = \frac{\ln Y_0}{1 - \mathrm{b}/t_0}$$

and applying the GADA to Eq. 7.28 with respect to x would result in a dynamic equation based on Schumacher's equation that provides polymorphic base-age invariant curves with variable asymptotes:

$$\ln Y(t, t_0, Y_0) = \ln Y_0 \frac{t_0(t - \mathrm{b})}{t(t_0 - \mathrm{b})} \tag{7.29}$$

The assignment of x to a means that given an objective measure of growth intensity, the upper production limit would be increasing with increasing innate growth potential. This relationship would result in variable asymptotes. The assignment of x to b means that the shapes of curves change with changing growth intensity which defines a polymorphic equation. Clearly, both variable asymptotes and polymorphism occur if x affects both a and b (Cieszewski and Bailey 2000).

7.8.2.2 GADA Extensions to More Complex Functions

Cieszewski and Bailey (2000) labeled as complex dynamic equations those that require in their derivations the roots of an equation in order to determine x; the solutions required may be difficult to obtain. The selection of the most appropriate expression for x may depend on the equation parameters, the data available, and the domain of applicable ages. For each base equation, several possible approaches may be used to derive the implicit dynamic equation.

Generic equations might be formulated in cases where there are no specific expectations about the final model form. Such equations can result from simple or complex derivations. However, Cieszewski and Bailey (2000) caution that development of generic equations can lead to over parameterization and model instability as well as difficulties with parameter estimation.

7.8.2.3 Applications of GADA

Since the introduction of the generalized algebraic difference approach, there have been a number of applications of the method. These applications have varied in the amount and quality of data available, the base functions evaluated, the site-specific parameters considered, and the fitting methods employed. A brief description of several applications of the GADA techniques follows.

Cieszewski (2001) used data generated from existing site-index curves for Douglas-fir (Monserud 1984) to demonstrate three approaches to derivation of dynamic equations that provide compatible height and site index values from one common equation. In another analysis, Cieszewski (2002) compared a fixed-base-age height growth equation with several dynamic equations derived by the GADA process. The focus of the comparison was on ability to exhibit concurrent polymorphism and multiple asymptotes. With data from subalpine fir, Cieszewski (2003) developed a dynamic site equation using a modified Hossfeld IV function. The proposed model performed better than other base-age specific and base-age-invariant models that were evaluated for both fit to the data and behavior in extrapolations.

Diéguez-Aranda et al. (2005a) used data from stem analysis of 161 trees to develop a model for predicting height growth of radiata pine plantations in Galicia (north-west Spain). Different base equations were evaluated for modeling the dominant height growth pattern and several variants of each base equation were developed using the simplified approach of mixed-effects model of Cieszewski (2003). Models with both one and two parameters specified to be site-specific were considered. Many seemingly different model forms and variants resulted in essentially the same curves. Thus, for simplicity, three base equations with variants were used as a basis for developing a final model (Table 7.1).

For convenience, Diéguez-Aranda et al. (2005a) let $\mathrm{a}_1, \mathrm{a}_2, \ldots, \mathrm{a}_n$ denote parameters in base models, and $\mathrm{b}_1, \mathrm{b}_2, \ldots, \mathrm{b}_m$ symbolize global parameters in subsequent GADA formulations. The GADA based models have the general implicit form of $Y = f(t, t_0, Y_0, \mathrm{b}_1, \mathrm{b}_2, \ldots, \mathrm{b}_m)$. The first group of models Diéguez-Aranda developed using the GADA were formulated on the basis of the function proposed by Bertalanffy. Three models were derived by applying the GADA to the Bertalanffy-Richards function while considering only one parameter to be site-specific (in this case, the GADA is equivalent to the ADA; dynamic equations (2)–(4) in Table 7.1). Both dynamic Equations (3) and (4) in Table 7.1 have a common asymptote. However, an important property of site equations that should be considered during modeling various growth trends is the ability to simulate concurrent polymorphism and multiple asymptotes, which requires that more than one parameter be site-specific. Therefore, in the next model derived using the GADA (Equation (5) in Table 7.1) both the asymptote and the shape parameter a_3 were assumed to be dependent on X. As the site variable X is arbitrary, the base equation was reparameterized to provide a mathematically tractable solution for X. When developing Equation (5), the solution for X involved finding roots of a quadratic equation and selection of the most appropriate root to substitute into the dynamic equation. As noted by Cieszewski and Bailey (2000), the selection of the most appropriate expression for X may depend on the equation parameters that in turn depend on the data and the domain of the applicable ages.

Table 7.1 Base models and GADA formulations considered for modeling dominant height growth of radiata pine (From Diéguez-Aranda et al. 2005a)

Base equation	Parameter related to site	Solution for X with initial values $(t_0 Y_0)$	Dynamic equation	
Bertalanffy-Richards:	$a_1 = X$	$X_0 = \frac{Y_0}{(1-\exp(-b_2 t_0))^{b_3}}$	$Y = Y_0\left(\frac{1-\exp(-b_2 t)}{1-\exp(-b_2 t_0)}\right)^{b_3}$	(2)
$Y = a_1(1-\exp(-a_2 t))^{a_3}$	$a_2 = X$	$X_0 = \frac{-\ln(1-(Y_0/b_1)^{1/b_3})}{t_0}$	$Y = b_1\left(1-\left(1-\left(\frac{Y_0}{b_1}\right)^{1/b_3}\right)^{t/t_0}\right)^{b_3}$	(3)
	$a_3 = X$	$X_0 = \frac{\ln(Y_0/b_1)}{\ln(1-\exp(-b_2 t_0))}$	$Y = b_1\left(\frac{Y_0}{b_1}\right)^{(\ln(1-\exp(-b_2 t)))/(\ln(1-\exp(-b_2 t_0)))}$	(4)
	$a_1 = \exp(X)$	$X_0 = \frac{1}{2}\left((\ln Y_0 - b_2 L_0) \pm \sqrt{(\ln Y_0 - b_2 L_0)^2 - 4b_3 L_0}\right)$	$Y = Y_0\left(\frac{1-\exp(-b_1 t)}{1-\exp(-b_1 t_0)}\right)^{(b_2+b_3/X_0)}$	(5)
	$a_3 = b_2 + b_3/X$	with $L_0 = \ln(1-\exp(-b_1 t_0))$		
Log-logistic (log-transformed):	$a_1 = b_1 + X$	$X_0 = \frac{1}{2}\left(Y_0 - b_1 \pm \sqrt{(Y_0-b_1)^2 + 4b_2 Y_0 t_0^{-b_3}}\right)$	$Y = \frac{b_1+X_0}{1+b_2/X_0 t^{-b_3}}$	(6)
$Y = \frac{a_1}{1+a_2 t^{-a_3}}$	$a_2 = b_2/X$			
	$a_1 = b_1 + X$	$X_0 = \frac{Y_0-b_1}{1-b_2 Y_0 t_0^{-b_3}}$	$Y = \frac{b_1+X_0}{1+b_2 X_0 t^{-b_3}}$	(7)
	$a_2 = b_2 X$			
Cieszewski (2003):	$a_1 = b_1 + X$	$X_0 = \frac{1}{2}\left(Y_0^3 t^{-1} - b_1 \pm \sqrt{(Y_0^3 t^{-1} - b_1)^2 + 2b_3 Y_0^3 t^{-b_2}}\right)$	$Y = Y_0\left(\frac{(b_3+t_0^{b_2-1}R_0)t^{b_2}}{(b_3+t^{b_2-1}R_0)t_0^{b_2}}\right)^{1/3}$	
$Y^3 = \frac{a_1 t^{a_2}}{a_3+t^{a_2-1}}$	$a_3 = \frac{1}{2}b_3/X$		Or if $b_2 \le 1$:	(8)
			$Y = Y_0\left(\frac{(b_3 t_0^{1-b_2}+R_0)t}{(b_3 t^{1-b_2}+R_0)t_0}\right)^{1/3}$	

The subsequent two models (Equations (6) and (7) in Table 7.1) may be viewed as special cases of the log-logistic class of models, which are equivalent to Hossfeld models (Cieszewski 2000). Within the different forms investigated by Diéguez-Aranda et al. (2005a), one model proposed by Cieszewski (2002) was found to perform particularly well. Its formulation is identical to the general solution of the differential equation developed by McDill and Amateis (1992), whose difference equation solved by parameter a_2 has been successfully used for modeling the height growth in other studies (Sharma et al. 2002a; Fontes et al. 2003). Both dynamic Equations (6) and (7) are polymorphic with variable asymptotes.

The last model evaluated (Equation (8) in Table 7.1) was the GADA based equation developed by Cieszewski (2003). The two forms of Equation (8) are equivalent except for the fact that only one of them is defined for $t = 0$, depending on the value of parameter b_2.

Parameter estimation for site models involves many different statistical considerations such as criteria of fitting, error structures and independence of errors, homogeneity of variance and balance of data panels. When estimating parameters, Diéguez-Aranda et al. (2005a) used the base age invariant (BAI) method by Cieszewski et al. (2000). In this method, the initial conditions are specified to be identical for all the measurements belonging to the same tree. The initial age can be, within limits, arbitrarily selected, even for each tree; however, age zero is not allowed. The height corresponding to the initial age is then simultaneously estimated for each tree along with all of the global model parameters during the fitting process. The dummy variables method recognizes that each measurement is made with error, and, therefore, the model is not forced through any given measurement. Instead, the curve is fitted to the observed individual trends in the data. With this method, all the data can be utilized and there is no need to make any arbitrary choice regarding measurement intervals (i.e., it is a BAI method). In the dummy variables method, a minimum of two measurements per tree is required, and the number of trees must be greater than the number of global parameters in the model.

Diéguez-Aranda et al. (2005a) used a second-order continuous-time autoregressive error structure to correct the inherent autocorrelation of the longitudinal data. The GADA formulation derived on the basis of the Bertalanffy-Richards model by considering the asymptote and the initial pattern parameters as related to site productivity (Equation (5) in Table 7.1) resulted in the best compromise between biological and statistical considerations, producing the most adequate site curves.

Using procedures similar to those applied when modeling radiata pine height growth, Diéguez-Aranda et al. (2006) evaluated four dynamic site equations derived with GADA methods to describe height development in loblolly pine plantations in the USA. The data came from a large, long-term study involving permanent sample plots. As in the study with radiata pine, the fittings were done in one stage using the base-age invariant dummy variables method. A first-order autoregressive error structure was used to correct the serial correlation in the longitudinal data. In this case, Cieszewski's (2002) model best described the data.

Table 7.2 Selected studies that have employed GADA-type approaches in the development of site curves. The data base (species, country), base model selected, and citation for the study are shown.

Data base	Base model selected	Citation for the study
Scots pine, Spain	Hossfeld	Palahi et al. (2004)
Pyrenean Oak, Portugal	Bertalanffy-Richards	Carvalho and Parresol (2005)
Radiata pine, Spain	Bertalanffy-Richards	Diéguez-Aranda et al. (2005a)
Loblolly pine, USA	Cieszewski	Diéguez-Aranda et al. (2006)
European beech, Denmark	Logistic	Nord-Larsen (2006)
Mediterranean maritime pine, Spain	Hossfeld	Bravo-Oviedo et al. (2007)
Red alder, USA	Schumacher	Weiskittel et al. (2009)
Longleaf pine, USA	Hossfeld	Lauer and Kush (2010)

Other examples of using a GADA-type approach to site index equation modeling include studies of Scots pine in Spain (Palahí et al. 2004), Pyrenean Oak stands in Portugal (Carvalho and Parresol 2005), European beech in Denmark (Nord-Larsen 2006), Mediterranean maritime pine in the Iberian Peninsula (Bravo-Oviedo et al. 2007, 2008), red alder plantations in the Pacific Northwest, USA (Weiskittel et al. 2009), thinned stands of even-aged naturally regenerated longleaf pine in the East Gulf region of the United States (Lauer and Kush 2010), and maritime pine in Portugal (Nunes et al. 2011). A tabulation of these studies, along with those of Diéguez-Aranda et al. (2005a, 2006), giving information on the data base used and the base model selected is provided in Table 7.2. The Hossfeld function was selected as the base model in three of the eight studies, the Bertalanffy-Richards function was chosen in two instances; the Cieszewski, logistic, and Schumacher functions were each selected in one case. At this juncture, it seems that there are multiple options for base functions and for specifying which parameter(s) are site-specific that can produce essentially equivalent results.

7.8.3 Estimating Parameters in ADA- and GADA-Type Formulations

In developing the ADA concept for site curve construction, Bailey and Clutter (1974) specified that it exhibit the base-age invariance property. Being invariant under choices of base age is dependent not only on model formulation but on parameter estimation methodology as well. To be base-age invariant the height-age model must be reparameterized as a function of site index and age and the global and site-specific parameters must be estimated with least squares. If the first requirement is met but not the second, biased parameter estimates result. The bias stems from forcing the equation through a particular data point (height at index age – i.e., site index) rather than assuming that height at index age also contains error as do all other height-age pairs. There is no generally-accepted terminology to distinguish between the so-called GADA models that are truly base-age invariant and those that are not. However, it is important that the parameter estimation methodology, as well as the model formulation, be noted when interpreting and applying results from studies that have used GADA-type approaches. Bailey and Cieszewski (2000) provide additional discussion on the mathematical definition of invariant. As part of the discussion, they demonstrated a simple means to evaluate whether or not a given site index model formulation with attendant parameter estimates is in fact base-age invariant.

Furnival et al. (1990) showed analytically that different methods of parameter estimation can, in the special case of a linear model, give identical estimates for the slope coefficient. These results do not, however, extend to models that are nonlinear in the parameters. Cao (1993) evaluated three different methods for estimating coefficients of five base-age-invariant site index models. Results indicate that in most cases coefficients of the site index models evaluated should be obtained from a height growth or difference equation.

Eight hypothetical data points (four growth series with two observations each) were used by Strub and Cieszewski (2006) to illustrate differences between base-age invariant stochastic regression (BAI) and the method of all base ages (BAA) for site index curve parameter estimation. The anamorphic form of the Schumacher height-age equation was used for the comparison. The two estimation techniques are equivalent for the linear case of the model but produce different parameter estimates for its nonlinear form. For both linear and nonlinear cases the BAI approach produced a lower root-mean-squared error than the BAA method. Further, the BAI approach produced unique estimates for the site-specific parameters of both model forms and both growth series.

Additional detail about the comparison of parameter estimation techniques published by Strub and Cieszewski (2006) follows. The Schumacher (1939) log-transform height-age equation was expressed as:

$$\ln(h_{domij}) = \mathrm{a}_j + \mathrm{b}/t_{ij},$$

where h_{domij} = dominant height for measurement i and series j; t_{ij} = age for measurement i; a and b = model parameters to be estimated.

The nonlinear form of this model was given as:

$$h_{domij} = \mathrm{c}_j \mathrm{e}^{\mathrm{b}/t_{ij}}$$

where $\mathrm{c}_j = \mathrm{e}^{\mathrm{a}j}$ and e = the base of the natural logarithm.

An anamorphic form of the site index curve results when b is the global parameter. If all growth series contain two observations, as in the example constructed by Strub and Cieszewski, the least-squares estimates for the site-specific parameters (a and c) suggested by the BAI approach are

$$\hat{\mathrm{a}}_j = \frac{\sum_{i=1}^{2}\left(\ln(h_{domij}) - \hat{\mathrm{b}}/t_{ij}\right)}{2} \tag{7.30}$$

$$\hat{c}_j = \frac{\sum_{i=1}^{2} h_{domij} e^{\hat{b}/t_{ij}}}{\sum_{i=1}^{2} e^{2\hat{b}/t_{ij}}} \tag{7.31}$$

With the BAI approach, the estimates for site-specific parameters are unique and use all of the data for a growth series rather than depending on the value of a single observation. Thus, the BAI estimates of the site-specific parameters are independent of any arbitrary choice of base age, and therefore they are base-age invariant.

Estimates for the site-specific parameters with the BAA approach are

$$\hat{a}_{ij} = \ln(h_{domij}) - \hat{b}/t_{ij} \tag{7.32}$$

$$\hat{c}_j = h_{domij}/e^{\hat{b}/t_{ij}} \tag{7.33}$$

For each of the BAA estimates ($\hat{a}_{ij}$ or $\hat{c}_{ij}$), an arbitrary choice for base age (i) must be selected before determining the estimate. In the Strub and Cieszewski example, either age 10 or age 40 was chosen as the base age. The estimates for each of the base ages are, therefore, not base-age invariant. Furthermore, even though the overall estimate of the BAA method averages the estimates for each of the base ages for the linear model, in the case of a nonlinear model the resulting average is not equal to the base-age invariant estimate. Although Eqs. 7.30 and 7.32 illustrate that, for the linear model the BAI estimate of $\hat{a}_j$ is equal to the average of the two BAA estimates ($\hat{a}_{1j}$ and $\hat{a}_{2j}$), Eqs. 7.31 and 7.33 illustrate that, for the nonlinear model the BAI estimate of $\hat{c}_j$ is different from the average of the two BAA estimates ($\hat{c}_{1j}$ and $\hat{c}_{2j}$).

Strub and Cieszewski (2006) provide a specific numeric example and conclude with the observation that both approaches BAI and BAA are based on least-squares estimation of the global parameter ($\hat{b}$), but the BAI method also relies on least-square estimates of local ($\hat{a}$ or $\hat{c}$) site parameters. Because the BAI estimates are based on least-square criteria and can assume any values for the site estimates, this method must have root-mean-square errors smaller than or equal to that of any other method.

Wang et al. (2008b) proposed a method to develop polymorphic base-age invariant models with multiple asymptotes by specifying the asymptotic parameter as the site-specific parameters (x) and one of the other parameters as related to x as a simple power function. Their approach is a constrained form of the generalized algebraic difference (GADA) and eliminates the requirement to obtain an explicit solution for x. In this constrained GADA approach, x is estimated along with other parameters of the model. This approach may be used to adapt base-age specific models to be base-age invariant. The Wang et al. approach was evaluated using data sets for Chinese fir, red alder, and beech with models being fitted using the dummy variable method in which each observed data series has its own specific x parameter. The base-age invariant models were superior to their comparable base-age specific models.

7.9 Mixed-Effects Models for Height Prediction

Mixed-effect models provide a flexible and efficient approach for estimating site-specific parameters in height-age equations. Traditional regression models are fixed effects models, containing only constants in the functional form and one random variant, the error term. Fixed-effects models describe a mean function. Mixed-effects arise when some of the model components are fixed while others are random. Mixed models contain at least one random effect as well as the random error term; as such, they describe a mean trend and deviations from the mean trend.

The standard linear regression model can be written in matrix form as:

$$y = X\beta + e$$
$$e \sim N_n(0, \sigma^2 I_n)$$

where $y = (y_1, y_2, \ldots, y_n)'$ is the response vector; X is the model matrix with typical row $x'_i = (x_{1i}, x_{2i}, \ldots x_{pi})$; $\beta = (\beta_1, \beta_2, \ldots, \beta_p)'$ is the vector of regression coefficients: $e = (e_1, e_2, \ldots, e_n)'$ is the vector of errors; N_n represents the n-variable multivariate-normal distribution; $\mathbf{0}$ is an $n \times 1$ vector of zeroes; and I_n is the order-n identity matrix. *Mixed-effect* models (or just *mixed models*) include additional random-effect terms, and are often used for representing clustered data resulting, for example, when observations are collected hierarchically, when measurements are taken on related individuals, or when observations are gathered over time on the same individuals. The mixed-effects model can be represented in matrix form:

$$y_i = X_i\beta + Z_i\mathbf{b}_i + e_i$$

$$\mathbf{b}_i \sim N_q(0, D)$$

$$e_i \sim N_{n_i}(0, \sigma^2 R_i)$$

where

y_i is the $n_i \times 1$ response vector for observations in the ith group.
X_i is the $n_i \times p$ model matrix for the fixed effects for observations in group i.
β is the $p \times 1$ vector of fixed-effect coefficients.
Z_i is the $n_i \times q$ model matrix for the random effects for observations in group i.
$\mathbf{b}_i$ is the $q \times 1$ vector of random-effect coefficients for group i.
e_i is the $n_i \times 1$ vector of errors for observations in group i.
D is the $q \times q$ covariance matrix for the random effects.
$\sigma^2 R_i$ is the $n_i \times n_i$ covariance matrix for the errors in group i.

For mixed models the $\mathbf{b}_i$ are random effects or coefficients with mean $\mathbf{0}$ and variance-covariance matrix D that vary across clusters and the e_i are the

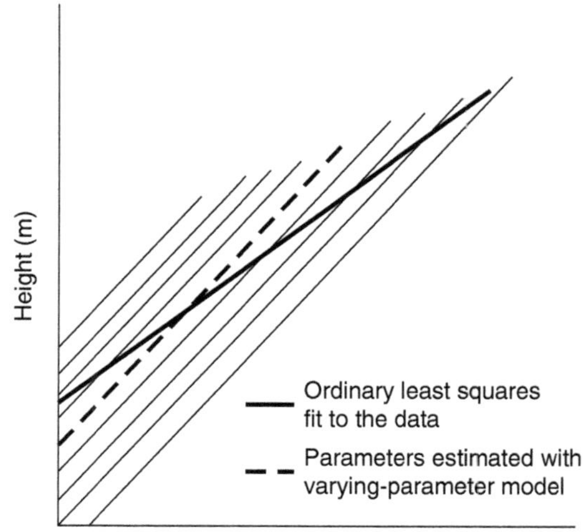

Fig. 7.4 Comparison of the regression surfaces derived from a varying-parameter model and a standard regression fit to pooled data (Adapted from Biging 1985)

within-cluster errors. The model contains a fixed-effect mean structure given by $\boldsymbol{X}_i\boldsymbol{\beta}$ and a random structure given by $\boldsymbol{Z}_i\mathbf{b}_i + \boldsymbol{e}_i$. In addition to linear mean functions, mixed modeling procedures can be applied when the mean function is nonlinear.

7.9.1 Varying Parameter Model

In many past applications data have been pooled and height has been predicted as a function of age or age and site. However, as pointed out by Biging (1985), if there is significant variation in individual tree growth the conditional mean height of the sample given specific values of the explanatory variables will differ from that of an estimator that allows coefficients in the growth model to vary by individual. It cannot be assumed that the mean regression derived by standard procedures represents height growth of the average tree. To allow for individual tree variation, Biging (1985) developed a varying parameter (random regression coefficient) approach using weighted least squares fitting of height growth models to the observations of individual trees over time to produce a height growth curve for the average tree.

Data for Biging's analysis consisted of four to six dominant trees chosen randomly on sample plots in the mixed conifer region of California, USA, and felled as site index trees for stem analysis. Between-tree heterogeneity was treated by using a varying-parameter model in which coefficients of height over age curves for individual trees are treated as random. Figure 7.4 illustrates how the varying-parameter approach can produce results that differ from those of standard regression fitting with pooled data.

7.9.2 Mixed-Effects Models with Multiple Random Components

Using data from slash pine plantations, Lappi and Bailey (1988) illustrated a procedure that explicitly describes the major random components in the variation of height curves. The average height of dominant and codominant trees in the population is expressed as a function of age. Then a model is developed for the variance-covariance structure of deviations from this average height curve due to stands and trees within stands. The Lappi-Bailey formulation takes into account the fact that, within a given stand, heights of trees at a specific age are correlated, and tree heights and average stand height are correlated over time. This special covariance structure is specified and estimated by including random stand and tree parameters in the height growth model. With the estimates of variances and covariances, and of the parameters of the function for average height, the height development of a given stand or single tree can be predicted using statistical prediction methods that utilize all available height measurements.

The varying-parameter approach of Biging (1985) did not take the stand level variation into account but rather treated tree parameters as independent for trees belonging to the same stand. Lappi and Bailey (1988) recognized three different height over age curve types: (1) the population average curve, (2) the stand average curve, and (3) the tree curve. The population height curve is the deterministic (fixed) average height curve for all dominant and codominant trees in the population. Stand and tree curves are assumed to be random, but predictable, fluctuations from the population curve. Specifying a model with random stand and tree parameters allows analysis of the within-stand and between-stand variance in the height of trees at different ages and the autocorrelation among heights over time. Therefore, in application the model allows all available data from a plot (stand) for all measurements to be used for predicting subsequent growth.

After estimating parameters of the population height curve and of the covariance structure, stand and tree heights can be predicted from any available height measurements using linear prediction methods. The main advantage of specifying stand-specific parameters as random is that the resulting predictor combines optimally (in the "best linear unbiased" sense) the available measurements and the prior knowledge of the variances and covariances of the random parameters (Lappi and Bailey 1988).

Lappi and Malinen (1994) further developed the modeling structure of Lappi and Bailey (1988) for the situation where stand-specific parameters are correlated with stand age. When this correlation exists, the population mean curve will be biased. (The bias that results when guide curve methods are applied and possible ways to mitigate against the bias are discussed in Sect. 7.4.) Lappi and Malinen (1994) showed how ordinary least squares estimates of tree-specific parameters can be used for computing unbiased estimates of the population mean parameters and for estimating the relationship between stand parameters and stand age.

In other applications of mixed-effects models for predicting height development, Fang and Bailey (2001) used a modified Richards growth function with nonlinear mixed effects for modeling slash pine dominant height growth in conjunction with different silvicultural treatments. Noting that there are strong site and genetic variations in eucalyptus stands, Calegario et al. (2005) used a nonlinear mixed-effects model to represent the height growth pattern of eucalyptus clonal stands in the Brazilian coastal region.

7.9.3 Accounting for Serial Correlation

The inclusion of random effects into mixed models is expected to bring about a reduction in serial correlation. However, random effects may not be sufficient to account for serial correlation; directly modeling the error structure to further reduce the correlation may be required.

Generalized least squares (GLS) procedures have been applied to model directly the error structure for various forest growth models. The mixed-effects modeling approach addresses the correlation issue by incorporating the random effects into models and/or by directly modeling the error structure. Including a function to account for serial correlation generally results in an improvement of the model's fit statistics (Gregoire et al. 1995). This improvement in fit may not, however, always result in improvement in the model's predictive ability. Fortin et al. (2007) reported an improved predictive ability of a model fitted using the GLS method when estimated correlation of the model was used to adjust predictions. Only one prior measurement was used to adjust the predictions.

Meng and Huang (2010) conducted a study to determine if accounting for serial correlation would result in improved predictive ability for three nonlinear mixed models calibrated using individual tree sectional data and repeated measurements from permanent sample plots. Results showed that accounting for the serial correlation using the spatial power (SP(POW)) or Toeplitz (TOEP(X)) functions resulted in a large reduction in serial correlation and improved the fit of the models. The improved model fits, however, did not necessarily translate into improved model predictions when evaluated under different scenarios. In many cases, the models with the simple independent and identically distributed structure outperformed the models with the SP(POW) or TOEP(X) structure in terms of the models' predictive ability.

Meng and Huang (2010) also examined the effect of adjusting predictions according to a theorem that indicates the estimated correlation of a model can be used to enhance the model predictions based on known prior measurements. It was shown that, in general, the adjusted predictions had lower errors than those without adjustment, but the differences were small in many cases. The adjustment with three prior measurements was better in predictions that the adjustment with only one or two prior measurements for the models with the TOEP(X) structure, but not for SP(POW).

In addition to serial correlation, heteroscedasticity is also often encountered. Various procedures can be applied to combat heteroscedasticity (such as transforming variables, applying weights), or it can be modeled directly. When fitting height-age functions, the heteroscedasticity is often modeled by fitting a power function with age as a covariate.

7.9.4 *Calibration of Nonlinear Mixed-Effects Models*

The calibration of nonlinear mixed effects models is central to making local predictions. Numerous authors (e.g. Hall and Bailey 2001; Nothdurft et al. 2006; Meng and Huang 2009) have addressed calibration issues. Meng and Huang (2009) observed that simplified calibration procedures for nonlinear mixed-effects models can distort local predictions. Using a nonlinear height growth model for lodgepole pine, Meng and Huang (2009) demonstrated procedures to obtain improved calibration of the height growth model. Calibration using the improved method reduced bias and variance of the errors.

Meng and Huang (2009) used two data sets to demonstrate the improved calibration of a nonlinear mixed-effects height growth model. A logistic-type function was adopted as the height growth model and appropriate procedures for calibrating nonlinear mixed models expanded via a first-order Taylor series expansion at zero and at EBLUP (the empirically best linear unbiased predictors) were applied. The improved calibration, based on a three-step iteration was compared to conventional calibration. The resultant predictions differed considerably between the two methods.

7.9.5 *Evaluation of Population-Averaged and Subject-Specific Predictions*

Meng et al. (2009) evaluated population-averaged (PA) and subject-specific (SS) approaches for modeling the dominant height of lodgepole pine. The population-averaged response was modeled using a set of covariates and fixed-effects parameters, that is:

$$\boldsymbol{h}_{domi} = f(\boldsymbol{X}_i, \boldsymbol{\beta}) + \boldsymbol{\varepsilon}_i \tag{7.34}$$

where $\boldsymbol{h}_{domi}$ is a vector of heights for site tree i; $\boldsymbol{X}_i$ is a matrix of covariates usually consisting of age, site index, and other predictors; $\boldsymbol{\beta}$ is a vector of fixed-effects parameters; and $\boldsymbol{\varepsilon}_i$ is a vector of error terms.

Top height was also modeled using a mixed-effects modeling approach; the mixed-effects model can be expressed in a general form as:

$$\boldsymbol{h}_{domi} = f(\boldsymbol{X}_i, \boldsymbol{\beta}, \boldsymbol{u}_i) + \boldsymbol{\varepsilon}_i \tag{7.35}$$

where $\boldsymbol{\beta}$ is the fixed-effects parameters common to all subjects (trees) and $\boldsymbol{u}_i$ is a vector of the random-effects parameters unique for subject *i*, which is assumed to follow a multivariate normal distribution with mean zero and a variance-covariance matrix $\boldsymbol{D}$.

To obtain a subject-specific response using model (7.35), a proportion of premeasured $\boldsymbol{X}_i$ and $\boldsymbol{h}_{domi}$ for subject *i* has to be acquired to predict $\boldsymbol{u}_i$, which is then used together with estimates of $\boldsymbol{\beta}$ and the variance-covariance matrix **D** to predict other $\boldsymbol{h}_{domi}$ values at given values of $\boldsymbol{X}_i$. This procedure differs from PA models, which require only $\boldsymbol{X}_i$ values to predict $\boldsymbol{h}_{domi}$. In cases where there is an insufficient proportion of premeasured $\boldsymbol{X}_i$ and $\boldsymbol{h}_{domi}$ to predict $\boldsymbol{u}_i$, then the estimated fixed-effects parameters of Eq. 7.35 can be used to predict a typical mean (TM) response. The TM response obtained using only the fixed-effects component of a mixed- effect model is a PA response only for linear mixed-effects models.

Using only the fixed-effects parameters of a linear mixed-effects model can yield a true PA response, and TM is equivalent to PA. However, for a nonlinear mixed-effects model because $\boldsymbol{u}_i$ enters the model in a nonlinear fashion, the random-effects parameters cannot be factored out of the integration as in linear mixed-effects models; therefore, the TM response generated using only the fixed-effects parameters is not equal to the PA mean.

Obtaining a PA response using only fixed-effects parameters for a nonlinear mixed-effects model is not appropriate; the true PA response should be generated from a separate fit of the nonlinear model without the inclusion of random effects (Meng et al. 2009). Six candidate models derived from the Chapman-Richards and logistic functions were included in the evaluation reported by Meng et al. (2009). The true PA response obtained from separate fits of the models was compared with the typical mean (TM) response computed using only the fixed-effects parameters of the mixed-effects models. Results showed that the TM response had higher prediction errors than the PA response, suggesting that a true PA response and not the TM response is needed to reflect the overall mean response of the model. The SS approach produced improved height prediction relative to the PA approach when evaluated using independent validation data. In addition, the logistic performed better than the Chapman-Richards function, regardless of whether the SS or PA approach was applied. Among the candidate models, the logistic function with the inclusion of site index gave the most accurate predictions. Three scenarios of calibrating the mixed-effects models on the validation data set were compared. The SS predictions obtained using only one premeasured observation per subject were not as good as those that used all observations, but they were still generally better than PA predictions.

7.10 Comparison of Subject-Specific Approaches for Modeling Dominant Height

Wang et al. (2008a) compared two subject-specific approaches to height modeling, the dummy variable method (fixed individual effects) and the mixed model method (random individual effects). Chapman-Richards type models were fitted to loblolly pine data from a designed experiment.

Table 7.3 lists the models considered by Wang et al. (2008a). The three ADA models (MI–M3) are examples of one-local-parameter models, and they are either anamorphic (M1) by choosing the asymptote parameter a to be site-specific or polymorphic with single asymptotes (M2 and M3) by choosing either parameter b or c as site-specific. GADA model M4 was developed by Cieszewski (2004) and has been used by Diéguez-Aranda et al. (2005a, 2006), and Nord-Larsen (2006). This model is an example of a one-varying-parameter model which is polymorphic with multiple asymptotes. Model M5 is a two-local-parameter example, which has been used in Lappi and Bailey (1988) and Hall and Bailey (2001). When attempting to fit a model with three random parameters, convergence could not be obtained; thus the most flexible model with two random parameters that was obtained is model M5.

Parameters in the candidate models were estimated using measurements taken at 3-year intervals starting at age six and extending to age 18. A brief description of the parameter estimation procedures employed by Wang et al. (2008a) follows.

Dummy variables were generated to indicate plot-specific parameters. For M1–M4, the dummy variable models in the general form are given as:

$$h_{domij} = f(t_{ij}; x, \beta) + \varepsilon_{ij}$$

$$x = \sum_{i=1}^{n} x_i D_i$$

where h_{domij} is the jth observed heights for the ith plot, t_{ij} is the age, ε_{ij} is the error term, n is the number of plots, D_i is the dummy variable with value of 1 for the ith

Table 7.3 Height models considered by Wang et al. (2008a)

No.	Model form	Derivation
M1	$h_{dom} = a_i(1-\mathrm{e}^{-bt})^c + \varepsilon$	$\chi = a_i$
M2	$h_{dom} = a(1-\mathrm{e}^{-b_i t})^c + \varepsilon$	$\chi = b_i$
M3	$h_{dom} = a(1-\mathrm{e}^{-bt})^{c_i} + \varepsilon$	$\chi = c_i$
M4	$h_{dom} = \mathrm{e}^{\chi}(1-\mathrm{e}^{-bt})^{c_1+c_2/\chi} + \varepsilon$	$\chi_1 = a_i = e^{\chi}, \chi_2 = c_i = c_1 + c_2/\chi$
M5	$h_{dom} = a_i(1-e^{-bt})^{c_i} + \varepsilon$	$\chi_1 = a_i, \chi_2 = c_i$

Where h_{dom} is dominant height measurements at age t, a, b, c, c_1, c_2 are common parameters; i indicates subject; a_i, b_i, c_i, and χ indicate local parameters plot and 0 otherwise, and x_i is the fixed plot-specific parameter. For M5, the dummy variable model was specified as

$$h_{domij} = \left(\sum_{i=1}^{n} \mathrm{a}_i D_i\right)(1 - e^{-bt})^{\sum_{i=1}^{n} \mathrm{c}_i D_i} + \varepsilon_{ij}$$

where a_i and c_i are the two fixed plot-specific parameters.

Nonlinear mixed models for M1–M4 can be expressed in the general form as:

$$\begin{aligned} h_{domij} &= f(t_{ij}; x, \beta) + \varepsilon_{ij} \\ x &\sim N(\mu_x, \sigma_x^2) \end{aligned}$$

where N denotes the bivariate normal distribution, μ_x and σ_x^2 are the fixed mean and variance parameters of the random subject-specific parameter x, respectively. For M5 the general form is:

$$h_{domij} = \mathrm{a}_i \left(1 - \mathrm{e}^{-\mathrm{b}t}\right)^{\mathrm{c}_i} + \varepsilon_{ij}$$

$$\begin{pmatrix} \mathrm{a}_i \\ \mathrm{c}_i \end{pmatrix} \sim N\left(\begin{pmatrix} \mu_\mathrm{a} \\ \mu_\mathrm{c} \end{pmatrix}, \begin{pmatrix} \sigma_\mathrm{a}^2 & \rho\sigma_\mathrm{a}\sigma_\mathrm{c} \\ \rho\sigma_\mathrm{a}\sigma_\mathrm{c} & \sigma_\mathrm{c}^2 \end{pmatrix}\right)$$

where N denotes the bivariate normal distribution, μ_a and μ_c are the fixed mean parameters and σ_a and σ_c are the variance parameters of the two random plot-specific parameters a_i and c_i, and ρ is the correlation between the two random parameters.

In the comparison of two subject-specific approaches to height modeling, Wang et al. (2008a) concluded that, for height growth description, the dummy variable method is preferred. In terms of height prediction, results showed that using single height measurements to recover single local parameters, the mixed predictor generally, but not always, performed better. When using multiple observations, there was no substantial difference in prediction between the least squares (LS) predictor and the mixed predictor, and when using the LS predictor both the dummy variable method and the mixed model method produced nearly identical height prediction. However, using single height measurements and recovering several local parameters through the mixed predictor did not perform better in terms of height prediction than recovering single local parameters.

Wang et al. (2008a) further concluded that with respect to height prediction given few measurements, the mixed model method may be more appropriate because it takes advantage of more information obtained from model building (i.e., that of local parameters and error parameter(s)). In this regard, the authors emphasized that the usefulness of mixed models in height modeling results largely from the fact that it provides a new local parameter predictor; proper use of mixed models for height modeling should include both "mixed" estimation of the common parameters and "mixed" prediction of local parameters.

7.11 Including Concomitant Information in Height-Age Models

In many instances developing site curves that exhibit varying asymptotes and/or shapes for differing levels of site index is sufficient. However, there is a plethora of genetic, edaphic and climatic variables that influence height development and, to improve predictive ability, it may be necessary to include such factors in site index models. Typically, in the past, differences in growing conditions have been taken into account by including categorical variables such as physiographic regions, soils groups, or habitat types in height growth models.

As an example, Burkhart and Tennent (1977a) divided data from a country-wide set of permanent plot installations in radiata pine in New Zealand into regions. Where differences warranted, a further division was made into soils groups. Eight region/soils groups were recognized and separate coefficients were computed for each group. The two-parameter Eq. 7.6 was fitted, resulting in the estimation of 16 parameters. The resultant site index curves are polymorphic, having different shapes by site index classes within each group and different shapes among the eight groups for land of the same site index. Figure 7.5 illustrates the variation in curve shape between the volcanic plateau and the Canterbury Plain of New Zealand for site index 24 m. Adding concomitant information accommodates nondisjoint site curves (i.e., curves of the same site index having differing values at ages other than index age).

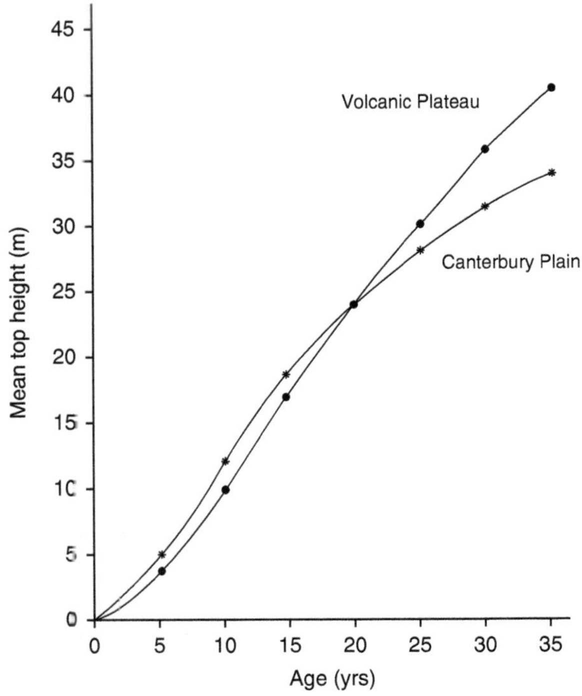

Fig. 7.5 Site index 24 m curve for volcanic plateau and Canterbury Plain in New Zealand (From Burkhart and Tennent 1977a)

Monserud (1984) found that forest habitat type was a useful concomitant variable when developing site index curves for inland Douglas-fir in the USA. Three habitat types (based on overstory and understory vegetation composition) proved important in determining the shape of the height growth and site index curves. In a follow-up comparison of Douglas-fir site index and height growth curves in the Pacific Northwest region of the USA, Monserud (1985) noted that differences in height growth pattern increased with increasing distance between sample areas. The trend in height growth divergence along an east-west gradient was attributed to the multiplicity of factors, both environmental and genetic, that vary across the region.

Amateis et al. (2006) used data from permanent plots established in stands planted throughout the native range of loblolly pine in the southeastern United States to model differences in dominant height development by physiographic region and geographic locale. Three physiographic regions (Atlantic Coastal Plain, Piedmont, and Gulf Coastal Plain) were recognized and geographic locale was represented by the longitude and latitude of each sample location. A significant reduction in the error sum of squares resulted from inclusion of both physiographic region and geographic coordinates. These results indicate that longitude and latitude, which are continuous variables that are easily obtained, may serve as useful surrogates for biologic, edaphic, and climatic factors that are difficult to measure, and, thus, provide more general predictive ability across broad areas.

In the case of naturally regenerated stands, and in particular with shade tolerant species, variability in early suppression of regeneration can have a profound effect on subsequent height development. Stage (1963) included the number of annual rings in a 1.5-in. (3.8 cm) radius from the pith at breast height as a variable when determining the appropriate height-age curves for grand fir in the Inland Empire region of the United States. Grand fir, a shade tolerant species with a high growth potential, responds well to release and the inclusion of diameter increment information improved ability to estimate height growth and site index.

Bravo-Oviedo et al. (2007) developed a polymorphic base-age invariant model with multiple asymptotes for Mediterranean maritime pine in the inland part of the Iberian Peninsula using a generalized algebraic formulation of Hossfeld's function. The selected model was used in a regional-based comparison of growth patterns. The parameter identified as 'global' in the first part of the study was expanded with dummy variables to assess the differences among regions. In follow-up studies (Bravo-Oviedo et al. 2008; Nunes et al. 2011), the applicability of height-age models across regions was improved by expressing parameters as a function of climate and/or soils variables.

7.12 Effect of Stand Density on Height Growth

When applying the site index method, an assumption that stand density does not affect height development of the dominant portion of the stand is generally invoked. While it is recognized that stand density affects average stand height, the evidence regarding effects on the dominant portion of the canopy is mixed. Stand density has been reported to affect dominant height growth of conifers in a number of studies (such as, Alexander et al. 1967, for lodgepole pine; Bennett 1975, for slash pine; Curtis and Reukema 1970, for Douglas-fir; Harms and Lloyd 1981, for loblolly pine; and Boyer 1983, for longleaf pine). However, there are instances in which no significant differences in dominant height due to stand density were detected, including the results of Pienaar and Shiver (1984) for slash pine and Harms et al. (1994, 2000) for two spacing trials with loblolly pine.

Stand density is confounded with site factors and it is difficult to assess its impact on height growth without designed experiments that are measured over long time periods. Antón-Fernández et al. (2011) analyzed density effects on height growth using rotation-age data from a loblolly pine spacing trial (intermediate results were reported by MacFarlane et al. 2000, and by Sharma et al. 2002b, c). These spacing trials were established at four locations (two on Atlantic Coastal Plain sites, two on Piedmont sites in the USA) with three replicates at each location and 16 plots per replicate. Initial planting densities ranged from 750 to 6,730 trees per hectare. Annual measurements from ages 1–25 were taken.

The Chapman-Richards equation was adopted as the base model to describe height development. Nonlinear mixed-effects modeling techniques were applied with random effects at the location, block, and plot levels. Average stand height as well as four different definitions of dominant height were modeled. In all instances, there were significant impacts of stand density on height development with the differences among initial planting densities being evident from age 6 and consistent to the end of the 25-year period of study. Height-age models that take into account the effect of spacing on average and dominant height were developed from the spacing trial data.

References

Aertsen W, Kint V, van Orshoven J, zkan K, Muys B (2010) Comparison and ranking of different modelling techniques for prediction of site index in Mediterranean mountain forests. Ecol Mod 221:1119–1130

Alexander RR, Tackle D, Dahms WG (1967) Site indexes for lodgepole pine, with corrections for stand density: methodology. USDA Forest Service, Rocky Mountain Forest and Range Experiment Station, Fort Collins, Research Paper RM-29

Amaro A, Reed D, Tomé M, Themido I (1998) Modeling dominant height growth: Eucalyptus plantations in Portugal. For Sci 44:37–46

Amateis RL, Burkhart HE (1985) Site index curves for loblolly pine plantations on cutover site-prepared lands. South J Appl For 9:166–169

Amateis RL, McDill ME (1989) Developing growth and yield models using dimensional analysis. For Sci 35:329–337

Amateis RL, Prisley SP, Burkhart HE, Liu J (2006) The effect of physiographic region and geographic locale on predicting the dominant height and basal area of loblolly pine plantations. South J Appl For 30:147–153

Antón-Fernández C, Burkhart HE, Strub MR, Amateis RL (2011) Effects of initial spacing on height development of loblolly pine. For Sci 57:201–211

Bailey RL, Cieszewski CJ (2000) Development of a well-behaved site-index equation: jack pine in north-central Ontario: comment. Can J For Res 30:1667–1668

Bailey RL, Clutter JL (1974) Base-age invariant polymorphic site curves. For Sci 20:155–159

Batho A, Garcia O (2006) De Perthuis and the origins of site index: a historical note. FBMIS 1:1–10

Beaulieu J, Raulier F, Prégent G, Bousquet J (2011) Predicting site index from climatic, edaphic, and stand structural properties for seven plantation-grown conifer species in Quebec. Can J For Res 41:682–693

Beck DE (1971) Height-growth patterns and site index of white pine in the Southern Appalachians. For Sci 17:252–260

Bennett FA (1975) Slash pine: some intriguing growth and yield relationships. USDA Forest Service, Southeastern Forest Experiment Station, Asheville, Research Paper SE-133

Biging GS (1985) Improved estimates of site index curves using a varying-parameter model. For Sci 31:248–259

Borders BE, Bailey RL, Ware KD (1984) Slash pine site index from a polymorphic model by joining (splining) nonpolynomial segments with an algebraic difference method. For Sci 30:411–423

Boyer WD (1983) Variations in height-over-age curves for young longleaf pine plantations. For Sci 29:15–27

Bravo-Oviedo A, Rio MD, Montero G (2004) Site index curves and growth model for Mediterranean maritime pine (*Pinus pinaster,* Ait.) in Spain. For Ecol Manage 201:187–197

Bravo-Oviedo A, Rio MD, Montero G (2007) Geographic variation and parameter assessment in generalized algebraic difference site index modeling. For Ecol Manage 247:107–119

Bravo-Oviedo A, Tomé M, Bravo F, Montero G, Rio MD (2008) Dominant height growth equations including site attributes in the generalized algebraic difference approach. Can J For Res 38:2348–2358

Brewer JA, Burns PY, Cao QV (1985) Short-term projection accuracy of five asymptotic height-age curves for loblolly pine. For Sci 31:414–418

Brickell JE (1968) A method of constructing site index curves from measurements of three age and height – its application to Inland Douglas-Fir. USDA Forest Service, Intermountain Forest and Range Experiment Station, Ogden, Research Paper INT-47

Brisco D, Klinka K (2002) Height growth models for western larch in British Columbia. West J Appl For 17:66–74

Bruce D (1981) Consistent height-growth and growth-rate estimates for remeasured plots. For Sci 27:711–725

Burkhart HE, Tennent RB (1977a) Site index equations for radiata pine in New Zealand. N Z J For Sci 7:408–416

Burkhart HE, Tennent RB (1977b) Site index equations for Douglas fir in Kaingaroa forest. N Z J For Sci 7:417–419

Calegario N, Daniels RF, Maestri R, Neiva R (2005) Modeling dominant height growth based on nonlinear mixed-effects model: a clonal *Eucalyptus* plantation case study. For Ecol Manage 204:11–20

Cao QV (1993) Estimating coefficients of base-age-invariant site index equations. Can J For Res 23:2343–2347

Carmean WH (1972) Site index curves for upland oaks in the Central States. For Sci 18:109–120

Carmean WH (1975) Forest site quality evaluation in the United States. Adv Agron 27:209–269

Carmean WH, Lenthall DJ (1989) Height-growth and site-index curves for jack pine in north central Ontario. Can J For Res 19:215–224

Carvalho JP, Parresol BR (2005) A site model for Pyrenean oak (*Quercus pyrenaica)* stands using a dynamic algebraic difference equation. Can J For Res 35:93–99

Cieszewski CJ (2000) Analytical site index solution for the generalized log-logistic height equation. For Sci 46:291–296

Cieszewski CJ (2001) Three methods of deriving advanced dynamic site equations demonstrated on inland Douglas-fir site curves. Can J For Res 31:165–173

Cieszewski CJ (2002) Comparing fixed- and variable-base-age site equations having single versus multiple asymptotes. For Sci 48:7–23

Cieszewski CJ (2003) Developing a well-behaved dynamic site equation using a modified Hossfeld IV function $Y^3 = (ax^m)/(c + x^{m-1})$, a simplified mixed model and scant subalpine fir data. For Sci 49:539–554

Cieszewski CJ (2004) GADA derivation of dynamic site equations with polymorphism and variable asymptotes from Richards, Weibull, and other exponential functions. University of Georgia, PMRC-TR 2000-5

Cieszewski CJ, Bailey RL (2000) Generalized algebraic difference approach: theory based derivation of dynamic site equations with polymorphism and variable asymptotes. For Sci 46:116–126

Cieszewski CJ, Bella IE (1989) Polymorphic height and site index curves for lodgepole pine in Alberta. Can J For Res 19:1151–1160

Cieszewski CJ, Strub M (2008) Generalized algebraic difference approach derivation of dynamic site equations with polymorphism and variable asymptotes from exponential and logarithmic functions. For Sci 54:303–315

Cieszewski CJ, Harrison M, Martin SW (2000) Practical methods for estimating non-biased parameters in self-referencing growth and yield models. University of Georgia, PMRC-TR 2000-7

Cieszewski CJ, Zasada M, Strub M (2006) Analysis of different base models and methods of site model derivation for Scots pine. For Sci 52:187–197

Clutter JL, Jones Jr EP (1980) Prediction of growth after thinning in old-field slash pine plantations. USDA Forest Service, Southeastern Forest Experiment Station, Asheville, Research Paper SE-217

Clutter JL, Lenhart JD (1968) Site index curves for old-field loblolly pine plantations in the Georgia Piedmont. Georgia Forest Research Council Report 22-Series 1

Corral RJ, Álvarez GJ, Ruiz González AD, Gadow Kv (2004) Compatible height and site index models for five pine species in El Salto, Durango (Mexico). For Ecol Manage 201:145–160

Curtis RO (1964) A stem-analysis approach to site-index curves. For Sci 10:241–256

Curtis RO, Reukema DL (1970) Crown development and site estimates in a Douglas-fir plantation spacing test. For Sci 16:287–301

Curtis RO, DeMars DJ, Herman FR (1974) Which dependent variable in site-index-height-age regressions? For Sci 20:74–87

Dahms WG (1963) Correction for a possible bias in developing site index curves from sectioned tree data. J For 61:25–27

Devan JD, Burkhart HE (1982) Polymorphic site index equations for loblolly pine based on a segmented polynomial differential model. For Sci 28:544–555

Diéguez-Aranda U, Burkhart HE, Soalleiro RR (2005a) Modeling dominant height growth of radiata pine (*Pinus radiata* D. Don) plantations in north-western Spain. For Ecol Manage 215:271–284

Diéguez-Aranda U, Gonzalez JGA, Anta MB, Alboreca AR (2005b) Site quality equations for *Pinus sylvestris* L. plantations in Galicia (northwestern Spain). Ann For Sci 62:143–152

Diéguez-Aranda U, Burkhart HE, Amateis RL (2006) Dynamic site model for loblolly pine (*Pinus taeda* L.) plantations in the United States. For Sci 52:262–272

Dyer ME, Bailey RL (1987) A test of six methods for estimating true heights from stem analysis data. For Sci 33:3–13

Ek AR (1971) A formula for white spruce site index curves. University of Wisconsin, Forest Research Notes. No. 161

Elfving B, Kiviste A (1997) Construction of site index equations for *Pinus sylvestris* L. using permanent plot data in Sweden. For Ecol Manage 98:125–134

Eriksson H, Johansson U, Kiviste A (1997) A site-index model for pure and mixed stands of *Betula pendula* and *Betula pubescens* in Sweden. Scand J For Res 12:149–156

Fabbio G, Frattegiani M, Manetti MC (1994) Height estimation in stem analysis using second differences. For Sci 40:329–340
Fang Z, Bailey RL (2001) Nonlinear mixed effects modeling for slash pine dominant height growth following intensive silvicultural treatments. For Sci 47:287–300
Farrelly N, Ní Dhubháin Á, Nieuwenhuis M (2011) Site index of Sitka spruce (*Picea sitchensis*) in relation to different measures of site quality in Ireland. Can J For Res 41:265–278
Feng Z, Stadt KJ, Lieffers VJ, Huang S (2006) Linking juvenile growth of white spruce with site index. For Chron 82:819–824
Fontes L, Tomé M, Coelho MB, Wright H, Luis JS, Savil P (2003) Modelling dominant height growth of Douglas-fir (*Pseudotsuga menziesii* (Mirb.) Franco) in Portugal. Forestry 76: 509–523
Fortin M, Daigle G, Chhun-Huor U, Bégin J, Archambault L (2007) A variance-covariance structure to take into account repeated measurements and heteroscedasticity in growth modeling. Eur J For Res 126:573–585
Furnival GM, Gregoire TG, Valentine HT (1990) An analysis of three methods for fitting site-index curves. For Sci 36:464–469
García O (1983) A stochastic differential equation model for the height growth of forest stands. Biometrics 39:1059–1072
García O (1998) Estimating top height with variable plot size. Can J For Res 28:1509–1517
García O (1999) Height growth of *Pinus radiata* in New Zealand. N Z J For Sci 29:131–145
García O (2005) Comparing and combining stem analysis and permanent sample plot data in site index models. For Sci 51:277–283
García O (2011) Dynamical implications of the variability representation in site-index modelling. Eur J For Res 130:671–675
García O, Batho A (2005) Top height estimation in lodgepole pine sample plots. West J Appl For 20:64–68
García O, Ruiz F (2003) A growth model for eucalypt in Galicia. Spain For Ecol Manage 173: 49–62
Goelz JCJ, Burk TE (1992) Development of a well-behaved site index equation: jack pine in north central Ontario. Can J For Res 22:776–784
Goelz JCG, Burk TE (1996) Measurement error causes bias in site index equations. Can J For Res 26:1585–1593
Golden MS, Meldahl R, Knowe SA, Boyer WD (1981) Predicting site index for old-field loblolly pine plantations. South J Appl For 5:109–114
Graney DL, Burkhart HE (1973) Polymorphic site index curves for shortleaf pine in the Ouachita Mountains. USDA Forest Service, Southern Forest Experiment Station, New Orleans, Research Paper SO-85
Gregoire TG, Schabenberger O, Barrett JP (1995) Linear modeling of irregularly spaced, unbalanced, longitudinal data from permanent-plot measurements. Can J For Res 25:137–156
Hägglund B (1981) Evaluation of forest site productivity. For Abstr 42:515–527
Hall DB, Bailey RL (2001) Modeling and prediction of forest growth variables based on multilevel nonlinear mixed models. For Sci 47:311–321
Harms WR, Lloyd FT (1981) Stand structure and yield relationships in a 20-year-old loblolly pine spacing study. South J Appl For 5:162–166
Harms WR, DeBell DS, Whitesell CD (1994) Stand and tree characteristics and stockability in *Pinus taeda* plantations in Hawaii and South Carolina. Can J For Res 24:511–521
Harms WR, Whitesell CD, DeBell DS (2000) Growth and development of loblolly pine in a spacing trial planted in Hawaii. For Ecol Manage 126:13–24
Heger L (1973) Effect of index age on the precision of site index. Can J For Res 3:1–6
Hu Z, García O (2010) A height-growth and site-index model for interior spruce in the sub-boreal spruce biogeoclimatic zone of British Columbia. Can J For Res 40:1175–1183
Johansson T (2011) Site index curves for poplar growing on former farmland in Sweden. Scand J For Res 26:161–170

Jones EA, Reed DD (1991) Improved site index curves for young red pine plantations in the Northern Lake States. North J Appl For 8 59–63

Kariuki M (2002) Height estimation in complete stem analysis using annual radial growth measurements. Forestry 75:63–74

King JE (1966) Site index curves for Douglas-fir in the Pacific Northwest. Weyerhaeuser Forestry Paper No. 8

Lappi J, Bailey RL (1988) A height prediction model with random stand and tree parameters: an alternative to traditional site index methods. For Sci 34:907–927

Lappi J, Malinen J (1994) Random-parameter height/age models when stand parameters and stand age are correlated. For Sci 40:715–731

Lauer DK, Kush JS (2010) Dynamic site index equation for thinned stands of even-aged natural longleaf pine. South J Appl For 34:28–37

Lenhart JD (1972) An alternative procedure for improving height/age data from stem analysis. For Sci 18:332

Lundgren AL, Dolid WA (1970) Biological growth functions describe published site index curves for Lake States timber species. USDA Forest Service, North Central Forest Experiment Station, St. Paul, Research Paper NC-36

MacFarlane DW, Green EJ, Burkhart HE (2000) Population density influences assessment and application of site index. Can J For Res 30:1472–1475

Magnussen S (1999) Effect of plot size on estimates of top height in Douglas-fir. West J Appl For 14:17–27

Magnussen S, Penner M (1996) Recovering time trends in dominant height from stem analysis. Can J For Res 26:9–22

McDill ME, Amateis RL (1992) Measuring forest site quality using the parameters of a dimensionally compatible height growth function. For Sci 38:409–429

Meng SX, Huang S (2009) Improved calibration of non-linear mixed-effects models demonstrated on a height growth function. For Sci 55:238–248

Meng SX, Huang S (2010) Incorporating correlated error structures into mixed forest growth models: prediction and inference implications. Can J For Res 40:977–990

Meng SX, Huang S, Yang Y, Trincado G, VanderSchaaf CL (2009) Evaluation of population-averaged and subject-specific approaches for modeling the dominant or codominant height of lodgepole pine trees. Can J For Res 39 1148–1158

Milner KS (1992) Site index and height growth curves for ponderosa pine, western larch, lodgepole pine, and Douglas-fir in western Montana. West J Appl For 7:9–14

Monserud RA (1984) Height growth and site index curves for inland Douglas-fir based on stem analysis data and forest habitat type. For Sci 30:943–965

Monserud RA (1985) Comparison of Douglas-fir site index and height growth curves in the Pacific Northwest. Can J For Res 15:673–679

Newberry JD (1991) A note on Carmean's estimate of height from stem analysis data. For Sci 37:368–369

Newnham RM (1988) A modification of the Ek-Payandeh nonlinear regression model for site index curves. Can J For Res 18:115–120

Nigh GD (1995) The geometric mean regression line: a method for developing site index conversion equations for species in mixed stands. For Sci 41:84–98

Nord-Larsen T (2006) Developing dynamic site index curves for European beech (*Fagus sylvatica* L.) in Denmark. For Sci 52:173–181

Northway SM (1985) Fitting site index equations and other self-referencing functions. For Sci 31:233–235

Nothdurft A, Kublin E, Lappi J (2006) A non-linear hierarchical mixed model to describe tree height growth. Eur J For Res 125:281–289

Nunes L, Patrício M, Tomé J, Tomé M (2011) Modeling dominant height growth of maritime pine in Portugal using GADA methodology with parameters depending on soil and climate variables. Ann For Sci 68:311–323

OMNR, Ontario Forest Research Institute (2009) An evaluation of site index models for young black spruce and jack pine plantations in a changing climate. Ontario Ministry of Natural Resources, Sault Ste Marie, pp 1–31
Palahí M, Tomé M, Pukkala T, Trasobares A, Montero G (2004) Site index model for *Pinus sylvestris* in north-east Spain. For Ecol Manage 187:35–47
Payandeh B, Wang Y (1994a) A site-index model remodified. Can J For Res 24:197–198
Payandeh B, Wang Y (1994b) Relative accuracy of a new base-age invariant site index model. For Sci 40:341–348
Payandeh B, Wang Y (1995) Preliminary site index equations for three planted species in Northern Ontario. North J Appl For 12:57–63
Pienaar LV, Shiver BD (1984) The effect of planting density on dominant height in unthinned slash pine plantations. For Sci 30:1059–1066
Raulier F, Lambert M-C, Pothier D, Ung C-H (2003) Impact of dominant tree dynamics on site index curves. For Ecol Manage 184:65–78
Rennolls K (1978) "Top height"; its definition and estimation. Commonw For Rev 57:215–219
Rennolls K (1995) Forest height growth modelling. For Ecol Manage 71:217–225
Richards FJ (1959) A flexible growth function for empirical use. J Exp Bot 10:290–300
Rivas JJC, González JGA, González ADR, Gadow Kv (2004) Compatible height and site index models for five pine species in El Salto, Durango (Mexico). For Ecol Manage 201:145–160
Salas C, García O (2006) Modelling height development of mature *Nothofagus obliqua*. For Ecol Manage 229:1–6
Schumacher FX (1939) A new growth curve and its application to timber-yield studies. J For 37:819–820
Sharma M, Amateis RL, Burkhart HE (2002a) Top height definition and its effect on site index determination in thinned and unthinned loblolly pine plantations. For Ecol Manage 168: 163–175
Sharma M, Burkhart HE, Amateis RL (2002b) Modeling the effect of density on the growth of loblolly pine trees. South J Appl For 26:124–133
Sharma M, Burkhart HE, Amateis RL (2002c) Spacing rectangularity effect on the growth of loblolly pine plantations. Can J For Res 32:1451–1459
Sharma M, Smith M, Burkhart HE, Amateis RL (2006) Modeling the impact of thinning on height development of dominant and codominant loblolly pine trees. Ann For Sci 63:349–354
Skovsgaard JP, Vanclay JK (2008) Forest site productivity: a review of the evolution of dendrometric concepts for even-aged stands. Forestry 81:13–31
Stage AR (1963) A mathematical approach to polymorphic site index curves for grand fir. For Sci 9:167–180
Strand L (1964) Numerical constructions of site-index curves. For Sci 10:410–414
Strub M, Cieszewski CJ (2006) Base-age invariance properties of two techniques for estimating the parameters of site index models. For Sci 52:182–186
Subedi N, Sharma M (2010) Evaluating height-age determination methods for jack pine and black spruce plantations using stem analysis data. North J Appl For 27:50–55
Thrower JS, Goudie JW (1992) Estimating dominant height and site index for even-aged interior Douglas-fir in British Columbia. West J Appl For 7:20–25
Trousdell KB, Beck DE, Lloyd FT (1974) Site index for loblolly pine in the Atlantic Coastal plain of the Carolinas and Virginia. USDA Forest Service, Southeastern Forest Experiment Station, Asheville, Research Paper SE-115
Upadhyay A, Eid T, Sankhayan PL (2005) Construction of site index equations of even aged stands of *Tectona grandis* (teak) from permanent plot data in India. For Ecol Manage 212:14–22
Vanclay JK (1992) Assessing site productivity in tropical moist forests: a review. For Ecol Manage 54:257–287
Walters DK, Gregoire TG, Burkhart HE (1989) Consistent estimation of site index curves fitted to temporary plot data. Biometrics 45:23–33
Walters DK, Burkhart HE, Reynolds MR Jr, Gregoire TG (1991) A Kalman filter approach to localizing height-age functions. For Sci 37:1526–1537

Wang Y, Payandeh B (1994a) A bi-segmental curve fitting approach to improve the accuracy of site index equations. For Ecol Manage 67:35–38

Wang Y, Payandeh B (1994b) Application of the Kalman filter model in site index equation construction. Can J For Res 24:1415–1418

Wang Y, Payandeh B (1995) A base-age invariant site index model for aspen stands in north central Ontario. For Ecol Manage 72:207–211

Wang Y, Huang S, Yang RC, Tang S (2004) Error-in-variable method to estimate parameters for reciprocal base-age invariant site index models. Can J For Res 34:1929–1937

Wang M, Borders B, Zhao D (2007) Parameter estimation of base-age invariant site index models: which data structure to use? For Sci 53:541–555

Wang M, Borders B, Zhao D (2008a) An empirical comparison of two subject-specific approaches to dominant heights modeling: the dummy variable method and the mixed model method. For Ecol Manage 255:2659–2669

Wang M, Rennolls K, Borders B (2008b) Base-age invariant site index models for a generalized algebraic parameter prediction approach. For Sci 54:625–632

Weiskittel AR, Hann DW, Hibbs DE, Tzeng YL, Bluhm AA (2009) Modeling top height of red alder plantations. For Ecol Manage 258:323–331

Yang Y, Huang S (2011) Estimating a multilevel dominant height-age model from nested data with generalized errors. For Sci 57:102–116

Zeide B (1999) Pattern of height growth for southern pine species. For Ecol Manage 118:183–196

Zeide B, Zakrzewski WT (1993) Selection of site trees: the combined method and its application. Can J For Res 23:1019–1025

Chapter 8
Quantifying Stand Density

8.1 Stocking and Stand Density

Forest stand dynamics are a function of the species present, the site quality, the degree to which the site is occupied, stand age in the case of even-aged stands or elapsed time from a specified initial condition for uneven-aged stands, and management treatments. Quantification of site occupancy (or stand density) is central to developing reliable models for predicting forest growth and yield. Response to silvicultural treatments and the amount and size-class distribution of volume produced over time are closely related to competition for site resources (light, water, nutrients); stand density measures that are simple and direct to determine yet highly correlated with tree and stand growth and mortality are needed. Because of the importance of quantitatively describing site occupancy and competition for growth resources, much attention has been devoted to the problem of developing stand density measures, but it remains a rather elusive and vexing component of forest growth and yield forecasting.

Although stocking and stand density are terms that are often applied interchangeably in forestry use, the two terms are not synonymous. *Stand density* denotes a quantitative measurement of the stand, whereas *stocking* refers to the adequacy of a given stand density to meet some management objective (Bickford et al. 1957). Accordingly, stands may be referred to as understocked, fully stocked, or overstocked. A stand that is "overstocked" for one management objective could be "understocked" for another.

Stand density is a quantitative term describing the degree of stem crowding within a stocked area; it can be expressed in absolute or relative terms. Absolute measures of density are determined directly from a given stand without reference to any other stand. For example, number of trees per hectare is an absolute measure that expresses the density of trees on an area basis. Relative density is based on a selected standard density. If, for instance, "fully stocked" is defined on a basal-area basis, the ratio of the measured basal area in a stand to that of the fully stocked

H.E. Burkhart and M. Tomé, *Modeling Forest Trees and Stands*,
DOI 10.1007/978-90-481-3170-9_8,

standard is a relative measure of stand density. The problem of what constitutes full stocking makes application of relative density measures difficult (Avery and Burkhart 2002).

The extent of competition in a stand is determined by the number of trees per unit area, their respective sizes and their spatial distribution. Spatial distribution is generally not explicitly considered in measures of stand density. For practical purposes of forest management and growth and yield prediction, measures of stand density should be objective, easily measured, and highly correlated with stand growth, yield and mortality.

Quantitative expressions of the degree of crowding and competition for site resources fall into two broad categories: (1) measures of average stand density, and (2) measures of point density (crowding at a particular location or for individual subject trees; point density measures, which are commonly designated competition indices, are the subject of Chap. 9). Some stand density measures can be computed directly from measurements in the stand of interest while others require reference to previously-determined relationships.

8.1.1 Trees Per Unit Area

Number of trees per unit area is a simple, easy to measure numeric expression for average stand density that does not rely on reference to other stands, agreed upon standards, or previously determined relationships. When employing fixed-area sampling units in the field, only a frequency count of stems is needed to determine trees per hectare. Although no information on tree sizes or spatial distribution of stems is included, trees per hectare, along with stand age and a measure of site quality, has been found highly useful and is commonly employed as a measure of stand density for deriving silvicultural prescriptions and predicting growth for planted stands of a single species.

8.1.2 Basal Area Per Unit Area

Basal area is the sum of the cross-sectional area at *dbh* of all trees or a specified portion of trees in the stand, expressed on a per unit area basis (m^2ha^{-1}). It, like trees per unit area, is simple, easily measured, and does not require information other than that from the stand of interest. When using point sampling methodology, basal area per hectare can be determined in the field with a simple frequency tally that does not require any tree measurements. Size information is incorporated as part of the basal area measure; basal area per hectare has been widely used in deriving silvicultural prescriptions and it is highly correlated with stand growth, yield and mortality when used in conjunction with stand age and a measure of site quality. Together basal area and trees per hectare specify average tree size. Hence both measures are sometimes combined to obtain an improved quantification of average stand density for input into planning silvicultural treatments and projecting stand growth and yield.

8.2 Size-Density Relationships

Forest biometricians have long devoted, and continue to devote, a great deal of attention to the problem of developing simple and effective indices of competition in forest stands. As individual trees grow in size their demands on site resources and growing space increase. When resources are no longer adequate to support additional growth of all of the trees present, self-thinning will be initiated and the number of trees per unit area will decrease. Quantifying self-thinning relationships is important for prescribing silvicultural treatments such as thinning and for predicting forest stand development. Consequently, a number of indices have been developed to study the influence of density on self-thinning; the indices combine an expression of the size of the average tree (diameter, height, volume, or biomass) with the number of trees per unit area. The best known and most commonly employed of these self-thinning or maximum size-density relationships are those of Reineke (1933), Yoda et al. (1963), and Hart (1926). Reineke's stand density index is based on the relationship between numbers of trees per unit area and the quadratic mean diameter of stands, whereas Yoda et al. related mean plant volume (or biomass) to numbers per unit area. Hart proposed a measure based on the average distance between trees and the average height of the dominant canopy.[1] Each of these size-density measures will be explained in more detail in the following sections and relationships among them will be shown.

8.2.1 Reineke's Stand Density Index

In fully stocked even-aged stands the limiting relationship between the number of trees per hectare (N) and the quadratic mean *dbh* ($\bar{d}_g$) is generally linear on the log-log scale (Fig. 8.1). For any given $\bar{d}_g$ there is a limit to the number of trees per unit that can be carried. Reineke (1933) noted that for a variety of species the slope of the limiting line was approximately −1.6 on the log-log scale, that is[2]:

$$\log N = -1.6 \log \bar{d}_g + \mathrm{k} \tag{8.1}$$

where k is a constant varying by species and log indicates logarithm.

---

[1]The concept of using height in a numerical expression of stand density was largely brought to the attention of foresters in the USA through an article by Wilson (1946).

[2]In his original paper Reineke (1933) used a value of −1.605 for the common slope; in our discussion the slope value has been rounded to −1.6. Also, in some treatments of Reineke's SDI an index diameter of 25.4 cm is used. We have used 25 cm for simplicity.

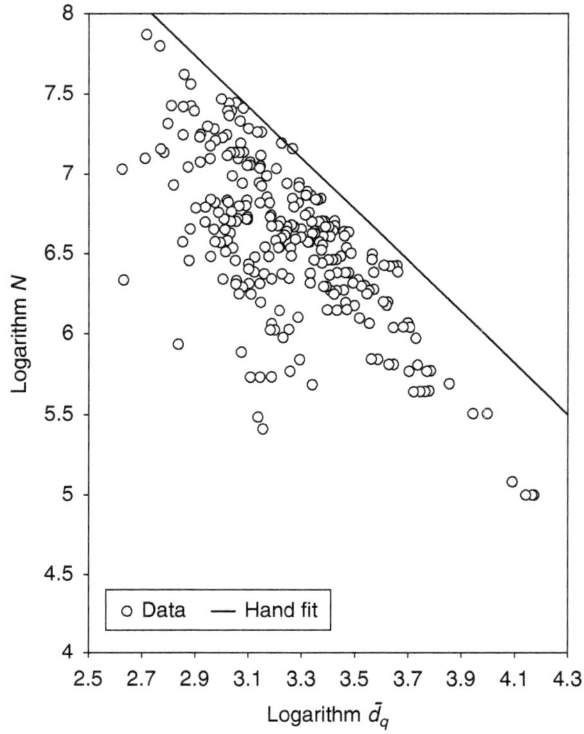

Fig. 8.1 The size density relationship for even-aged eastern white pine by hand-fitting a line with slope equal to −1.6 (Adapted from Zhang et al. 2005)

Reineke defined the limiting number of trees when $\bar{d}_g$ was equal to 10 in. to be the stand density index (*SDI*). In metric units the index $\bar{d}_g$ value is generally taken to be 25 cm. Noting that the limiting relationship can be written as

$$N = a\bar{d}_g^{\mathrm{b}}$$

and that stands at the limiting density with a quadratic mean diameter of 25 cm define stand density index, that is:

$$SDI = a25^{\mathrm{b}}$$

Thus, for any stand of known N and $\bar{d}_g$, the stand density index can be computed as

$$SDI = N(25/\bar{d}_g)^{\mathrm{b}} \tag{8.2}$$

where b is often taken to be −1.6.

8.2.2 3/2 Rule of Self-thinning

The so-called 3/2 rule of self-thinning, like Reineke's stand-density index, is based on the concept of a mean size-density relationship. In the case of the 3/2 rule of self-thinning the logarithm of mean tree volume or weight is plotted against the logarithm of the number of trees per unit area. For pure, even-aged stands that are sufficiently crowded such that competition-induced mortality ("self-thinning") is occurring the slope of the line of logarithm of mean volume (or weight) versus logarithm of trees per unit area has been found to be approximately −3/2, but the intercept varies by species. That is,

$$\log \bar{v} = -3/2 \log N + \mathrm{a}$$

where

$\bar{v}$ = mean tree volume
a = a constant varying with species

Obviously, the 3/2 rule of self-thinning is closely related to the stand-density index – in fact, the two can be shown to be mathematically equivalent. Reineke's stand-density index was developed with log N on the left-hand side of the Eq. 8.1. Rearranging the 3/2 relationship with log N on the left-hand side gives

$$-3/2 \log N = \log \bar{v} - \mathrm{a} \tag{8.3}$$

Assuming that mean tree volume is proportional to the diameter of the tree of average basal area raised to the power of 2.4 (Bredenkamp and Burkhart 1990), that is,

$$\bar{v} = \mathrm{c}\,\bar{d}_g^{2.4}$$

where c is a constant and substituting the definition for $\bar{v}$ into Eq. 8.3 and simplifying one obtains

$$\log N = -1.6 \log \bar{d}_g + \mathrm{k}$$

which is the stand-density index reference line.

Although the two concepts are, for the stated assumptions, mathematically equivalent, due to measurement considerations and historical precedent, the stand density index has been applied widely in forestry, whereas the 3/2 rule of self-thinning has been utilized prevalently by plant ecologists.

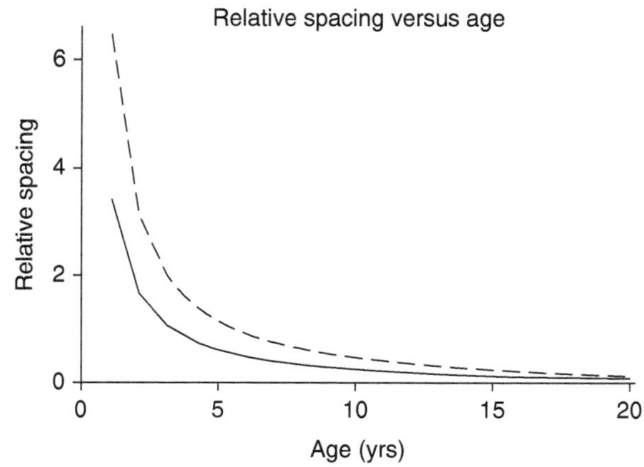

Fig. 8.2 Relative spacing trends for the 1.8 × 1.8 m (*solid line*) and the 3.6 × 3.6 m (*dashed line*) plots in a loblolly pine spacing trial (Sharma et al. 2002 present a full description of the spacing trial)

8.2.3 Relative Spacing

The average distance between trees divided by the average height of the dominant canopy has been termed *relative spacing*. Assuming square spacing, the average distance between trees can be computed as the square root of the number of m^2 per ha (10,000) divided by the number of trees per ha. This average distance between trees in m is then divided by the average height of the dominant canopy in m to compute relative spacing as:

$$R_S = \frac{\sqrt{10{,}000/N}}{h_{dom}} \tag{8.4}$$

where h_{dom} = average height of the dominant canopy, m

For even-aged stands, relative spacing initially drops rapidly; then it levels off at a lower limit (Fig. 8.2). After reaching the lower limit, R_S will increase somewhat if the stand is carried to an advanced age. The lower limit of relative spacing is fairly constant for a given species regardless of the site quality and the initial density.

Although it may not be immediately obvious, relative spacing is closely related to stand-density index. Height has been found to be proportional to diameter raised to the power of 0.8 (Curtis 1970), that is,

$$h = \mathrm{a}\, d^{0.8}$$

Assuming that the height of the dominant canopy (h_{dom}) can be related to the quadratic mean diameter ($\bar{d}_g$) by the preceding relationship, h_{dom} in the relative spacing formula can be replaced by $\mathrm{a}\, d^{0.8}$, giving

$$R_S = \frac{\sqrt{\mathrm{b}/N}}{\mathrm{a}\, \bar{d}_g^{0.8}}$$

where b is an appropriate constant (10,000). Combining the constants b and a into a single constant denoted c, squaring both sides, taking the logarithm of both sides and rearranging terms gives

$$\log N = \log c - 2\log R_S - 1.6\log \bar{d}_g$$

If R_S is at or near its lower limit, it can be assumed constant, and the terms $\log c - 2\log R_S$ can be set equal to a constant called k, giving

$$\log N = -1.6\log \bar{d}_g + \mathrm{k}$$

which is the stand-density index reference line.

8.3 Methods for Fitting Maximum Size-Density Relationships

8.3.1 Data Screening

The maximum size-density relationship that results from statistical analysis depends on how the data are screened to determine which observations are at the maximum and what fitting procedure is applied. There is no established statistical procedure for selecting appropriate plots to include (Bi and Turvey 1997). Consequently a visual inspection of the data is often employed to determine which observations to include. The extent to which the slope of the maximum size density line is affected by the data screening step depends, to some extent, on the fitting algorithm employed. For instance when applying ordinary least squares, the slope of the self-thinning line can be greatly affected by inclusion of plots that diverge from the stage of maximum size for a given number per unit area.

8.3.2 Free Hand Fitting

Reineke (1933) noted that when using data from fully stocked stands, plotting numbers of trees per unit area over quadratic mean diameter on double logarithmic cross-section paper resulted in a straight-line relationship. For many species the slope of the logarithmic straight-line graph was constant, but the level of the line varied by species. Consequently he proposed the general relationship

$$\log N = -1.6\log \bar{d}_g + \mathrm{k}$$

In which the constant k varies by species. Plotting data and then positioning a line with a fixed slope by hand in such a way that all, or essentially all (with the exception of "outliers"), points are below the line is quite straight forward (as in Fig. 8.1). However, as computing capability and statistical methodology advanced so did interest in deriving more objective ways of estimating size-density relationships. The vast majority of this work has focused on the stand density index or the −3/2 self-thinning concepts. A summary of the primary methods proposed for statistically estimating maximum size-density relationships follows.

8.3.3 Reduced Major Axis Regression

Leduc (1987) pointed out that there are two primary reasons for estimating an equation to describe the trend of a set of bivariate data: (i) the estimation of conditional means of one variable for given values of the other, and (ii) the description of the functional relationship between two variables. While least-squares regression is generally regarded as best for the first purpose, an alternative fitting procedure may be preferred when interpretation of functional relationships is the main objective.

Leduc describes a number of techniques (Bartlett's three-group methods, Schnute's trend line, the general structural relationship, major axis regression, and reduced major axis regression) that might be used when the primary interest is in the values of the equation parameters themselves. After comparing the relative merits of six methods, he concluded that reduced major axis regression is most applicable because of its desirable properties and ease of estimation.

Solomon and Zhang (2002) established maximum size-density relationships for three mixed-softwood forest types (hemlock-red spruce, spruce-fir, and cedar-black spruce) in the northeastern USA. Plots with relative density index (number of trees per unit area present divided by the maximum number for the mean tree volume of a given plot) greater than 0.7 were selected for model development. Reduced major axis regression was used to fit the coefficients of the self-thinning lines.

8.3.4 Frontier Functions

Frontier functions, which originated in economics, are aimed at estimating maximum possible values for given inputs. In all frontier production functions the slope parameter β can be consistently estimated by ordinary least squares (OLS); the intercept parameter α can be consistently estimated in a frontier function by shifting the OLS line upward so that the largest residual is zero thus, the corrected ordinary least squares (COLS) estimation of the intercept is:

$$a_{COLS} = a + \max \varepsilon_i$$

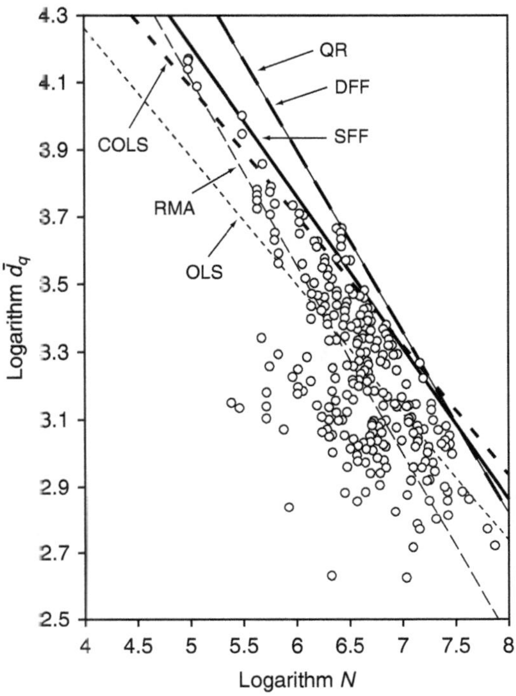

Fig. 8.3 Maximum size-density lines obtained from six modeling methods. *COLS* corrected ordinary least squares, *RMA* reduced major axis, *OLS* ordinary least squares, *QR* quantile regression, *DFF* deterministic frontier function, *SFF* stochastic frontier function (From Zhang et al. 2005)

In a deterministic frontier model, the output is bounded from above by a deterministic function. By contrast, a stochastic frontier model specifies that maximum output is determined both by the production function and by random factors. Zhang et al. (2005) used data from an even-aged white pine stand to show that quantile regression (QR), deterministic frontier function (DFF) and stochastic frontier function (SFF) methods can potentially produce an upper limiting boundary line above all plot values for the maximum size-density relationship without subjectively selecting a subset of data points based on predetermined criteria. On the other hand, Zhang et al. observed that ordinary least squares (OLS), corrected ordinary least squares (COLS), and reduced major axis (RMA) methods are sensitive to the data selected for model fitting and may produce self-thinning lines with inappropriate slopes. Statistical inference is difficult with the DFF and QR methods, and Zhang et al. recommended the SFF approach due to the satisfactory results achieved and the ease of producing statistics for inference on the model coefficients. A comparison of results produced by the six methods applied to the eastern white pine data is displayed in Fig. 8.3.

Bi (2001) developed a self-thinning surface that defines a density-dependent upper frontier of stand biomass over a gradient of site productivity for a given species. The equation was formulated for parameter estimation as a stochastic frontier function with two error components that have different distributional properties. Weiskittel et al. (2009) applied stochastic frontier functions to test the influence of additional covariates when fitting self-thinning boundary lines. Likelihood ratio tests indicated that site index, stand origin (natural versus planted), and composition (proportion of basal area in the primary species) significantly influenced the species self-thinning boundary line intercept for all of the species examined (Douglas-fir, western hemlock, and red alder). In Douglas-fir and western hemlock stands the slope of the self-thinning boundary was also dependent on stand origin; in addition site index was related to the slope of the boundary in the case of Douglas-fir.

8.3.5 Mixed Models

VanderSchaaf and Burkhart (2007a) fitted maximum size-density relationships (MSDR) for a range of planting densities using data from a loblolly pine spacing trial. Three methods for estimating the slope of the MSDR species boundary line were compared: ordinary least squares, first-difference approach, and linear mixed-effects model. In the linear mixed effects model formulation the intercept (b_0) and the slope (b_1) terms were assumed to contain random effects; that is:

$$\ln N = (b_0 + u_{0i}) + (b_1 + u_{1i}) \ln \bar{d}_g + \varepsilon \tag{8.5}$$

where u_{0i}, u_{1i} are cluster-specific random effects to be predicted and assumed to be $N(0, \sigma_0^2)$ and $N(0, \sigma_1^2)$, respectively. A cluster is an individual plot (indexed by i). ε is the random error where it is assumed $\varepsilon \sim N(0, \sigma^2 \mathrm{I})$.

The three methods of parameter estimation were compared on the basis of stability of parameter estimates, where stability refers to the extent to which parameter estimates do not change when the range of planting densities in the fitting data set vary. Results from fitting with data from a spacing trial consisting of planting densities ranging from 6,727 to 747 seedlings per ha showed that mixed-effects models produced the most stable estimates of the slope while OLS resulted in the least stable.

When developing self-thinning models for even-aged stands of Scots pine, Norway spruce, and birch, Hynynen (1993) used mixed-effects analysis to estimate the MSDR species boundary line slopes.

8.3.6 Curvilinear Size-Density Boundaries

In a detailed analysis of the 3/2 power rule of self-thinning, Zeide (1987) asserted that the limiting line does not necessarily have a constant slope of −3/2 on the log-log scale, or any other constant slope. Rather his analysis led to the conclusion that the self-thinning line is a curve concave down.

Cao et al. (2000) used data from direct-seeded slash pine stands to develop a model to describe the trajectory of maximum quadratic mean diameter after

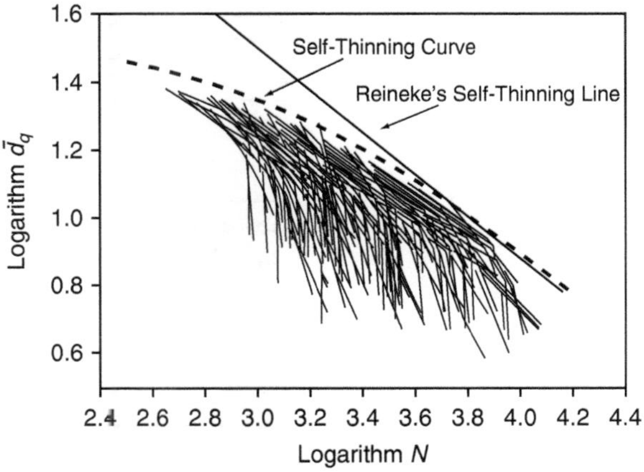

Fig. 8.4 Relationship between quadratic mean diameter at breast height and stand density in direct-seeded slash pine stands. The *self-thinning curve* is from Eq. 8.6 (From Cao et al. 2000)

self-thinning begins when the trajectory departs from Reineke's size-density line. Noting that Reineke's relationship between quadratic mean diameter and tree density can be written as

$$\bar{d}_g = \mathrm{b}_1 N^{\mathrm{b}_2}$$

where b_2 was taken as $1/(-1.605) = -0.623$ Cao et al. proposed a new relationship

$$d_{\max} = \bar{d}_g \left[1 - \exp(\mathrm{b}_3 N^{b_4})\right]$$

or

$$d_{\max} = b_1 N^{-0.623} \left[1 - \exp(\mathrm{b}_3 N^{\mathrm{b}_4})\right] \tag{8.6}$$

where $d_{\max}$ is the maximum stand diameter at tree density N. The values of the expression within the square brackets in Eq. 8.6 ranges from 0 to 1 for negative values of b_3. The result of estimating the self-thinning curve (8.6) using data from direct-seeded slash pine stands is shown in Fig. 8.4.

8.3.7 Segmented Regression

Size-density trajectories on the logarithmic scale are generally considered to consist of two major stages. The first stage, often referred to as density-independent, is represented by precanopy conditions. Within the self-thinning stage segments of a size-density trajectory can be represented by a non-linear approach to a linear portion, a linear portion, and a divergence from the linear (Fig. 8.5). Fully-stocked stands are thought to be in phase II (the linear portion of density-dependent mortality on a log-log scale) of this overall size-density trajectory and most of the past effort

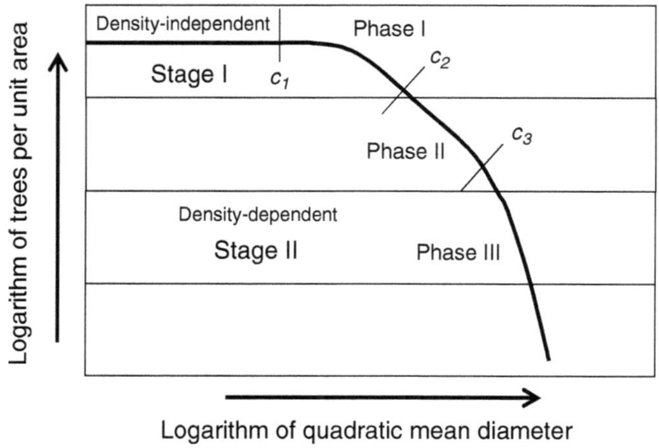

Fig. 8.5 Depiction of a size-density trajectory for an individual stand. Two stages of stand development are shown: density-independent mortality and density-dependent mortality. Within the density-dependent mortality stage three phases of stand development are shown. The join points (c_1, c_2, and c_3,) used in Eq. 8.7 to differentiate stages and phases of stand development in size-density trajectories are indicated (From VanderSchaaf and Burkhart 2008)

to model maximum size-density relationships has focused on this phase. Long-term remeasurement data have, however, provided an opportunity to model the size-density trajectory over the full range of stages and phases of stand development.

VanderSchaaf and Burkhart (2008) used data from a loblolly pine spacing trial to demonstrate the use of segmented regression techniques for estimating stages and phases of self-thinning. A segmented regression model to represent the two stages of stand development and the three phases of self-thinning illustrated in Fig. 8.5 can be written as

$$\ln N = (\mathrm{b}_1)J_1 + \left[\mathrm{b}_1 + \mathrm{b}_2\left(\ln \bar{d}_g - \mathrm{c}_1\right)^2\right] J_2 + \left[\mathrm{b}_1 + \mathrm{b}_2(\mathrm{c}_2 - \mathrm{c}_1)^2 + \mathrm{b}_3\left(\ln \bar{d}_g - \mathrm{c}_2\right)\right] J_3$$
$$+ \left[\mathrm{b}_1 + \mathrm{b}_2(\mathrm{c}_2 - \mathrm{c}_1)^2 + \mathrm{b}_3(\mathrm{c}_3 - \mathrm{c}_2) + \mathrm{b}_4 \ln\left(\bar{d}_g - \mathrm{c}_3\right)\right] J_4 \tag{8.7}$$

where $\bar{d}_g$ is the quadratic mean diameter (cm), J_1, J_2, J_3 and J_4 are indicator variables for the stages and phases of stand development, $J_1 = 1$ if $\ln \bar{d}_g$ is within the density-independent mortality stage of stand development (Stage I in Fig. 8.5) and 0 otherwise, $J_2 = 1$ if $\ln \bar{d}_g$ is within the curved approach to the MSDR dynamic thinning line phase of self-thinning (Phase I of Stage II in Fig. 8.5) and 0 otherwise, $J_3 = 1$ if $\ln \bar{d}_g$ is within the MSDR dynamic thinning line phase of self-thinning (Phase II of Stage II in Fig. 8.5) and 0 otherwise, and $J_4 = 1$ if $\ln \bar{d}_g$ is within the divergence phase of self-thinning (Phase III of Stage II in Fig. 8.5) and 0 otherwise and other variables are as defined previously.

Estimates of $\ln \bar{d}_g$ and $\ln N$ where the linear segment begins and ends were obtained from the segmented regression analyses and used as response variables predicted as a function of planting density. Predicted values of $\ln \bar{d}_g$ and $\ln N$ can then be used to develop self-thinning line trajectories for varying initial planting densities.

Cao and Dean (2008) used data from direct-seeded slash pine stands in the southeastern USA when fitting segmented regression functions to describe the trajectory of tree density and quadratic mean diameter through time for individual stands. Two-segment and three-segment models were fitted; the authors concluded that, for their data, a two segment model was sufficient. Because the overall trend lines from Cao and Dean are the result of two quadratic curves, they are curvilinear throughout.

8.4 Applying Maximum Size-Density Concepts to Complex Stand Structures

Stand density index has been widely applied to assess and contrast stand structure and growing stock in single-species, even-aged stands. The usefulness of *SDI*, which is based on quadratic mean diameter and number of trees in a subject stand, for uneven-aged and multi-species stands has been questioned because of the need to partition the index into additive components to describe the relative stocking of a stand by species or size classes. Stage (1968) showed that *SDI* can be partitioned to allow for the contribution of various classes of trees in the stand towards the total index value. The summation method involves calculating *SDI* for each tree (or by diameter class) and then summing for the total stand value as

$$SDI = \sum N_i \left(\frac{D_i}{25} \right)^{1.6} \tag{8.8}$$

Where D_i is the midpoint of the ith diameter class (cm) and N_i is the number of trees per hectare in the diameter class. In the case of even-aged monocultures, where the diameter distribution is typically unimodal and fairly symmetric (i.e. is approximately normal), the application of Eq. 8.2 with overall quadratic mean diameter and number of trees will result in an overall value of *SDI* that is very close to that computed by the summation methods given in (8.8). However, for uneven-aged stands with a large number of small trees and an inverse shaped diameter distribution, then the total *SDI* value can be considerably different when computed by Eqs. 8.2 and 8.8.

A number of authors (Long and Daniel 1990; Long 1996; Shaw 2000) have advocated computing *SDI* for uneven-aged stands by applying Eq. 8.8 in order to avoid the bias that use of quadratic mean diameter alone introduces for highly skewed diameter distributions and to provide a way for partitioning the growing stock among the various size classes. Woodall et al. (2003) found that the *SDI* summation method was biased for apportionment of relative density across diameter classes in uneven-aged ponderosa pine stands because a greater relative density was assigned to small trees than to larger ones. Thus, *SDI* by summation may over-predict site occupancy for reverse *J*-shaped diameter distributions with relatively large numbers of small trees and it may underpredict occupancy for stands with nonreverse *J*-shaped diameter distributions.

Using data from even- and uneven-aged mixed species stands in northern Idaho and northwestern Montana, USA, Sterba and Monserud (1993) showed that the slope of Reineke's maximum density line depended on the skewness of the $dbh^{1.5}$ distribution, which in turn was correlated with structural stand characteristics like "unevenagedness" and species mixture. In an effort to improve application of *SDI* in uneven-aged mixed species stands, Woodall et al. (2005) examined the relationship between a stand mean specific gravity of component trees and maximum *SDI*. Results indicated that the maximum *SDI* that any particular species can attain is affected to varying degrees by species composition of subject stands. A strong relationship was found between the mean specific gravity of all trees in a stand and the 99th percentile of the observed distribution of stand *SDI* values by classes of mean stand specific gravity. A model was proposed whereby the mean specific gravity of individual trees in a stand may serve as a predictor of a stand's maximum stocking potential, regardless of the stand's diameter distribution and species composition.

Although differences between *SDI* calculated using an "average" stand diameter and by summation of individual trees or diameter classes are generally not large, a considerable amount of attention has been given to a "best" method of computation for complex stand structures (Long and Daniel 1990; Long 1996; Shaw 2000). Ducey and Larson (2003) have, however, argued that additivity may not always be required and that theory alone cannot determine if one index is superior to another. They opined that direct tests of the predictive ability of stand density indices are required to provide concrete guidance on what density index is best for a given species and management context.

One disadvantage of stand density indices is that they are more difficult to determine in the field than simpler measures such as trees and basal area per hectare. However, Ducey and Valentine (2008) derived a direct method of establishing *SDI* using point sampling methodology.

8.5 Incorporating Size-Density Relationships in Models of Stand Dynamics

Size-density or self-thinning relationships have been used in various ways when developing models of forest stand dynamics, growth and yield. Some modelers (e.g. Smith and Hann 1984, 1986; Lloyd and Harms 1986; Tang et al. 1994) have used the self-thinning boundary as a central component that governs stand dynamics, while others (e.g. Yang and Titus 2002; Monserud et al. 2005) have imposed self-thinning constraints on a specific component, namely on the mortality function for growth and yield prediction. Turnblom and Burk (2000) and Pittman and Turnblom (2003) formulated the mean size-density relationship as a coupled system in order to allow for mean size to affect mortality and mortality to affect mean size.

Smith and Hann (1984) developed a model of maximum mean biomass for a given number of stems per unit area. In order to incorporate the third dimension of time they subsequently (Smith and Hann 1986) modeled numbers of stems as a function of time. Thus, the self-thinning analytical model was converted into a growth model for self-thinning stands.

The growth model of Lloyd and Harms (1986) for pure, even-aged stands is asymptotically bounded above by the self-thinning rule that relates maximum plant size to plant density. Their model characterizes accretion in mean size as a deviation from the limiting size, through a function relating mean size to time and density and a companion survival model. The growth model is obtained by substituting the survival model for density in the mean size relationship.

A stand growth model was developed by Tang et al. (1994) by combining the self-thinning model with a basal-area increment model. Their stand growth model can be used to estimate the average diameter and stand density at any given stand age with any initial stand density.

Yang and Titus (2002) used a maximum size-density relationship as a constraint on an existing individual tree mortality model in a stand simulator for boreal mixedwood species in Alberta, Canada. For a given stand, its quadratic mean diameter and density can be calculated at any point during growth projection. If the combination of quadratic mean diameter and density exceeds the maximum size-density line tree mortality is accelerated to maintain the stand on or below the maximum size-density boundary.

Monserud et al. (2005) examined the question of whether or not a tree-specific mortality model can elicit expected forest density dynamics without imposing stand-level constraints such as Reineke's maximum stand density index (SDI_{max}) or the −3/2 rule of self-thinning. This question was investigated using a stand simulator that does not use stand density constraints to determine individual tree mortality rates. Initial conditions were obtained from research plots that were established in young pure stands of Norway spruce and Scots pine in Austria. From the results for Norway spruce they concluded that stand-level density constraints are not necessary to obtain Reineke's maximum size-density relations. Results from simulation of Scots pine also displayed reasonable and stable stand dynamics, except that they greatly exceeded Reineke's maximum stand density index determined empirically from the literature. This result argues for stand-level constraints (such as specifying SDI_{max}) to ensure that the appropriate intercept for the maximum density line is applied. Thus, the authors were left with ambiguous results. First, that a density-dependent individual-tree mortality model, developed on an adequate data set, is sufficient for the desired stand-level behavior of Reineke to be exhibited. Second, that stand-level constraints on SDI_{max} need to be imposed if the underlying mortality modeling database is not adequate.

Turnblom and Burk (2000), studied the density-mean size trajectory using simultaneous differential equations. Growth and mortality were interrelated and bounded by the maximum density-mean tree size line. Using this modeling framework they found that initial stand density in red pine plantations in the Lake States (Minnesota, Michigan, and Wisconsin) had a large impact on the level of the self-thinning boundary. Site quality (as measured by site index) chiefly affected the rate at which stand dynamics progressed. Pittman and Turnblom (2003) extended the two-dimensional system of Turnblom and Burk (2000) by the inclusion of a measure of vertical growth, h_{dom} (height of the dominant stand). After calibrating this model with data from Douglas-fir permanent sample plots located throughout western Washington and western Oregon, USA, and southwest British Columbia, Canada, they compared model predictions quantitatively with data and then qualitatively with the −3/2 power rule. Results indicated that allometric relationships are interdependent and the authors concluded that it is desirable to model the mean size-density relationship using a coupled dynamical system rather than with uncoupled algebraic equations.

8.6 Other Proposed Measures of Stand Density

Numerous measures of stand density have been proposed. The similarity of SDI, the −3/2 self-thinning rule, and relative spacing was shown in Sect. 8.2. Before proceeding with categorizing the various measures and showing analogies among them, two additional measures will be described, namely the tree-area ratio of Chisman and Schumacher (1940) and the crown competition factor of Krajicek et al. (1961).

8.6.1 Tree-Area Ratio

As described by Clutter et al. (1983), tree-area ratio is a measure of stand density based on the assumption that the land area A occupied by any given tree in a stand can be represented by the equation

$$A = \beta_0 + \beta_1 d + \beta_2 d^2 \tag{8.9}$$

where d is tree *dbh*. The total area occupied by n trees on a unit area of land is then

$$\sum_{j=1}^{n} A_j = \beta_0 n + \beta_1 \sum_{j=1}^{n} d_j + \beta_2 \sum_{j=1}^{n} d_j^2 \tag{8.10}$$

where the summation is made over the n trees growing on a specified area (generally on an acre or a ha). Suppose N sample plots of a given size were established in a population of interest, and on each plot the values $n_i, \sum_{j=1}^{n_i} d_{ij}$ and $\sum_{j=1}^{n_i} d_{ij}^2$ ($i = 1,2, \ldots ,N$) were obtained, where d_{ij} is the *dbh* of the jth tree on the ith sample plot and n_i is the number of trees on sample plot i. Estimates of the parameters β_0, β_1, and β_2 can be obtained by minimizing

$$\sum_{j=1}^{n} \left(1 - \hat{\beta}_0 n_i - \hat{\beta}_1 \sum_{j=1}^{n_j} d_{ij} - \hat{\beta}_2 \sum_{j=1}^{n_j} d_{ij}^2 \right)^2 \tag{8.11}$$

These least squares estimates can then be used to evaluate Eq. 8.10 for any given stand with observed $n, \sum_{j=1}^{n} d_j$ and $\sum_{j=1}^{n} d_j^2$. The computed tree-area ratio ($\sum_{j=1}^{n} A_j$) is a measure of stand density relative to the average relationship in the original sample, which is typically restricted to "fully stocked" stands. While the tree-area is not restricted to be less than 1.0, it generally will be except for stands of exceptionally high density.

Tree-area ratio is a measure that relies on a predetermined relationship. It has been applied in both even and uneven-aged stands. However, it has not been used as extensively in growth and yield prediction as simpler measures such as basal area and number of trees per unit area.

8.6.2 Crown Competition Factor

Developed by Krajicek et al. (1961), crown competition factor (*CCF*) reflects the area available to the average tree in a stand in relation to the maximum area it could use if it were open-grown. To compute *CCF* values, the crown-width/*dbh* relationship for open-grown trees of the species of interest must be established. Generally, a simple linear regression of the form

$$c_w = \mathrm{b}_0 + \mathrm{b}_1 d \tag{8.12}$$

suffices to establish this relationship where c_w is crown width in m and d is diameter at breast height in cm.

Assuming that the crowns of open-grown trees are circular in shape the maximum crown area (*MCA*), expressed as percent of a ha, that can be occupied by the crown of a tree with a specific bole diameter is computed as

$$MCA = \frac{\pi (c_w)^2 (100)}{(4)(10{,}000)} = 0.0078 (c_w)^2 \tag{8.13}$$

CCF for a plot or stand is computed from a stand table by summing the *MCA* values for each diameter class and dividing by the area in ha. In formula form, the expression for *CCF* is

$$CCF = \frac{1}{\mathrm{a}}\left[\mathrm{a}'\sum n_i + \mathrm{b}'\sum D_i n_i + \mathrm{c}'\sum D_i^2 n_i\right] \tag{8.14}$$

where

a = plot or stand size, ha
n_i = number of trees in i^{th} *dbh* class
D_i = midpoint of ith *dbh* class, cm, and
a′, b′, c′ are constants that vary depending on the value of the regression coefficient
(b_0 and b_1) in (8.12).

A predetermined relationship between crown width and *dbh* of open-grown trees is required to apply the *CCF* measure of stand density. Although not as widely used as some of the other measures of stand density, it has been found to be well correlated with growth and yield for a number of species.

8.7 Similarity of Stand Density Measures

Curtis (1970) observed that many stand density measures can be regarded as expressions of average area available to trees in a given stand, relative to that occupied by trees growing under a standard density condition and comparable in *dbh* or another measure of size. Either the open-grown condition (as for *CCF*) or the "fully stocked" stand (as for tree-area ratio and a number of other measures) can be used as the standard and lead to similar results. Differences are introduced by use of stand diameter, height, or volume as alternative bases for relating observed stands to a standard condition.

Although the various measures of stand density often seem to be regarded as distinct and separate entities, Curtis (1970) provided a unifying view of a number of common relative measures of stand density as expressions of the same basic relationship, which differ mainly in details of algebraic form and methods of estimation of the constants. He concluded that "most common measures appear to be practically equivalent."

In a subsequent paper, Curtis (1971) derived equations for a number of alternative relative stand density measures for Douglas-fir. A simple sum of diameters, without reference to other stand characteristics (i.e., $\sum d_i^{\mathrm{b}}$, where b is approximately 3/2), provided an index of density closely related to tree-area ratio. This sum was also closely related to measures based on ratios of observed numbers of trees or basal area to that expected for a given stand diameter.

West (1983) compared stand density measures in an even-aged eucalypt forest in Tasmania, Australia. Most measures for even-aged forests provide estimates of the degree of approach of a stand to a "maximum density" condition. West (1983) computed 17 different density measures for stands of eucalypts and presented the various measures in mathematically equivalent forms so that the relationships among them can be easily seen. The 17 density measures were found to fall into four groups: (i) stand basal area, (ii) other measures based on sums of tree diameters, (iii) measures based on sums of tree volumes, and (iv) measures based on sums of tree height. The measures within each group did not seem to differ appreciably; of particular note was the lack of difference among the diameter-based measures. Many of the more complex measures differed little from their simpler counterparts in the way they represented stand density and the author concluded that little gain seems to have been achieved with developing the more complex expressions of stand density.

8.8 Efficacy of Various Stand Density Measures for Growth and Yield Prediction

Stand density measures are used to guide silvicultural prescriptions and to predict stand dynamics, growth and yield. Choosing an expression for stand density generally depends on the application at hand, the types of stands of interest, and the preferences of the individual analyst. When the purpose is to include a measure of stand density for predicting growth and yield a choice can be made on the basis of selecting a measure that is highly correlated with the growth and yield response variables of interest. More precisely, it is the multiple correlation between the growth and yield variables of interest and stand age, site quality, and stand density that is of primary interest when modeling even-aged stands. One might, for instance, select the stand density measure that results in the highest R^2 value for a fitted growth or yield relationship for even-aged stands that includes stand age, site quality, and density as predictor variables.

Nelson and Brender (1963) tested total stand basal area, SDI, and percent full stocking measures in conjunction with age and site index for ability to account for variation in growth of loblolly pine stands via the following equation form

$$\Delta V_m = \mathrm{b}_0 + \mathrm{b}_1 t^{-1} + \mathrm{b}_2 S + \mathrm{b}_3 SD_i^2 + \mathrm{b}_4 (S)(SD_i) + \mathrm{b}_5 S^2 \qquad (8.15)$$

where

ΔV_m = merchantable periodic net annual growth in cubic volume per unit area
t = stand age
S = site index
SD_i = stand density using the ith measure at the beginning of the growth period

Basal area, *SDI*, and percent full stocking were successively substituted for SD_i in Eq. 8.15 and the R^2 value of each fitted equation was noted. The R^2 values for fitting Eq. 8.15 using the three density measures with age and site index were very similar, ranging from 0.77 to 0.78; Nelson and Brender recommended use of basal area per unit area because of the ease of computation and because it is a direct measure that does not rely on having or establishing any prior standards or relationships.

In a similar study Larson and Minor (1968) evaluated nine measures of stand density for predicting 10-year basal area growth of ponderosa pine in the southwestern USA. Multiple correlation coefficients were best when including stocking percent, stand basal area, or crown competition factor, with values being essentially identical for these three measures.

Allen and Duzan (1981) included basal area, MacKinney and Chaiken's (1935) modification of *SDI*, percent full stocking, and number of trees in three growth models to determine which measure performed best for estimating gross growth in untreated loblolly pine plantations. The coefficients of determination for the simple liner regression

$$\Delta V = \mathrm{b}_0 + \mathrm{b}_1 SD_i \quad (8.16)$$

where

ΔV = 5-year growth in cubic volume per unit area
SD_i = the ith stand density measure

were 44.6%, 45.0%, 45.0% and 16.8%, respectively, for basal area, modified SDI, percent full stocking, and number of trees per unit area. When site index was included along with one of the measures of stand density, coefficients of determination increased to 65.7%, 67.0%, 67.7% and 34.1% for basal area, modified SDI, percent full stocking, and number of trees, respectively. From those results, Allen and Duzan recommended that basal area be used when predicting growth in untreated stands because it is easy to determine and effective in predicting growth.

A generally-accepted model for yield estimation in even-aged stands (Chap. 11) was fitted with four different measures of stand density (total basal area per ha, *CCF*, SDI, trees per ha) to evaluate ability to predict yield of natural and planted stands of loblolly pine (Burkhart et al. 1982):

$$\ln V = \mathrm{b}_0 + \mathrm{b}_1 t^{-1} + \mathrm{b}_2 S + \mathrm{b}_3 \ln SD_i \quad (8.17)$$

where

V = total cubic volume per ha for all trees
t = stand age
S = site index
SD_i = the ith measure of stand density

There was little difference in the coefficients of determination among *SDI*, *CCF*, and basal area (0.970, 0.984, 0.992, respectively) for fitting (8.17) to data from planted stands, but all three were superior to using trees per ha ($R^2 = 0.937$). For plot data from natural stands the R^2 values were 0.974 for *SDI* and *CCF*, 0.990 for basal area and 0.872 for trees per unit area. Thus, there was little difference among *SDI*, *CCF*, and basal area, but basal area was best and all were superior to trees per ha, especially for natural stands.

Results from comparing the efficacy of various stand density measures for reducing the error sum of square after inclusion of age and site index in growth and yield relationships have consistently shown that basal area per unit area is fully as effective as more complex measures. These results are not surprising, given the near equivalency of many commonly employed stand density measures. Making such comparisons, however, presents a dilemma. In past applications the "base" model has been held constant and different measures are successively substituted for the stand density term(s). This process avoids confounding model specification and density measure, but, due to the different algebraic constructs of stand density measures, it is possible that different models should be specified, depending on the stand density measure to be included in the empirical fitting.

As an example, consider the yield function (8.17). If the measure of stand density (SD_i) is basal area, the coefficient b_3 is expected to be near 1.0. Setting $b_3 = 1.0$ and taking the antilogarithm of both sides gives:

$$V = e^{b_0} e^{b_1 t^{-1}} e^{b_2 S} G \qquad (8.18)$$

where *V* is cubic volume and *G* is basal area per unit area. Given that specifying age (t) and site index (S) is equivalent to specifying height of the dominant stand (h_{dom}) and setting $e^{b_0} = F$, the stand form factor, Eq. 8.18 can be written:

$$V = (F)(h_{dom})(G) \qquad (8.19)$$

which is, by definition, an expression for total stand cubic volume. While the result that basal area as the measure of stand density results in the best fit of (8.17) is expected, it should be noted that basal area has consistently produced good results for a variety of growth and yield prediction objectives. These empirical results, coupled with the objective definition of basal area per ha and the ease of its measurement, make basal area a highly attractive measure of density for stand modeling purposes.

In highly uniform stands – such as planted monocultures – number of trees per unit area is quite adequate for growth and yield estimation. Trees per ha is sometimes combined with basal area per ha in order to obtain a more complete assessment of site occupancy and stand structure.

In some of the literature on stand density measurement, the desirability of developing measures that can be applied without reference to stand age or site index is expressed. While assessment of density for stands that are at particular points of development (e.g. actively self-thinning) may be possible without reference to age or site quality, many forest management situations require that site occupancy be determined throughout all stages of stand development from establishment to crown closure to self-thinning and beyond. In these situations stand density must be modeled as a function of species, stand age, and site index and often silvicultural treatments as well. These circumstances require that site occupancy be regarded as dynamic and that stand density functions be incorporated as a component in a system of inter-related equations for modeling stand development through time. In subsequent chapters which describe stand-level models for even-aged structures, we will deal primarily with basal area and stems per hectare as measures of site occupancy and we will consider these measures to be dynamic with regard to age and site index.

8.9 Evaluation of Concepts Underlying Stand Density Measures

While empirical correlations between growth and yield values and a number of stand density expressions do not vary greatly, that does not imply that all are equally desirable or sound for modeling purposes. Having sound supporting biological and ecological rationale for measures of stand density provides generality in their application and use. In a series of papers, Zeide (1985, 1987, 1991, 1995, 2002, 2005, 2010) analyzed self-thinning and stand density using reasoning and empirical evidence. A brief summary of the arguments and conclusions presented in Zeide's papers follows.

The simplest measure of density is number per unit area, but to be useful in quantifying crowding or competition for space, tree size is an essential component of stand density measures. Number of trees is necessary but not sufficient to adequately describe stand density. A large number of stand density measures that include tree size have been advanced. Commonly-used measures of size are tree-stem volume, diameter, and height. Relying on knowledge of forest tree characteristics and stand development when comparing self-thinning relationships, Zeide (2010) concluded that measures based on diameter are to be preferred.

Basal area has the same components as Reineke's index (number of trees and stem diameter), but the power of diameter is 2 for basal area and 1.6 for the index. Empirical evidence indicates that numbers of trees in self-thinning stands is related to mean diameter expressed as a power less than 2.

Tree stem volume is a principal variable for many forestry purposes, but it is not the best representation of crowding (Zeide 2010). Crowding depends on the space trees occupy, which is closely related to crown, not stem, size. The size of crowns increases with stem diameter but, given trees of the same diameter, decreases with tree height within a stand. As elaborated by Zeide (2002):

> This means that taller trees have smaller crowns than shorter trees with the same diameter. On the other hand, stem volume increases with both stem diameter and height. This reasoning explains why stem diameter, the easiest tree dimension to measure, has a closer relationship with crown size than stem mass or volume.

Zeide (2010) noted that in dense stands with complete crown closure, the number of trees is inversely related to the square of average crown diameter. Therefore, he asserted, the variable most closely related to crown diameter will be the best predictor of self-thinning. Average height is not as highly correlated with crown width as average diameter, and stem volume, which involves both diameter and height, is intermediate in correlation.

Similar correlation of stand density measures in empirical fitting of growth and yield equations does not necessarily mean equivalent behavior when applied for a variety of purposes and across a spectrum of conditions. Based on reasoning advanced by Zeide and on empirical relationships between crown width and tree variables, one would expect measures using stem diameter to be "best" followed by those employing stem volume and measures involving height. However, it is important to note that quantifying stand density has been and remains one of the most vexing problems in forest modeling and a completely satisfactory measure has not yet been advanced.

References

Allen HL, Duzan HW (1981) What measure of stand density is best for growth predictions in loblolly pine plantations. In Barnett JP (ed) Proceedings of the first biennial southern silvicultural research conference. USDA Forest Service, Southern Forest Experiment Station, New Orleans, pp 175–178, General Technical Report SO-34

Avery TE, Burkhart HE (2002) Forest measurements, 5th edn. McGraw-Hill, New York

Bi H (2001) The self-thinning surface. For Sci 47:361–370

Bi H (2004) Stochastic frontier analysis of a classic self-thinning experiment. Aust Ecol 29: 408–417

Bi H, Turvey ND (1996) Competition in mixed stands of *Pinus radiata* and *Eucalyptus obliqua*. J Appl Ecol 33:87–99

Bi H, Turvey ND (1997) A method of selecting data points for fitting the maximum biomass-density line for stands undergoing self-thinning. Aust J Ecol 22:356–359

Bi H, Wan G, Turvey ND (2000) Estimating the self-thinning boundary line as a density-dependent stochastic biomass frontier. Ecology 81:1477–1483

Bickford CA, Baker F, Wilson FG (1957) Stocking, normality, and measurement of stand density. J For 55:99–104

Bredenkamp BV, Burkhart HE (1990) An examination of spacing indices for *Eucalyptus grandis*. Can J For Res 20:1909–1916

Buford MA (1989) Mean stem size and total volume development of various loblolly pine seed sources planted at one location. Can J For Res 19:396–400

Burkhart HE, Selph CE, Amateis RL (1982) Preliminary results from a comparison of stand density measurements for loblolly pine. Unpublished research report, 8pp

Cao QV, Dean TJ (2008) Using segmented regression to model the density-size relationship in direct-seeded slash pine stands. For Ecol Manage 255:948–952

Cao QV, Dean TJ, Baldwin VC (2000) Modeling the size-density relationship in direct-seeded slash pine stands. For Sci 46:317–321

Chisman HH, Schumacher FX (1940) On the tree-area ratio and certain of its applications. J For 38:311–317

Clutter JL, Fortson JC, Pienaar LV, Brister GH, Bailey RL (1983) Timber management: a quantitative approach. Wiley, New York

Curtis R (1970) Stand density measures: an interpretation. For Sci 16:403–414
Curtis RO (1971) A tree area power function and related stand density measures for Douglas-fir. For Sci 17:146–159
Curtis RO (1982) A simple index of stand density for Douglas-fir. For Sci 28:92–94
Curtis RO (2011) Effect of diameter limits and stand structure on relative density indices: a case study. West J Appl For 25:169–175
de Montigny L, Nigh G (2007) Density frontiers for even-aged Douglas-fir and western hemlock stands in Coastal British Columbia. For Sci 53:675–682
Dean TJ, Baldwin VC (1996a) Growth in loblolly pine plantations as a function of stand density and canopy properties. For Ecol Manage 82:49–58
Dean TJ, Baldwin VC (1996b) The relationship between Reineke's stand-density index and physical stem mechanics. For Ecol Manage 81:25–34
DeBell DS, Harms WR, Whitesell CD (1989) Stockability: a major factor in productivity differences between *Pinus taeda* plantations in Hawaii and the southeastern United States. For Sci 35:708–719
Drew TJ, Flewelling JW (1977) Some recent Japanese theories of yield-density relationships and their application to Monterey pine plantations. For Sci 23:517–534
Drew TJ, Flewelling JW (1979) Stand density management: an alternative approach and its application to Douglas-fir plantations. For Sci 25:518–532
Ducey MJ (2009) The ratio of additive and traditional stand density indices. West J Appl For 24: 5–10
Ducey MJ, Knapp RA (2010) A stand density index for complex mixed species forests in the northeastern United States. For Ecol Manage 260:1613–1622
Ducey MJ, Larson BC (1999) Accounting for bias and uncertainty in nonlinear stand density indices. For Sci 45:452–457
Ducey MJ, Larson BC (2003) Is there a correct stand density index? An alternative interpretation. West J Appl For 18:179–184
Ducey MJ, Valentine HT (2008) Direct sampling for stand density index. West J Appl For 23:78–82
Gingrich SF (1967) Measuring and evaluating stocking and stand density in upland hardwood forests in the Central States. For Sci 13:38–53
Harms WR, Whitesell CD, DeBell DS (2000) Growth and development of loblolly pine in a spacing trial planted in Hawaii. For Ecol Manage 126:13–24
Hart HMJ (1926) Stem density and thinning: pilot experiment to determine the best spacing and thinning method of teak. Proefsta. Boschwesen, Batavia, Meded. 21
Hasenauer H, Burkhart HE, Sterba H (1994) Variation in potential volume yield of loblolly pine plantations. For Sci 40:162–176
Honer TG (1972) A height-density concept and measure. Can J For Res 2:441–447
Hynynen J (1993) Self-thinning models for even-aged stands of *Pinus sylvestris*, *Picea abies* and *Betula pendula*. Scand J For Res 8:326–336
Jack SB, Long JN (1996) Linkages between silviculture and ecology: an analysis of density management diagrams. For Ecol Manage 86:205–220
Keim RF, Dean TJ, Chambers JL, Conner WH (2010) Stand density relationships in baldcypress. For Sci 56:336–343
Krajicek JE, Brinkman KA, Gingrich SF (1961) Crown competition – a measure of density. For Sci 7:35–42
Kurinobu S, Arisman H, Hardiyanto E, Miyaura T (2006) Growth model for predicting stand development of *Acacia mangium* in South Sumatra, Indonesia, using the reciprocal equation of size-density effect. For Ecol Manage 228:91–97
Larson FR, Minor CO (1968) A comparison of stand density measurements for ponderosa pine in the Southwest. Northern Arizona University, Arizona Forestry Notes, No 4
Le Goff N, Ottorini J-M, Ningre F (2011) Evaluation and comparison of size-density relationships for pure even-aged stands of ash (*Fraxinus excelsior* L.), beech (*Fagus silvatica* L.) oak (*Quercus petraea* Liebl.), and sycamore maple (*Acer pseudoplatanus* L.). Ann For Sci 68: 461–475

Leduc DJ (1987) A comparative analysis of the reduced major axis technique of fitting lines to bivariate data. Can J For Res 17:654–659

Lloyd FT, Harms WR (1986) An individual stand growth model for mean plant size based on the rule of self-thinning. Ann Bot 57:681–688

Long JN (1996) A technique for the control of stocking in two-storied stands. West J Appl For 11:59–61

Long JN, Daniel TW (1990) Assessment of growing stock in uneven-aged stands. West J Appl For 5:93–96

Long JN, Smith FW (1984) Relation between size and density in developing stands: a description and possible mechanisms. For Ecol Manage 7:191–206

Lynch TB, Wittwer RF, Stevenson DJ, Huebschmann MM (2007) A maximum size-density relationship between Lorey's mean height and trees per hectare. For Sci 53:478–485

MacKinney AL, Chaiken LE (1935) A method of determining density of loblolly pine stands. USDA Forest Service, Appalachian Forest Experiment Station, Asheville, Technical Note 15

Monserud RA, Ledermann T, Sterba H (2005) Are self-thinning constraints needed in a tree-specific mortality model? For Sci 50:848–858

Nelson TC, Brender EV (1963) Comparison of stand density measures for loblolly pine cubic-foot growth prediction. For Sci 9:8–14

Newton PF (1997) Stand density management diagrams: review of their development and utility in stand-level management planning. For Ecol Manage 98:251–265

Pittman SD, Turnblom EC (2003) A study of self-thinning using coupled allometric equations: implications for coastal Douglas-fir stand dynamics. Can J For Res 33:1661–1669

Poage NJ, Marshall DD, McClellan MH (2007) Maximum stand-density index of 40 western hemlock-Sitka spruce stands in southeast Alaska. West J Appl For 22:99–104

Pretzsch H, Biber P (2005) A re-evaluation of Reineke's rule and stand density index. For Sci 51:304–320

Puettmann KJ, Hann DW, Hibbs DE (1993) Evaluation of the size-density relationships for pure red alder and Douglas-fir stands. For Sci 39:7–27

Reineke L (1933) Perfecting a stand-density index for even-aged forests. J Agric Res 46:627–638

Río MD, Montero G, Bravo F (2001) Analysis of diameter–density relationships and self-thinning in non-thinned even-aged Scots pine stands. For Ecol Manage 142:79–87

Schütz J-P, Zingg A (2010) Improving estimations of maximal stand density by combining Reineke's size-density rule and the yield level, using the example of spruce (*Picea abies* (L.) Karst.) and European Beech (*Fagus sylvatica* L.). Ann For Sci 67:507p1–507p12

Sharma M, Burkhart HE, Amateis RL (2002) Modeling the effect of density on the growth of loblolly pine trees. South J Appl For 26:124–133

Shaw JD (2000) Application of stand density index to irregularly structured stands. West J Appl For 15:40–42

Smith NJ (1989) A stand-density control diagram for western red cedar, *Thuja plicata*. For Ecol Manage 27:235–244

Smith NJ, Hann DW (1984) A new analytical model based on the −3/2 power rule of self-thinning. Can J For Res 14:605–609

Smith NJ, Hann DW (1986) A growth model based on the self-thinning rule. Can J For Res 16: 330–334

Solomon DS, Zhang L (2002) Maximum size–density relationships for mixed softwoods in the northeastern USA. For Ecol Manage 155:153–170

Stage AR (1968) A tree-by-tree measure of site utilization for grand fir related to stand density index. USDA Forest Service, Intermountain and Forest Range Experiment Station, Ogden, Research Note INT-77

Stahelin R (1949) Thinning even-aged loblolly and slash pine stands to specified densities. J For 47:538–540

Sterba H (1987) Estimating potential density from thinning experiments and inventory data. For Sci 33:1022–1034

Sterba H, Monserud RA (1993) The maximum density concept applied to uneven-aged mixed-species stands. For Sci 39:432–452
Tang S, Meng CH, Meng F-R, Wang YH (1994) A growth and self-thinning model for pure even-age stands: theory and applications. For Ecol Manage 70:67–73
Tang S, Meng F-R, Meng CH (1995) The impact of initial stand density and site index on maximum stand density index and self-thinning index in a stand self-thinning model. For Ecol Manage 75:61–68
Turnblom EC, Burk TE (2000) Modeling self-thinning of unthinned Lake States red pine stands using nonlinear simultaneous differential equations. Can J For Res 30:1410–1418
Vanclay JK (2009) Tree diameter, height and stocking in even-aged forests. Ann For Sci 66: 702–702
Vanclay JK, Sands P (2009) Calibrating the self-thinning frontier. For Ecol Manage 259:81–85
Vanderschaaf CL (2010) Estimating individual stand size-density trajectories and a maximum size-density relationship species boundary line slope. For Sci 56:327–335
VanderSchaaf CL, Burkhart HE (2007a) Comparison of methods to estimate Reineke's maximum size-density relationship species boundary line slope. For Sci 53:435–442
VanderSchaaf CL, Burkhart HE (2007b) Relationship between maximum basal area carrying capacity and maximum size-density relationships. Mod Appl Sci 1(4):3–6
VanderSchaaf CL, Burkhart HE (2008) Using segmented regression to estimate stages and phases of stand development. For Sci 54:167–175
Weiskittel A, Gould P, Temesgen H (2009) Sources of variation in the self-thinning boundary line for three species with varying levels of shade tolerance. For Sci 55:84–93
West PW (1983) Comparison of stand density measures in even-aged regrowth eucalypt forest of southern Tasmania. Can J For Res 13:22–31
Wilson FG (1946) Numerical expression of stocking in terms of height. J For 44:758–761
Wilson FG (1979) Thinning as an orderly discipline: a graphic spacing schedule for red pine. J For 77:483–486
Woodall CW, Fiedler CE, Milner KS (2003) Stand density index in uneven-aged ponderosa pine stands. Can J For Res 33:96–100
Woodall CW, Miles PD, Vissage JS (2005) Determining maximum stand density index in mixed species stands for strategic-scale stocking assessments. For Ecol Manage 216:367–377
Xue L, Feng H, Chen F (2010) Time-trajectory of mean component weight and density of self-thinning *Pinus densiflora* stands. Eur J For Res 129:1027–1035
Yang Y, Titus SJ (2002) Maximum size-density relationship for constraining individual tree mortality functions. For Ecol Manage 168:259–273
Yoda K, Kira T, Ogawa H, Hozumi K (1963) Self-thinning in overcrowded pure stands under cultivated and natural conditions. J Biol Osaka City Univ 14:107–129
Zeide B (1985) Tolerance and self-tolerance of trees. For Ecol Manage 13:149–166
Zeide B (1987) Analysis of the −3/2 power rule of plant self-thinning. For Sci 33:517–537
Zeide B (1991) Self-thinning and stand density. For Sci 37:517–523
Zeide B (1995) A relationship between size of trees and their number. For Ecol Manage 72: 265–272
Zeide B (2002) Analysis of a concept: stand density. J Sustain For 14(4):51–62
Zeide B (2005) How to measure stand density. Trees 19:1–14
Zeide B (2010) Comparison of self-thinning models: an exercise in reasoning. Trees 24:1117–1126
Zhang L, Bi H, Gove JH, Heath LS (2005) A comparison of alternative methods for estimating the self-thinning boundary line. Can J For Res 35:1507–1514

Chapter 9
Indices of Individual-Tree Competition

Growth of individual trees on particular sites is influenced by a number of factors such as age, size, micro-environment, genetic characteristics, and competitive status. Past growing conditions and genetic potential to grow account for actual characteristics of the tree, such as size and vigor, which are usually introduced in a tree growth model by initial tree size and age. The influence of other factors may be separated into the following three components:

- Micro-environmental and genetic influences, represented by a ratio of some dimension of the tree to the mean or maximum value of this dimension in the stand;
- General environment of competition, which is usually taken into account using stand density measures such as total basal area per ha or others presented in Chap. 8;
- Average growth potential of individual trees as modified by the influence of local neighbors.

Competition may be defined as an interaction between individuals brought about by a shared requirement for a resource in limited supply, and leading to a reduction in the survival, growth and/or reproduction of the individual concerned (Begon et al. 1986). The effect of competition on growth of individual trees has long been studied in an attempt to more accurately predict tree increment and mortality.

The effect of local neighbors is usually expressed by some mathematical formulation – commonly referred to as a "competition index" – representing how much each tree is affected by its neighbors. Functions used to quantify competition range from simple formulations expressing the hierarchical position of the tree within the stand or plot to more complex indices that express the size of, distance to, and number of local neighbors. Munro (1974) classified competition indices as distance-independent (not requiring individual tree locations) and distance-dependent (requiring tree coordinate locations for computation).

H.E. Burkhart and M. Tomé, *Modeling Forest Trees and Stands*,
DOI 10.1007/978-90-481-3170-9_9,

Subsequently, Stage and Ledermann (2008) and Ledermann (2010) presented a class of competition measures that they termed "semi-distance-independent" indices.

This chapter focuses on the quantification of competition by local neighbors and its influence on individual tree growth.

9.1 Distance-Independent Indices

Distance-independent indices do not require individual tree coordinates, since they are simple functions of stand level variables and/or dimensions of the subject tree in relation to the average or maximum tree value of the stand. A brief description of representative examples of these indices by category or type follows.

Relative dimensions

These indices are mathematical formulations that measure the hierarchical position of the subject tree within the stand. Examples are ratios of a tree's dimension to the dimension of the average tree (Glover and Hool 1979; Daniels et al. 1986), larger tree, or the average dimension of the dominant trees (Alder 1979) in the stand:

$$Rx_m = \frac{x_i}{x_m} \quad Rx_{max} = \frac{x_i}{x_{max}} \quad Rx_{dom} = \frac{x_i}{x_{dom}} \tag{9.1}$$

where x is a tree variable such as diameter, height or some crown variable (Biging and Dobbertin 1995) and the subscripts m, max and dom indicate, respectively, the stand average, maximum and the average size of dominant trees.

Area proportional to relative tree basal area

This measure, designated APg_i, consists of dividing the plot area among the individual trees according to their dimension (for instance basal area) in relation to the dimension of the average tree of the stand (Tomé and Burkhart 1989):

$$APg_i = \frac{10{,}000}{N} \frac{g_i}{\bar{g}} \tag{9.2}$$

Crown ratio

Crown ratio (crown length divided by total tree height) has also been used to express the past competition undergone by each individual tree (Daniels et al. 1986; Soares and Tomé 2003):

$$c_r = c_l / h$$

Measures based on trees larger than the subject tree

The basal area of the trees greater than the subject tree ($G_{>di}$) was first proposed by Wykoff et al. (1982) and has been largely used since then either in its absolute value or relative to stand basal area ($G_{>di}/G$). Later on, Schröder and Gadow (1999)

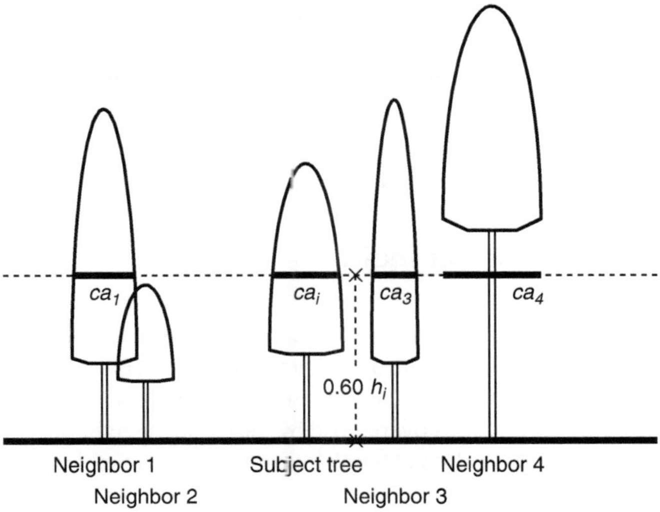

Fig. 9.1 Computation of a distance-independent index based on crown cross sectional areas calculated at a reference height equal to $p\%$ of the height of the subject tree (Adapted from Biging and Dobbertin 1995)

proposed a related competition measure ($G_{>dimod}$) that combines $G_{>di}$ with relative spacing (R_s):

$$G_{>dimod} = \frac{1}{R_s}\left(1 - \left(\frac{G_{>di}}{G}\right)\right) \tag{9.3}$$

Ritchie and Hann (1986) employed the crown competition factor of trees larger than the subject tree to model individual tree height growth of Douglas-fir in Oregon; Biging and Dobbertin (1995) used similar indices based on tree crown variables including crown cross-sectional area, crown volume, and crown surface area.

Measures based on crown variables evaluated at a certain percentage of crown length

Biging and Dobbertin (1995) proposed a distance independent measure of competition that utilized crown areas computed at a reference height equal to $p\%$ of the height of the subject tree (h_p). If the base of the crown of a competitor is above this height, the full crown area is used instead; whereas if the tree height is below the reference height the tree is not considered (Fig. 9.1):

$$CC_p = \frac{1}{S_a}\sum_{i=1}^{n} ca_{p_i} \tag{9.4}$$

where CC_p is crown cover computed at h_p, S_a is plot surface, ca_p is the crown cross sectional area of one tree at the same height (= 0 if the tree is smaller than this height and to the crown area if the tree crown base is higher than h_p).

The authors also used the same index with crown volume and crown surface above h_p instead of crown cross-sectional area.

Distance-independent indices are easy to calculate and less demanding in data and computer time, which make them preferable to distance-dependent indices in some applications.

9.2 Distance-Dependent Indices

According to Opie (1968), Staebler presented in 1951 perhaps the first individual tree competition index. He assumed each tree had a circular area of influence expressed as a function of its size and measured competitive stress as the degree to which this influence area was overlapped by those of its neighbors.

Later on, measures of local stand basal area around individual trees were utilized to quantify the local inter-tree competition and its relationship with tree growth (Lemmon and Schumacher 1962a, b; Spurr 1962). Two years later, Newnham (1964) used a competition index in the development of a growth model for Douglas-fir in British Columbia. Opie (1968) formulated an area-overlap index, and since that time a wide variety of competition measures have been developed.

The computation of distance-dependent measures of point density, or competition indices, involves two main steps: (i) selection of competitors, and (ii) computation of an index that synthetize the degree to which the subject tree has to share resources with its competitors. With the exception of a few indices, implementation of these two steps is not necessarily linked; therefore the following sections will present them separately.

9.2.1 Selection of Competitors

The selection of the neighbors that affect growth of a subject tree is of crucial importance when modeling inter-tree competition. While various methods for choosing competing neighbors have been proposed, the most common approaches include those listed below.

Using a fixed area or a fixed number of trees

Examples are:

- All trees within a circle of fixed radius centered at the subject tree (Hegyi 1974; Pukkala and Kolström 1987)
- Search radii as a multiple of the mean crown radius of overstory trees or as a multiple of the crown width of the subject tree (Lorimer 1983)
- A fixed number of nearest neighbors (Soares and Tomé 1999; Rivas et al. 2005)

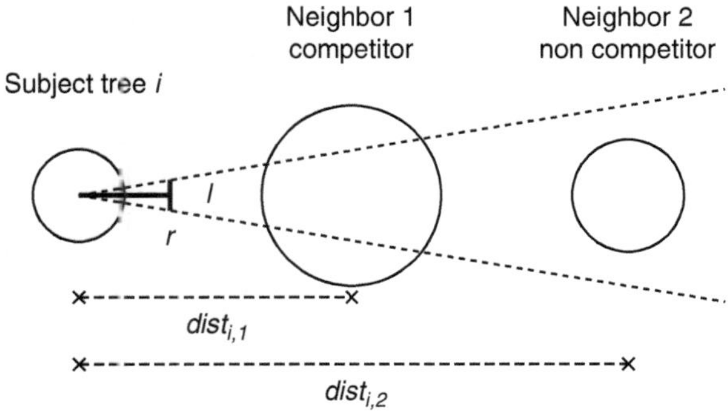

Fig. 9.2 Selection of competitors using Bitterlich's angle count method

Trees selected by angle count sampling

- Trees selected according to Bitterlich's angle count sampling centered at the subject tree were used by Hamilton (1969) and Daniels (1976).

The condition for a neighbor j to be considered as a competitor when applying Bitterlich's angle count method is that:

$$dist_{ij} < dist_{\lim} = kd_j$$

where $dist_{ij}$ is the distance between subject tree i and neighbor j, d_j is the diameter of tree j and k is a function of the basal area factor used (Fig. 9.2). Soares and Tomé (1999) noted that the selection of competitors using a basal area factor, which results in a linear relationship between the search radius and the tree dimension, did not result in a realistic number of competitors. To overcome this problem they tested several alternative rules among which several were based on asymptotic functions of the tree dimension (Fig. 9.3); the best results were obtained by use of the Richards function with an asymptote equal to 7 m.

Areas of influence overlap

Areas of influence overlap have been found useful when quantifying local competition. A measure of size (e.g. *dbh*) determines each tree's zone of influence. In Fig. 9.4, $dist_{ij}$ is the distance between the subject tree i and the neighbor j, with the radius of the respective areas of influence R_i and R_j. The neighbor j is a competitor if:

$$dist_{ij} < R_j + R_i$$

Competition elimination angle

Competing trees may be selected according to the competition elimination angle (proposed by Lee and Gadow and referred in Gadow and Hui 1999). The method is based on a fixed search radius to select the neighbors of a given subject tree. Each neighbor may be an active or a passive competitor, based on a competition elimination sector defined by a specific elimination angle. The magnitude of the angle is fixed a priori. The nearest neighbor is first selected as a competitor and all

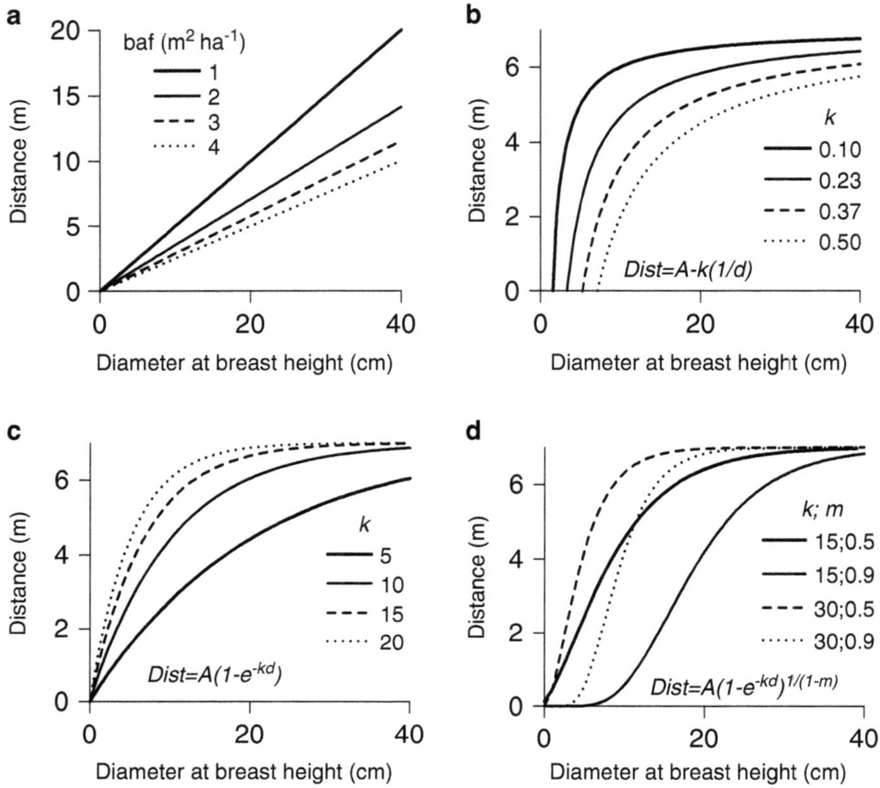

Fig. 9.3 Relationship between the distance of competition and tree size: (**a**) basal area factors of 1–4 m^2 ha^{-1}; (**b**) hyperbolic function; (**c**) monomolecular function; (**d**) Richards function (an asymptote (A) of 7 m was assumed); k and m are shape parameters (Adapted from Soares and Tomé 1999)

the trees located within the angle with the vertex at the subject tree and centered at the neighbor are considered as passive competitors and discarded from the selection procedure; the nearest neighbor outside the elimination angle is then selected and the respective passive competitors identified; the procedure ends when all the active competitors have been identified.

Selection using a vertical search cone

Trees can be selected using a vertical search cone. An upside-down search cone is set up at a certain height of the subject tree (stem base, base of the crown or some point within the crown) and all the trees whose crowns overlap the search cone are considered as competitors (Pukkala and Kolström 1987; Biging and Dobbertin 1992; Pretzsch 2009). Using the notation in Fig. 9.5 (search cone, with an opening angle of β, centered at a height hc_i of the subject tree), a neighbor is selected as a competitor if:

$$h_j > hc_i + dist_{ij} \, tan\,(90 - \beta/2)$$

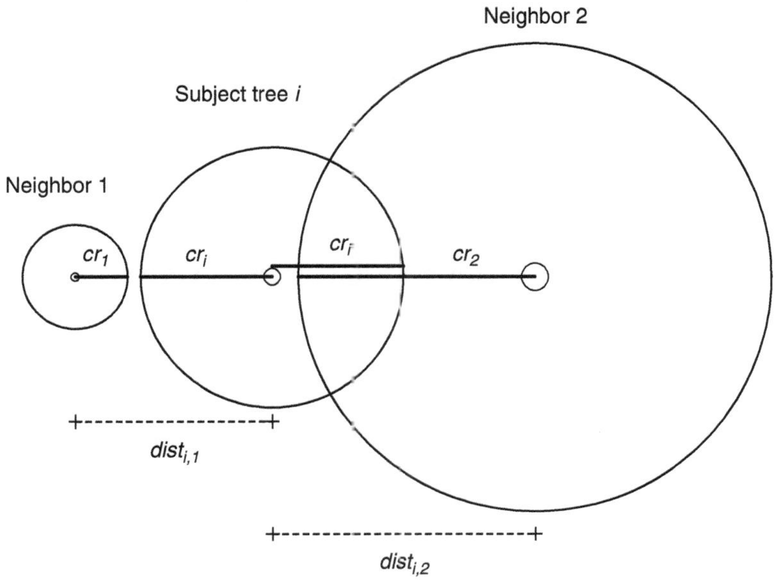

Fig. 9.4 Selection of competitors in area overlap competition indices ($cr_1 + cr_i < dist_{i,1}$ therefore neighbor 1 is not a competitor to subject tree *i*; $cr_2 + cr_i > dist_{i,2}$ therefore neighbor 2 is a competitor to subject tree *i*)

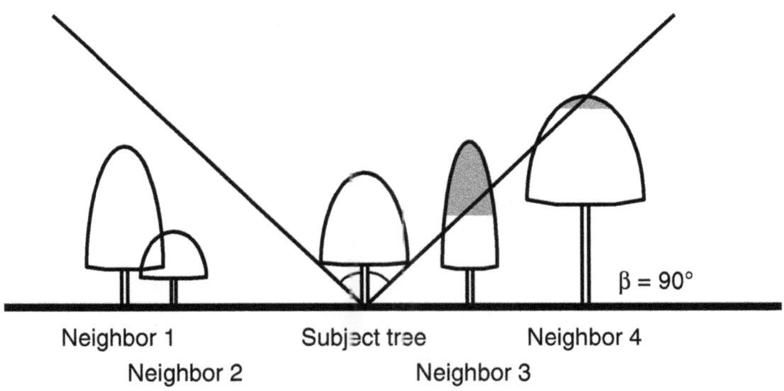

Fig. 9.5 Selection of competitors using a vertical search cone with an opening angle $\beta = 90°$. Neighbors 3 and 4 are selected as competitors. The *shaded areas* in the crowns of the competitors are used in the competition indices CI_{cvha} and CI_{cvsha} proposed by Biging and Dobbertin (1992)

9.2.2 Formulation of the Competition Index

Distance-dependent competition indices include, directly or indirectly, the size of the neighbors and their distance to the subject tree. The competitive influence of a neighboring tree should be a decreasing function of the distance between the neighbor and the subject tree and an increasing function of the neighbor's size

(Weiner and Solbrig 1984). Different methods have been applied to synthetize the information from the neighbors' size and distance from the subject tree into an index. According to the principles in which they are based, the competition indices can be classified into several types that will be further described in the next sections.

9.2.2.1 Point Density Indices

Lemmon and Schumacher (1962a, b) pointed out that basal area measurements with an angle gauge, using the center of study trees as a sampling point, would provide an assessment of the density of competing trees around the tree being studied.

Spurr (1962) adapted point basal area for use as a point density measure (CI_{PD}). Whereas in the Bitterlich angle-count technique a limiting angle is arbitrarily chosen and the number of trees exceeding the angle when viewed from a center point are counted (Fig. 9.2), in the angle-summation method, an angle is chosen so as to define a one-tree plot and an estimate of basal area is computed; a second angle is chosen to create a two-tree plot and a second estimate of basal area results, etc.; the process is continued until the desired number of trees has been measured. The final estimate of basal area is the mean of the individual estimates, each based upon the angle subtended by a separate individual tree in the plot. The one-tree plot is based on the tree that subtends the largest angle from the viewing point. This tree may not be the closest or the largest tree in the vicinity. In the ordered sequence, the second tree is the one subtending the second largest angle, which is not necessarily the second closest or second largest tree near the sampling point.

Spurr (1962) presented two options for this index, which consist of excluding (CI_{PD1}) and including (CI_{PD2}) the subject tree:

$$CI_{PD1i} = \frac{2500}{n}\left[\sum_{j=1}^{n}(j-0.5)\left(\frac{d_j}{dist_{ij}}\right)\right] \tag{9.5a}$$

$$CI_{PD2i} = \frac{2500}{n}\left[\sum_{j=1}^{n}(j+0.5)\left(\frac{d_j}{dist_{ij}}\right)\right] \tag{9.5b}$$

where n is the number of competitors, d_j is diameter at breast height of competitor j and $dist_{ij}$ is the distance between subject tree i and competitor j.

9.2.2.2 Area Overlap Indices

Area overlap indices (AO), the first distance-dependent indices to be developed, are based on the sharing of the areas of influence of the subject tree and its competitors. The area of influence, or influence zone, can be defined as an area over which the tree obtains or competes for site resources (Opie 1968). Competition between trees is assumed to occur when their zones overlap. All trees whose area of influence overlaps the area of influence of the subject tree are considered to be its competitors. Various definitions of area of influence, measure of overlap, and the use of weights in summing area overlaps lead to different area overlap indices.

The most common definitions of area of influence involve linear functions of *dbh* (e.g. Opie 1968; Alemdag 1978; Tomé and Burkhart 1989) or of open-grown tree crown radius (e.g. Bella 1971; Arney 1974; Alemdag 1978; Ek and Monserud 1974).

According to Bella (1971), Staebler's area overlap index used linear overlap within competition circles. Angles subtended by overlapping crowns were used by Newnham (1964), but area overlap (Opie 1968; Bella 1971; Arney 1974; Ek and Monserud 1974) has been the most frequently applied measure.

Most area overlap indices can be included under the generalized formula:

$$CI_{AOi} = \sum_{j=1}^{n} \frac{ao_{ij}}{AI_i} \left(R_{ji}\right)^m \tag{9.6}$$

where ao_{ij} is the area overlap between the subject tree i and the competitor j; R_{ji} is the ratio between the dimensions of the competitor j and the subject tree i; m is an exponent; and AI_i is the area of influence of subject tree i and other symbols are as before.

Holmes and Reed (1991), recognizing the importance of root competition, developed several variants of a root/crown index that include not only crown overlap but also root overlap.

Pretzsch (2009) describes the "lateral crown restriction" (ε) as the area overlap of the potential crown diameter of the subject tree with the crowns of the adjacent trees computed at the height at which the crown of the subject tree is larger (Fig. 9.6). This methodology has the advantage of computing lateral restriction. Using Fig. 9.6 as an example, in horizontal projection the crown of the subject tree overlaps the crowns of neighbors 1, 2 and 3. However, at the height of the widest part of the subject tree's crown, just the crowns of trees 1 and 3 overlap its potential crown.

9.2.2.3 Indices Based on the Size and Distances of the Neighbors Within a Search Radius

The best-known indices of this type fall under the category of distance-weighted size ratio (CI_{DR}) indices, first used by Hamilton (1969) and Hegyi (1974) and defined as the sum of the ratios between the dimensions of each competitor to the subject tree, weighted by a function of the inter-tree distance. In Hegyi's (1974) index all trees within a fixed radius were considered competitors. Hamilton (1969) defined as competitors all trees selected by a fixed angle gauge sweep centered at the subject tree (Fig. 9.2) and this definition of competitors has been preferred in most of the following applications (Daniels 1976; Alemdag 1978; Daniels et al. 1986;

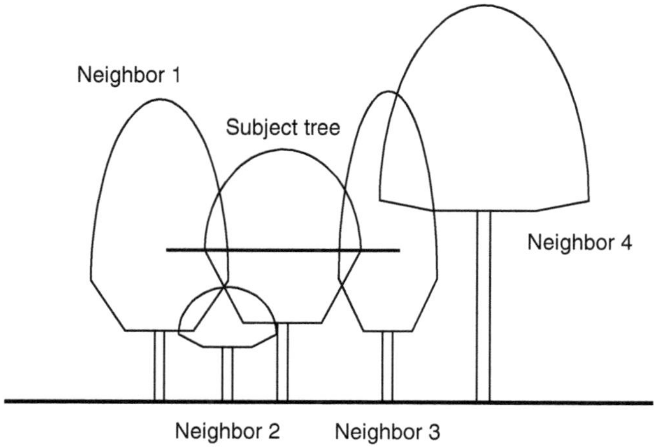

Fig. 9.6 Area overlap index computed according to the "lateral crown restriction" proposed by Pretzsch (2009). In horizontal projection the potential crown of the subject tree (represented by the *bold line*) overlaps the crowns of neighbors 1, 2 and 3. However, at the height of its larger crown just the crowns of trees 1 and 3 overlap its potential crown

Tomé and Burkhart 1989). However the index can be applied with any other rule for the selection of competitors. CI_{DR} type indices have the advantage of being easy to compute, while explaining variation in growth with precision similar to other indices.

Indices of this type can be generalized by:

$$CI_{DRi} = \sum_{j=1}^{n} R_{ji} f\left(dist_{ij}\right) \tag{9.7}$$

where n and R_{ji} are as previously defined and $f(dist_{ij})$ is a function of the distance between the subject tree i and competitor j.

Lorimer (1983) pointed out a limitation to these types of indices, namely that their numerical values decrease in a given stand over time even when the stocking level remains constant. To overcome this problem, Lorimer proposed using a constant search radius defined as a multiple of the average crown radius of the overstory and weighting the distance function with the inverse of the search radius:

$$CI_{DRi} = \sum_{j=1}^{n} R_{ji} \frac{1}{dist_{ij}/dist_s} \tag{9.8}$$

where $dist_s$ represents a fixed search radius, which is a multiple of the mean crown radius of overstory trees and other symbols are as defined before.

Different variables have been used in the ratio R_{ji} including tree diameter or basal area at breast height, tree height, and crown variables. Biging and Dobbertin (1992) used the crown cross-sectional area of the competitors and the subject tree computed at a reference height defined as $p\%$ of the height of the subject tree in the computation of R_{ji}. Other crown variables, such as crown surface area or crown volume computed above the reference height, have also been used.

Biging and Dobbertin (1992) proposed two indices that did not utilize distance as a weighting factor. They used the crown volume or crown surface area of the competitor tree above the point where a vertical angle from the base of the subject tree cuts the axis of the stem of the competitor (Fig. 9.5) relative to the crown volume or crown surface area of the subject tree:

$$CI_{cvha} = \sum_{j=1}^{n} \frac{cvha_j}{c_{vi}} \tag{9.9a}$$

$$CI_{csha} = \sum_{j=1}^{n} \frac{csha_j}{c_{si}} \tag{9.9b}$$

where $cvha_j$ and $csha_j$ are crown volume and crown surface area of the competitor j above the height at which the vertical angle cuts its stem axis (ha_j) and c_{vi} and c_{si} are crown volume and crown surface area of the subject tree i. In these indices the weighting function for distance is not used as the distance between the subject tree and each competitor is already embedded in the crown variable itself.

Cole and Lorimer (1994) also used crown measures in size ratio indices and found the variable exposed crown surface area to perform well. Exposed crown radius was measured as the portion of the total radius estimated to be free from overlap from above by the branches of neighboring trees.

Several authors formulated indices similar to those just described but using just the size of the neighbors instead of the ratio of the sizes of the competitor and the subject tree (Lorimer 1983; Newton and Jolliffe 1998; Canham et al. 2004; Zhao et al. 2006). Also, variants on the way the distance to competing neighbors is taken into account have been proposed (see, for instance, Martin and Ek 1984; Newton and Jolliffe 1998).

9.2.2.4 Indices Based on Horizontal or Vertical Angles Centered at the Subject Tree

Pukkala and Kolström (1987) and Rouvinen and Kuuluvainen (1997) used the sum of the horizontal/vertical angles from the subject tree to all the neighbors within a fixed search radius (competitors) as a competition measure:

$$CI_{Sang} = \sum_{j=1}^{n} \sum_{k=1}^{na} \alpha_{jk} \tag{9.10}$$

where α_{jk} is the horizontal/vertical angle subtended by some dimension x_{jk} of the neighbor. They tested vertical angles centered at the base of the subject tree or at a certain height within its crown and used different dimensions of the neighbor, such as diameter at breast height and crown width, for horizontal angles; total tree height or height above the insertion point of the vertical angle was used for vertical angles. Rouvinen and Kuuluvainen (1997) proposed using the widest horizontal angle from a subject tree without neighbors, called the widest free angle, in a competition measure formulation.

When selecting competitors with a search cone of magnitude β, a possible measure of competition is given by the sum of the angles γ between the surface line of the search cone and the line connecting the tip of the competitor tree j with the cone apex on the subject tree i. The closer and taller the competitor compared to the subject tree, the greater the angle γ and the competitive strength of this neighbor. Pretzsch (2009) describes a competition measure he developed that also includes the ratio of the competitor and subject tree crown cross-sectional areas at the height of the search cone insertion (if the height of the maximum crown width of the competitor is higher than the cone insertion height, the crown projection area is used instead) and a species-specific light transmission coefficient:

$$CI_{Sang_i} = \sum_{j=1}^{n} \gamma_j \frac{ca_{hcj}}{ca_{hci}} lt_j = \sum_{j=1}^{n} \left[arctan \left(\frac{h_j - (p/100)\, h_i}{dist_{ij}} \right) - \beta \right] \frac{ca_{hcj}}{ca_{hci}} lt_j \tag{9.11}$$

where ca_{hc} indicates the crown cross-sectional area at height of cone insertion hc, γ_j is the angle defined by the surface line of the search cone and the line between the insertion point of the cone and the top of the competitor tree and lt_j is the light transmission coefficient for competitor j (dependent on species) and other symbols are as before.

9.2.2.5 Growing Space and Area Potentially Available

Alemdag (1978) proposed an index defined on the basis of the growing space for the subject tree that may also be included under the distance-weighted size ratio indices. The stand surrounding a central tree i is divided into as many imaginary circle segments as there are competitors. Each circle has a radius proportional to the size of the subject tree in relation to the sum of its size and that of the corresponding competitor. The sum of the area of these n segments, assumed to be the area available for tree growth, is used as the competition index:

$$CI_{GS} = \sum_{j=1}^{n} \underbrace{\pi \left[dist_{ij} \frac{d_i}{d_i + d_j} \right]^2}_{\text{area of the circle}} \underbrace{\frac{d_j}{dist_{ij}} \Bigg/ \frac{\sum_{j=1}^{n} d_j}{\sum_{j=1}^{n} dist_{ij}}}_{\text{opening width of the segment}} \tag{9.12}$$

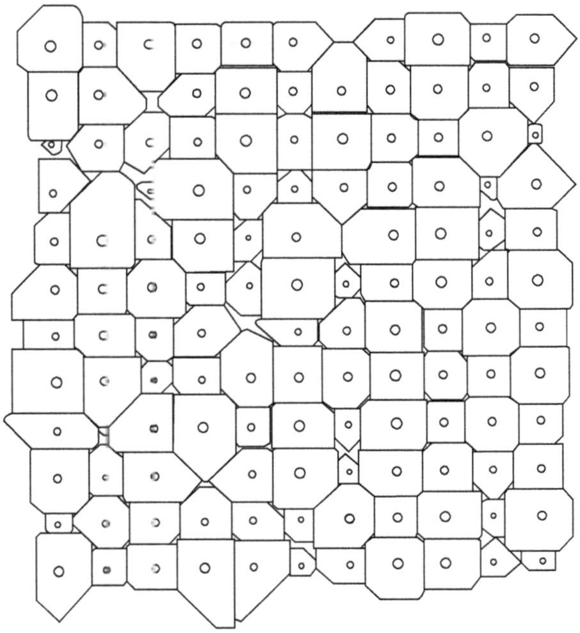

Fig. 9.7 Area potentially available to each tree in a permanent plot established in a *Eucalyptus globulus* plantation

The first term in Eq. 9.12 defines the area of the circle corresponding to each competitor and the second term defines the opening width of the segment. The radius of the segments increases with the diameter of the subject tree and with its distance to the neighbor and its width is proportional to the diameter/distance ratio of the competitor in relation to the same ratio for all the competitors.

An area potentially available index (*APA*) was first defined by Brown (1965) as a measure of point density. The area available to each tree is calculated as the area of a polygon defined by the bisectors to the inter-tree lines (example in Fig. 9.7). Essentially the *APA* for each tree is derived from a Dirichlet tessellation (or Voronoi diagram) of the point pattern of trees in a stand. Moore et al. (1973) modified Brown's index by bisecting the inter-tree lines proportionally to the subject tree and competitor sizes; this version was used with success by Pelz (1978), Daniels et al. (1986), and Tomé and Burkhart (1989).

Pelz (1978) extended the concept of tree polygons to the third dimension by defining a tree growing space represented by a geometric solid obtained by multiplication of the tree polygon area by the height of the subject tree. However, simple correlations of this three-dimensional *APA* with tree basal area growth were not higher than those obtained with the traditional *APA*.

Nance et al. (1988) constrained the distance to a polygon side by a function of the radius of an open-grown tree of the same diameter as the subject tree to prevent polygon areas from becoming excessively large. This variant of the *APA* index may be useful in stands with irregular spatial patterns.

9.2.2.6 Indices Based on Ecological Field Theory (EFT) and on Field of Neighborhood (FON)

In the last two decades ecologists have given particular attention to the so-called individual-based models in which inter-plant competition plays an essential role (e.g. Burton 1993; Berger and Hildenbrandt 2000; Grimm and Railsback 2005; Berger et al. 2008). These individual-based models are similar in concept to distance-dependent, individual-tree growth and yield models. Berger et al. (2008) identified two types of spatially explicit individual-based models describing plant competition: (i) site-based neighborhood models, and (ii) individual-based models. The first class, often referred to as "cellular automata" (CA) or as "grid-based models", refers to models that represent spatial relationships on regular (hexagonal or squared) lattices or grids, i.e., space is discretized. The second class includes models that consider plant positions in continuous space. Distance-dependent models described in Sects. 9.2.2.1, 9.2.2.2, 9.2.2.3, 9.2.2.4, and 9.2.2.5 are part of this class although the individual-based literature uses different designations for the indices. Indices based on the size of and distances to neighbors within a search radius have been used with a fixed search radius and designated as fixed-radius neighborhood models, area overlap indices as zone of influence models, and area potentially available as tessellation models.

Following is a brief description of two competition models that have been advocated in the literature on individual-based models: models applying ecological field theory (*EFT*), and those based on the field of neighborhood (*FON*) model.

The *EFT* model (Wu et al. 1985) uses classical field theory concepts to outline a general theoretical approach to describe spatial interference among plants, assuming that the combined effects of a plant's crown, stem and roots on the resources available to other plants in the area constitute a field of influence. The field of influence of each local plant is first defined for each resource (water, nutrients and light) as a continuous function scaled from 0 to 1 (0 represents the most favorable environment) and the effects of the plant on the different resources are then integrated to compute a general measure of resource availability around each single isolated plant. An iterative procedure is used to compute resource availability at each point (x, y) resulting from the influences of all the individuals in the plant community. In this way it is possible to use an integrated measure of the resources available to each plant when modeling its growth. One problem with this type of model, other than the complexity of the associated computation, is that the influence of a single tree on resource availability cannot be observed in the field because each point is influenced simultaneously by more than one tree. Pukkala (1989) used *EFT* to model competition among Scots pine trees in a 70-year-old stand located in a rather poor site of North Carelia, Finland. Miina and Pukkala (2002) developed *EFT*-based competition indices to model growth of Scots pine and Norway spruce in Finland. In both cases, *EFT*-based competition indices performed slightly better than indices based on the sum of vertical angles.

Berger and Hildenbrandt (2000) developed the concept of field of neighborhood (*FON*) that is related to *EFT*. The basic idea is also the definition of a tree's field of influence that takes the value 1 at the tree location and decreases exponentially across the radius of the zone of influence. The method assumes that the *FON*'s of all trees superimpose, therefore at a given location (*x, y*), the aggregate field strength of all trees, *F(x, y)*, is given by the sum of their single field intensities. The specific, dimensionless value of *F(x, y)* depends on the local configuration of all neighboring trees and represents the local neighborhood situation taking into account the number, the weighted distance and the individual size of the neighboring trees. When describing the neighborhood situation of a particular tree, the authors assume that the tree 'perceives' the aggregate field strength *F* on its entire zone of influence. Consequently, the mean value *FA* of the aggregate field strength *F* produced by all other trees on the tree's area is computed and used as a competition index. Similarly, Pommerening et al. (2011) used the idea of the shot-noise field from Illian et al. (2008) to derive the competition load of a tree by additively aggregating the competition effects of all other trees. The authors noted that *EFT* and *FON* are conceptually similar, with the difference being that competition effects are aggregated in a multiplicative manner in the *EFT* model and in an additive fashion in the *FON* and shot-noise field approaches.

9.2.2.7 Indices Based on the Estimation of Shading or Light Interception

Among the resources limiting growth of individual trees in forest stands, light (i.e., photosynthetically active radiation, *PAR*) is often assumed to be the most important (e.g., Cannell and Grace 1993). Therefore indices that estimate light interception or the degree of shading from neighbors seem useful to model competition for light. The simulation of light interception within a forest stand has been a topic of research in ecology with the objective of better understanding stand production and dynamics. Brunner (1998) presents an exhaustive review of such light models. However, some recent studies have used simplified models of light interception as a way to include inter-tree competition in modeling tree growth. Examples of these types of indices are described by Brunner and Nigh (2000), Pretzsch (2009), and Canham et al. (2004).

9.2.3 Asymmetric/One-Sided Versions of the Competition Indices

Competition processes have been defined according to two basic models: symmetric/asymmetric and one-sided/two-sided competition (Brand and Magnussen 1988; Weiner 1985, 1986, 1990). In two-sided competition, resources are shared (equally or proportionally to size) by all the trees while in one-sided competition larger trees are not affected by smaller neighbors. When there is perfect sharing relative to size, competition is symmetric. One-sided competition may be considered as an extreme case of asymmetric competition, and two-sided competition can be symmetric or asymmetric according to whether or not the sharing of resources is proportional to the size of the individuals. It has been shown for some species that, in the early stages of stand development, competition for light may not be present, although the effects of competition for water and nutrients are evident. Additionally, even when competition for light is the main factor impacting individual plant growth, two-sided competition for water and nutrients also has an effect.

Depending on the respective formulation, competition indices implicitly assume an asymmetric or symmetric partitioning of resources among neighboring trees. In this section we analyze how the different types of competition indices previously identified have been used to reflect symmetry or asymmetry in the partitioning of resources among trees. The term asymmetric competition will be used throughout this section to indicate that the neighbors bigger than the subject tree are assumed to use the resources more than proportionally to their size, eventually in a unilateral way.

All the distance-independent competition indices that are based on the trees larger than the subject tree implicitly assume asymmetric competition and therefore reflect mainly competition for light. In distance-dependent indices the distinction between these two models relates also to the selection of competitors.

Lorimer (1983) included, among the competition indices that he compared, several formulations that reflect asymmetric competition by restricting the competitors to those neighbors that belong to certain crown classes or to those that belong to a crown class higher than that of the subject.

Most of the competition indices used by Pukkala and Kölstrom (1987) expressed asymmetric competition as they restricted the neighbors to those larger (in dbh and/or height or some function of height) than the subject tree. Rouvinen and Kuuluvainen (1997) also used both symmetric and asymmetric variants of the competition indices.

When studying maize, Yoda et al. (1957) found that within a row of plants, once a difference has been triggered, it is progressively exaggerated (Harper 1977). Accordingly, Tomé and Burkhart (1989) expressed the competitive status between each tree and its neighbors in eucalyptus plantations taking into account their relative dimensions. It was assumed that neighbors larger than the subject tree place the subject tree at a competitive disadvantage, whereas those smaller put it at a competitive advantage. This leads to competition indices that are sums of positive and negative values. Dominant neighbors make a positive contribution to the index while suppressed neighbors subtract from the index. Moreover, Tomé and Burkhart assumed that in the competition between two trees of different sizes, the contribution of the larger tree to the smaller tree competition index is equal in absolute value, although opposite in sign, to the contribution of the smaller tree to the larger tree competition index. When applying these competition indices, dead neighbors were included as a special kind of suppressed neighbor. The size of a dead tree was set as $d_{j0} = min(d_0/(2ny), d_{min})$ where d_0 is the *dbh* of the tree before dying, *ny* is the number of years since it died, and d_{min} is the *dbh* of the smallest tree in the stand. When selecting neighbors, dominant competitors were selected using the search radius centered at the subject tree while suppressed competitors were selected by centering the search radius at the neighbor. The number of competitors selected in this way was, in most cases, larger than if dominance/suppression relationships were not considered. Also some modification was made in the calculation of ratios of dimensions between the competitors and the subject tree. For dominant neighbors the ratio was neighbor/subject tree as usual, but for suppressed neighbors the ratio was subject tree/neighbor (with a minus sign) in order to ensure a decrease in the competition index directly proportional to subject tree dimension and inversely proportional to suppressed neighbor dimension. This computational algorithm guarantees that the contribution of every neighbor to a subject tree's competition index is equal, but opposite in sign, to the subject tree's contribution to its own index. Taking distance-weighted size ratio indices as an example (the index for area overlap indices is developed in a similar way), the index expression is specified as:

$$\begin{aligned} CI_{ADR_i} = & \sum_{j=1}^{n_1} \frac{d_j}{d_i} f\left(dist_{ij}\right) \text{ dominant neighbors} \\ & - \sum_{j=1}^{n_2} \frac{d_i}{d_j} f\left(dist_{ij}\right) \text{ suppressed neighbors} \\ & - \sum_{j=1}^{m} \frac{d_i}{d_{j0}} f\left(dist_{ij}\right) \text{ dead neighbors} \end{aligned} \tag{9.13}$$

where n_1, n_2, and m are, respectively, the number of dominant, suppressed and dead neighbors. Based on the same philosophy the authors also developed an asymmetric distance-weighted size differences index:

$$\begin{aligned} CI_{ADD_i} = & \sum_{j=1}^{n_1} \left(d_j - d_i\right) f\left(D_{ij}\right) \text{ dominant and suppressed neighbors} \\ & + \sum_{j=1}^{m} \left(d_{j0} - d_i\right) f\left(D_{ij}\right) \text{ dead neighbors} \end{aligned} \tag{9.14}$$

where the symbols are as before. Canham et al. (2004) also used an asymmetric version of the indices they tested based on the differences of the sizes of the competitor and the subject tree.

Soares and Tomé (1999) compared different competition indices, including the asymmetric variants from Tomé and Burkhart (1989) and the corresponding unilateral variants (only trees larger than the subject tree considered as competitors), with data from permanent plots and spacing trials established in eucalypt plantations. In general, the unilateral and asymmetric versions of the indices performed better than the other measures included in the comparison.

Wimberly and Bare (1996) used asymmetric/unilateral versions of the *DR* and *AO* competition indices they tested by imposing the constraint that only neighbors with a basal area larger than the subject tree would be included as competitors.

They also presented a layered variant of *APA* index (*APAL*) that computes the tessellation separately for each of three crown classes – dominant and co-dominant combined into one class, intermediate and suppressed.

9.2.4 Interspecific Competition

Models for complex, multi-species and multi-aged forest structures are often required. Hence, quantifying competition in such stands with the identity of the species of neighbors has been an area of research emphasis.

The competition index used by Pretzsch in the SILVA model (Pretzsch et al. 2002), presented in Eq. 9.11, includes a species-specific light transmission coefficient. Additionally, the authors found that height and diameter growth of spruce, pine and beech are significantly affected by the type of competitor, and that this effect is not covered by the light transmission coefficients, therefore an additional differentiation between deciduous and coniferous trees was introduced by using a tree-type specific value that is calculated as the ratio between the sum of crown surface area of the coniferous competitors in relation to that of all competitors.

Canham et al. (2004), when defining competition indices for application in forests of northern, interior British Columbia dominated by western hemlock and western red cedar, used a parameter (λ_k) as shown in (9.15) to account for tree species:

$$CI_{DN} = \sum_{k=1}^{s}\sum_{j=1}^{n} \lambda_k \frac{d_{jk}^{\alpha}}{dist_{ijk}^{\beta}} \tag{9.15}$$

where d_{jk} is the diameter of competitor j of species k, located at a distance d_{ijk} from subject tree i and α and β are parameters.

Zhao et al. (2006) modeled growth and survival of individual trees in a natural temperate species-rich forest taking species groups neighborhood effects into account. They used distance-weighted size-ratio and distance-weighted neighbors' size indices that were computed separately according to the species group of the neighbors. The species group specific competition indices were tested in individual-based diameter growth and survival models.

Miina and Pukkala (2000) used optimization techniques combined with regression analysis to specify the parameters of five competition index types for a mixture of Scots pine and Norway spruce. The best model included an index computed from vertical angles formed by a horizontal plane and the tops of competitors. The elevation of the horizontal plane was computed with a species-specific linear regression model using height of the subject tree as the predictor. Pine competitors nearer than 6 m and spruce competitors nearer than 9–10 m were included in the optimal competition index.

Kaitaniemi and Lintunen (2010) studied the potential importance of competitive species interaction for tree growth prediction in mixed stands of silver birch, Scots pine, and Siberian larch by selecting subject trees with a high local abundance of a single dominant neighboring species. The dominant neighboring species was defined as the one with the sum of basal areas being over half (typically close to 80%) of the total sum of basal areas of all the neighboring trees. The importance of neighbors identity was assessed by fitting models of tree growth – average annual height increment, diameter growth, shoot length, branch numbers per unit crown length, and whorl distances – that included a competition index as a covariate, neighboring species as a fixed factor and site as a random factor. Interspecific neighbors influenced annual height increment, shoot length, and branch number per unit crown length, especially in Scots pine. Silver birch and Siberian larch were predominantly affected by the level of competition alone, as estimated with competition indices. A simple extrapolation of individual tree growth to the stand level suggested that Scots pine and silver birch may grow faster in mixed than in pure stands. Siberian larch showed negative growth responses to interspecific neighbors, but the effects may be counterbalanced at the stand level by a corresponding increase in pine or birch growth.

9.2.5 Clumping, Differentiation and Mingling

When modeling structurally irregular stands, it is often important to account for spatial characteristics such as clumping, differentiation and mingling in the competition indices. Clumping means that the neighbors are not uniformly distributed around the subject tree, differentiation expresses the variability in the size of the neighbors, and mingling measures the spatial structure of the species mixture.

The effect of the directional distribution of the neighbors was described by Pukkala (1989) using an index aimed at expressing the distance of the center of the competition from the subject tree. The index was computed in two stages. First, a weighted average of the x- and y-coordinates was computed for neighbors nearer than the search radius:

$$\bar{x}_i = \sum_{\substack{j=1 \\ j \neq i}}^{n} w_j x_j \Bigg/ \sum_{\substack{j=1 \\ j \neq i}}^{n} w_j$$

$$\bar{y}_i = \sum_{\substack{j=1 \\ j \neq i}}^{n} w_j y_j \Bigg/ \sum_{\substack{j=1 \\ j \neq i}}^{n} w_j$$

where $\bar{x}_i$ and $\bar{y}_i$ are the means of the x- and y-coordinate of the competitors of subject tree i, (x_i, y_i) are the coordinates of subject tree i, w_j is a weight variable for tree j and n is the number of neighbors. The studied weights depended on the distance and diameter of the neighbor and the ratio between the heights of the neighbor and the subject tree.

Second, the distance of the subject tree to the center of competition was calculated by

$$distc_i = \sqrt{(x_i - \bar{x}_i)^2 + (y_i - \bar{y}_i)^2} \qquad (9.16)$$

This index was used in a tree basal area growth model, jointly with a competition index based on the sum of horizontal angles.

The SILVA simulator (Pretzsch et al. 2002) uses the index $distc_i$ but expresses it in relation to the average distance r_i that would result if all trees inside the search cone used to derive the competition index $CI_{sang\beta}$ were randomly distributed:

$$r_i = \frac{1}{2\sqrt{Mc_i / Ac_i}}$$

where Ac_i represents the projection area of the search cone at the subject tree i and Mc_i the trees inside the area Ac_i. The normalized distance of the subject tree to the center of competition is then:

$$Ndistc_i = \frac{distc_i}{\frac{1}{2\sqrt{Mc_i/Ac_i}}} = 2\ distc_i \sqrt{Mc_i/Ac_i} \qquad (9.17)$$

Gadow and Hui (1999) describe several indices to quantify size differentiation and species mingling in the neighborhood of the subject tree. Other structural indices are delineated in Pommerening and Stoyan (2008).

The differentiation index (T_i) is defined as (Gadow and Hui 1999):

$$T_i = 1 - \frac{1}{m} \sum_{j=1}^{m} \left[\frac{\min(d_i, d_j)}{\max(d_i, d_j)} \right] \qquad (9.18)$$

where m is the number of competitors and d_i and d_j are the diameters of the subject tree i and neighbor j, respectively. In plantations or mature stands managed for future crop trees differentiation is low and T_i approaches 0; maximum differentiation corresponds to T_i values close to 1.

The mingling index (M_i) is defined as the proportion of neighbors of another species (Gadow and Hui 1999):

$$M_i = \frac{1}{m} \sum_{j=1}^{m} v_{ij} \qquad (9.19)$$

where the variable v_{ij} is $= 0$ if neighbor j is the same species as the central tree i and $= 1$ if the neighbor j is a species different from the subject tree i. The index M_i varies between 0, when all the neighbors are the same species as the subject tree, and 1, when all the neighbors are another species.

Canham et al. (2004) added the effect of clumping by including a clumping index (δ, presented in Zar 1996) in Eq. 9.15. The clumping index is calculated as a function of the angles from the target tree to each neighbor and ranges from 0 when the neighbors are uniformly distributed around the target tree to 1 when they are tightly clumped:

$$CI_{DN} = \left[\sum_{k=1}^{s} \sum_{j=1}^{n} \lambda_k \frac{d_{jk}^{\alpha}}{dist_{ijk}^{\beta}} \right] (1 - \delta) \tag{9.20}$$

where the variables and parameters are as described for Eq. 9.15.

9.2.6 Using Change in Competition Indices to Model Thinning Effects

Modeling thinning effects remains a challenge for forest analysts (see Chap. 16), and competition indices can potentially improve ability to simulate reaction of individual trees to thinning.

Based on the assumption that individual tree growth following thinning should be correlated with the amount of new growing space made available by the treatment, Wimberley and Bare (1996) proposed that this new growing space can be measured in terms of the change in the competitive status of the residual trees. For competition indices that exhibit a positive correlation with competitive stress (all except *APA*, *APAL* and the relative size indices), the corresponding thinning index was computed as:

$$\Delta CI = 1 - \frac{CI_{at}}{CI_{bt}}$$

where ΔCI is the thinning index based on competition index *CI*, CI_{at} is the competition index measured immediately following thinning and CI_{bt} is the competition index measured before thinning. For competition indices that exhibit a negative correlation with competitive stress (*APA*, *APAL* and the relative size indices), the thinning index was computed as:

$$\Delta CI = \frac{CI_{at}}{CI_{bt}} - 1$$

Both forms of this index equal zero when no thinning treatment is applied and they increase with the amount of removal.

The change in the competition index value is used both in the height and diameter increment models of the SILVA simulator to express the reaction of single tree growth to varying pressure from local neighborhood (Pretzsch et al. 2002). Thinning will have a strong impact on the competition index in the year thinning occurs and this effect is therefore taken into account when predicting tree growth.

9.2.7 Edge Bias in Competition Indices Computation

As described in the previous sections, the computation of local or distance-dependent measures of competition implies knowledge of the size and location of all trees located in the neighborhood of each tree within a plot. Bias can be introduced if neighbors of trees near the edge of measured plots are ignored. The edge effect is especially problematic in forest growth predictions when the long-term development of a stand is based on a small sample area without information about the stand structure around the plot border. Several methods that have been proposed to correct for plot edge bias are briefly described in the following paragraphs.

A very simple method for avoiding edge bias is the use of a boundary strip or "buffer" zone, of width R_b, inside the measurement plot. This method has been used by many researchers when dealing with the computation of competition indices (e.g., Soares and Tomé 1999; Miina and Pukkala 2002; Canham et al. 2004). In the computation of competition indices only the trees in an inner plot are used as subject trees, namely those that are at a distance larger than R_b from the plot boundary. The width of the buffer zone should be large enough so that all relevant interaction between trees in the reduced plot is accounted for within the plot. Determining the optimal width of the buffer zone is difficult; if it is too small residual edge effects will remain; if it is too large valuable data will be discarded unnecessarily. The use of a "buffer" zone takes the actual border trees into account, rather than applying a correction method.

The border zone outside the plot can be simulated by utilizing the structure of the measured plot. Pommerening and Stoyan (2008), for instance, developed methods for reconstructing plot structure using modeling of the observed point patterns.

Edge bias can be reduced by creating an artificial stand structure in the border zone of the plot. Assuming that the border zone has a spatial structure similar to the one observed inside the plot, the border zone can be created by one of two methods: reflection or translation. Reflection assumes a symmetric distribution of the trees on both sides of a reflection line that may pass through the plot border, the border tree closest to the plot border, or at a fixed distance from the plot border. The plot is first reflected right and left and then the enlarged plot is reflected up and down (Fig. 9.8). This method results in large trees near the border competing with equal sized neighbors (i.e. themselves in reflection) only a short distance away. Additionally the method implies some periodicity at tree level as seen in Fig. 9.8. Translation involves copying the plot and placing it around the plot boundaries until the border zone is complete (Fig. 9.8). This method also has a propensity for introducing periodicity at plot level (Monserud and Ek 1974) as is apparent in Fig. 9.8. Reflection and translation methods also pose problems when using circular plots, although it is possible to reflect or translate circles by first "transforming" them into squares by filling the empty areas using the reflection or translation of a square circumscribed inside the circle.

The linear expansion method for edge bias correction, developed by Martin et al. (1977), is applicable to a wide variety of plot shapes and sizes and is reported to

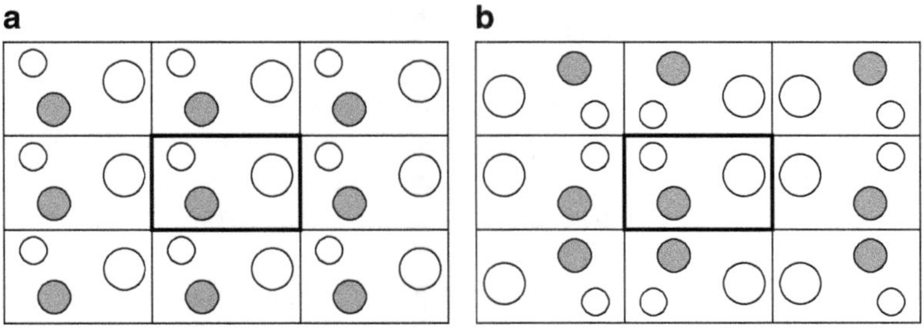

Fig. 9.8 Illustration of the translation (**a**) and reflection (**b**) edge-correction methods. The observation window, *W* is the center with boundaries in *bold* (From Pommerening and Stoyan 2006)

be unbiased under general assumptions about the forest tree spatial pattern. When comparing the linear expansion method with plot translation techniques, Martin et al. (1977) found that both methods performed equally well for square plots 0.08 ha in size. As the plot size diminished or its shape deviated from square, the linear expansion method provided greater accuracy and lower bias than translation.

9.2.8 Modeling and Simulating the Spatial Pattern of Forest Stands

It is important to note that measurements of tree coordinates are not required in order to apply distance-dependent individual-tree models for many purposes. One of the components of most individual-based tree growth models is a stand structure simulator. These simulators use information measured in study plots to obtain the variables needed to simulate a stand. Information about the spatial structure of the stand can be given by directly providing the *x, y* coordinates of each tree, in case they are measured, or by providing descriptive information about stand structure: random, regular, or aggregated (different levels of aggregation may be provided). Measurements of tree coordinates are essential for the development of distance-dependent models but not for all applications in forestry practice. Distance-dependent growth models can be applied to simulate response to alternative management treatments, but they are generally not used for inventory updating because it is not practical to record tree locations for the large number of samples typically obtained in forest inventories

Descriptions of methodologies useful for simulating spatially-specific stands are provided by a number of authors including Tomppo (1986), Biging et al. (1994), Pretzsch (1997), Paulo et al. (2002), and Pommerening and Stoyan (2006). (See also Sect. 14.5.)

9.3 Evaluation and Comparison of Competition Measures

There has been considerable debate in the literature as to whether tree spatial information improves prediction of individual tree growth. Conceptually, one would expect some improvement in precision when going from distance-independent to distance-dependent indices; however, most of the comparisons of competition do not report large differences.

Competition indices have often been compared by computing simple correlation coefficients with tree growth and determining significance of the competition index when added to a tree growth model in which the influence of tree size and stand density is already accounted for in the base model. In some instances, a distance-independent competition index has been included along with tree size and overall stand density before testing for significance of adding a distance-dependent measure.

9.3.1 Simple Correlations with Tree Growth or Models with the Competition Index as the Unique Independent Variable

Daniels (1976) compared area overlap and distance weighted size ratio indices for predicting diameter and height growth of trees in plantations of loblolly pine and found similar correlation coefficients with tree growth for the best variants of both types of indices tested (Ek and Monserud 1974 for area overlap indices and a variant of Hegyi's index for distance weighted size ratio).

Using data from three even-aged temperate hardwood stands, Lorimer (1983) tested several formulations of distance-weighted size-ratio indices with indices that did not include inter-tree distances, concluding that the correlation between competition and growth is optimal over a wide range of competition radii and that the inclusion of inter-tree distances in the formulation of the index is of little value despite considerable small-scale variability in the stocking level around individual trees. Highest correlations were obtained when competitors were defined to be only those trees of equal or higher crown class than the subject tree, suggesting that asymmetric competition was dominant in these hardwood stands. Another conclusion was that inclusion of the size of the subject tree as well as that of the competitors was necessary for obtaining reasonable correlations with growth.

Pukkala and Kolström (1987), using data from three plots with areas from 0.10 to 0.19 ha, observed simple correlation coefficients between competition indices and tree growth, at plot level, between 0.40 and 0.77. Generally, correlation increased with increasing length of the search radius, but after 4–5 m the change became negative. The indices based on the sum of horizontal and vertical angles, as well as Hegyi's index, provided the highest correlation coefficients. When the three plots were combined, the sum of horizontal angles to the neighbors with a size larger than the subject tree, selected within a search radius of 5 m, had the highest correlation.

The use of crown characteristics did not increase the correlation with growth above that obtained with diameter and/or height.

Holmes and Reed (1991) compared competition indices to predict annual diameter growth in mixed species northern hardwood stands in upper Michigan, USA. Their study used simple correlation coefficients between annual diameter growth and each of the competition measures studied. Generally, size-ratio indices performed well, but Spurr's point density and *APA* indices did not. The root/crown indices tested showed good correlations with growth but did not outperform the size ratio indices. An interesting conclusion from the study was that area overlap indices (e.g., Bella 1971; Arney 1974) using species-specific search radii to differentiate competition levels among tolerance classes proved useful.

9.3.2 Contribution of Competition Indices to Tree Growth Models in Which Tree Size and/or Stand Variables Are Already Included

Daniels et al. (1986) compared several distance-independent indices with variants of area overlap, distance-weighted size-ratio, Spurr's point density and *APA* indices on the basis of simple correlations and multiple correlations in the presence of other tree and stand attributes with growth of planted loblolly pine trees. The best distance-dependent indices had little if any advantage, either in simple or multiple correlation, over the best distance-independent indices. However, the point density measure of Spurr and especially the *APA* contributed significantly to growth prediction even in the presence of tree size, stand density, and the distance-independent relative size and crown ratio indices. Further, *APA* had the highest partial correlation when all variables were included in this multiple correlation.

Tomé and Burkhart (1989) evaluated the contribution of several distance-dependent competition measures when added to a linear model for tree diameter growth that already included the influence of tree size, a measure of stand density and a distance-independent index. Generally, the contribution of competition indices was significant, even if the differences in multiple correlation values were not very large in magnitude. The contributions of *APA* and asymmetric versions of distance-weighted size ratios were significantly higher than that of other indices.

Biging and Dobbertin (1995) evaluated several competition indices in individual tree diameter squared and height growth models in multi-aged mixed-species conifer stands in California, USA. The assumption was that if distance-dependent competition indices are superior to distance-independent measures it should become evident in heterogeneous stands, such as those examined in this study, where individual tree competition varied greatly. Based on the reduction in mean square error, they found that the distance-independent indices based on crown variables evaluated at a specified percentage of crown length performed as well as or slightly better than the best distance-dependent competition indices.

Wimberly and Bare (1996) compared distance-independent and distance-dependent competition measures as regressors in a basal area growth equation for Douglas-fir and western hemlock, using data from plots measured in a single stand that had been subjected to a range of thinning and fertilization treatments. Performance of competition indices was analyzed as their contribution to a model of basal area growth that already included tree size and age, crown class and a thinning index. The distance independent index $G_{>di}$ was significant for both species but much more important for Douglas-fir than for hemlock. The best distance-dependent competition measure was the *APAL*, but its contribution was small for both species, although larger than that of the distance-independent competition variable ($G_{>di}$) in the western hemlock model.

The study of Rouvinen and Kuuluvainen (1997) evaluated the contribution of competition indices for prediction of crown base height and crown width. A distance-dependent competition index was significant in the models for both variables. In the height to crown base model the best competition index was an asymmetric variant of a sum of vertical angles centered at 80% of the height of the subject tree while the crown width model included a distance weighted size index and an asymmetric variant of the widest free angle.

Moravie et al. (1999) compared several competition measures, including non-spatial indices of tree vigor such as crown ratio and *h/d* ratio, tree position in the canopy (tree social status), and several point density measures. Assessment of the contribution of the various competition measures to a model of *dbh* growth, in which *dbh* was present, showed that tree social status and all of the distance-dependent competition indices were significantly related to diameter increment.

Soares and Tomé (1999) compared different competition indices, including the asymmetric variants from Tomé and Burkhart (1989), using data from permanent plots and spacing trials established in eucalypt plantations covering different stages of stand development that were defined following a methodology proposed by Perry (1985). In a first stage, in which competition is not yet evident, small trees have relative growth rate (*RGR*) values greater than those for larger trees. In an intermediate stage, *RGR* differs little among social classes. Finally, competition effects are clearly visible and trees in the lower diameter classes are suppressed exhibiting smaller *RGR* than the bigger trees. The contribution of the tested indices to a multiple linear regression equation to predict the annual increment of tree basal area in which tree size, stand density and a distance independent index were present was analyzed separately for the three stages of stand development. The asymmetric formulations of the distance weighted size ratio indices exhibited the best performance in the first two stages of competition. When asymmetric competition was clearly evident performance of these indices was still good but the area potentially available indices showed the best performance. An interesting result of this study was that the simple correlations of point density, distance weighted size ratio (all variants) and *APA* were all relatively high and increasing with the three stages of stand development; the differences among the indices became evident only when their contribution to a model for tree basal area prediction was considered.

Pukkala (1989) and Miina and Pukkala (2002) compared the performance of two competition indices, one based on the sum of angles and another one based on *EFT*.

The performance of the indices was assessed when they were added to a tree growth model that already included tree size. The index based on *EFT* exhibited a slightly better performance.

Canham et al. (2004) considered the separation of two effects of competition: shading by the neighbors and crowding that reflects both belowground competition and physical aboveground inhibition of crown development. They assumed that traditional competition indices are a measure of crowding and tested the performance of these two types of competition measures in a model of annual radial growth of individual trees in forests of northern, interior British Columbia, Canada, dominated by western hemlock and western redcedar. The effects of crowding and shading were subtracted from the potential radial growth. For both species, the most parsimonious regression models included terms for the effects of tree size, crowding, and shading and separate competitive effects of four different groups of competing species. The models explained 33–59% of the variation in radial growth of the two species. For both species, growth declined more steeply as a function of crowding than shading. The measure of crowding that gave the best fit did not include the dimension of the subject tree. There was striking asymmetry in the strength of interspecific competition between hemlock and redcedar, with crowding by hemlock having a strong effect on redcedar, while crowding by redcedar had relatively little effect on the radial growth of hemlock.

Rivas et al. (2005) compared the contribution of a large list of distance-independent, area overlap, and distance-weighted size indices when predicting tree basal area growth in mature even-aged stands of Cooper pine in Mexico. The distance-dependent indices were computed using several methods for the selection of competitors. Based on the mean square error reduction when a competition index was added to an individual tree basal area growth model in which site index, stand density, basal area and the ratio between crown width and tree height had already been included, the distance-independent competition indices $G_{>di}$ and the corresponding variant modified by Schröder and Gadow (1999) performed as well as the best distance-dependent competition indices.

Zhao et al. (2006) developed individual-based spatially-explicit diameter growth and survival models for each species group in a natural temperate species-rich forest in the southeastern USA. Their models explicitly partition the competitive effects of different species groups of neighbors. The individual-based approach proved effective in detecting density-dependent relationships and understanding the ecological processes of the mixed-species stand. From the results, the authors concluded that the competitive effects among different species are unequal and asymmetric and that the identity of neighboring species was important.

Contreras et al. (2011) evaluated 16 measures of tree competition in terms of their effectiveness as growth predictors for three important conifer tree species in western Montana, USA. Strong correlations were exhibited between several competition indices and tree growth for all three species. The best distance-dependent indices explained a larger proportion of the variation in growth than the best distance-independent indices (64% vs. 56%). Competition indices derived from light interception models performed poorly in terms of predicting tree growth.

The low correlations between light-value indices and growth, the authors opined, suggest that trees in the semi-arid conditions of their study area are not competing primarily for light.

9.3.3 Distance-Independent Versus Distance-Dependent Competition Indices

A wide array of competition indices have been developed and no single index or class of indices has been found to be universally superior. Performance varies according to forest type and forest conditions. There remain, however, many opportunities for refinement and improvement on past methods for quantifying local competition in forest stands, including a better understanding of resource depletion (symmetric competition) and resource preemption (asymmetric competition). Better definition of the zone of influence (and hence search radius) of each tree may also bring about improvements in predictive ability. The impact of sample plot size on the relative performance of distance-dependent and distance-independent indices also needs to be elucidated.

Although superiority for distance-dependent indices of point density for growth predictions has not been consistently exhibited, there are many management and ecological considerations that require spatially-explicit forest growth models. Plantation management questions, such as the influence of spacing rectangularity (ratio of distance between rows to distance between trees in rows) of planting stock on tree and stand growth cannot be evaluated without spatially-explicit models. Simulating a wide range of thinning options, including the tradeoff between maximizing the quality of trees in the residual stand versus aiming for a relatively even distribution of the residual growing stock, requires growth response functions that include spatial information. Furthermore, evaluations of alternative silvicultural systems and stand management options to maintain diverse stand composition and structure involve spatially-explicit considerations. Models involving stand density measures and distance-independent competition indices will provide the same predictions of tree and stand growth response regardless of differences in the spatial distribution resulting from various treatments. While the presently-available density-dependent point density measures may not provide improved precision over density-independent indices for longer-term forecasts, they can provide information on relative differences likely to result for a number of management alternatives that involve spatial differences.

References

Alder D (1979) A distance-independent tree model for exotic conifer plantations in East Africa. For Sci 25:59–71

Alemdag IS (1978) Evaluation of some competition indexes for the prediction of diameter increment in planted white spruce. Canadian Forestry Service, Department of the Environment. Forest Management Institute Information Report FMR-X-108

Arney JD (1974) An individual tree model for stand simulation in Douglas-fir. In: Fries J (ed) Growth models for tree and stand simulation. Royal College of Forestry, Stockholm, pp 38–46, Research Notes 30

Begon M, Harper JL, Townsend CR (1986) Ecology. Blackwell Science, Oxford

Bella IE (1971) A new competition model for individual trees. For Sci 17:364–372

Berger U, Hildenbrandt H (2000) A new approach to spatially explicit modelling of forest dynamics: spacing, ageing and neighbourhood competition of mangrove trees. Ecol Mod 132:287–302

Berger U, Piou C, Schiffers K, Grimm V (2008) Competition among plants: concepts, individual-based modelling approaches, and a proposal for a future research strategy. Perspect Plant Ecol Evol Syst 9(3–4):121–135

Biging GS, Dobbertin M (1992) A comparison of distance-dependent competition measures for height and basal area growth of individual conifer trees. For Sci 38:695–720

Biging GS, Dobbertin M (1995) Evaluation of competition indexes in individual tree growth models. For Sci 41:360–377

Biging GS, Robards AT, Turnblom CE, Deusen PCV (1994) The predictive models and procedures used in the forest stand generator (STAG). Hilgardia 61(1):1–36

Brand DG, Magnussen S (1988) Asymmetric, two-sided competition in even-aged monocultures of red pine. Can J For Res 18:901–910

Brown GS (1965) Point density in stems per acre. Forest Research Institute, New Zealand Forest Research Notes No. 38

Brunner A (1998) A light model for spatially explicit forest stand models. For Ecol Manage 107:19–46

Brunner A, Nigh G (2000) Light absorption and bole volume growth of individual Douglas-fir trees. Tree Physiol 20:323–332

Burton PJ (1993) Some limitations inherent to static indices of plant competition. Can J For Res 23:2141–2152

Canham CD, Finzi AC, Pacala SW, Burbank DH (1994) Causes and consequences of resource heterogeneity in forests: interspecific variation in light transmission by canopy trees. Can J For Res 24:337–349

Canham CD, Coates KD, Bartemucci P, Quaglia S (1999) Measurement and modeling of spatially explicit variation in light transmission through interior cedar-hemlock forests of British Columbia. Can J For Res 29:1775–1783

Canham CD, LePage PT, Coates KD (2004) A neighborhood analysis of canopy tree competition: effects of shading versus crowding. Can J For Res 34:778–787

Cannell MGR, Grace J (1993) Competition for light: detection, measurement, and quantification. Can J For Res 23:1969–1979

Cole WG, Lorimer CG (1994) Predicting tree growth from crown variables in managed northern hardwood stands. For Ecol Manage 67:159–175

Contreras MA, Affleck D, Chung W (2011) Evaluating tree competition indices as predictors of basal area increment in western Montana forests. For Ecol Manage 262:1939–1949

Daniels RF (1976) Simple competition indexes and their correlation with annual loblolly-pine tree growth. For Sci 22:454–456

Daniels RF, Burkhart HE, Clason TR (1986) A comparison of competition measures for predicting growth of loblolly-pine trees. Can J For Res 16:1230–1237

De Luis M, Raventós J, Cortina J, Moro MJ, Bellot J (1998) Assessing components of a competition index to predict growth in an even-aged *Pinus nigra* stand. New For 15:223–242

Ek AR, Monserud RA (1974) FOREST: a computer model for simulating the growth and reproduction of mixed species forest stands. University of Wisconsin, School of Natural Resources, Research Report R2635

Gadow Kv, Hui G (1999) Modelling forest development. Kluwer, Dordrecht

Glover GR, Hool JN (1979) A basal area ratio predictor of loblolly pine plantation mortality. For Sci 25:275–282

Grimm V, Railsback S (2005) Individual-based modeling and ecology. Princeton University Press, Princeton

Hamilton GJ (1969) The dependence of volume increment of individual trees on dominance, crown dimensions, and competition. Forestry 42:133–144

Harper JL (1977) Population biology of plants. Academic, London

Hegyi F (1974) A simulation model for managing jack-pine stands. In: Fries J (ed) Growth models for tree and stand simulation. Royal College of Forestry, Stockholm, pp 74–90, Research Notes 30

Holmes MJ, Reed DD (1991) Competition indexes for mixed species northern hardwoods. For Sci 37:1338–1349

Hynynen J, Ojansuu R (2003) Impact of plot size on individual-tree competition measures for growth and yield simulators. Can J For Res 33:455–465

Illian J, Penttinen A, Stoyan H, Stoyan D (2008) Statistical analysis and modelling of spatial point patterns. Wiley, Chichester

Kaitaniemi P, Lintunen A (2010) Neighbor identity and competition influence tree growth in Scots pine, Siberian larch, and silver birch. Ann For Sci 67:604p1–604p7

Ledermann T (2010) Evaluating the performance of semi-distance-independent competition indices in predicting the basal area growth of individual trees. Can J For Res 40:796–805

Ledermann T, Stage AR (2001) Effects of competitor spacing in individual-tree indices of competition. Can J For Res 31:2143–2150

Lemmon PE, Schumacher FX (1962a) Stocking density around ponderosa pine trees. For Sci 8:397–402

Lemmon PE, Schumacher FX (1962b) Volume and diameter growth of ponderosa pine trees as influenced by site index, density, age, and size. For Sci 8:236–249

Lorimer CG (1983) Tests of age-independent competition indices for individual trees in natural hardwood stands. For Ecol Manage 6:343–360

Martin GL, Ek AR (1984) A comparison of competition measures and growth-models for predicting plantation red pine diameter and height growth. For Sci 30:731–743

Martin GL, Ek AR, Monserud RA (1977) Control of plot edge bias in forest stand growth simulation models. Can J For Res 7:100–105

Miina J, Pukkala T (2000) Using numerical optimization for specifying individual-tree competition models. For Sci 46:277–283

Miina J, Pukkala T (2002) Application of ecological field theory in distance-dependent growth modelling. For Ecol Manage 161:101–107

Monserud RA, Ek AR (1974) Plot edge bias in forest stand growth simulation models. Can J For Res 4:419–423

Moore JA, Budelsky CA, Schlesinger RC (1973) A new index representing individual tree competitive status. Can J For Res 3:495–500

Moravie MA, Durand M, Houllier F (1999) Ecological meaning and predictive ability of social status, vigour and competition indices in a tropical rain forest (India). For Ecol Manage 117:221–240

Munro DD (1974) Forest growth model – a prognosis. In: Fries J (ed) Growth models for tree and stand simulation. Royal College of Forestry, Stockholm, pp 7–21, Research Notes 30

Nance WL, Grissom JE, Smith WR (1988) A new competition index based on weighted and constrained area potentially available. In: Ek AR, Shifley SR, Burk TE (eds) Forest growth modelling and prediction. USDA Forest Service, North Central Forest Experiment Station, St. Paul, pp 134–142, General Technical Report NC-120

Newnham RM (1964) The development of a stand model for Douglas fir. Ph.D. thesis, University of British Columbia, 201pp

Newton PF, Jolliffe PA (1998) Assessing processes of intraspecific competition within spatially heterogeneous black spruce stands. Can J For Res 28:259–275

Opie JE (1968) Predictability of individual tree growth using various definitions of competing basal area. For Sci 14:314–323

Pacala SW, Canham CD, Saponara J, Silander JA, Kobe RK, Ribbens E (1996) Forest models defined by field measurements: estimation, error analysis and dynamics. Ecol Monogr 66(1): 1–43

Paulo MJ, Stein A, Tomé M (2002) A spatial statistical analysis of cork oak competition in two Portuguese silvopastoral systems. Can J For Res 32:1893–1903

Pelz DR (1978) Estimating individual tree growth with tree polygons. In: Fries J, Burkhart HE, Max TA (eds) Growth models for long term forecasting of timber yields. School of Forestry and Wildlife Resources, Virginia Polytechnic Institute and State University, Blacksburg, pp 172–178, Pub. FWS-1-78

Perry DA (1985) The competition process in forest stands. In: Jackson JE, Cannel MGR (eds) Attributes of trees as crop plants. Institute of Terrestrial Ecology, Abbots Ripton, pp 481–506

Pommerening A (2002) Approaches to quantifying forest structures. Forestry 75:305–324

Pommerening A, Stoyan D (2006) Edge-correction needs in estimating indices of spatial forest structure. Can J For Res 36:1723–1739

Pommerening A, Stoyan D (2008) Reconstructing spatial tree point patterns from nearest neighbour summary statistics measured in small subwindows. Can J For Res 38:1110–1122

Pommerening A, LeMay V, Stoyan D (2011) Model-based analysis of the influence of ecological processes on forest point pattern formation – a case study. Ecol Mod 222:666–678

Pretzsch H (1997) Analysis and modeling of spatial stand structures. Methodological considerations based on mixed beech-larch stands in Lower Saxony. For Ecol Manage 97:237–253

Pretzsch H (2009) Forest dynamics, growth and yield. Springer, Berlin

Pretzsch H, Biber P, Ďurský J (2002) The single tree-based stand simulator SILVA: construction, application and evaluation. For Ecol Manage 162:3–21

Pukkala T (1989) Methods to describe the competition process in a tree stand. Scand J For Res 4:187–202

Pukkala T, Kolström T (1987) Competition indices and the prediction of radial growth in Scots pine. Silva Fennica 21:55–67

Radtke PJ, Burkhart HE (1998) A comparison of methods for edge-bias compensation. Can J For Res 28:942–945

Richards M, McDonald AJS, Aitkenhead MJ (2008) Optimisation of competition indices using simulated annealing and artificial neural networks. Ecol Mod 214:375–384

Ritchie MW, Hann DW (1986) Development of a tree height growth model for Douglas-fir. For Ecol Manage 15:135–145

Rivas JJC, Álvarez-González JG, Aguirre O, Hernández FJ (2005) The effect of competition on individual tree basal area growth in mature stands of *Pinus cooperi* Blanco in Durango (Mexico). Eur J For Res 124:133–142

Rouvinen S, Kuuluvainen T (1997) Structure and asymmetry of tree crowns in relation to local competition in a natural mature Scots pine forest. Can J For Res 27:890–902

Sabatia CO, Burkhart HE (2012) Competition among loblolly pine trees: does genetic variability of the trees in a stand matter? For Ecol Manage 263:122–130

Schröder J, Gadow Kv (1999) Testing a new competition index for Maritime pine in northwestern Spain. Can J For Res 29:280–283

Soares P, Tomé M (1999) Distance-dependent competition measures for eucalyptus plantations in Portugal. Ann For Sci 56:307–319

Soares P, Tomé M (2003) GLOBTREE: an individual tree growth model for Eucalyptus globulus in Portugal. In: Amaro A, Reed D, Soares P (eds) Modelling forest systems. CABI Publishing, Wallingford, pp 97–110

Spurr SH (1962) A measure of point density. For Sci 8:85–96

Stage AR, Ledermann T (2008) Effects of competitor spacing in a new class of individual-tree indices of competition: semi-distance-independent indices computed for Bitterlich versus fixed-area plots. Can J For Res 38:890–898

Tomé M, Burkhart HE (1989) Distance-dependent competition measures for predicting growth of individual trees. For Sci 35:816–831

Tomppo E (1986) Models and methods for analysing spatial patterns of trees. Commun Inst For Fenn 138:1–65

Weiner J (1985) Size hierarchies in experimental populations of annual plants. Ecology 66: 743–752

Weiner J (1986) How competition for light and nutrients affects size variability in *Ipomoea tricolor* populations. Ecology 67:1425–1427

Weiner J (1990) Asymmetric competition in plant populations. Trends Ecol Evol 5(11):360–364

Weiner J, Solbrig OT (1984) The meaning and measurement of size hierarchies in plant populations. Oecologia 61:334–336

Welden CW, Slauson WL (1986) The intensity of competition versus its importance: an overlooked distinction and some implications. Q Rev Biol 61:23–46

Welden C, Slauson WL, Ward RT (1988) Competition and abiotic stress among trees and shrubs in northwest Colorado. Ecology 69:1566–1577

Wimberly MC, Bare BB (1996) Distance-dependent and distance-independent models of Douglas-fir and western hemlock basal area growth following silvicultural treatment. For Ecol Manage 89:1–11

Wu HI, Sharpe PJH, Walker J, Penridge LK (1985) Ecological field theory: a spatial analysis of resource interference among plants. Ecol Mod 29:215–243

Wykoff WR, Crookston NL, Stage AR (1982) User's guide to the stand prognosis model. USDA Forest Service, Intermountain Forest and Range Experiment Station, Ogden, General Technical Report INT-133

Yoda K, Kira T, Hozimu K (1957) Intraspecific competition among higher plants, IX. Further analysis of the competition interaction between adjacent individuals. J Inst Polytech Osaka City Univ 8:161–178

Zar JH (1996) Biostatistical analysis, 3rd edn. Prentice Hall, Upper Saddle River

Zhao D, Borders B, Wilson M, Rathbun SL (2006) Modeling neighborhood effects on the growth and survival of individual trees in a natural temperate species-rich forest. Ecol Mod 196: 90–102

Chapter 10
Modeling Forest Stand Development

10.1 Need for Stand Models

Growth and yield models are essential for informed forest management decision making, and a great deal of emphasis has been placed on developing reliable models for predicting stand characteristics (volume, basal area, numbers of trees per unit area, and height and diameter distributions) at selected times in the production cycle. Growth and yield forecasts may be required for a short-term or long-term basis, for the overall stand volume or volume by product and size classes. With the wide variety of existing stand conditions and the diverse objectives and needs of users of growth and yield models, as expected, numerous approaches have been proposed. These approaches range from models that provide only a specified aggregate stand volume to models with information about individual trees. Regardless of the structural complexity and amount of stand detail provided, all growth and yield models have a common purpose: to produce estimates of stand characteristics at specified points in time.

While growth and yield forecasts enter into virtually all stand management decisions, the primary uses of growth and yield information can be categorized as: production forecasting, inventory updating, evaluation of silvicultural alternatives, management planning, and harvest scheduling.

Questions arise with regard to whether it is more efficient to develop multipurpose models that can be used for a variety of forecasting objectives as opposed to developing a number of limited-purpose models for specific tasks. In choosing a growth and yield model users must be concerned with the stand detail needed for the particular decision(s) of interest and the efficiency in providing the required information. When predictions are required for a broad range of management decisions, it would be desirable to have a system of growth and yield models capable of providing logical and consistent estimates for varying levels of stand detail (whole stand values, size class data, or individual tree information), thus allowing users to efficiently compute estimates with stand detail appropriate to the use of the information.

H.E. Burkhart and M. Tomé, *Modeling Forest Trees and Stands*,
DOI 10.1007/978-90-481-3170-9_10,

10.2 Approaches to Modeling Forest Stands

Numerous approaches to modeling forest stand development have been taken and there is no universally accepted means of categorizing the various techniques. One possibility is to classify according to the modeling entity (stand, size class, tree) involved. Alternatively, growth and yield models might be grouped according to the level of resolution or detail about stand structure provided (overall stand values, size-class information, tree-level data).

In the first classification alternative, models that provide disaggregation of whole-stand values into size-class information would be called "whole-stand" models, whereas in the second approach they would be labeled "size-class models". The most commonly-applied disaggregation approach involves using a continuous distribution function to describe the diameter at breast height distribution. Although only stand-level variables are projected, size-class information is developed and yield is estimated from the resulting stand table. Due to the wide-spread application of this so-called "diameter-distribution" method, we have recognized it as a separate category in our classification scheme (Table 10.1) and have presented details of the approach in Chap. 12. Thus, to some extent, we have applied somewhat of a "hybrid" between two alternative groupings when organizing the presentation of growth and yield models into chapters for this book. The categories recognized are arbitrary and not exhaustive but they provide a framework for presenting the key features of the most commonly used modeling structures. Table 10.1 contains a summary to the key components of the most prevalent growth and yield modeling systems. In practice, elements from more than one modeling approach may be used when developing an operational growth and yield forecasting system.

Although no classification system of modeling approaches is fully satisfactory, categorizing provides a succinct way of identifying the general features involved rather than having to provide a detailed description of each model's structure. While all "whole-stand", "size-class", "diameter-distribution", and "individual-tree" models are not exactly the same – indeed no two are exactly the same – they share certain common characteristics. And the various modeling approaches can be linked. Developing linkages allows for efficiencies and consistency in providing different levels of detail on stand structure. Linkages can consist of disaggregation from stand to size class to individual tree, or conversely of aggregation of individual tree to size class to whole stand. There are limitless ways that these linkages might be structured, each with its own advantages and disadvantages. Relating different levels of resolution of growth and yield models can be an aid to understanding relative merits of alternative approaches as well as insuring consistency in predictions at varying levels of detail. Figure 10.1 illustrates relationships among the most commonly used growth and yield modeling methods and shows how the alternate structures are related.

In the whole-stand approach (Chap. 11), quantities such as volume, basal area, and/or number of trees per unit area are forecast. The input or predictor variables for these models for even-aged stands are generally age, site index, and stand density

Table 10.1 Key components of commonly-used structures for modeling growth and yield of forest stands (modified and adapted from Davis et al. 2001).

Model type	Model equations: primary relationships and variables (growth equations in bold)
Whole stand models	
Direct prediction of growth	$\begin{cases} iV_{1-2} = f(t_1, t_2, S, SD_1) \\ \mathrm{V}_2 = \mathrm{V}_1 + \mathrm{iV}_{1-2} \end{cases}$
Direct prediction of future volume	$\begin{cases} SD_2 = f(t_1, t_2, S, SD_1) \\ V_2 = f(t_2, S, SD_2) \end{cases}$
Prediction from stand density projection	$\begin{cases} SD_2 = f(t_1, t_2, S, SD_1) \\ \mathrm{V}_2 = \mathrm{f}(\mathrm{t}_2, \mathrm{S}, \mathrm{D}_2) \end{cases}$
Diameter distribution models	
Implicit prediction	$\begin{cases} SD_2 = f(t_1, t_2, S, SD_1) \\ \mathrm{pdf}(\mathrm{d}_2) = \mathrm{f}(\mathrm{t}_2, \mathrm{S}, \mathrm{D}_2) \\ \mathrm{v}_{k2} = \mathrm{f}(\mathrm{d}_{k2}) \\ \mathrm{V}_2 = \Sigma_k \mathrm{v}_{k2}\mathrm{n}_{k2} \end{cases}$
Size class models	
Diameter class growth models	$\begin{cases} nk_2 = f(nk_1, t_{2-1}, S, SD_1) \\ \mathrm{v}_{k2} = \mathrm{f}(\mathrm{d}_{k2}) \\ \mathrm{V}_2 = \Sigma_k \mathrm{vk}_2\mathrm{nk}_2 \end{cases}$
Matrix models	$\begin{cases} n_{k2} = n_{k1} P d_{kj} \\ \mathrm{v}_{k2} = \mathrm{f}(\mathrm{d}_{k2}) \\ \mathrm{V}_2 = \Sigma_k \mathrm{v}_{k2}\mathrm{n}_{k2} \end{cases}$
Individual tree models	
Distance-dependent	$\begin{cases} \mathrm{CI}_{kl} = \mathrm{f}\left[(\mathrm{d}_k,\mathrm{h}_k,\mathrm{c}_k, \mathrm{XY}_k)_1, (\mathrm{d}_{\Sigma j},\mathrm{h}_{\Sigma j},\mathrm{c}_{\Sigma j}, \mathrm{XY}_{\Sigma j})_1\right] \\ (d_k, h_k, c_k)_2 = f\left[t_1, t_2, S, SD_1, CI_{k1}, (d_k, h_k, c_k, XY_k)_1\right] \\ \mathrm{v}_{k2} = \mathrm{f}(\mathrm{d}_k,\mathrm{h}_k)_2 \\ \mathrm{V}_2 = \Sigma_k \mathrm{v}_{k2}\mathrm{n}_{k2} \end{cases}$
Distance-independent	$\begin{cases} \mathrm{CI}_{k1} = \mathrm{f}\left[\mathrm{S}, \mathrm{SD}_1, (\mathrm{d}_k,\mathrm{h}_k,\mathrm{c}_k)_1\right] \\ (d_k, h_k, c_k)_2 = f\left[t_1, t_2, S, SD_1, CI_{k1}, (d_k, h_k, c_k)_1\right] \\ \mathrm{v}_{k2} = \mathrm{f}(\mathrm{d}_k,\mathrm{h}_k)_2 \\ \mathrm{V}_2 = \Sigma_k \mathrm{v}_{k2}\mathrm{n}_{k2} \end{cases}$

S = site index; t_i = stand age at time t_i; $\mathrm{SD_i}$ = stand density (number of trees per ha and/or stand basal area) at time t_i; V_i = stand volume at time t_i; iV_{1-2} = volume growth in the period between t_1 and t_2; *pdf(d)* = diameter distribution function; n_{ki} = number of trees in diameter class k at time t_i; d_{ki}, h_{ki}, c_{ki} = diameter, height and crown size of tree k at time t_i; v_k = volume of a tree with d_k; $\mathrm{Pd_{kj}}$ = probability of moving from state k to j; CI_{ki} = competition index for tree k at time t_i; XY_k = spatial coordinates of tree k; subscript Σ_j indicates that all competitors j are included in the computation

(numbers of trees planted per unit area for plantations; initial basal area for natural stands). Often only aggregated volume growth and/or yield is predicted for the total stand. As a variation on this approach, several researchers have applied probability density functions to estimate the number of trees by *dbh* (diameter at breast height)

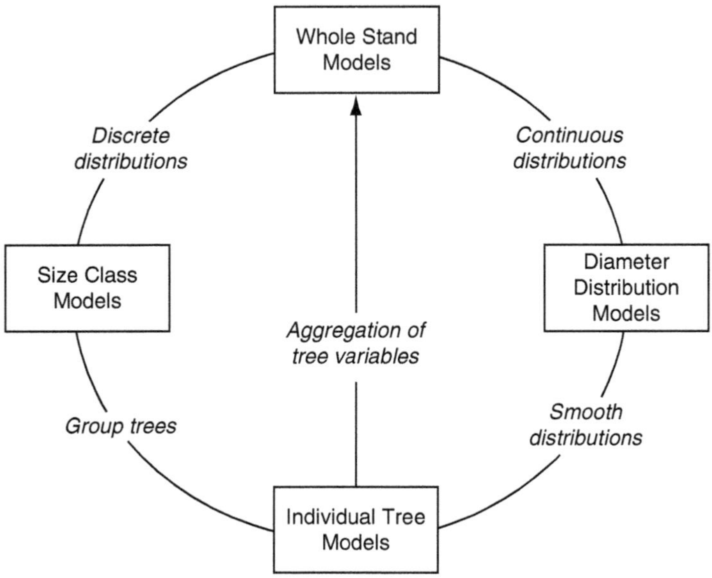

Fig. 10.1 General relationships among commonly applied approaches to modeling forest stand development

classes, given that an estimate of the total number of trees per unit area is available. This approach, commonly termed the "diameter distribution approach," still relies on overall stand values as the basic modeling unit (Chap. 12).

Size-class models typically involve projecting the stand table (numbers of trees by diameter class) forward for a specified time period (Chap. 13). Stand tables can be converted to stock tables (volume or weight by diameter class) through application of tree taper functions, volume or weight prediction equations (Chaps. 2, 3, and 4); the difference between the current stock table and the estimated future stock table (after allowance for mortality) is the periodic growth. Stand tables (or tree lists by *dbh* classes) can be projected via a system of equations for upgrowth (trees moving into larger diameter classes) and ingrowth (trees growing into the smallest measured diameter class), or by means of a matrix model (Chap. 15) with transition probabilities. The so-called "absorbing states" (states that once entered cannot be changed) for mortality and for harvested trees are generally included in matrix models.

Approaches to predicting stand growth and yield that use individual trees as the modeling entity are referred to as "individual tree models." The components of tree growth (e.g., diameter increment, height increment) in these models are commonly linked through a computer program that simulates the growth of each tree and then aggregates these values to provide estimates of stand growth and yield. Individual tree models are divided into two classes, distance-independent and distance-dependent, depending on whether or not individual tree locations are used. Distance-independent models project tree growth either individually or by size classes, usually as a function of present size and stand-level variables (e.g., age, site index, number of trees per unit area). In distance-dependent models, initial stand conditions are input or generated and each tree is assigned spatial coordinates. The growth of each tree is predicted as a function of its attributes, the site quality, and a measure of competition from neighbors.

Modeling methodology continues to evolve and to become increasingly sophisticated as forest biometricians bring new biological rationale, advanced statistical techniques, and powerful computing technology to bear on growth and yield prediction problems. Levins (1966) argued that modelers of population biology strive to maximize simultaneously three desirable properties of a given model: generality, reality and precision. Generality refers to applicability to a range of instances, reality might be thought of as conformity of model assumptions and relationships to the real system, and precision indicates the degree of exactness in predictions. In any one model, Levins asserted, developers may sacrifice one of these desired properties to achieve a higher level of the other two. In traditional growth and yield models, generality is sacrificed for increased reality and precision (with the primary emphasis being on precision). Given the usual objectives of growth and yield modelers, this is a reasonable strategy. However, due to rapidly changing management and environmental conditions, there is increased interest in enhancing the generality of growth and yield models.

10.3 Prediction, Parsimony and Noise

To have high utility, models must be accurate. Model accuracy can be improved by collecting more data, improving the quality of the data obtained, or by applying more sophisticated modeling techniques to existing data. Collecting data is expensive relative to performing analyses; thus analysts must make the best possible use of the data available. When planning data collection efforts, it is important that a wide range of site and stand conditions be included in the sample.

It is commonly believed that a model can be no more accurate than the data on which it is based. This is not necessarily true, however, because models can amplify patterns and discard unwanted noise, they can be more accurate than the data used to build them. Gauch (1993) emphasized that models being more accurate than the data available is dependent on: (i) the precise question being asked of the model, (ii) the design of the experiment, and (iii) the quantity and accuracy of the available data.

Typical modeling efforts attempt to enhance prediction by amplifying pattern and discarding noise. Ordinarily, most of the pattern in a data set is recovered quickly with relatively simple models. Patterns usually depend on a few main causal factors that can be summarized readily. Noise, on the other hand, is recovered slowly as a model's complexity increases. The accuracy of prediction, therefore, increases quickly as parameters are added to relatively simple models. Predictive ability tends to peak rather quickly (this point is sometimes referred to as 'Ockham's Hill')

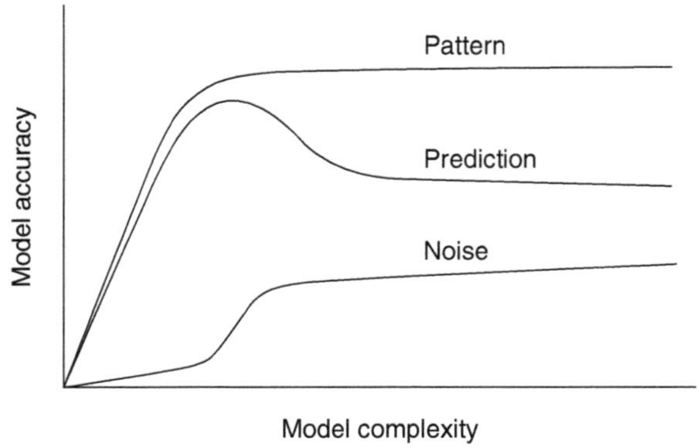

Fig. 10.2 Pattern is generally determined by a few factors, meaning that relatively simple models recover much of it in a data set; noise is more idiosyncratic and complex and is thus recovered more slowly as model complexity increases; predictive ability arises from pattern recovery minus noise recovery so it is expected to increase quickly in simple models but peak on 'Ockham's Hill' and then decrease in increasingly complex models (Adapted from Gauch 1993)

and then decrease with increasingly complex models (Gauch 1993). Figure 10.2 illustrates how model accuracy (in terms of predictive ability) is affected by the recovery of pattern and noise through increasing model complexity. Modeling thus offers its greatest benefits when a parsimonious, simple model captures the essence of the data's pattern in a large, noisy data set (Burkhart 2003).

10.4 Level for Modeling Forest Stands

The level at which forest stands can be modeled is often dictated by the data available. If, for instance, individual trees are not numbered and identified, individual-tree-based approaches are not possible. Permanent plots established in the past have sometimes had limited usefulness because of inadequacies in the measurements taken. To allow for flexibility in modeling approaches, in permanent plots established for growth estimation purposes, tree measurements should include, as a minimum, *dbh*, height, crown measures, stem quality assessment and tree spatial coordinates.

Whether one should model at the tree level and aggregate for stand estimates or model at an aggregated level depends on the specific objectives for modeling. The use for which a growth model is intended, it is generally argued, should determine the resolution level at which one should operate. However, as Leary (1979) discussed, another consideration is the relationship between dimensionality of the model (or resolution level) and the time horizon over which projections are to be made. The following relationship from Kahne (1976) has been found useful in a number of large-scale modeling efforts:

$$m_d = \frac{k_\alpha}{t_h}$$

where t_h is the time horizon over which projections are to be made, m_d is the dimension of the model state vector and k_α is a constant for a given accuracy or precision level. This relationship indicates that the model dimension should be reduced for long-term projections and increased for short-term projections, to give the same level of accuracy.

The relationship between model dimensionality and projection length has been fairly well accepted by scientists working in various fields with large-scale models, but it has not received much attention by researchers involved in forest projection. Forest projections are often made for any time horizon of interest without regard to the dimensionality of the model.

Insight into the influence of dimensionality of growth and yield models and performance over increasing projection lengths can be gained from the study of Shortt and Burkhart (1996). They evaluated a whole-stand (Sullivan and Clutter 1972) and an individual-tree, distance-independent (Amateis et al. 1989) growth and yield model for loblolly pine for the purpose of updating forest inventory data. The growth and yield models were evaluated at varying projection periods by using permanent plots measured at 0, 3, 6 and 9 years after initial plot establishment. Evaluations were based solely on the capability of each model to predict merchantable volume. The individual-tree model produced the best result until the 6-year period, at which time it was approximately equal to the whole-stand model. After 6 years, the whole-stand model produced more reliable results. Both models displayed increasing error of prediction with increasing projection length. As expected from the general relationship of model dimension to projection length, for short-term projections the more detailed individual-tree model performed best, but for long-term projections the simple whole-stand model performed best.

Although the science of modeling forests has advanced greatly, and continues to advance, there is still a great deal of judgment and intuition involved. One cannot overemphasize the necessity of having clearly stated objectives for modeling, because no approach can satisfy all purposes. Regardless of the modeling objective, one should strive to: (i) select as parsimonious a model as possible to describe the population trends of interest, and (ii) adjust the dimensionality of the model to suit the projection length (Burkhart 2003).

10.5 Field Data for Growth and Yield Modeling

The typical approach taken in past growth and yield studies has involved defining a population of interest, obtaining a sample from the defined population (the sample could consist of temporary plots, permanent plots, or both), and estimating coefficients (invariably with least squares) in specified equation forms. This approach produces satisfactory prediction tools for many purposes, but it may not be adequate in circumstances where forest management practices and objectives are changing rapidly. Given that growth and yield models are used to project the present forest resource and to evaluate alternative treatment effects, data both of the inventory type (which describe operational stands of interest) and of the experimental or research type (which describe response to treatment) are needed.

As an example, when the primary interest is in evaluating silvicultural alternatives, designed-experiment type data with the relevant silvicultural treatments included would be required. The model structure would likely be quite detailed in terms of the underlying equations and the types of output produced so that the full range of treatments could be evaluated under varying assumptions. If, on the other hand, one were primarily interested in inventory updating, the data for equation fitting should be obtained from a representative sample of the population of stands to which the model is going to be applied. The input to the model would necessarily need to be consistent and compatible with the inventory data definitions and quantities available. The fundamental equations should be as simple and straightforward as possible for producing output (updated stand statistics) that is needed for and consistent with the inventory data base. What is "best" depends primarily on the objective(s) for developing the model; obviously, the objectives should be specified clearly before data collection and analyses are initiated (Adlard 1995).

Growth and yield information is used for a variety of purposes; no single data base or modeling approach can be optimal for all applications. Köhl et al. (1995), in a comparison of Swiss growth and yield plots and forest survey plots, noted that sample plots from forest inventories are representative of the total population but give only limited information on site conditions and management history. On the other hand, growth and yield plots contain detailed information on site conditions and management history but are not representative of the total population.

Empirical growth and yield models are based on field plot measurements. Ideally, one would like to have *permanent plots* that have been established, maintained, and measured regularly over a full rotation (that is, a complete time series). This, of course, is not always feasible. More commonly, permanent plots are established in stands of varying initial conditions (i.e. differing ages, site qualities, and stand densities) and measured over one or more growth periods. Such *interval plots* represent a partial time series for each stand and they are an effective and efficient means for obtaining growth and yield data. In some instances where permanent plots are not available and results are needed quickly, *temporary plots* are established in stands of varying ages, site qualities, and densities. Equations for estimating yield can be fitted to temporary plot data and growth can be estimated by differencing the yield function. Diameter and height measurements from temporary plots are sometimes supplemented with information from a sample of increment cores or stem analysis trees. While temporary plots may be satisfactory for some purposes, they generally do not provide fully satisfactory information for growth and yield modeling. Figure 10.3 illustrates these three types of field plots.

Many permanent plot installations in forest types of interest for growth and yield modeling consist of plots with different thinning treatments plus an unthinned

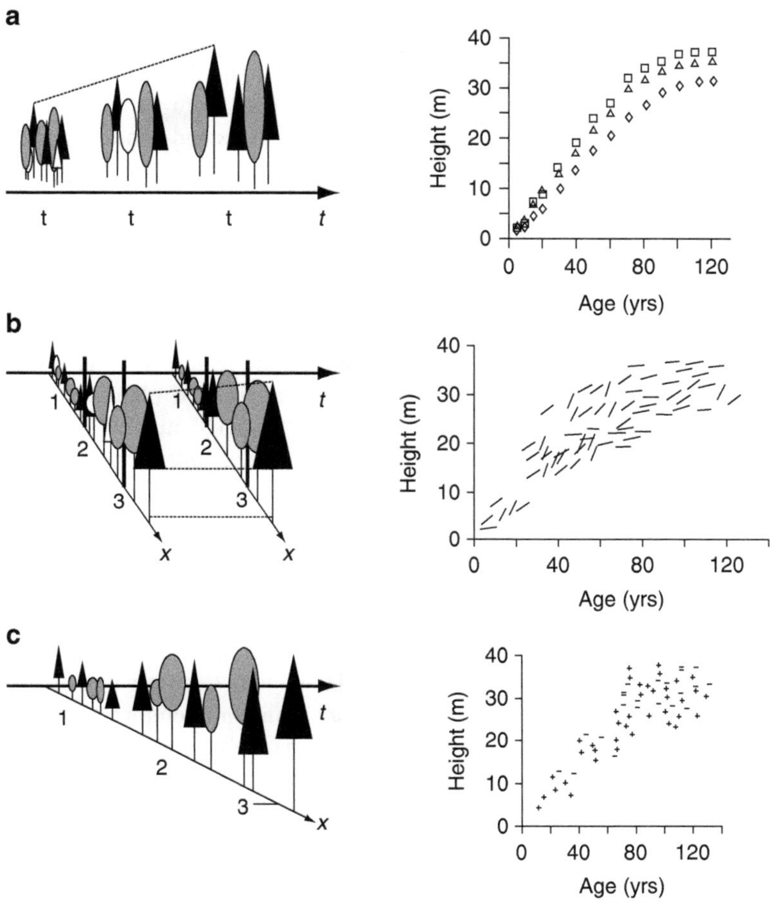

Fig. 10.3 Schematic representation of (**a**) permanent plots, (**b**) interval plots, and (**c**) temporary plots established for growth and yield estimation. (**a**) *Left*: A permanent plot with three successive measurements (white trees are removed during thinning operations; the time axis is designated t). *Right*: Hypothetical data series derived from three permanent plots. (**b**) *Left*: Three interval plots; white trees are removed during thinning operations. *Right*: Interval data for obtaining rates of change of observed state variables. (**c**) *Left*: Three temporary plots of varying age; the x-axis signifies the tree position; the symbol t indicates the time axis. *Right*: Independent height-age data obtained from temporary plots (Adapted from Gadow and Hui 1999)

control. The ability to predict response to thinning is often seen as a central part of growth and yield prediction systems. Vanclay et al. (1995) recommended that permanent plot installations for developing growth models include experimental observations that are manipulated to provide data on a wide range of stand densities and thinning treatments. In addition to thinning, for intensively managed forest types, managers may also need to evaluate response to fertilizer applications, vegetation control, and other treatments that could be applied at various times (stand establishment, during the rotation) in the production cycle and in various combinations. It is not possible to apply all treatments of interest to a single installation of permanent plots. Consequently, data from a variety of silvicultural experiments are often used in the modeling process. Response data from various sources can be incorporated into an overall modeling framework; stand modeling serves to integrate and synthesize growth response information, identify knowledge gaps, and provide forecasting capability for managers. (Chap. 16 includes additional detail on modeling stand response to thinning, vegetation control, fertilizer applications, and genetic enhancement.)

As an illustration of the need for data from designed experiments as well as from plots in operational stands, consider the problem of determining an appropriate initial planting density. One of the most important factors that is within the direct control of forest managers is tree spacing at time of planting. Economic considerations dictate that managers strive for the "optimal" number of trees. This optimum will, of course, vary widely depending on the management objective. Data from surveys of existing stands rather than from designed experiments with the independent variables controlled at specified levels result in imprecise estimates of some parameters and place restrictions on the type of analyses that can be performed. One cannot legitimately treat density (number of trees) as a controlled variable when analyzing the yield response surfaces unless density was a controlled variable in the data set. Hence designed spacing trials are essential for providing definitive answers to the important question "How many trees per unit area should be planted?" Designed spacing experiments are commonly employed to augment permanent plot data from stands in order to develop reliable information on optimal density for specified product objectives.

10.6 Looking Ahead

Growth and yield models provide an effective means for summarizing and integrating information from a variety of sources, and they are essential decision-support tools for forest managers. The next four chapters provide an overview of the basic structure of three important modeling approaches for even-aged stand structures (Chap. 11, whole-stand models; Chap. 12, diameter-distribution models; Chap. 13, size-class models; Chap. 14, individual-tree models).

Special considerations and approaches for modeling uneven-aged stands are covered in Chap. 15. Methods for incorporating silvicultural treatments and wood quality in growth and yield models are presented in Chaps. 16 and 17, respectively. In the final chapter (Chap. 18) we provide information on evaluating and implementing forest stand models.

References

Adlard PG (1995) Myth and reality in growth estimation. For Ecol Manage 71:171–176
Amaro A, Reed D, Soares P (eds) (2003) Modelling forest systems. CABI Publishing, Cambridge, MA
Amateis RL, Burkhart HE, Walsh TA (1989) Diameter increment and survival equations for loblolly pine trees growing in thinned and unthinned plantations on cutover, site-prepared lands. South J Appl For 13:170–174
Avery TE, Burkhart HE (2002) Forest measurements, 5th edn. McGraw-Hill, New York
Burkhart HE (1987) Data collection and modeling approaches for forest growth and yield prediction. In: Chappell HN, Maguire DA (eds) Predicting forest growth and yield: current issues, future prospects. Institute of Forest Resources, University of Washington, Seattle, pp 3–16, Contribution No. 58
Burkhart HE (1990) Status and future of growth and yield models. In: LaBau VJ, Cunia T (eds) State-of-the-art methodology of forest inventory: a symposium proceedings. USDA Forest Service, Pacific Northwest Research Station, Portland, pp 409–414, General Technical Report PNW-GTR-263
Burkhart HE (1999) Development of empirical growth and yield models. In: Amaro A, Tomé M (eds) Proceedings IUFRO conference on empirical and process-based models for forest tree and stand growth simulation. Edicoes Salamandra, Lisboa, pp 53–60
Burkhart H (2003) Suggestions for choosing an appropriate level for modeling forest stands. In: Amaro A, Reed D, Soares P (eds) Modelling forest systems. CAB International, Wallingford, pp 3–10
Chappell HN, Maguire DA (eds) (1987) Predicting forest growth and yield: current issues, future prospects. Institute of Forest Resources, University of Washington, Seattle, Contribution No. 58
Clutter JL, Fortson JC, Pienaar LV, Brister GH, Bailey RL (1983) Timber management: a quantitative approach. Wiley, New York
Curtis RO, Hyink DM (1985) Data for growth and yield models. In: Hooser DD, Van Pelt N (eds) Proceedings – growth and yield and other mensurational tricks: a regional technical conference. USDA Forest Service, Intermountain Forest and Range Experiment Station, Ogden, pp 1–5, General Technical Report INT-GTR-193
Davis LS, Johnson KN, Bettinger PS, Howard TE (2001) Forest management, 4th edn. McGraw-Hill, New York
Dennis B, Brown BE, Stage AR, Burkhart HE, Clark S (1985) Problems of modeling growth and yield of renewable resources. Am Stat 39:374–383
Gadow Kv, Hui G (1999) Modelling forest development. Kluwer, Dordrecht
Gauch HG (1993) Prediction, parsimony and noise. Am Sci 81:468–478
Hasenauer H (ed) (2006) Sustainable forest management: growth models for Europe. Springer, Berlin/Heidelberg
Kahne S (1976) Model credibility for large-scale systems. IEEE Trans Syst Man Cyber 6(8):53–57
Köhl M, Scott CT, Zingg A (1995) Evaluation of permanent sample surveys for growth and yield studies: a Swiss example. For Ecol Manage 71:187–194
Leary RA (1979) Design. In: A generalized forest growth projection system applied to the Lake State region. USDA Forest Service, North Central Forest Experiment Station, St. Paul, pp 5–15, General Technical Report NC-49
Levins R (1966) The strategy of model building in population biology. Am Scientist 54:421–431
Mohren GMJ, Burkhart HE (1994) Contrasts between biologically-based process models and management-oriented growth and yield models. For Ecol Manage 69:1–5
Pretzsch H (2009) Forest dynamics, growth and yield. Springer, Berlin/Heidelberg
Shortt JS, Burkhart HE (1996) A comparison of loblolly pine plantation growth and yield models for inventory updating. South J Appl For 20:15–22
Sullivan AD, Clutter JL (1972) A simultaneous growth and yield model for loblolly pine. For Sci 18:76–86

Thompson JR (1989) Empirical model building. Wiley, New York
Vanclay JK (1994) Modelling forest growth and yield: applications to mixed tropical forests. CABI Publishing, Cambridge, MA
Vanclay JK, Skovsgaard JP, Pilegaard Hansen C (1995) Assessing the quality of permanent sample plot databases for growth modelling in forest plantations. For Ecol Manage 71:177–186
Walstad JD, Kuch PJ (eds) (1987) Forest vegetation management for conifer production. Wiley, New York

Chapter 11
Whole-Stand Models for Even-Aged Stands

11.1 Background

Yield prediction for even-aged stands began with the development of normal yield tables. Temporary plots were located in fully stocked or "normal" density portions of a sample of stands of varying ages representing various site qualities. These plot observations of volume per unit area were sorted into site-quality classes, and volume values were plotted over age. A volume-age curve was drawn through the points for each site-quality class by using graphical techniques. Values were read from the curve for selected site-quality classes and ages to compile a normal yield table. Many normal yield tables also contain auxiliary information, such as basal area and number of trees per unit area and diameter distributions, as well as volume per unit area.

Normal yield tables were constructed in an era when only two variables could be included readily by graphical techniques. Thus analysts eliminated the variable of density by holding it constant at fully stocked or "normal" levels.

As a variation on the normal yield table approach, so-called "empirical yield tables" were sometimes developed. An empirical yield table is similar to a normal yield table except that it supposedly applies to "average" rather than full, or normal, stocking. Thus the problem of defining normal stocking is eliminated, but an empirical yield table applies only to the average density levels found on the sample plots included.

Growth and yield equations that do not rely on either "normal" or "average" stand density concepts, but, rather, include density as a dynamic part of the stand-projection system are commonly termed variable-density models. The focus of this chapter is on variable-density growth and yield equations that use whole-stand values as the basic modeling unit.

H.E. Burkhart and M. Tomé, *Modeling Forest Trees and Stands*,
DOI 10.1007/978-90-481-3170-9_11,

11.2 Growth and Yield Relationships

Before proceeding further, we define the terms growth and yield. *Growth* is the increase (increment) over a given period of time. *Yield* is the total amount available for harvest at a given time. Thus yield can be regarded as the summation of the annual increments. To be meaningful, growth and yield values must be qualified with regard to the part of the tree and the portion of the stand being considered. Further, the unit of measurement being used and, for growth, the time period involved, must be specified.

The factors most closely related to growth and yield of forest stands, apart from management treatments, are (1) the point in time in stand development, (2) the site quality, and (3) the degree to which the site is occupied. For even-aged stands, these factors can be expressed quantitatively through the variables of stand age, site index, and stand density, respectively. The measure of stand density most commonly used in growth and yield models for natural stands has been basal area per unit area, whereas most models for planted stands have employed number of trees per unit area.

For a given site index and initial stand-density level, yield (volume per unit area) plotted over stand age results in a sigmoid curve. The growth curve (often referred to as current annual growth or current annual increment), which is the first derivative of the yield function, increases up to the inflection point of the yield curve and decreases thereafter. Another important quantity is the mean annual growth or increment, defined as the yield at any given time divided by the total number of years (age) required to achieve that yield. The current annual growth curve crosses the mean annual growth curve at its highest value (Fig. 11.1).

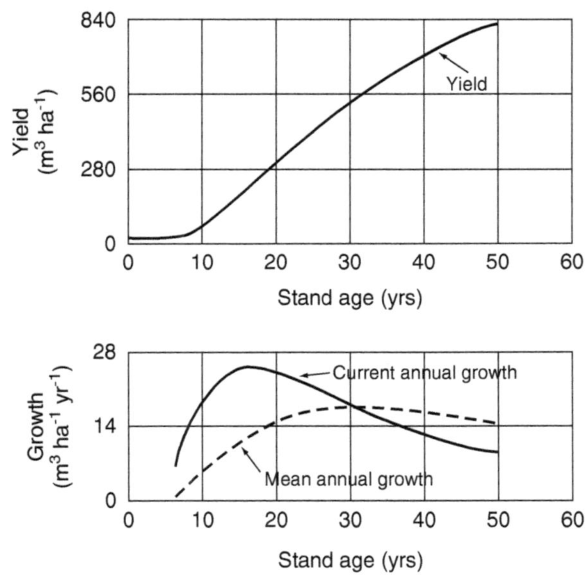

Fig. 11.1 Relationship between yield, current annual growth, and mean annual growth for even-aged stands with a specified site index and initial stand density. The yield equation is $V = e^{7.34-32t^{-1}}$, age of maximum current annual growth = 16, age of maximum mean annual growth = 32 years (Adapted from Avery and Burkhart 2002)

11.3 Variable-Density Growth and Yield Equations

11.3.1 Schumacher-Type Equations

MacKinney et al. (1937) suggested the use of multiple regression to construct variable-density yield equations. Subsequently, MacKinney and Chaiken (1939) used a multiple regression approach to construct a yield equation for natural stands of loblolly pine. Their prediction model was:

$$\log V = \mathrm{b}_0 + \mathrm{b}_1 t^{-1} + \mathrm{b}_2 S + \mathrm{b}_3 \log SDI + \mathrm{b}_4 C$$

where

$\log V$ = logarithm of yield (total cubic volume per unit area of loblolly pine)
t^{-1} = reciprocal of stand age
S = site index
$\log SDI$ = logarithm of stand-density index
C = composition index (basal area per unit area of loblolly pine divided by total stand basal area)

The measure of density used was Reineke's stand-density index (Chap. 8) and a "composition index" was included, because not all of the sample plots were pure loblolly pine. Since MacKinney and Chaiken's milestone publication, many investigators have used multiple regression techniques to predict growth and/or yield for total stand values or for some merchantable portion of stands. Stand-level variables, such as age, site index, basal area or number of trees per unit area, are used to predict some specified aggregate stand volume. The resultant equations from this approach are commonly referred to as *whole-stand models.*

The variable forms used in subsequent analyses have generally been similar to those employed by MacKinney and Chaiken. Logarithmic transformation of yield is usually made prior to equation fitting to stabilize variance and thus to conform to the assumptions customarily made in linear regression analysis. Furthermore, the use of the logarithm of yield as the dependent variable is a convenient way to mathematically express the interaction of the independent variables in their effect on yield. The effects of age, site index, and stand density are additive for the logarithm of yield but multiplicative for yield. In most yield analyses, stand age has been expressed as a reciprocal to allow for the "leveling off" or asymptotic effect of yield with increasing age. Site index often is not transformed prior to fitting, but logarithmic or reciprocal transformations are sometimes employed. The measure of stand density is commonly subjected to logarithmic transformation – particularly in models employing basal area – but the exact form in which density is included is quite variable, especially for models fitted to data from planted stands that use number of trees per unit area as a predictor variable.

The rationale for the logarithm of yield-reciprocal of age stand yield function was laid out by Schumacher (1939). Subsequently yield models of the basic form

$$\log Y = \mathrm{b}_0 + \mathrm{b}_1 t^{-1} + \mathrm{b}_2 f(S) + \mathrm{b}_3 g(SD)$$

where Y is a measure of yield per unit area, t equals stand age, $f(S)$ is some function of site index, and $g(SD)$ is some function of stand density have frequently been referred to as "Schumacher yield models". The Schumacher or logarithm of yield-reciprocal of age function is a special case of the Lundquist-Korf equation described in Chap. 6.

11.3.2 Chapman-Richards Equations

Various growth functions have been fitted to stand growth and yield data. Besides the Schumacher-type equations, one of the most commonly-fitted equation forms is the Chapman-Richards generalization of Bertalanffy's growth model. Bertalanffy formulated a hypothesis which expressed rate of volume growth of an organism as the difference between anabolic growth rate (constructive metabolism) and catabolic growth rate (destructive metabolism). The anabolic rate was assumed to be proportional to surface area of the organism raised to the power 2/3, while the catabolic rate was assumed proportional to size or volume. Thus, expressing the hypothesis in mathematical terms yields

$$dV/dt = aV^{2/3} - \mathrm{b}V \qquad (11.1)$$

where dV/dt is the volume growth rate and V is the size or volume, a and b are parameters and 2/3 is an allometric constant.

Richards (1959) in studying plant growth and Chapman (1961) in analyzing growth of fish argued that Bertalanffy's allometric constant of 2/3 was too restrictive, leading to the modified growth function:

$$dV/dt = aV^{m} - \mathrm{b}V \qquad (11.2)$$

where the value of m is to be estimated for each organism and environment.

The integration of Eq. 11.2 and simplification of the resultant expressions for the constants leads to the form in which the so-called Chapman-Richards function has generally been fitted to forestry yield data:

$$Y = a_0(1 + \exp(a_1 t))^{a_2} \qquad (11.3)$$

where Y is the cumulative value (or yield) of interest and t is time (stand age in the context of modeling even-aged stands). The parameters a_0, a_1, a_2, are readily interpreted as the asymptote, rate and shape for the fitted function. While Eq. 11.3 has been found to be flexible in assuming sigmoid shapes exhibited by individual-tree and stand-level characteristics for even- aged stands, it has been most prevalently applied in fitting height-age (site index) relationships.

Pienaar and Turnbull (1973) provide a comprehensive discussion of the origin and generalization of Bertalanffy's growth model and give examples of its application to tree growth. They further extended the Chapman-Richards generalization of Bertalanffy's growth model for basal area growth and yield of even-aged stands.

Equations, such as Bertalanffy's, that are based on biological rationale are sometimes termed "theoretical" and are often times thought to be superior to strictly empirical fits to data. Typically a number of equations exist that will approximate the observed data points with nearly equal fidelity. The principal advantages of employing functional forms with underlying biological rationale often rest with the ability to interpret the parameter estimates and to extrapolate the fitted functions with more confidence than is generally the case with results from strictly data-based curve fitting.

11.4 Compatible Growth and Yield Equations

11.4.1 Analytic Compatibility

Early work did not attempt to relate growth analyses to yield analyses, although the biological relationships can be readily expressed mathematically. Buckman (1962) and Clutter (1963) were the first researchers in the United States to explicitly recognize the mathematical relationships between growth and yield in their analyses. When deriving compatible growth and yield models for loblolly pine, Clutter started with a particular form of the Schumacher yield function, namely:

$$\ln V = \mathrm{b}_0 + \mathrm{b}_1 S + \mathrm{b}_2 \ln G + \mathrm{b}_3 t^{-1} \tag{11.4}$$

where ln V = logarithm of cubic volume per unit area of all pine stems

S = site index
G = basal area per unit area
t = stand age in years

Differentiating (11.4) with respect to age (noting that G is a function of age) gives the total derivative

$$(dV/dt)/V = \mathrm{b}_2 G^{-1}(dG/dt) - \mathrm{b}_3 t^{-2} \tag{11.5}$$

where

dV/dt = instantaneous rate of cubic volume growth
dG/dt = instantaneous rate of basal area growth with respect to age

To estimate cubic volume growth one must specify the relationship between basal area growth and age, site index, and basal area. Plotting of data indicated that the following would be appropriate as a basal area yield model:

$$\ln G = a_0 + a_1 S + a_2 t^{-1} + a_3 \ln (G_{20}) t^{-1} + a_4 S t^{-1} \qquad (11.6)$$

where G_{20} = basal area at age 20.

Differentiating (11.6) with respect to age and algebraic rearrangement yields:

$$dG/dt = t^{-1} G \left[a_0 + a_1 S - \ln G\right] \qquad (11.7)$$

The growth rate equation can be integrated to obtain the difference equation:

$$\ln G_2 = \left(\frac{t_1}{t_2}\right) \ln G_1 + a_0 \left(1 - \frac{t_1}{t_2}\right) + a_1 S \left(1 - \frac{t_1}{t_2}\right) \qquad (11.8)$$

where

t_1 = initial age
t_2 = projection age
G_1 = initial basal area (age t_1)
G_2 = predicted basal area at age t_2

11.4.2 Ensuring Numeric Consistency

Clutter's (1963) system of compatible growth and yield equations was derived to ensure that a derivative-integral relationship existed between the analytic expressions for stand volume and basal area. In extending and building upon this analytic framework, Sullivan and Clutter (1972) noted that the parameters in any one equation are not independent of those in other equations of the system. This dependence must be recognized in the parameter estimation process in order for the equations to be numerically consistent. As a means of ensuring numeric consistency in Clutter's compatible system of growth and yield equations, a single linear model relating projected stand volume to initial stand age, projected age, site index, and initial basal area was developed. If projected age is equal to initial age, the model simplifies to a conventional yield equation.

The models selected by Sullivan and Clutter (1972) can be written:

$$E(\ln y_1) = \beta_0 + \beta_1 S + \beta_2 t_1^{-1} + \beta_3 \ln G_1 \qquad (11.9)$$

$$E(\ln y_2) = \beta_0 + \beta_1 S + \beta_2 t_2^{-1} + \beta_3 \ln G_2, \text{ and} \qquad (11.10)$$

$$E(\ln G_2) = (t_1/t_2) \ln G_1 + \alpha_1 (1 - t_1/t_2) + \alpha_2 (1 - t_1/t_2) S \qquad (11.11)$$

where

y_i = stand volume per unit area at age t_i
S = site index
G_i = basal area per unit area at age t_i
t_i = stand age at time i (t_1 = present stand age, t_2 = projected stand age)
ln denotes logarithm to the base e, and E denotes expected value.

Replacing $\ln G_2$ in Eq. 11.10 by the functional form of its expected value (Eq. 11.11) and simplifying the resulting expression results in:

$$E(\ln y_2) = \beta_0 + \beta_1 S + \beta_2 t_2^{-1} + \beta_3(t_1/t_2)\ln G_1 + \beta_3\alpha_1(1 - t_1/t_2) + \beta_3\alpha_2(1 - t_1/t_2)S \tag{11.12}$$

By letting $\beta_4 = \beta_3\alpha_1$ and $\beta_5 = \beta_3\alpha_2$, Eq. 11.12 can be written as:

$$E(\ln y_2) = \beta_0 + \beta_1 S + \beta_2 t_2^{-1} + \beta_3(t_1/t_2)\ln G_1 + \beta_4(1 - t_1/t_2) + \beta_5(1 - t_1/t_2)S \tag{11.13}$$

Equation 11.13 is a model for projected volume as a function of site index, initial age, projected age, and initial basal area. When $t_2 = t_1$, i.e., the projection period is zero years, Eq. 11.13 reduces to the yield model (11.9). Thus Eq. 11.13 is simultaneously a yield model for observations at the initial time and a projection or growth model for subsequent times.

Sullivan and Clutter (1972) showed that Eq. 11.13 is invariant for different combinations of projection lengths. That is, if yield is projected from t_1 to t_2 and then from t_2 to t_3, the result is identical with a projection from t_1 to t_3.

A second desirable property is that substituting projected basal area into the volume yield equation to calculate yield at t_2 should be equivalent to using initial basal area in the volume projection equation. To ensure this equivalency property, Sullivan and Clutter (1972) noted that the definitions $\beta_4 = \beta_3\alpha_1$ and $\beta_5 = \beta_3\alpha_2$ in the derivation of their simultaneous growth and yield model imply that at $\alpha_1 = \beta_4/\beta_3$ and $\alpha_2 = \beta_5/\beta_3$. Given estimates for the parameters of (11.13), they suggested that, although not the most efficient estimators, estimates of α_1 and α_2 in (11.11) could be computed as $\hat{\alpha}_1 = \hat{\beta}_4/\hat{\beta}_3$ and $\hat{\alpha}_2 = \hat{\beta}_5/\hat{\beta}_3$. While this procedure ensures the desired equivalency property, the estimates of α_1 and α_2, and thus the basal area projection equation, are dependent on the volume units and merchantability limits chosen for the dependent variable in (11.13).

Burkhart and Sprinz (1984) selected the models specified by Sullivan and Clutter (1972) for fitting with data from thinned loblolly pine plantations because these models have the desired properties of analytic compatibility between growth and yield, invariance for projection length, and numeric equivalency between alternative applications of the equations. In applying these models, however, Burkhart and Sprinz sought to develop more efficient and stable (i.e., not dependent on volume units or merchantability standards) estimators for the basal area equation.

Rather than deriving coefficients for the basal area projection Eq. 11.11 from parameter estimates in the volume projection Eq. 11.13, both functions were fitted simultaneously. When defining the criterion or loss function to be minimized in the simultaneous fitting, equal weights were assigned to volume and basal area projection. This led to the standardized loss function:

$$L = \frac{\Sigma(y_i - \hat{y}_i)^2}{\sigma_y^2} + \frac{\Sigma(G_i - \hat{G}_i)^2}{\sigma_G^2} \tag{11.14}$$

where y_i and $\hat{y}_i$ are the observed and predicted volume logarithms, G_i and $\hat{G}_i$ are the observed and predicted basal area logarithms, and $\hat{\sigma}_y^2$ and $\hat{\sigma}_G^2$ are estimates of the variance around the regression lines for volume and basal area, respectively. These variance estimates, $\hat{\sigma}_y^2$ and $\hat{\sigma}_G^2$, were computed as the mean square error from ordinary least squares regression fits of Eqs. 11.13 and 11.11, respectively.

Loss function (11.14) was used when fitting projection equations for total stand volume, volume of the pulpwood portion of the stand, and volume of the sawtimber sized trees. In all cases basal area included all stems. By using the simultaneous fitting procedure with the imposed restrictions that $\hat{\alpha}_1 = \hat{\beta}_4/\hat{\beta}_3$ and $\hat{\alpha}_2 = \hat{\beta}_5/\hat{\beta}_3$, coefficients in the basal area projection equations were nearly identical regardless of the merchantability definitions employed in the volume equation. Further, with the imposed restrictions, this process results in a system of equations that gives numerically consistent results. The three sets of parameter estimates for the basal area equations were quite close to those computed by ordinary least squares. As expected, there was a slight increase in the sum of squared error (SSE) for volume when fitting by the simultaneous approach. However, this small increase in SSE for volume was more than offset by the decrease in the SSE for basal area in the simultaneous fit versus deriving basal area coefficients from the volume equation.

While the method of Burkhart and Sprinz (1984) proved effective, it does not use information on correlation between residuals of the component equations and the statistical properties of the estimators are not known. VanDeusen (1988) showed that the system of equations fitted by Burkhart and Sprinz (1984) can be solved in the framework of seemingly unrelated regression (SUR). In SUR no analytical relationships are implied between variables to be predicted. Relationships are conceptual and take the form of correlations between error terms of component equations. VanDeusen (1988) pointed out that, if the requirements for SUR are met, the estimates are consistent and asymptotically efficient. When VanDeusen fitted the equation system (11.11) and (11.13) using loss function (11.14) and using SUR with data from unthinned plots of slash pine, both methods gave similar results. Thus, it seems that including across-equation constraints brings about large gains while additional gains from alternative parameter estimation techniques are not likely to be large.

11.5 Growth Models Based on Annual Increments

Compatibility between growth and yield is highly desirable because it ensures consistent results. Imposing compatibility constraints may, however, limit the types of models that can be fitted, thus reducing options and perhaps accuracy of predictions. Ochi and Cao (2003) advanced an alternative approach based on modeling annual stand growth. Although such models are not necessarily conceptually compatible, they can provide estimates that are invariant for different combinations of projection lengths. Furthermore, models of annual increment are not subject to restrictions resulting from compatibility constraints; consequently they provide increased flexibility.

Ochi and Cao (2003) evaluated two compatible models (Sullivan and Clutter 1972; Pienaar and Harrison 1989) and an annual increment growth model for predicting survival, basal area, and volume in loblolly pine stands. The annual increment model consistently performed better than the two compatible models for short (4–7 years), medium (10–12 years), and long (15–17 years) projection lengths. The model based on annual growth included both current measures of stand density (number of trees and basal area) as independent variables, whereas the compatible models did not. Due to restrictions of compatible growth and yield models, which might be thought of as special cases of annual increment models, one of the two measures of stand density could not be predicted from the other.

11.6 Simultaneous Systems of Growth and Yield Equations

Growth and yield of volume in even-aged stands is ordinarily viewed as a function of site quality, stand age, and stand density. The stand density measure is usually considered a function of site quality, age, and initial density. Site quality, expressed as site index, is frequently quantified as a function of age only. These equations form an interdependent system in which all relationships are assumed to hold simultaneously. Ordinary least squares procedures have customarily been used to estimate the parameters separately for each equation within the system. But independent, sequential fitting of the components of such a system of equations will often lead to biased and inconsistent estimates.

Furnival and Wilson (1971) suggested that techniques commonly used to fit simultaneous systems of equations in econometrics may be applicable to forest growth and yield modeling. Subsequently, Murphy and Sternitzke (1979) used three-stage least square estimation procedures to fit the Sullivan and Clutter (1972) volume and basal area projection system (with a modified basal area equation) with data from loblolly pine stands; Murphy and Beltz (1981) applied similar procedures for shortleaf pine data.

The simultaneous system fitted by Borders and Bailey (1986) will be used to illustrate an interdependent, multi-equation system of growth and yield equations.

As a first step, Borders and Bailey (1986) specified models for predicting volume and basal area for planted stands:

$$\ln V = \mathrm{b}_0 + \mathrm{b}_1 \ln h_{dom} + \mathrm{b}_2 \ln G \tag{11.15}$$

$$\ln G = a_0 + a_1 \ln h_{dom} + a_2 \ln N_p \tag{11.16}$$

where

V = volume per unit area
G = basal area per unit area
h_{dom} = dominant stand height
N_p = number of trees planted per unit area
b_0, b_1, b_2, a_0, a_1, a_2 are unknown parameters to be estimated.

Equations 11.15 and 11.16 are yield equations. These equations can be differentiated to define analytically compatible growth equations. Differentiating (11.15) and (11.16) with respect to age one obtains:

$$(dV/dt)/V = (\mathrm{b}_1/h_{dom})(dh_{dom}/dt) + (\mathrm{b}_2/G)(dG/dt) \tag{11.17}$$

and

$$(dG/dt)/G = (a_1/h_{dom})(dh_{dom}/dt) \tag{11.18}$$

Relative volume growth is dependent upon relative changes in both dominant height and basal area while relative growth rate of basal area is dependent upon relative growth rate of dominant height. Therefore height-age prediction and projection components are also needed. The following height-age model was adopted:

$$\ln h_{dom} = \mathrm{c}_0 + \mathrm{c}_1 t^{-1} \tag{11.19}$$

The first derivative of (11.19) with respect to age is

$$(dh_{dom}/dt)/h_{dom} = -\mathrm{c}_1 t^{-2} \tag{11.20}$$

Equations 11.17, 11.18, and 11.20 were solved as separable linear differential equations and integrated from t_1 to t_2 to produce the following growth (projection) equations (Borders and Bailey 1986):

$$\ln V_2 = \ln V_1 + \mathrm{b}_1 (\ln h_{dom2} - \ln h_{dom1}) + \mathrm{b}_2 (\ln G_2 - \ln G_1) \tag{11.21}$$

$$\ln G_2 = \ln G_1 + a_1 (\ln h_{dom2} - \ln h_{dom1}) \tag{11.22}$$

$$\ln h_{dom2} = \ln h_{dom1} + \mathrm{c}_1 \left(t_2^{-1} - t_1^{-1}\right) \tag{11.23}$$

where

V_i = volume per unit area at age t_i
G_i = basal area per unit area at age t_i
h_{domi} = dominant height at age t_i
t_i = age in years at time i
$i = 1,2$

and b_1, b_2, a_1, c_1 are defined in the yield models (11.15, 11.16 and 11.19).

In systems of interdependent equations, such as (11.15, 11.16, 11.21, 11.22 and 11.23), variables that occur on the left-hand sides are customarily referred to as endogenous variables; they are assumed to be determined by the structure of the model. The endogenous variables may also appear as predictor variables on the right-hand sides of equations within the system. All other variables are normally referred to as exogenous variables. Application of OLS estimation techniques to this equation system will yield biased and inconsistent parameter estimates because of correlations between explanatory variables and error terms. Alternatives to OLS estimation of simultaneous system parameters include two-stage least squares (2SLS) and three-stage least squares (3SLS). These alternatives will yield consistent, although not necessarily unbiased, parameter estimates.

Borders and Bailey (1986) fitted the five equations individually using OLS. These same models were also fitted as a system using unrestricted 3SLS and restricted 3SLS. In the restricted 3SLS case the cross-equation restrictions: (1) b_1 and b_2 must be the same in (11.15) and (11.21), and (2) a_1 must be the same in (11.16) and (11.22) were imposed. It was assumed that the observations used to estimate parameters were independently and identically distributed (iid); however, different equations in the system may exhibit contemporaneous correlations (these contemporaneous correlations are used in the 3SLS procedure to obtain more efficient estimates). Results showed that parameter estimates were similar for all three cases, but the 3SLS estimates generally had smaller predicted standard errors. Multistage procedures provide consistent estimators and allow for ready incorporation of constraints to insure that compatible results are achieved. Further, interval, as well as point, estimates are possible.

Borders (1989) suggested an alternative parameter estimation procedure to two-stage and three-stage least squares that can be applied to any number of sequentially related linear or nonlinear equations.

11.7 Mixed-Effects Models for Growth and Yield Prediction

Mixed-effects models have become important tools for modeling forest trees and stands. Most applications to date have focused on modeling particular components or variables such as tree height (Chap. 7) or tree taper (Chap. 2), but there have been some investigations aimed at yield prediction systems.

Gregoire et al. (1995) outlined a linear mixed-effects model that accounts for the covariance among repeated measurements and for random plot effects. Their formulation includes a continuous-time autocorrelation error structure that permits the model to be applied to irregularly spaced, unbalanced data. When fitted to two permanent-plot data bases, the model showed improvement when compared with models that do not account for the error structure.

Hall and Clutter (2004) developed a multivariate multilevel nonlinear mixed effects model for describing several plot-level volume characteristics simultaneously. Their system consists of models for height of dominant trees (h_{dom}) basal area (G) and numbers of trees per ha (N). Using trivariate three-level nonlinear mixed models they obtained predictions of the plot-level variables h_{dom}, G, and N at a future age of interest. The authors proposed to use these predictions as input into a known or a fitted model for volume as a function of the three variables. As an example, they fitted the following volume model to the same data to which the trivariate nonlinear mixed models for h_{dom}, G, and N were applied:

$$V = \mathrm{e}^{\mathrm{b}_1} h_{dom}^{\mathrm{b}_2} G^{\mathrm{b}_3} N^{\mathrm{b}_4} \tag{11.24}$$

where the parameters b_i, $i = 1, 2, 3, 4$ were assumed to be functions of age. After expressing each of the b_i's as a polynomial in the reciprocal of age, model (11.24) was linearized by taking logarithms and fitted as a linear mixed effects model. The Hall and Clutter (2004) mixed models approach provides an alternative to other methods that have been applied for developing simultaneous systems of linear equations for growth and yield prediction. They listed the primary advantages of their methodology as: (1) the ability to model and predict timber growth and yield at the plot, stand, and population level, and (2) the availability of a prediction variance estimator, which allows for quantification of uncertainty in yield predictions.

11.8 State Space Models

Forest growth models predict future values of product volumes per ha for given inputs such as site index, stand density, age, and silvicultural treatments. Considering that both inputs and outputs are functions of time, and that outputs depend on the entire past history of the stand, García (1984, 1994) adopted a state space approach to stand modeling. The idea is to characterize the state of the system at any point in time so that given the present state the future state does not depend on the past. For example, an even-aged monoculture might be characterized by its basal area, stems per ha, and dominant stand height. It is then assumed that two stands with the same values for these variables will behave in essentially the same manner regardless of how they happened to reach that state. In other words, the state of the stand at any given time is assumed to be the necessary and sufficient information needed to determine its future behavior.

García (1994) laid out the general structure of the state space approach as follows:

Let the state at a given time, t, be specified by a list of n numbers (state variables), that is, by an n-dimensional state vector $\mathbf{x}(t)$. The inputs and outputs are finite-dimensional vectors denoted $\mathbf{u}(t)$ and $\mathbf{y}(t)$, respectively. Then the behavior of the system is described by a transition function:

$$\mathbf{x}(t) = \mathrm{F}\left[\mathbf{x}(t_0), \mathbf{U}, t - t_0\right] \tag{11.25}$$

and an output function:

$$\mathbf{y}(t) = g\left[\mathbf{x}(t)\right] \tag{11.26}$$

Equation 11.25 gives the state at any time t as a function of the state at some other time t_0, of the inputs (as a function of time, denoted by $\mathbf{U}$), and of the elapsed time between t_0 and t. The output function (11.26) gives the current outputs as a function of the current state.

A transition function must possess the properties:

1. No change for zero elapsed time:

$$\mathbf{F}\left[\mathbf{x}(t),\ \mathbf{U},\ 0\right] = \mathbf{x}(t) \text{ for all } t, \mathbf{x}(t), \mathbf{U}$$

2. The result of projecting the state first from t_0 to t_1, and then from t_1 to t_2, must be the same as that of the one-step projection from t_0 to t_2:

$$\mathbf{F}\left[\mathbf{F}\left[\mathbf{x}(t_0), \mathbf{U}, t_1 - t_0\right], \mathbf{U}, t_2 - t_1\right] = \mathbf{F}\left[\mathbf{x}(t_0), \mathbf{U}, t_2 - t_0\right] \text{ for any } t_0 \le t_1 \le t_2$$

3. A change of state can only be influenced by inputs within the relevant time interval:

$$\mathbf{F}\left[\mathbf{x}(t_0), \mathbf{U}_1, t_1 - t_0\right] = \mathbf{F}\left[\mathbf{x}(t_0), \mathbf{U}_2, t_1 - t_0\right]$$

$$\text{if } \mathbf{u}_1(t) = \mathbf{u}_2(t) \text{ for } t_0 \le t \le t_1$$

The next step is to exploit the fact that transition functions generated by integration of differential equations (or summation of difference equations when using discrete time) automatically satisfy these conditions. The model can then be stated as:

$$\mathrm{d}\mathbf{x}/\mathrm{d}t = \mathrm{f}(\mathbf{x}, \mathbf{u}) \tag{11.27}$$

$$\mathbf{y} = \mathrm{g}(\mathbf{x}) \tag{11.28}$$

In a discrete-time model $\Delta\mathbf{x}$ is substituted for $d\mathbf{x}/dt$. Integration of the local transition function (11.27) between t_0 and t gives the global transition function (11.25).

To illustrate the state space approach, consider a simple system for pure, even-aged stands. The two state variables to be modeled are dominant stand height and basal area per unit area. For a given site, the rate of change of h_{dom} can be modeled as a function of the current height, namely:

$$dh_{dom}/dt = f_{1,1}(h_{dom}) \text{ or } \Delta h_{dom} = f_{1,2}(h_{dom}) \tag{11.29}$$

Stand basal area increment is considered a function of h_{dom} and current basal area (G), giving:

$$dG/dt = f_{2,1}(h_{dom}, G) \text{ or } \Delta G = f_{2,2}(h_{dom}, G) \tag{11.30}$$

The output (volume per unit area, V) can be computed by employing a stand volume equation:

$$V = f_3(h_{dom}, G) \tag{11.31}$$

In most applications three to five state variables have been used to characterize the stand systems. In addition to the transition functions describing growth and mortality, auxiliary relationships to estimate instantaneous change in the state variables caused by silvicultural intervention and to estimate volumes of various products for given states may be included. Forecasting is done by integrating the transition functions or through accumulation.

Examples of applications of the state space approach to growth modeling for even-aged stands include the work of García (1984) for radiata pine in New Zealand, the model of García and Ruiz (2003) for eucalypt coppice stands in Spain, the stand growth model for European beech in Denmark by Nord-Larsen and Johannsen (2007), the model for interior spruce in British Columbia, Canada, by García (2011), and a model for loblolly pine in the Piedmont physiographic region of the United States by García et al. (2011).

References

Avery TE, Burkhart HE (2002) Forest measurements, 5th edn. McGraw-Hill, New York

Beekhuis J (1966) Prediction of yield and increment in *Pinus radiata* stands in New Zealand. New Zealand Forest Service, Wellington, Forest Research Institute Technical Paper 49

Borders BE (1989) Systems of equations in forest stand modeling. For Sci 35:548–556

Borders BE, Bailey RL (1986) A compatible system of growth and yield equations for slash pine fitted with restricted three-stage least squares. For Sci 32:185–201

Bruce D (1977) Yield differences between research plots and managed forests. J For 75:14–17

Buckman RE (1962) Growth and yield of red pine in Minnesota. USDA Forest Service, Washington, DC, Technical Bulletin 1272

Buckman RE, Bishaw B, Hanson TJ, Benford FA (2006) Growth and yield of red pine in the Lake States. USDA Forest Service, North Central Forest Experiment Station, St. Paul, General Technical Report NC-271

Burkhart HE (1986) Fitting analytically related models to forestry data. In: Proceedings invited papers, XIIIth international biometric conference, Seattle, WA, 15pp

Burkhart HE, Sprinz PT (1984) Compatible cubic volume and basal area projection equations for thinned old-field loblolly pine plantations. For Sci 30:86–93

Chapman DG (1961) Statistical problems in dynamics of exploited fisheries populations. In: Proceedings of 4th Berkeley Symposium Mathematical Statistics and Probability. University of California Press, Berkeley, pp 153–168

Clutter JL (1963) Compatible growth and yield models for loblolly pine. For Sci 9:354–371

Coble DW (2009) A new whole-stand model for unmanaged loblolly and slash pine plantations in east Texas. South J Appl For 33:69–76

Curtis RO (1967) A method of estimation of gross yield of Douglas-fir. For Sci Monograph 13: 1–24

Curtis RO (1972) Yield tables past and present. J For 70:28–32

Curtis RO, Clendenen GW, DeMars DJ (1981) A new stand simulator for coast Douglas-fir: DFSIM user's guide. USDA Forest Service, Pacific Northwest Forest Experiment Station, Portland, General Technical Report PNW-128

Furnival GM, Wilson RW Jr (1971) Systems of equations for predicting forest growth and yield. In: Patil GP, Pielou EC, Waters WE (eds) Statistical ecology, vol 3. Pennsylvania State University Press, University Park, pp 43–55

García O (1984) New class of growth models for even-aged stands: *Pinus radiata* in Golden Downs Forest. N Z J For Sci 14:65–88

García O (1988) Growth modelling – a (re)view. N Z J For 33(3):14–17

García O (1994) The state-space approach in growth modelling. Can J For Res 24:1894–1903

García O (2011) A parsimonious dynamic stand model for interior spruce in British Columbia. For Sci 57:265–280

García O, Ruiz F (2003) A growth model for eucalypt in Galicia. Spain For Ecol Manage 173: 49–62

García O, Burkhart HE, Amateis RL (2011) A biologically-consistent stand growth model for loblolly pine in the Piedmont physiographic region. USA For Ecol Manage 262:2035–2041

Gregoire TG (1987) Generalized error structure for forestry yield models. For Sci 33:423–444

Gregoire TG, Schabenberger O, Barrett JP (1995) Linear modelling of irregularly spaced, unbalanced, longitudinal data from permanent-plot measurements. Can J For Res 25:137–156

Hall DB, Clutter M (2004) Multivariate multilevel nonlinear mixed effects models for timber yield predictions. Biometrics 60:16–24

Huuskonen S, Miina J (2007) Stand-level growth models for young Scots pine stands in Finland. For Ecol Manage 241:49–61

MacKinney AL, Chaiken LE (1939) Volume, yield, and growth of loblolly pine in the mid-Atlantic coastal region. USDA Forest Service, Appalachian Forest Experiment Station, Asheville, Technical Note 33

MacKinney AL, Schumacher FX, Chaiken LF (1937) Construction of yield tables for nonnormal loblolly pine stands. J Agric Res 54:531–545

Murphy PA (1983) A nonlinear timber yield equation system for loblolly pine. For Sci 29:582–591

Murphy PA, Beltz RC (1981) Growth and yield of shortleaf pine in the West Gulf region. USDA Forest Service, Southern Forest Experiment Station, New Orleans, Research Paper SO-169

Murphy PA, Sternitzke HS (1979) Growth and yield estimation for loblolly pine in the West Gulf. USDA Forest Service, Southern Forest Experiment Station, New Orleans, Research Paper SO-154

Murphy PA, Lawson ER, Lynch TB (1992) Basal area and volume development of natural even-aged shortleaf pine stands in the Ouachita Mountains. South J Appl For 16:30–34

Nord-Larsen T, Johannsen VK (2007) A state-space approach to stand growth modelling of European beech. Ann For Sci 64:365–374

Ochi N, Cao QV (2003) A comparison of compatible and annual growth models. For Sci 49: 285–290

Pienaar LV, Harrison WM (1989) Simultaneous growth and yield prediction equations for *Pinus elliottii* plantations in Zululand. South African For J 149:48–53

Pienaar LV, Shiver BD (1986) Basal area prediction and projection equations for pine plantations. For Sci 32:626–633

Pienaar LV, Turnbull KJ (1973) The Chapman-Richards generalization of von Bertalanffy's growth model for basal area growth and yield in even-aged stands. For Sci 19:2–22

Reed DD, Jones EA, Bottenfield TR, Trettin CC (1986) Compatible cubic volume and basal area equations for red pine plantations. Can J For Res 16:416–419

Richards FJ (1959) A flexible growth function for empirical use. J Exp Bot 10:290–300

Sadiq RA (1983) Estimation of stand basal area growth and yield with a reverse logistic function. Can J For Res 13:289–297

Schultz EB, Iles JC, Matney TG, Ezell AW, Meadows JS, Leininger TD, Booth WC, Jeffreys JP (2010) Stand-level growth and yield component models for red oak-sweetgum forests on mid-south minor stream bottoms. South J Appl For 34:161–175

Schumacher FX (1939) A new growth curve and its application to timber-yield studies. J For 37:819–820

Sullivan AD, Clutter JL (1972) A simultaneous growth and yield model for loblolly pine. For Sci 18:76–86

Sullivan AD, Reynolds MR (1976) Regression problems from repeated measurements. For Sci 22:382–385

Tang S, Meng CH, Meng F-R, Wang YH (1994) A growth and self-thinning model for pure even-age stands: theory and applications. For Ecol Manage 70:67–73

Van Deusen PC (1988) Simultaneous estimation with a squared error loss function. Can J For Res 18:1093–1096

West PW, Ratkowsky DA, Davis AW (1984) Problems of hypothesis testing of regressions with multiple measurements from individual sampling units. For Ecol Manage 7:207–224

Zhang Y, Borders BE (2001) An iterative state-space growth and yield modeling approach for unthinned loblolly pine plantations. For Ecol Manage 146:89–98

Chapter 12
Diameter-Distribution Models for Even-Aged Stands

12.1 Estimating Yields by Size Class Using a Distribution Function Approach

Overall stand volume is sufficient for many purposes, but effective forest management and planning often requires information about the distribution of volume by size and product classes. In the past, stand (numbers of trees by diameter class) and stock (volume by diameter class) tables were sometimes included as auxiliary information to yield tables. Because of the importance of stand and stock table information, a great deal of attention has been focused on modeling diameter distributions for even-aged stands. In typical applications, the total number of trees per unit area is distributed through the use of a probability density function (pdf), which provides the relative frequency of trees by diameters. Mean total tree heights are predicted for trees of given diameters growing under specified conditions. Volume per diameter class is calculated by substituting the predicted mean tree heights and the diameter class midpoints into tree volume equations. Yield estimates are obtained by summing the diameter classes of interest. Although only overall stand values (such as age, site index, and number of trees per ha) are needed as input, detailed stand distributional information is obtainable as output.

The various diameter distribution models differ chiefly in the function used to describe the diameter distribution. Regardless of the probability density function used, the procedure that is commonly referred to as "parameter prediction" involves estimating the pdf parameters for each plot in the data set (typically by maximum likelihood, the method of moments, or percentile estimators) and then developing regression equations to relate these parameter estimates to stand characteristics such as age, site index, and number of trees per unit area.

Functions for relating pdf parameters to stand characteristics have not been fully satisfactory; furthermore, predictions of basal area and overall stand volume from the "parameter prediction" approach to stand modeling will not be compatible with direct prediction of these stand variables. Consequently alternative methods

H.E. Burkhart and M. Tomé, *Modeling Forest Trees and Stands*,
DOI 10.1007/978-90-481-3170-9_12,

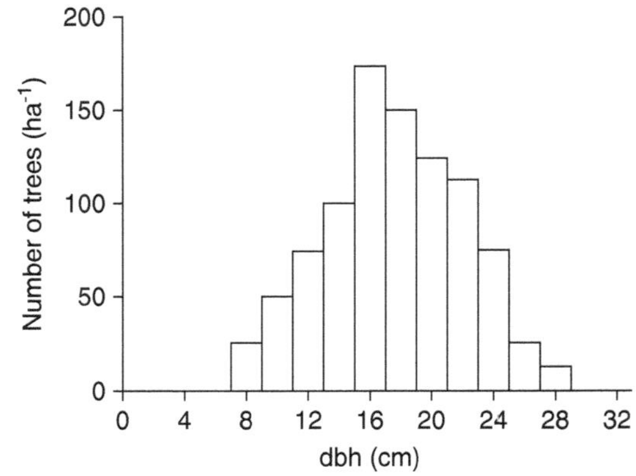

Fig. 12.1 A typical *dbh* distribution for pure, even-aged stands (Adapted from Avery and Burkhart 2002)

commonly referred to as "parameter recovery", have been developed. The so-called parameter recovery method consists of forecasting overall stand attributes (such as stand mean diameter and basal area per ha) and solving for the parameters of a theoretical distribution model that will give rise to the overall stand attributes included. Such an approach allows for a direct mathematical link between the predicted overall stand characteristics and a diameter distribution that is consistent with those characteristics.

A typical *dbh* distribution for pure, even-aged stands is shown by the histogram in Fig. 12.1. As a rule diameter distributions for even-aged stands are unimodal and slightly skewed. Curves can be fitted to such distributions by a variety of mathematical functions, including the beta, Weibull, gamma, Johnson S_B and, lognormal distributions.

12.1.1 Selecting a Distribution Function

When fitting diameter distribution data, the choice of a statistical distribution function to characterize the probabilities of interest is critical. Criteria for choosing a distribution function include ease of parameter estimation, flexibility to describe a broad spectrum of shapes, simplicity of integration methods for estimating proportions in various size classes, and accuracy in fitting the observed data. Hafley and Schreuder (1977) advocated using the skewness coefficient, $\sqrt{\beta_1}$, and kurtosis coefficient, β_2, of candidate statistical distributions as a means of evaluating their flexibility to reproduce a variety of shapes. Skewness measures asymmetry; negative values indicate a distribution with a long tail to the left and positive values a long tail to the right. Kurtosis is generally considered a relative measure of flatness or peakedness of a distribution, with the larger the value of β_2 the more peaked the distribution.

A graph of $\beta_1 - \beta_2$ space is useful to demonstrate the range of skewness and kurtosis that can be assumed by various statistical distributions. Such a graph can be

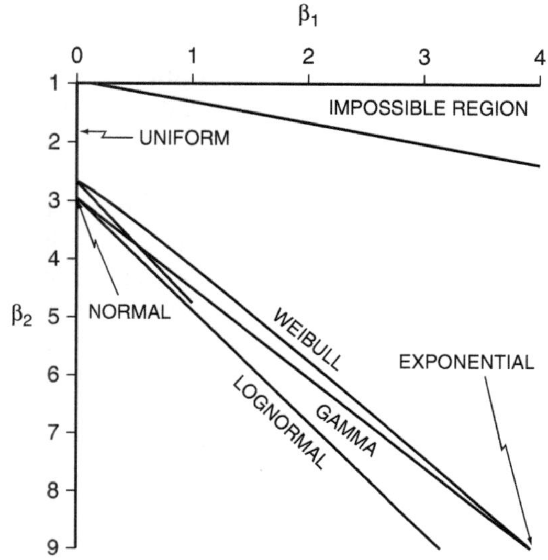

Fig. 12.2 The $\beta_1 - \beta_2$ space showing the plots of possible values for selected distributions (From Hafley and Schreuder 1977)

helpful to narrow the search for distribution functions that might accurately describe forest stand diameter and/or height distributions. Figure 12.2 shows the $\beta_1 - \beta_2$ space for a number of statistical distributions, including several functions that have been employed for modeling diameter distributions.

The 'impossible region' indicates combinations of β_1 and β_2 that are mathematically impossible. Further, by tradition, the ordinate scale is plotted upside down. One use of such a graph is to suggest distributions which might fit a set of data based on sample estimates of β_1 and β_2.

The normal, exponential, and uniform are all represented by points in the space, showing that they all have but one shape. The other three distributions shown in the figure are more flexible in terms of their ability to approximate a broader segment of the space. The three distributions gamma, lognormal, and Weibull are represented by lines in the $\beta_1 - \beta_2$ space, demonstrating their capability to assume a variety of shapes. The fact that these lines fall rather close to each other helps to explain why approximately comparable fits to sets of data are often achieved with these three distributions. However, as indicated in the graph, the Weibull distribution is somewhat more flexible than the gamma and lognormal; as expected, empirical results have repeatedly shown the Weibull function to result in a somewhat better fit to forestry data.

A further distinction between these three distributions is their ability to represent different types of skewness. Both the lognormal and gamma distributions are limited to shapes that have positive skewness, while the Weibull distribution has the ability to describe both positive and negative skewness. Since Fig. 12.2 presents β_1 as the square of the skewness coefficient, the positive and negative skewness aspect of a set of data is not readily discernable from the graph. When considering the sign of $\sqrt{\beta_1}$ the lower line of the Weibull plot in Fig. 12.2 is generated by negatively skewed shapes. The beta distribution covers the entire region between the gamma distribution line, the impossible region, and the β_2 axis. Hence, the beta distribution covers a broad spectrum of shapes and is quite flexible, fitting both positively and negatively skewed data. Johnson's S_B system of distributions, based on transformations of a standard normal variate, spans the $\beta_1 - \beta_2$ space. Johnson's (1949a,b) system consists basically of three distributions identified as S_B, S_L, and S_U. (Sometimes the normal distribution, which is a special case of the three, is included and denoted by S_N.) The S_L distribution is a three-parameter lognormal distribution with one parameter being the lower limit, the S_B distribution covers the region above the lognormal line in Fig. 12.2, and the S_U distribution covers the region below the lognormal line. Hence, Johnson's S_B distribution provides somewhat more flexibility in skewness and kurtosis than the beta distribution (Hafley and Schreuder 1977).

For demonstration purposes, Hafley and Schreuder (1977) fitted the beta, S_B, Weibull, gamma, lognormal, and normal distributions to 21 data sets of diameter and height measurements for even-aged stands. Based on the log likelihood criterion the best fit was achieved by the S_B distribution followed by the beta, Weibull, gamma, lognormal, and normal distributions for describing diameter distributions. For the height distribution data, the ranking was S_B, beta, Weibull, normal, gamma, and lognormal.

Mateus and Tomé (2011) found that the Johnson S_B was the only pdf of those evaluated that could assume the range of (β_1, β_2) values observed in their data from eucalyptus plantations.

12.1.2 Characterizing Diameter Distributions Using Parameter Prediction

Various functions, including the lognormal (Bliss and Reinker 1964) and gamma (Nelson 1964), have been used to characterize diameter distribution in even-aged forest stands. It was, however, not until Clutter and Bennett (1965) utilized the beta distribution to model diameter class frequencies, and their subsequent application to produce stand structure and yield estimates for slash pine plantations (Bennett and Clutter 1968), that the potential for this approach to stand modeling began to be thoroughly explored and developed. Following on Clutter and Bennett's (1965) work on slash pine diameter distribution estimation, Beck and Della-Bianca (1970) published yields, based on the beta distribution, for naturally-regenerated stands of yellow-poplar. Lenhart and Clutter (1971) and Burkhart and Strub (1974) applied the beta distribution when developing stand yield models for loblolly pine plantations.

For a given pdf $f(x;\theta)$, where θ is a vector of parameter values, integration between d_1 and d_2 gives the proportion of trees in that diameter interval. Since a pdf has the property that its integral over the range on which it is defined equals 1, the sum of the relative proportions of all defined diameter classes times the total number of trees present must equal the total. Often the cumulative distribution function (CDF) is employed and the proportion in a given diameter interval is computed as $F(d_2) - F(d_1)$ where $F(d_i)$ represents the cumulative area under the CDF from $-\infty$ to d_i. Given a function for number of trees per unit area and a site index equation and assuming that the paramaters of the chosen pdf can be related to the stand characteristics of age, number of trees, and site index, a stand table can be generated. A stock table and current yield can be produced through application of a height-diameter relationship and appropriate tree taper, volume or weight equations. Future yields are forecast by advancing age, predicting future numbers of trees, and updating the parameters of the pdf. Growth is estimated by differencing yield values.

Since Bailey and Dell (1973) suggested the Weibull function (Weibull 1951) for modeling forest stand diameter distributions it has been widely adopted and applied. Desirable attributes cited for the Weibull distribution include flexibility in assuming a variety of shapes, ease of estimating the distribution parameters by a number of methods, and efficiency in computing relative frequencies by diameter class due to the Weibull CDF being in closed form. The Weibull is a three-parameter distribution defined by the probability density function:

$$\begin{aligned} f(x) &= \frac{\mathrm{c}}{\mathrm{b}}\left(\frac{x-\mathrm{a}}{\mathrm{b}}\right)^{\mathrm{c}-1} \exp\left[-\left(\frac{x-\mathrm{a}}{\mathrm{b}}\right)^{\mathrm{c}}\right] \qquad (\mathrm{a} \le x < \infty) \\ &= 0 \text{ otherwise} \end{aligned} \tag{12.1}$$

The parameter a is referred to as the location parameter, b is the scale parameter, and c is the shape parameter. The b and c parameters must be positive while in general a can be positive, zero, or negative, for diameter distribution applications it must be nonnegative. For c less than 1 the distribution assumes the inverse *j*-shapes found in uneven-aged stands. When c equals 1, the negative exponential distribution results. Mound shape curves typical of even-aged stands are produced for c greater than 1. When c is equal 3.6 the Weibull approximates a normal distribution. Right-skewed curves are defined for c less than 3.6, and left-skewed curves for c greater than 3.6. As c approaches infinity the distribution approaches a spike over a single point. One can interpret a + b as that diameter where approximately 63% of all trees are smaller in diameter. The location parameter is directly related to the minimum diameter in a stand.

Integrating the pdf produces the cumulative distribution function for the Weibull distribution:

$$\begin{aligned} F(X) &= 1 - \exp\{-[(x-\mathrm{a})/\mathrm{b}]^{\mathrm{c}}\} \qquad (\mathrm{a} \le x < \infty) \\ &= 0 \text{ otherwise} \end{aligned} \tag{12.2}$$

When developing a diameter-distribution based yield model sample plot data consisting of age, total number of trees per ha, site index (or, equivalently, average dominant stand height), numbers of trees per ha by *dbh* classes, and height and *dbh* for selected sample trees are required. A separate Weibull distribution is fitted, often using maximum likelihood methods, to the *dbh* class frequency data from each plot. The data are then used to develop regression equations to predict the Weibull parameters, that is:

$$\mathrm{a} = f_1(t, N, S)$$
$$\mathrm{b} = f_2(t, N, S)$$
$$\mathrm{c} = f_3(t, N, S)$$

where $t =$ stand age, $N =$ total number of trees per ha, $S =$ site index (dominant stand height in conjunction with t may be used). Typically the functions for predicting a, b and c account for only a small part of the variation. The R^2 value for predicting the location parameter, a, which is equivalent to predicting the minimum diameter of the stand, is generally around 0.3. The shape parameter, c, is especially difficult to model with R^2 values often being close to 0.1. In most applications, the equations for estimating pdf parameters have been fitted independently, but Newton et al. (2005) employed stepwise regression and seemingly unrelated regression (SUR) techniques when modeling maximum likelihood estimates for the location, scale and shape parameters of the Weibull pdf fitted to data for black spruce plantations.

The information on heights and *dbh* of a set of sample trees is used to fit a height-*dbh* relationship that is specific to given stand conditions, that is:

$$h = f(d, t, S, N)$$

where $h =$ total tree height, $d =$ tree *dbh*, and t, S, N remain as previously defined.

Tree diameters and predicted heights are substituted into tree volume weight or taper equations for generating yield estimates. The tree survival and site index equations needed to implement the system may be developed from the same data set as the one used to estimate the pdf parameters or the functions may come from other sources. Examples of applications using the Weibull function and the parameter prediction approach for developing yield models include Smalley and Bailey (1974a, b), Dell et al. (1979), and Feduccia et al. (1979).

12.1.3 *Characterizing Diameter Distributions Using Parameter Recovery*

For an appropriate density function, Strub and Burkhart (1975) presented a class-interval-free method for obtaining yield estimates over specified diameter class limits. The general equation form is given by:

$$V_m = N \int_l^u g(d) f(d) dd$$

where

V_m = expected volume per unit area for a specified portion of the stand
N = number of trees per unit area,
$d = dbh$,
$g(d)$ = individual tree volume equation,
$f(d)$ = pdf for d, and
l, u = lower and upper merchantability limits, respectively, for the volume described by $g(d)$.

Using attributes from a whole stand model and the relationship given by the class-interval-free equation presented by Strub and Burkhart (1975), Hyink (1980) and Hyink and Moser (1983) introduced a method of solving for the parameters of a pdf approximating the diameter distribution. The approach involves predicting stand average attributes of interest for a specified set of stand conditions and using these estimates as a basis to "recover" the parameters of the underlying diameter distribution. When the number of stand attributes included is equal to the number of parameters in the pdf employed, a system of simultaneous equations results that can be solved for the pdf parameters, thus defining a "parameter recovery" method. The resultant diameter distribution will provide identical estimates of the stand average value(s) (e.g. average diameter, basal area) employed in the parameter recovery procedure.

12.1.3.1 Using Moments for Diameter Distribution Parameter Recovery

Use of moments to recover parameters for the pdf selected to represent the diameter distribution will be illustrated with the growth and yield model for yellow-poplar developed by Knoebel et al. (1986). As is often the case, the lower-bound of the diameter distribution is set equal to a constant or is predicted outside the parameter recovery system. In some instances a lower limit for tree size measurements is imposed in the field data collection; in cases where the recovery system is to be applied to thinned stands users generally wish to have the ability to set or predict the lower limit. The Knoebel et al. (1986) model consists of stand-level prediction plus diameter distribution information recovery using the Weibull function. The two parameter Weibull density was used, namely:

$$f(x;\mathrm{b},\mathrm{c}) = \left(\frac{\mathrm{c}}{\mathrm{b}}\right)\left(\frac{x}{\mathrm{b}}\right)^{\mathrm{c}-1} \exp\left[-\left(\frac{x}{\mathrm{b}}\right)^{\mathrm{c}}\right] \tag{12.3}$$

and the location parameter, a, was set to a constant or predicted external to the recovery of the b and c parameters.

With the general diameter distribution yield function,

$$Y_i = N \int_l^u g_i(x) f(x;\theta) dx \tag{12.4}$$

where

Y_i = total per unit area value of the stand attribute defined by $g_i(x)$
$g_i(x)$ = stand attribute as a function of x
$f(x;\theta)$ = pdf for x
N = number of trees per unit area
l, u = lower and upper diameter limits, respectively, for the attribute described by $g_i(x)$

integration over the range of diameters, x, for any $g_i(x)$, gives the total per unit area value of the stand attribute defined by $g_i(x)$. Average diameter, basal area per ha, and total cubic volume per ha are examples of such stand attributes. The number of stand attribute equations must equal the number of parameters to be estimated in order to solve the system of equations for recovery of the pdf parameters.

Letting $g_i(x)$ equal x^i, one obtains the ith noncentral moment of X as

$$E(X_i) = \int_{-\infty}^{\infty} X^i f(x;\theta)dx$$

and the parameter recovery system utilizes the method of moments technique of pdf parameter estimation. In the case of forest diameter distributions, the first noncentral moment, $E(X)$, is estimated by

$$\sum x_i/N = \bar{x},$$

the arithmetic mean diameter of the stand, and the second noncentral moment, $E(X^2)$, is estimated by

$$\sum x_i^2/N = \overline{x^2}$$

the square root of which is the quadratic mean diameter of the stand. Hence, the first two moments of the diameter distribution have stand-level interpretations that are common in forestry practice.

Stand average estimates of the first k moments produce a system of k equations with k unknown parameters which can be solved to obtained estimates of the pdf parameters while ensuring compatibility between whole stand and diameter distribution estimates of the stand attributes described by the moment equations.

Stand level basal area was estimated using the equation form of Sullivan and Clutter (1972). To ensure the required relationship $\overline{d^2} - \bar{d}^2 \geq 0$, the logarithm of the difference, $\ln(\overline{d^2} - \bar{d}^2)$, was predicted and, given a value for $\overline{d^2}$ obtained from the stand basal area prediction and the overall number of trees, $\bar{d}$ was determined algebraically. The minimum diameter (d_{min}) in the stand was specified as a constant or predicted and the location parameter was set to $0.5d_{min}$. The equations for the two parameter recovery system are:

$$\bar{x} = \int_0^\infty x f(x; \mathrm{b}, \mathrm{c}) dx = \mathrm{b}\Gamma(1 + 1/\mathrm{c}) \quad (12.5)$$

$$\overline{x^2} = \int_0^\infty x^2 f(x; \mathrm{b}, \mathrm{c}) dx = \mathrm{b}^2\Gamma(1 + 2/\mathrm{c}) \quad (12.6)$$

the estimated variance of the distribution is given by

$$s^2 = \overline{x^2} - \bar{x}^2 = \mathrm{b}^2 \left[\Gamma(1 + 2/\mathrm{c}) - \Gamma^2(1 + 1/\mathrm{c})\right] \quad (12.7)$$

and the coefficient of variation (CV) is estimated by

$$CV = \frac{s}{\bar{x}} = \frac{\left[\Gamma(1 + 2/\mathrm{c}) - \Gamma^2(1 + 1/\mathrm{c})\right]^{\frac{1}{2}}}{\Gamma(1 + 1/\mathrm{c})} \quad (12.8)$$

Given estimates of $\bar{x}$ and $\overline{x^2}$, the coefficient of variation is a function of c alone, thus reducing the order of the system. Under this formulation, there exists a unique solution for c, and simple iterative techniques for solving one equation in one unknown can be used to obtain a value for c. With c known, b is solved from $\bar{x} = \mathrm{b}\Gamma(1 + 1/\mathrm{c})$, and a is estimated with a constant or an equation external to the system. When applying the system, the same stand-level basal area equation is used when deriving diameter distributions and when estimating overall stand basal area in order to ensure compatibility between the two levels of stand detail.

Knoebel et al. (1986) used a set of randomly selected plots to evaluate the correspondence between observed and predicted diameter distributions and to check for logical consistencies that should be exhibited between stand tables for thinned and unthinned conditions. Although the predicted distributions closely approximated the observed distributions, some discrepancies were present among the stand tables of thinned and unthinned plots. Predicted numbers of trees increased in some diameter classes after thinning, and in some instances, the stand table after thinning had a larger maximum stand diameter and/or a smaller minimum stand diameter than those in the corresponding stand table prior to thinning. It was apparent that the diameter distribution predictions before and after a thinning from below could not be carried out independently, but had to be conditioned such that inconsistencies could not occur.

As an alternative to two independent predictions, the diameter distribution prior to thinning was predicted, as before, then a proportion of the basal area in each diameter class was removed to simulate the thinning. With this procedure the number of trees in a given class cannot increase as trees can only be removed from a class. Consequently, if any change occurs, minimum diameter can only increase and maximum diameter can only decrease.

A function was defined specifying the amount of basal area to be removed from each diameter class. The equation form relating the proportion of basal area removed in a diameter class to the ratio of the midpoint diameter of the class squared to the average squared diameter of the stand was:

$$P_j = \exp\left[\mathrm{b}_1\left(d_j^2/\bar{d}^2\right)^{\mathrm{b}_2}\right] \quad (12.9)$$

where

P_j = proportion of basal area removed from diameter class j
d_j = midpoint diameter of class j
$\bar{d}^2$ = average squared diameter of stand, and
b_1, b_2 = coefficients estimated from data.

As the yellow-poplar plot data were taken from stands thinned from below, the removal function reduces tree numbers more heavily in the smaller diameter classes than in the larger diameter classes. Equation 12.9, when fitted, represents the average removal pattern in the data used to estimate the parameters.

After defining basal area removal function (12.9), a thinning algorithm was implemented as follows:

1. Predict the diameter distribution prior to thinning from the Weibull distribution.
2. Starting with the smallest diameter class, remove the proportion of basal area specified by the removal function.
3. Proceed through the diameter classes until the desired level of basal area to be removed is achieved.
4. If the required basal area removal is not obtained after the largest diameter class is reached, return to the smallest diameter class and remove the remaining basal area in that class. Proceed in this manner through the diameter classes until the desired level of basal area removal is attained.

In another approach to insuring consistency between stand tables before and after thinning, Matney and Farrar (1992) and Farrar and Matney (1994) used a combination of parameter recovery and weighted constrained least squares procedures. Prior to the first thinning, diameter distributions are approximated by recovering the parameters of a three-parameter Weibull distribution so that its expected arithmetic- and quadratic-mean diameter equals predicted arithmetic- and quadratic-mean stand diameters. At first thinning, a list of tree diameters and the numbers per unit area are generated from the Weibull distribution. After generating the tree list, any specified thinning can be applied to the list; weighted constrained least squares procedures are employed to allocate mortality and diameter growth to the list. The mortality allocation function minimizes the weighted sum of squared differences between tree list elements before and after mortality, subject to constraints on the adjusted tree list, namely (1) the total number of trees per unit area represented equals the predicted total number of trees per unit area surviving to time $t_0 + \Delta t$, and (2) the arithmetic- and quadratic-mean diameters (first-and second-order diameter moments) equal the values for trees at time t_0 that will survive to time $t_0 + \Delta t$. Similarly, the least squares diameter growth procedure generates a tree diameter growth allocation function by minimizing the weighted sum of squared differences between tree list elements at time $t_0 + \Delta t$, subject to the constraint that the adjusted tree list has arithmetic- and quadratic-mean diameters equal to their projected values (Matney and Farrar 1992).

A number of growth and yield simulators use the moment-based parameter recovery method with the Weibull function, among them being models for unthinned slash pine plantations (Matney, et al. 1987), thinned loblolly pine plantations (Cao et al. 1982), and unthinned natural stands of loblolly pine (Burk and Burkhart 1984). In most applications of the parameter recovery technique, diameter distributions that are compatible with stand predicted values for mean diameter and basal area are produced, height is predicted as a function of diameter (Sect. 12.2), and yields are computed using volume, weight or taper functions. Using these procedures, the resultant volume per unit area from the diameter distribution analysis will not equal that obtained from a standard yield function. An exception is the model of Matney and Sullivan (1982) for compatible stand and stock tables for thinned and unthinned stands. They first developed compatible equations for projecting per hectare values of numbers of trees, basal area, and total cubic volume. Three-parameter Weibull distribution representations of diameter distributions were then calculated so that when integrated for per hectare basal area and cubic volume the result is the same as predicted.

12.1.3.2 Using Percentiles for Diameter Distribution Parameter Recovery

Percentile estimates provide another approach to recovering parameters of a specific distribution function. In the case of the Weibull function, percentile estimators are relatively easy to implement. The basic concept is similar to that of using moments for recovery of pdf parameters. If three sample percentiles are known, each can be equated to its corresponding Weibull cumulative distribution function value and the three equations can be solved iteratively for estimates of *a, b* and *c*.

Given the Weibull cumulative distribution function

$$F(X) = 1 - \exp\{-[(x - \mathrm{a})/\mathrm{b}]^{\mathrm{c}}\}$$

and letting X_p represent the p-percentile value in the sample (that is the value such that 100p-percent of the sample values are less that X_p), then

$$p = 1 - \exp\left\{-[(X_p - \mathrm{a})/\mathrm{b}]^{\mathrm{c}}\right\} \tag{12.10}$$

Expression (12.10) can be solved for X_p to give

$$X_p = \mathrm{a} + \mathrm{b}[-\ln(1 - p)]^{\frac{1}{\mathrm{c}}} \tag{12.11}$$

Selected percentiles of diameter distributions are computed for a series of sample plots and related to the overall plot characteristics of age, total number of trees per unit area, and site index (or, alternatively height of the dominant stand). A predicted diameter distribution for any specified combination of t, N and S (or h_{dom}) values can be obtained by solving the system of percentile estimators for the Weibull distribution parameters. Three percentiles can be used to develop a simultaneous system of equations to be solved but, as with the use of moment estimators, the location parameter a is often predicted directly or related to the predicted value for the minimum diameter and the scale and shape parameters (b and c, respectively) are recovered through the solution of two equations with two unknowns.

Although many different approaches to the application of percentile estimators are possible, the Weibull recovery procedure presented by Bailey et al. (1989) has been widely applied (examples include Brooks et al. 1992; Knowe 1992, and Knowe et al. 1994). The Bailey et al. (1989) parameter recovery procedure is based on the 0th, 25th, 50th and 95th diameter percentiles (denoted D_0, D_{25}, D_{50} and D_{95}). Assuming that c = 3, the location parameter, a, is obtained by the minimum (D_0) and median (D_{50}) diameters and sample size (n):

$$\mathrm{a} = \frac{n^{0.3333} D_0 - D_{50}}{n^{0.3333} - 1}$$

Negative values for a are set to zero. The shape parameter is determined by using the estimate for the location parameter (a) and D_{95} and D_{25}:

$$\mathrm{c} = \frac{2.343088}{\ln(D_{95} - \mathrm{a}) - \ln(D_{25} - \mathrm{a})}$$

Then the scale parameter, b, is obtained by solving the second moment of the Weibull distribution for the positive root using the estimates for c and $\bar{d}_g^2$:

$$\mathrm{b} = -\frac{\mathrm{a}\Gamma_1}{\Gamma_2} + \sqrt{\left(\frac{\mathrm{a}}{\Gamma_2}\right)^2 (\Gamma_1^2 - \Gamma_2) + \frac{\bar{d}_g^2}{\Gamma_2}}$$

where

$\Gamma_1 = \Gamma\left(1 + \frac{1}{c}\right)$
$\Gamma_2 = \Gamma\left(1 + \frac{2}{c}\right)$
Γ is the gamma function, and
$\bar{d}_g$ is the quadratic mean diameter

This procedure does not rely entirely upon percentiles but includes information on d_g. The system ensures that d_g in the predicted diameter distribution is the same as the quadratic mean diameter implied by prediction equations for stand basal area and number of trees.

12.1.4 Evaluations of Alternative Distributions and Parameter Estimation Methods

A number of comparisons have been made involving alternative methods for estimating pdf parameters from forest diameter distribution data. Zarnoch and Dell (1985) contrasted maximum likelihood and percentile estimators for estimating the three-parameter Weibull distribution. Employing computer simulations and field data comparisons, they found that maximum likelihood estimators had smaller bias and lower mean square error but larger variance than the percentile estimators. Comparisons of resultant distributions indicated that either maximum likelihood or percentile estimators should perform adequately when modeling pine plantation diameter distributions. The authors further noted that because the parameters are correlated, various combinations of parameter values can lead to similar distributions.

Shiver (1988) evaluated three methods of parameter estimation over four sample sizes for characterizing diameter distributions using the Weibull function. Maximum likelihood, moment (Burk and Newberry 1984; Garcia 1981) and percentile (Zanakis 1979) estimators were contrasted by generating samples of 30, 50, 75, and 100 trees using data from a simulator developed to provide realistic diameter distributions for unthinned slash pine plantations. Maximum likelihood estimation provided the best estimates of known distribution parameters. Under the assumption of no specific underlying parameters, both moments and percentile procedures reproduced the underlying distribution of diameters as well as or better than maximum likelihood estimation. All three methods required samples of approximately 50 trees to reproduce distributions with acceptable accuracy. In agreement with results of Zarnoch and Dell (1985), Shiver (1988) found that percentile estimators were biased but had smaller variances than maximum likelihood estimators. Percentile estimators were also more biased but with smaller variances than moment-based estimators. Maximum likelihood estimators proved best when the goal was to reproduce a given set of Weibull parameters. However, from a practical point of view, differences in results from different estimation methods were small and the alternate methods performed as well at reproducing diameter distributions.

The investigations by Zarnoch and Dell (1985) and Shiver (1988) were aimed at evaluating methods for estimating parameters of the Weibull function for the purpose of reproducing forest stand diameter distributions. In order to implement a diameter distribution based yield model, the parameters must be related to overall stand characteristics such as age, site index and trees per unit area. Liu et al. (2004) used parameter prediction, moment-based parameter recovery, and percentile-based parameter recovery methods for estimating parameters of the three-parameter Weibull pdf and tested their applicability for predicting diameter distributions of unthinned black spruce plantations. Using stepwise regression analyses in combination with seemingly unrelated regression techniques, the three methods were calibrated using commonly measured prediction variables (stand age, dominant height, site index, and stand density). Results indicated that, although all three methods were successful in predicting the diameter frequency distributions within the sample stands, the percentile-based recovery was superior in terms of prediction error. Specifically, the percentile-based method had the lowest mean error index (80.98), followed by the moment-based recovery (82.73) and the parameter prediction method (83.98). Consequently, among the three methods assessed, the percentile method was considered the most suitable for describing unimodal diameter distributions using the three-parameter Weibull probability density function within unthinned black spruce plantations.

Cao (2004) used data from a loblolly pine planting thinned to different residual densities to evaluate six methods for predicting parameters of Weibull functions applied to model diameter distributions. The following general form for regression equations was adopted:

$$y = \exp \mathrm{b}_1 + \mathrm{b}_2 R_s + \mathrm{b}_3 \ln N + \mathrm{b}_4 \ln h_{dom} + \mathrm{b}_5 t^{-1} \quad (12.12)$$

where:

y = a specific Weibull parameter, diameter percentile, or mean diameter;
R_s = relative spacing, $(10{,}000/N)^{0.5}/h_{dom}$, which is the ratio of average distance between trees (assuming square spacing) and dominant height;
N = number of trees per ha;
h_{dom} = dominant height in meters.
t = stand age in years; and
$\mathrm{b}_1, \ldots, \mathrm{b}_5$ = regression parameters.

Methods used to obtain the Weibull parameters were:

1. ***Parameter prediction***
   The system consisted of three regression equations, each in the form of Eq. 12.12. The dependent variables of these equations were Weibull parameters, a, b and c, which had previously been estimated for each plot measurement using the method of maximum likelihood. Because cross-equation correlations existed among error components of these equations, they were treated as a system of nonlinear, seemingly unrelated equations.
2. ***Moment estimation***
   The Weibull location parameter was computed as $0.5\hat{D}_0$, the predicted minimum diameter in the stand. The b and c parameters were recovered from the first two moments of the diameter distribution

$$\mathrm{b} = (\hat{\bar{d}} - \mathrm{a})/\Gamma_1$$

$$\hat{\bar{d}}_g^2 + \mathrm{a}^2 2\mathrm{a}\hat{\bar{d}} - \mathrm{b}^2\Gamma_2 = 0$$

where $\hat{\bar{d}}$ and $\hat{\bar{d}}_g$ are predicted arithmetic and quadratic mean diameters, respectively, and $\Gamma_i = \Gamma(1 + i/\mathrm{c})$. Borders' (1989) method was used to simultaneously estimate regression parameters of equations to predict D_0, $\bar{d}$, and $\bar{d}_g$.

3. ***Percentiles***
The method of Bailey et al. (1989) was used to compute the Weibull parameters from the quadratic mean diameter ($\hat{\bar{d}}_g$), minimum diameter ($\hat{D}_0$), 25th diameter percentile ($\hat{D}_{25}$), 50th percentile ($\hat{D}_{50}$), and 95th percentile ($\hat{D}_{95}$):

$$\mathrm{a} = (n^{0.3333}\hat{D}_0 - \hat{D}_{50})/(n^{0.3333} - 1)$$

$$\mathrm{c} = 2.343088/\left[\ln(\hat{D}_{95} - \mathrm{a}) - \ln(\hat{D}_{25} - \mathrm{a})\right]$$

$$\mathrm{b} = -\mathrm{a}\Gamma_1/\Gamma_2 + \left[(\mathrm{a}/\Gamma_2)^2(\Gamma_1^2 - \Gamma_2) + \hat{\bar{d}}_g^2/\Gamma_2\right]^{0.5}$$

where n is the number of trees in the plot. This method was not strictly percentiles as $\bar{d}_g$ was used. Regression parameters of equations to predict $\bar{d}_g$, D_0, D_{25}, D_{50}, and D_{95} were simultaneously estimated.
4. ***Moment-percentile method***
As in Method 2, the Weibull location parameter was computed from $\mathrm{a} = 0.5\hat{D}_0$. Baldwin and Feduccia (1987) developed this hybrid method in which the b and c parameters are recovered from two values: a moment (the quadratic mean diameter) and a percentile (the 93rd diameter percentile or $\hat{D}_{93}$):

$$\mathrm{b} = (\hat{D}_{93} - \mathrm{a})/2.65926^{1/\mathrm{c}}$$

$$\mathrm{a}^2 - \hat{d}_g^2 + 2\mathrm{a}(\hat{D}_{93} - \mathrm{a})\Gamma_1/2.65926^{1/\mathrm{c}} + \mathrm{a}(\hat{D}_{93} - \mathrm{a})^2\Gamma_2/2.65926^{2/\mathrm{c}} = 0$$

where $2.65926 = -\ln(1-0.93)$. The system of equations to predict, $\bar{d}_g$, D_0, and D_{93} was fitted simultaneously.
5. ***Maximum likelihood estimator (MLE) regression***
The Weibull location parameter was computed from $\mathrm{a} = 0.5\hat{D}_0$. The following two equations were used for predicting the scale parameter (b) and shape parameter (c):

$$\mathrm{b} = \exp\left[\mathrm{b}_1 + \mathrm{b}_2 R_s + \mathrm{b}_3\ln(N) + \mathrm{b}_4\ln(h_{dom}) + \mathrm{b}_5 t^{-1}\right]$$

$$\mathrm{c} = \exp\left[\mathrm{c}_1 + \mathrm{c}_2 R_s + \mathrm{c}_3\ln(N) + \mathrm{c}_4\ln(h_{dom}) + \mathrm{c}_5 t^{-1}\right]$$

The coefficients b_i and c_i were iteratively searched to maximize the sum of the log-likelihood values from all plots.
6. ***Cumulative Distribution Function (CDF) Regression***
This method is similar to Method 5, except that the coefficients b_i and c_i were iteratively searched to minimize the following function:

$$\sum_{i=1}^{p}\sum_{j=1}^{n_i}(F_{ij} - \hat{F}_{ij})^2/n_i$$

where

F_{ij} = observed cumulative probability of tree j in the ith plot-age combination;
$\hat{F}_{ij} = 1 - \exp\{-[(x_{ij} - \mathrm{a})/\mathrm{b}]^c\}$ = or the value of the CDF of the Weibull distribution evaluated at x_{ij};
x_{ij} = *dbh* of tree j in the ith plot-age combination; and
n_i = number of trees for the ith plot-age combination.

Results from three measures of goodness of fit were very similar. Parameter prediction (Method 1) was the poorest performer, ranking last for both fit and validation data. The "parameter recovery" group, which involved recovery of the Weibull parameters from moments (Method 2), percentiles (Method 3), or a combination of both (Method 4), produced similar evaluation statistics. Methods 5 and 6 ranked best among the alternatives evaluated, with Method 6 (CDF regression) consistently producing the lowest goodness-of-fit statistics for both fit and validation data.

When developing a diameter distribution prediction model for longleaf pine plantations, Jiang and Brooks (2009) evaluated the percentile-based parameter recovery method of Bailey et al. (1989) (Method 3 in Cao's comparison) and the CDF regression presented as Method 6 by Cao (2004) for estimating parameters of the Weibull function. In contrast to the results of Cao (2004), the percentile-based parameter recovery method consistently produced the better goodness-of-fit statistics for both fit and validation data sets of longleaf pine. Liu et al. (2009) compared six methods for predicting parameters of the Weibull function when describing diameter distributions in unthinned white spruce plantations in eastern Canada. Weibull parameters, moments, or diameter percentiles were related to stand age, site index, dominant height and relative spacing using stepwise linear regression analysis to identify functional forms. Results indicated that the percentile method performed best. Maximum likelihood regression, cumulative distribution function regression, and moment-based parameter recovery all performed about equally well and all were superior to parameter prediction. The two hybrid-based (moment-percentile) methods tested showed the poorest performance in predicting diameter frequency distributions.

The Johnson system of distributions was used by Zhao and McTague (1996) to fit both diameter and height data from ponderosa pine and mixed-conifer forest types in New Mexico and Arizona, USA. Five parameter estimation methods (percentile, Knoebel-Burkhart, mode, maximum likelihood, and a new linear regression approach developed by the authors) were compared and evaluated. For the sample data available, the linear regression method proved best for estimating parameters of S_B distributions for both diameter and height.

In a study aimed at developing tree diameter distributions for biomass estimation in four boreal forest types (trembling aspen, jack pine, black spruce, and mixed forest), Chen (2004) contrasted parameter prediction and parameter recovery methods using the Weibull, Johnson's S_B, and lognormal functions. The three distribution functions ranked Johnson's S_B first, lognormal second and Weibull third, but the differences in goodness of fit among them were minor. The parameter recovery method performed well in comparison to two parameter prediction methods that were evaluated.

Zhang et al. (2003) compared four commonly used estimation models to fit the three-parameter Weibull and Johnson's S_B distributions to pure and mixed stands of red spruce-balsam fir stands. The results indicated that the Weibull and the Johnson's S_B distributions were generally equally suitable for modeling the diameter frequency distributions. However, the relative performance depended on the parameter estimation method used. The observed diameter distributions of the whole plots were typically positively skewed, reverse-J, or mound shapes. The relative merits of the Weibull and Johnson's S_B distributions depended not on the inherent flexibility of the probability density functions themselves, but rather on the accuracy of methods employed to estimate the parameters. As more information from the tree list was used to fit the distributions, a more accurate representation of the tree diameter distribution was obtained.

12.1.5 Characterizing Diameter Distributions of Mixed-Species Stands

In addition to monocultures, diameter distributions for mixed-species stands have been modeled. Little (1983) reported that the three-parameter Weibull function met specified standards for goodness of fit as a model for the diameter distributions of mixed stands of western hemlock and Douglas- fir. Bowling et al. (1989) developed diameter distributions for five species groups of Appalachian hardwoods. Equations were specified to predict stand attributes by species group and for the whole stand to provide the inputs for moment-based recovery of parameters of Weibull functions. In another application, Maltamo (1997) used Weibull functions to model distributions of mixed Scots pine and Norway spruce stands. Tham (1988) investigated the structure of mixed Norway spruce and two birch species and found that the Johnson's S_B distribution fit well to all three species separately and to the entire stand.

The diameter distribution of the entire stand consisting of multiple species is often basically unimodal. When species are considered individually, multiple modes may be exhibited; neither the Johnson S_B nor the Weibull distribution can accurately represent bimodal distributions. Various solutions to the problem of modeling highly irregular tree-size distributions have been investigated including use of segmented distributions (Cao and Burkhart 1984), distribution-free models (Borders et al. 1987), and nonparametric statistical methods (Droessler and Burk 1989; Haara et al. 1997; Maltamo and Kangas 1998).

Liu et al. (2002) suggested using a finite mixture model (FMM) for characterizing the diameter distributions of mixed-species stands. A frequency distribution made up of two or more component distributions is defined as a "mixture" distribution.

In their study, Liu et al. compared a finite mixture of two Weibull distributions against two traditional methods: (1) fitting the Weibull function to the entire plot, and (2) fitting the Weibull function to each species separately. For comparison purposes, plots composed of two species were used. Results showed that the finite mixture model was capable of fitting irregular, multimodal, or highly skewed diameter distributions. Compared with traditional methods in which a single Weibull function is fitted to either the whole plot or each species component separately, the finite mixture model produced smaller root mean square error and bias. The finite mixture approach simultaneously estimates the proportion and component diameter distributions of different tree species in mixed-species stands. Liu et al. judged the finite mixtures modeling approach to be promising but pointed out that it has both advantages and disadvantages. Advantages listed were: (1) it is more flexible than the traditional Weibull function fit either to the whole plot or to individual species separately, (2) it is not necessary to classify the components of a multimodal distribution a priori during data collection, and (3) the proportions of each species component or group can be estimated when the information is not available in the data. A disadvantage of the FMM approach is that it may not predict each species component as accurately as fitting the component data separately.

12.1.6 Bivariate Approach

12.1.6.1 Modeling Diameter and Height Distributions

As an alternative to predicting mean tree heights by diameter class, Schreuder and Hafley (1977) suggested a bivariate distribution approach for describing height-diameter data from even-aged stands. Based on the comparatively good performance of the marginal S_B distribution for fitting both diameter and height data (Hafley and Schreuder 1977), Schreuder and Hafley (1977) asserted that the bivariate extension, commonly denoted S_{BB}, could provide more usable information than approaches that involve fitting marginal distributions for diameter and estimating mean heights by diameter class. The S_{BB} allows for generation of bivariate frequencies for diameter and height. In addition, the S_{BB} implies a height-diameter relationship that is comparable in fit to commonly-used regression models.

Hafley and Buford (1985) used the S_{BB} distribution for describing stand structure (height and diameter distributions) in thinned and unthinned stands of loblolly pine. By using a bivariate distribution approach height, as well as diameter, can be included in the thinning algorithm. In the Hafley-Buford application, a stand is described at any instance using the S_{BB} distribution on the assumption that the stand is a bivariate population of diameters and heights. The model predicts changes over time in nine stand characteristics and uses those characteristics to derive the nine parameters of the S_{BB} distribution. Stand yield at any desired point in time is determined from the S_{BB} distribution whose parameters are defined by the stand characteristics, and the number of surviving trees per hectare at that point. The quadratic mean diameter, basal area, average height, dominant height, and cubic volume are all obtained from moments of the respective marginal S_B distributions and the S_{BB} distribution using numerical integration. Integrating the bivariate distribution over the two-way cell boundaries gives two-way stand tables of frequency by diameter and height class.

12.1.6.2 Modeling Diameter Distributions at Two Points in Time

Knoebel and Burkhart (1991) presented a bivariate distribution approach to modeling forest diameter distributions at two points in time. There are a number of shortcomings associated with the usual parameter prediction and parameter recovery methods by which a given diameter distribution is projected to a future point in time. These procedures obtain initial and future diameter distribution parameter estimates for a specified stand either directly or indirectly from initial and future stand characteristics. The future diameter distribution does not directly depend on the initial diameter distribution; the initial and future diameter distributions are treated as independent entities and the dependencies that exist between them are not taken into account. Intuitively, the projected future distribution should be related to, or depend on, the current initial distribution. These considerations led Knoebel and Burkhart to hypothesize that the incorporation of the initial distribution information into a projection method would improve the prediction of the future diameter distribution. As initial and future diameters measured on the same tree represent correlated observations, and future tree diameter given current diameter is a random variable, an investigation of a bivariate distribution approach to modeling distributions at two points in time was undertaken. Such an approach assumes a dependency between the initial and future diameter distributions and allows future tree diameter, given current diameter, to be treated as a random variable.

Rather than independently predicting the initial and future diameter distributions, the bivariate distribution approach suggests the following: given an initial distributions and the implied relationships between initial and future tree diameters as defined by the particular bivariate distribution, determine or specify the future distribution. The first phase in development of the bivariate distribution approach involved the selection of an appropriate bivariate density capable of describing the diameter distributions. Criteria for selection included flexibility and ease of parameter estimation. A bivariate distribution was considered to be flexible if the marginal distributions were capable of describing the univariate diameter distributions while implying a biologically reasonable relationship between the initial and future diameter random variates. Many statistical distribution functions have been used to describe diameter distributions in forest stands with varying degrees of success. The selected bivariate distribution should be able to accommodate both positive and negative skewness, as well as symmetry for either marginal variate.

Ease of parameter estimation and computations was desired as was limiting the number of parameter to be estimated while ensuring sufficient flexibility in approximating a range of shapes. Estimation of proportions in various size classes should also be relatively simple. A bivariate distribution that satisfied all criteria set forth was the bivariate Johnson distribution or the S_{BB}.

For the S_{BB} distribution the median regression takes a relatively simple form and is bounded by the limits of the marginal distributions. Hence, the S_{BB} distribution was selected as an appropriate bivariate distribution to model tree diameter distributions at two points in time.

Simply fitting a bivariate S_B distribution to tree diameters from sample plots and predicting the distribution parameters from fitted functions relating them to stand characteristics would not establish a means of determining the future distribution given the initial distribution. While directly fitting the bivariate distribution will determine whether the initial and future diameter distributions can be described by such a bivariate distribution, this fitting requires knowledge, or specification, of both the initial and future marginal distributions, and the future distribution is assumed to be unknown.

For the purpose of modeling diameter distributions at two points in time, it was of interest to explicitly define the future distribution based on the current distribution. The general approach taken, which was possible due to the form of the median regression equation of the S_{BB}, was as follows. First specify the initial diameter distribution, described by an S_B density function. Next, define the median regression equation, which represents the relationship between initial and future tree diameters. From the parameters of the initial distribution and median regression, obtain the future diameter distribution parameters. The bivariate S_B distribution model presented by Knoebel and Burkhart was compared to several "standard" methods for predicting future diameter distributions including the parameter prediction, parameter recovery, and percentile prediction methods. Observed present and future diameter distributions were obtained from permanent plot installations in yellow-poplar stands. Comparisons of projected diameter distributions to observed values showed that the bivariate S_B distribution approach gave results as good as, and in some cases better than, the alternative methods.

12.2 Modeling Height-Diameter Relationships

In most applications, distribution functions have been used to estimate stem frequencies by *dbh* class. Height information is then generated via a regression equation fitted for predicting mean heights of trees of a given *dbh* growing in stands of specified age, site index (or average height of the dominant stand), and total number of stems per hectare. The midpoint of diameter classes, along with the predicted tree heights, are then used with tree taper, volume or weight functions (Chaps. 2, 3, and 4) to compute the per unit area values of interest. Heights, as well as diameters, are measured on all, or often times a sample, of the trees comprising the sample plots used to develop the diameter distribution information.

Because of their importance for a number of forest stand modeling applications, height-diameter equations have received considerable attention. In addition to predicting average heights associated with *dbh* classes in diameter distribution systems, height-diameter relationships are also employed in stand-table projection (Chap. 13) and individual-tree growth and yield simulators (Chap. 14).

Curtis (1967) compared a number of height-diameter and height-diameter-age equations for Douglas-fir and found that most gave similar results within the range of the observed data but some forms were more stable than others when extrapolation was involved. In an evaluation of model forms for estimating height-diameter relationships in loblolly pine plantations, Arabatzis and Burkhart (1992) found the form

$$\ln h = \mathrm{b}_0 + \mathrm{b}_1 d^{-1} \tag{12.13}$$

where h is total tree height for a tree of *dbh* equal to d to be most satisfactory.

In the context of diameter distribution models developed for growth and yield prediction, most have relied on the basic form (12.13) when predicting mean tree heights by diameter class. Bennett and Clutter (1968), in their original publication on yield prediction based on diameter distribution analyses, specified the height prediction model

$$\ln h = \mathrm{b}_0 + \mathrm{b}_1 S + \mathrm{b}_2 N + \mathrm{b}_3 t^{-1} + \mathrm{b}_4 d^{-1} \tag{12.14}$$

which reduces to $\ln h = \mathrm{b}_0 + \mathrm{b}_1 d^{-1}$ for specified stand variables of *S, N* and *t*.

In a subsequent analysis, Lenhart and Clutter (1971) adopted the form

$$\ln(h/h_{dom}) = \mathrm{b}_0 + \mathrm{b}_1(d^{-1} - d_{\max}^{-1}) + \mathrm{b}_2 t^{-1}(d^{-1} - d_{\max}^{-1}) + \mathrm{b}_3(d^{-1} - d_{\max}^{-1}) \ln N \tag{12.15}$$

where $d_{\max}$ is defined as the midpoint or the upper limit of the largest occupied diameter class. Formulation (12.15) also reduces to $\ln h = \mathrm{b}_0 + \mathrm{b}d^{-1}$ for given stand variables but it has the desirable property that predicted height for the larger *dbh* classes will be close to the specified average height of the dominant stand (h_{dom}). Variants of equation form (12.15) have been incorporated in a number of diameter-distribution-based models including Smalley and Bailey (1974a, b), Dell et al. (1979), Feduccia et al. (1979), and Amateis et al. (1984).

Most height-diameter relationships are aimed at predicting individual tree heights given stand age, site index, stand density, and *dbh*. However, as Lynch and Murphy (1995) pointed out, it is reasonable to expect that the current total height of a tree if known (either by direct measurement or from the current height-*dbh* relationship in the stand) should also be useful when predicting tree heights. Since "current" (previous to age of prediction) tree heights are not always known, a height prediction system which also retains the desirable features of traditional height-diameter relationships may also be required.

Accordingly, Lynch and Murphy (1995) developed a system of two compatible equations that can be used to predict individual tree heights at various ages. One equation uses known values of current tree height to project future heights, while the other provides height predictions that depend only on *dbh* and stand-level variables. The Lynch-Murphy system, when applied to data for trees in natural, even-aged shortleaf pine stands, showed that future total heights were more precisely predicted when previous values of total height were known. Seemingly unrelated nonlinear regression techniques were applied to the system of two equations, which had unequal sample sizes and restrictions on equation parameters.

Due to the importance of estimating heights from tree diameters, a number of evaluations of model forms and/or efficacy of adding additional stand-level predictor variables have been conducted (examples include Arabatzis and Burkhart 1992; Huang et al. 1992; Huang et al. 2000; Staudhammer and LeMay 2000; Temesgen et al. 2007; and Fast and Ducey 2011; Paulo et al. 2011). The addition of stand variables after *dbh* generally improves fit statistics and extends the generality of height-diameter models. Lei and Parresol (2001) listed the following desirable characteristics for functions used to model heigh-diameter relationships: (i) increase monotonically, (ii) have an upper asymptote, and (iii) have an inflection point. Paulo et al. (2011) questioned the third requirement. Both diameter and height growth curves have an inflection point, but this may not be necessarily so for the relationship between height and diameter. Although the fitting data set used by Paulo et al. included cork oak trees with large variability of diameter values (corresponding to young and old individuals), no evidence of an inflection point was found when plotting height against diameter.

Several relatively recent studies (for instance, Mehtätalo 2004; Calama and Montero 2004; Sharma and Parton 2007; Trincado et al. 2007; Budhathoki et al. 2008; Temesgen et al. 2008; and Coble and Lee 2011; and Paulo et al. 2011) have applied mixed-effects modeling techniques when developing height-diameter curves. Mixed models allow prediction of a typical response, using only fixed effects, and a calibrated response where random effects are predicted and included in the model using measurements of heights from a sample of trees.

12.3 Predicting Unit-Area Tree Survival

When applying distribution functions to estimate stand tables and yields, the total number of trees per unit area is assumed to be known. Given that a suitable site index equation is available, future yields and growth estimates (computed as the differences between yield values at two points in time) rely on projection of stand age and overall number of trees. The ability to estimate the number of stems per unit area over time and for varying stand characteristics is a key component in the distribution-based approach to growth and yield prediction.

The prediction of tree survival (or mortality) is difficult, and a number of different approaches have been advanced. In the context of diameter-distribution-based models, the basic structure

$$\ln(N_p/N_s) = t\ f(t, S(\text{or}\ h_{dom}), N_p) \tag{12.16}$$

where N_p represents the number of trees planted, N_s is the number surviving at age t, S is site index (alternatively, h_{dom} height of the dominant stand at age t) has been employed in several models, including those of Smalley and Bailey (1974a, b), Dell et al. (1979), Feduccia et al. (1979), and Amateis et al. (1984). The merits of form (12.16) are that when $t = 0$, $\ln(N_p/N_s) = 0$ and, thus, $N_s = N_p$ and the equation predicts 100% survival.

Tree survival prediction equations are commonly developed using data from at least one remeasurement of permanent sample plots. Typically, the data consists of numbers of stems (N_1) at an initial age (t_1), number (N_2) at a subsequent age (t_2), and site index of the stand. Such data allow fitting an equation to predict N_s as a function of t_1, t_2, and N_1; site index may also be included in the prediction equation, but it has often been found not to contribute significantly to the fitted relationship. Certain properties are desired in equations for estimating future numbers of trees in even-aged stands, including: (i) if t_2 equals t_1,N_2 should equal N_1, (ii) if t_2 is greater than t_1, N_2 should be less than or equal to N_1, (iii) as t_2 increases without bound, N_2 should approach 0; and (iv) if N_2 is predicted at age t_2 and t_2 and N_2 are then used to predict N_3 at age $t_3 (t_3 > t_2 > t_1)$, the result should be the same as that obtained by a single projection from t_1 to t_3.

Experience has shown that integration of mortality rate equations is often an effective means of expressing a difference equation model with desirable properties for estimating future numbers of trees. For example, one might assume that instantaneous mortality rate is constant, that is:

$$\frac{dN/dt}{N} = \text{k} \tag{12.17}$$

where N = number of trees per ha at age t, dN/dt = instantaneous mortality rate at age t, and k = constant.

Integration of equation (12.17) with the initial condition that $N = N_1$ when $t = t_1$ gives the difference equation:

$$N_2 = N_1 e^{\text{k}(t_2 - t_1)} \tag{12.18}$$

Equation 12.18, would be appropriate for populations where proportional mortality rate is constant at all ages, site indexes, and stand densities. Such an assumption is not realistic for most cases involving forest stands, but rather mortality rate is generally related to stand variables.

Assuming that mortality rate is proportional to stand age raised to a power gives:

$$\frac{dN/dt}{N} = \alpha t^{\beta} \tag{12.19}$$

Integrating Eq. 12.19 results in the difference equation model

$$\ln N_2 = \ln N_1 + b_1 \left(t_2^{b_2} - t_1^{b_2} \right) \tag{12.20}$$

which has been fitted by Piennar and Shiver (1981), Pienaar et al. (1990), and Amateis et al. (1997).

While difference Eq. 12.20 may be suitable for certain stages of stand development, an assumption that mortality rate is proportional to stand age and density is often more appropriate leading to the mortality rate model:

$$\frac{dN/dt}{N} = \alpha t^{\beta} N^{\gamma} \tag{12.21}$$

Solving (12.21) as a difference equation yields

$$N_2 = \left[N_1^{b_1} + b_2 \left(t_2^{b_3} - t_1^{b_3} \right) \right]^{(1/b_1)} \tag{12.22}$$

Model (12.22) exhibits several desirable properties, such as, when $t_1 = t_2$, $N_2 = N_1$ and as t_2 approaches infinity N_2 approaches zero. Clutter and Jones (1980) used model (12.22) when projecting number of trees in thinned slash pine plantations. In an evaluation of four equations for predicting mortality after thinning in loblolly pine plantations, Lemin and Burkhart (1983) found that the difference equation form (12.22) performed best.

Although site index frequently does not contribute significantly after including age and number of trees when predicting mortality, it is occasionally also included. Under the assumption that mortality rate is proportional to a function of age, number of trees, and site index one can write the differential equation as:

$$\frac{dN/dt}{N} = \alpha t^{\beta} N^{\gamma} S^{\delta} \tag{12.23}$$

Integrating (12.23) results in the difference equation:

$$N_2 = \left[N_1^{b_1} + b_2 S^{b_3} \left(t_2^{b_4} - t_1^{b_4} \right) \right]^{(1/b_1)} \tag{12.24}$$

Obtaining convergence when fitting (12.24) with nonlinear least squares can be difficult. A slightly different form was fitted by Clutter et al. (1984) when predicting mortality in loblolly pine plantations, namely:

$$N_2 = \left[N_1^{b_1} + \left(b_2 + b_3 S^{-1} \right) \left(t_2^{b_4} - t_1^{b_4} \right) \right]^{(1/b_1)} \tag{12.25}$$

Somers et al. (1980) demonstrated the use of the censored Weibull distribution for modeling survival in young, even-aged natural stands of loblolly pine. Setting parameters "c" of the Weibull distribution to a common value for all densities and relating parameter "b" to initial numbers of trees resulted in adequate predictions from ages 3 to 14.

In a study aimed at developing methods for predicting stand survival over much of the life of a plantation, Amateis et al. (1997) defined three distinct phases. The first phase occurs during the first year after planting. Survival during the first year tends to be highly variable and depends on many factors, including care of the planting stock at the nursery and at the planting site, storage time of seedlings, planting crew practices, and first year climatic factors (such as the amount and distribution of rainfall during the growing season).

The second survival phase occurs from year 1 to sometime beyond crown closure. Mortality during this period can be attributed to factors other than intraspecific competition. Important factors which may influence survival in phase 2 include levels of interspecific competition (both woody and herbaceous), stand establishment practices, and certain stochastic elements such as insect or disease attacks.

The third survival phase occurs from the onset of intraspecific competition-induced mortality to rotation age. During this period, the effects of intraspecific competition are the dominant forces affecting survival. Factors to be considered when modeling survival during this phase are age, stand density, and site index. These factors will also affect the time when competition-induced mortality begins (denser stands on better sites are expected to enter this stage of stand development sooner). In addition, intermediate silvicultural treatments applied to plantations during this phase will affect survival patterns. Thinning, in particular, has a direct effect on stand survival because it alters the amount and distribution of the growing stock, and it changes the overall vigor of the stand by removing smaller, slower growing trees (when thinning is from below) and provides additional growing space for the residual stand. Survival models for this period of intraspecific competition include natural changes in stand density due to self-thinning and perhaps artificial changes due to thinning and other silvicultural treatments as well.

Amateis et al. (1997) presented two equations which can be used to project survival for the second and third phases of stand development. Owing to the highly-variable nature of first-year survival, prediction equations could not be developed for this stage. Managers are required to specify first-year survival when applying the system. Model form (12.20) was used for estimating survival in young stands (phase 2) and a post-crown closure (phase 3) model was specified to complete the system.

Woollons (1998) suggested a two-step strategy for modeling mortality in even-aged stands. In the first step a logistic regression is developed to predict the probability of mortality (or, alternatively, of all trees surviving). A second step consists of fitting a mortality equation utilizing only plots where some tree death occurred over the observation period. Representative examples of applications of this two-step idea include models of mortality for even-aged stands fitted by Eid and Øyen (2003) using data from the Norwegian National Inventory, equations for predicting tree number decline in planted stands of Scots pine in northwestern

Spain developed by Diéguez-Aranda, et al. (2005), and mortality models for loblolly pine plantations in the southeastern United States published by Zhao et al. (2006).

In an effort to avoid specifying separate equations for estimating survival during different phases of the life span of plantations, Rose et al. (2004) derived whole-stand survival models that are capable of modeling complex underlying hazard functions. Knowledge of the empirical hazard function was used to limit selection to appropriate functions for survival estimation. Integrating selected functions results in initial condition difference equations that, when fitted to data, provided biologically reasonable whole-stand survival predictions and adequately represented the underlying hazard function.

12.4 Alternatives to the Distribution Function Approach

The use of unimodal probability density functions to apportion the total number of trees into a stand table has been widely applied and generally successful. These distributions are, however, somewhat restricted in the shapes they can assume and there are situations where a wide range of diameter distributions occur and need to be approximated. Hence, alternatives to the pdf approach have been proposed for describing diameter distributions and computing yields for various portions of the stand.

12.4.1 Percentile-Based Distributions

The classical "diameter distribution" approach to yield estimation involves defining a functional form (pdf or CDF) to approximate the tree frequencies by *dbh* class. In a departure from this approach, Borders et al. (1987)—rather than relying on a predefined functional form—characterized the diameter distribution by defining an empirical probability density function with 12 percentiles. Their percentile-based methods permit a variety of distribution shapes to be represented, including inverse-j, unimodal, and multimodal.

Borders et al. (1987) noted that a system of regression equations for estimating the ordered percentiles from the stand attributes can be constructed as

$$P_n = f(\text{stand attributes})$$

where P_n = the nth percentile of the distribution.

To fit the system of equations by least squares methods, it is formulated in general terms around a percentile selected to be the driver, as

$$\begin{aligned} P_D &= f\ (\text{stand attributes}) \\ P_{D+i} &= f(P_D, \text{stand attributes}) \end{aligned}$$

where

$i = -j, -j + 1, \ldots, -1, 1, 2, \ldots, k,$
P_D = percentile selected to drive the system (D for driver), generally the percentile that can be predicted with the greatest confidence,
P_{D+i} = other percentiles of the system, which, due to constraints, are directly or indirectly related to P_D such that there are j percentiles less than P_D and k percentiles greater than P_D.

The individual equations should be constructed such that percentile estimates are mathematically consistent across the system (i.e., percentiles should be monotonic with P_{D-j} the minimum and P_{D+k} the maximum). Given a set of percentiles as defined above, a stand table can be generated with any desired size classes (i.e., 1 cm, 2 cm, or other merchantability classes). An empirical probability density function can be defined by assuming trees are uniformly distributed between adjacent percentiles, that is:

$$f(d) = \begin{cases} N_i/N/(P_i - P_{i-1}) & P_{i-1} < d < P_i \text{ elsewhere} \\ 0 & \end{cases} \tag{12.26}$$

where

$d = dbh$
N_i = number of trees between percentiles P_i and P_{i-1},
N = number of trees per ha,
P_{i-1}, P_i are adjacent percentiles (Note $P_i > P_{i-1}$).

From (12.26) one can define

$$N_i = (p_i - p_{i-1})N \tag{12.27}$$

where

p_i = proportion of trees per ha with d less than P_i
p_{i-1} = proportion of trees per ha with d less than P_{i-1}, and all else is as defined above.

Thus (12.26) can be written

$$f(D) = \begin{cases} (p_i - p_{i-1})/(P_i - P_{i-1}) & P_{i-1} < d < P_i \\ 0 & \end{cases}$$

elsewhere

The number of trees in the Kth size class is defined as

$$N_K = N\left\{\int_{-\infty}^{{}^U D_K} f(d)dd - \int_{-\infty}^{{}^L D_K} f(d)dd\right\} \tag{12.28}$$

where

N_K = number of trees per unit area in the Kth diameter size class,
$_L D_K$ = lower limit of the Kth diameter size class,
$_U D_K$ = upper limit of the Kth diameter size class, and all else is as defined previously.

Note that

$$\int_{-\infty}^{_U D_K} f(d)dd = \int_{-\infty}^{P_1} f(d)dd + \int_{P_1}^{P_2} f(d)dd + \cdots + \int_{P_{j-1}}^{P_j} f(d)dd$$

$$+ \int_{P_j}^{_U D_K} f(d)dd = p_j - p_1 + (p_{j+1} - p_j)(_U D_K - P_j)/(P_{j+1} - P_j)$$

where
$P_1, P_2, \ldots, P_{j+1}$ are the first $j+1$ percentiles (with P_1 the minimum *dbh* and $P_j <_U D_K < P_{j+1}$).

Thus the closed form solution for (12.28), the number of trees per unit area in the Kth size class, is

$$N_K = \left\{ \frac{P_i - _L D_K}{P_i - P_{i-1}}(p_i - p_{i-1}) + (p_j - p_i) + (p_{j+1} - p_j)\frac{_U D_K - P_j}{P_{j+1} - P_j} \right\} N \tag{12.29}$$

where

$P_{i-1} <_L D_K < P_i$
$P_j <_U D_K < P_{i+1}$

The percentile-based method was tested by Borders et al. (1987) using 12 percentiles defined across the stand table: the 0th percentile (or minimum diameter), the 5th continuing to the 95th by increments of 10 percentile points, and the 100th percentile (or maximum diameter). The 65th percentile was used as the driver of the system and was estimated as a function of quadratic mean diameter and age. All other percentiles were estimated as functions of adjacent percentiles and quadratic mean diameter, except maximum diameter which was a function of an adjacent percentile and age. Data from a slash pine stand density study, were used to fit the system of equations using seemingly unrelated regression (SUR) techniques. From this empirical test, the authors concluded that 12 percentiles produced an adequate representation of the stand tables.

In a follow-up study, Borders and Patterson (1990) compared three yield projection systems: (1) a probability distribution approach based on the Weibull function, (2) the percentile-based method of Borders et al. (1987), and a stand table projection system reported by Pienaar and Harrison (1988). (The Pienaar and Harrison (1988) model is described in Chapter 13.) This comparison, based on two data sets, showed that the stand table projection method generally reproduced stand

and stock tables with less bias and with greater precision than the other methods. The percentile-based method was overall superior to the Weibull distribution based approach. The ordering – from stand table projection to percentile-based to Weibull-based – is in accordance to expectation because differing amounts of information are used in each instance. The stand table projection used the observed stand table when using projections, the percentile method used 12 percentiles from the observed stand table, and the Weibull method used only four percentiles when making projections.

12.4.2 Ratio Approach

A stand-level ratio model which uses stand attributes age, site index, basal area, and number of surviving trees per unit area to portion total stand yield to any specified top diameter and/or threshold *dbh* limit was presented by Amateis et al. (1986). In addition a model was developed which distributes the total number of trees by diameter such that basal area is consistent with total stand basal area. Comparisons with two Weibull-based diameter distribution approaches showed the ratio approach to be a satisfactory alternative.

Amateis et al. (1986) selected the following equation form for predicting merchantable yields to various top diameters and threshold *dbh* limits:

$$V_m = V e^{b_1 (d_t/\bar{d}_g)^{b_2} + b_3 N^{b_4} (d_T/\bar{d}_g)^{b_5}} \tag{12.30}$$

where

V_m = merchantable yield (m^3/ha) for trees d_T cm and above to a d_t cm top diameter limit
V = total yield (m^3/ha)
$\bar{d}_g$ = quadratic mean *dbh* (cm)
N = number of loblolly pine surviving (per ha)
d_t = top diameter merchantability limit (cm)
d_T = threshold *dbh* limit (cm)
e = base of the natural logarithm
$b_1 - b_5$ = parameters to be estimated

Equation 12.30 is conditioned such that when both d_t and d_T equal zero, merchantable yield equals total yield. When d_t equals zero, the merchantable portion of the stand is determined by the threshold *dbh* limit (d_T) alone. When d_T equals zero, the merchantable portion of the stand is determined only by the top diameter d_t merchantability limit. Equation 12.30 can be used to estimate the merchantable cubic yield for individual diameter classes as well as for groups of diameter classes by subtraction.

In order to use Eq. 12.30 it is necessary to estimate total cubic yield (V). Since estimates of both the number of loblolly pine per ha surviving (N) and the loblolly pine basal area per ha (G) are necessary for computing the quadratic mean *dbh* ($\bar{d}_g$), these variables were included as:

$$\ln V = c_0 + c_1(1/t) + c_2(h_{dom}/t) + c_3(t \ln N) + c_4(\ln G) \tag{12.31}$$

where

V = total cubic yield
t = years since planting
h_{dom} = average height of dominant and codominant trees
N = number of planted pine (ha^{-1})
G = basal area of planted pine (m^{2}ha^{-1})
ln = natural logarithm
$c_0 - c_4$ = coefficients to be estimated.

Equations 12.30 and 12.31 provide estimates of cubic yield for any portion of the stand. However, it is useful to know how the yield is distributed with regard to number of trees and basal area. Equation 12.32, which is of similar form to Eq. 12.30, can be used to distribute the total number of trees across the diameter distribution:

$$N_m = N e^{-f_1 G^{f_2} (d_T/\bar{d}_g)^{f_3}} \tag{12.32}$$

where

N_m = trees per ha larger than d_T cm
N = number of planted pine surviving (ha^{-1})
G = basal area of planted pine (m^{2}ha^{-1})
$\bar{d}_g$ = quadratic mean *dbh* (cm)
d_T = threshold *dbh* limit (cm)
e = base of the natural logarithm
$f_1 - f_3$ = parameters to be estimated.

By rearranging Eq. 12.32

$$N_m/N = e^{-f_1 G^{f_2} (d_T/\bar{d}_g)^{f_3}} \tag{12.33}$$

and noting that the form of the 2-parameter Weibull distribution is

$$P(X > x) = e^{-\left(\frac{x}{b}\right)^c}$$

it can be seen that Eq. 12.33 is in the form of the 2-parameter Weibull distribution where $c = f_3$ and $b = \frac{\bar{d}_g}{(f_1 G^{f_2})^{1/f_3}}$. Recognizing this, and that for the Weibull distribution,

$$\bar{d}_g^2 = b^2 \Gamma(1 + 2/c)$$

where Γ is the gamma function, then

$$b^2 = \frac{\bar{d}_g^2}{\Gamma(1+2/c)} \quad \text{or} \quad b = \frac{\bar{d}_g}{\Gamma^{\frac{1}{2}}(1+2/c)}.$$

Equating the two expression for b, one obtains:

$$f_1 = \frac{\Gamma^{f_3/2}(1+2/f_3)}{G^{f_2}}$$

By substituting this value of f_1 into Eq. 12.32, Eq. 12.34 is obtained which portions the total number of trees by *dbh* limit and is conditioned such that the sum of both the number of trees and the corresponding basal area across the diameter distribution is equivalent to the total stand values:

$$N_m = N e^{-\Gamma^{f_3/2(1+2/f_3)(d_T/\bar{d}_g)^{f_3}}} \tag{12.34}$$

where

f_3 = parameter to be estimated and all other variables are as previously defined.

The stand ratio model was contrasted and compared with two different parameter recovery methods (Matney and Sullivan 1982; Burk and Newberry 1984) using the Weibull distribution when fitting data from two sets of sample plots in loblolly pine stands. Results showed that the Weibull-based value recovery method (Matney and Sullivan 1982) was slightly better, in terms of smaller average absolute value of residuals, than either the Weibull-based moment recovery method (Burk and Newberry 1984) or the ratio method for predicting merchantable yields. However, there was little overall differences among the three techniques and any of the three methods should provide satisfactory estimates of yield to any specified top diameter or threshold *dbh* limit.

12.4.3 Functional Regression Tree Method

Because forest stand diameter distributions can adopt a wide variety of shapes that may not be adequately represented by a single parametric family, Lane et al. (2010) investigated the functional regression tree (FRT) method for modeling probability density functions. The FRT approach can be used to estimate stand diameter distributions without making assumptions about the functional form. Comparing the functional regression tree method with a parameter prediction and percentile method showed favorable results. The FRT approach was found suitable for diameter distributions that are multimodal and excessively skewed – situations that are not easily dealt with in a parametric context.

References

Affleck DLR (2006) Poisson mixture models for regression analysis of stand-level mortality. Can J For Res 36:2994–3006

Amateis RL, Burkhart HE, Knoebel BR, Sprinz PT (1984) Yields and size class distributions for unthinned loblolly pine plantations on cutover site-prepared lands. Virginia Polytechnic Institute and State University, Blacksburg, Pub. FWS-2–84

Amateis RL, Burkhart HE, Burk TE (1986) A ratio approach to predicting merchantable yields of unthinned loblolly pine plantations. For Sci 32:287–296

Amateis RL, Burkhart HE, Liu J (1997) Modeling survival in juvenile and mature loblolly pine plantations. For Ecol Manage 90:51–58

Arabatzis AA, Burkhart HE (1992) An evaluation of sampling methods and model forms for estimating height-diameter relationships in loblolly pine plantations. For Sci 38:192–198

Avery TE, Burkhart HE (2002) Forest measurements, 5th edn. McGraw-Hill, New York

Bailey RL (1980) Individual tree growth derived from diameter distribution models. For Sci 26:626–632

Bailey RL, Dell TR (1973) Quantifying diameter distributions with the Weibull function. For Sci 19:97–104

Bailey RL, Burgan TM, Jokela EJ (1989) Fertilized mid-rotation-aged slash pine plantations- stand structure and yield prediction models. South J Appl For 13:76–80

Baldwin VC Jr, Feduccia DP (1987) Loblolly pine growth and yield prediction for managed West Gulf plantations. USDA Forest Service, Southern Forest Experiment Station, New Orleans, Research Paper SO-236

Beck DE, Della-Bianca L (1970) Yield of unthinned yellow-poplar. USDA Forest Service, Southeastern Forest Experiment Station, Asheville, Research Paper SE-58

Bennett FA, Clutter JL (1968) Multiple-product yield estimates for unthinned slash pine plantations – pulpwood, sawtimber, gum. USDA Forest Service, Southeastern Forest Experiment Station, Asheville, Research Paper SE-35

Bliss CI, Reinker KA (1964) A lognormal approach to diameter distributions in even-aged stands. For Sci 10:350–360

Borders BE (1989) Systems of equations in forest stand modeling. For Sci 35:548–556

Borders BE, Patterson WD (1990) Projecting stand tables: a comparison of the Weibull diameter distribution method, a percentile-based projection method, and a basal area growth projection method. For Sci 36:413–424

Borders BE, Souter RA, Bailey RL, Ware KD (1987) Percentile-based distributions characterize forest stand tables. For Sci 33:570–576

Bowling EH, Burkhart HE, Burk TE, Beck DE (1989) A stand-level multispecies growth model for Appalachian hardwoods. Can J For Res 19:405–412

Brooks JR, Borders BE, Bailey RL (1992) Predicting diameter distributions for site-prepared loblolly and slash pine plantations. South J Appl For 16:130–133

Budhathoki CB, Lynch TB, Guldin JM (2008) A mixed-effects model for the dbh-height relationship of shortleaf pine (*Pinus echinata* Mill.). South J Appl For 32:5–11

Bullock BP, Boone EL (2007) Deriving tree distributions using Bayesian model averaging. For Ecol Manage 242:127–132

Bullock BP, Burkhart HE (2005) Juvenile diameter distributions of loblolly pine characterized by the two-parameter Weibull function. New For 29:233–244

Burk TE, Burkhart HE (1984) Diameter distributions and yields of natural stands of loblolly pine. Virginia Polytechnic Institute and State University, Blacksburg, Pub FWS-1–84

Burk TE, Newberry JD (1984) A simple algorithm for moment-based recovery of Weibull distribution parameters. For Sci 30:329–332

Burkhart HE (1971) Slash pine plantation yield estimates based on diameter distribution: An evaluation. For Sci 17:452–453

Burkhart HE, Strub MR (1974) A model for simulation of planted loblolly pine stands. In: Fries J (ed) Growth models for tree and stand simulation. Royal College of Forestry, Stockholm, pp 128–135, Research Notes 30

Calama R, Montero G (2004) Interregional nonlinear height-diameter model with random coefficients for stone pine in Spain. Can J For Res 34:150–163

Cao QV (2004) Predicting parameters of a Weibull function for modeling diameter distribution. For Sci 50:682–685

Cao QV, Burkhart HE (1984) A segmented distribution approach for modeling diameter frequency data. For Sci 30:129–137

Cao QV, Burkhart HE, Lemin RC Jr (1982) Diameter distributions and yields of thinned loblolly pine plantations. Virginia Polytechnic Institute and State University, Blacksburg, Pub FWS-1–82

Chen W (2004) Tree size distribution functions of four boreal forest types for biomass mapping. For Sci 50:436–449

Clutter JL, Bennett FA (1965) Diameter distributions in old-field slash pine plantations. Georgia Forest Research Council, Macon. Report 13

Clutter JL, Jones EP Jr (1980) Prediction of growth after thinning in old-field slash pine plantations. USDA Forest Service, Southeastern Forest Experiment Station, Asheville, Research Paper SE-217

Clutter JL, Harms WR, Brister GH, Rheney JW (1984) Stand structure and yields of site-prepared loblolly pine plantations in the lower coastal plain of the Carolinas, Georgia, and north Florida. USDA Forest Service, Southeastern Forest Experiment Station, Asheville, General Technical Report SE-27

Coble DW, Lee Y-J (2011) A mixed-effects height-diameter model for individual loblolly and slash pine trees in East Texas. South J Appl For 35:12–17

Crecente-Campo F, Tomé M, Soares P, Diéguez-Aranda U (2010) A generalized nonlinear mixed-effects height-diameter model for *Eucalyptus globules* L. in northwestern Spain. For Ecol Manage 256:943–952

Curtis RO (1967) Height-diameter and height-diameter-age equations for second-growth Douglas-fir. For Sci 13:365–375

Dell TR, Fedducia DP, Campbell TE, Mann WF Jr, Polmer BH (1979) Yields of unthinned slash pine plantations on cutover sites in the west Gulf region. USDA Forest Service, Southern Forest Experiment Station, New Orleans, Research Paper SO-147

Devine JO, Clutter JL (1985) Prediction of survival in slash pine plantations infected with fusiform rust. For Sci 31:88–94

Diéguez-Aranda U, Castedo-Dorado F, Alvarez-González JG, Rodríguez-Soalleiro R (2005) Modelling mortality of Scots pine (*Pinus sylvestris* L.) plantations in the northwest of Spain. Eur J For Res 124:143–153

Droessler TD, Burk TE (1989) A test of nonparametric smoothing of diameter distributions. Scand J For Res 4:407–415

Eid T, Øyen H (2003) Models for prediction of mortality in even-aged forest. Scand J For Res 18:64–77

Ek AR, Issos JN, Bailey RL (1975) Solving for Weibull diameter distribution parameters to obtain specified mean diameters. For Sci 21:290–292

Farrar RM Jr, Matney TG (1994) A dual growth simulator for natural even-aged stands of longleaf pine in the South's East Gulf region. South J Appl For 18:147–155

Fast AJ, Ducey MJ (2011) Height-diameter equations for select New Hampshire tree species. North J Appl For 28:157–160

Feduccia DP, Dell TR, Mann WF Jr, Campbell TE, Polmer BH (1979) Yields of unthinned loblolly pine plantations on cutover sites in the west Gulf region. USDA Forest Service, Southern Forest Experiment Station, New Orleans, Research Paper SO-148

Garcia O (1981) Simplified method-of-moments for the Weibull distribution. N Z J For Sci 11: 304–306

Gertner G, Cao X, Zhu H (1995) A quality assessment of a Weibull based growth projection system. For Ecol Manage 71:235–250

Green EJ, Burkhart HE, Clason TR (1984) A model for basal area distribution in loblolly pine. For Sci 30:617–628

Haara A, Maltamo M, Tokola T (1997) The k-nearest-neighbor method for estimating basal area diameter distribution. Scand J For Res 12:200–208

Hafley WL, Buford MA (1985) A bivariate model for growth and yield prediction. For Sci 31: 237–247

Hafley WL, Schreuder HT (1977) Statistical distributions for fitting diameter and height data in even-aged stands. Can J For Res 7:481–487

Huang S, Titus SJ, Wiens DP (1992) Comparison of nonlinear height-diameter functions for major Alberta tree species. Can J For Res 22:1297–1304

Huang S, Price D, Titus S (2000) Development of ecoregion-based height-diameter models for white spruce in boreal forests. For Ecol Manage 129:125–141

Hyink DM (1980) Diameter distribution approaches to growth and yield modeling. In: Brown KM, Clarke FR (eds) Forecasting forest stand dynamics. Lakehead University School of Forestry, Thunderbay, pp 138–163

Hyink DM, Moser JW Jr (1983) A generalized framework for projecting forest yield and stand structure using diameter distributions. For Sci 29:85–95

Jiang L, Brooks JR (2009) Predicting diameter distributions for young longleaf pine plantations in southwest Georgia. South J Appl For 33:25–28

Johnson NL (1949a) Systems of frequency curves generated by methods of translation. Biometrika 36:149–176

Johnson NL (1949b) Bivariate distributions based on simple translation systems. Biometrika 36:297–304

Kangas A, Maltamo M (2000) Calibrating predicted diameter distribution with additional information. For Sci 46:390–396

Knoebel BR, Burkhart HE (1991) A bivariate distribution approach to modeling Forest diameter distributions at two points in time. Biometrics 47:241–253

Knoebel BR, Burkhart HE, Beck DE (1986) A growth and yield model for thinned stands of yellow-poplar. For Sci Monogr 27:1–62

Knowe SA (1992) Basal area and diameter distribution models for loblolly pine plantations with hardwood competition in the Piedmont and upper coastal plain. South J Appl For 16:93–98

Knowe SA, Stein WI (1995) Predicting the effects of site preparation on development of young Douglas-fir plantations. Can J For Res 25:1538–1547

Knowe SA, Foster GS, Rousseau RJ, Nance WL (1994) Eastern cottonwood clonal mixing study: predicted diameter distributions. Can J For Res 24:405–414

Krug AG, Nordheim EV, Giese RL (1984) Determining initial values for parameters of a Weibull model: case study. For Sci 30:573–581

Lane SE, Robinson AP, Baker TG (2010) The functional regression tree method for diameter distribution modelling. Can J For Res 40:1870–1877

Lappi J (1997) A longitudinal analysis of height/diameter curves. For Sci 43:555–570

Leduc DJ, Matney TG, Belli KL, Baldwin VC Jr (2001) Predicting diameter distributions of longleaf pine plantations: a comparison between artificial neural networks and other accepted methodologies. USDA Forest Service, Southern Research Station, Asheville, Research Paper SRS-25

Lei Y, Parresol BR (2001) Remarks on height-diameter modeling. USDA Forest Service, Southern Research Station, Asheville, Research Note SRS-10

Lemin RC Jr, Burkhart HE (1983) Predicting mortality after thinning in old-field loblolly pine plantations. South J Appl For 7:20–23

Lenhart JD (1988) Diameter-distribution yield-prediction system for unthinned loblolly and slash pine plantations on non-old-fields in East Texas. South J Appl For 12:239–242

Lenhart JD, Clutter JL (1971) Cubic-foot yield tables for old-field loblolly pine plantations in the Georgia Piedmont. Georgia Forest Research Council, Macon, Report 22 – Series 3

Little SN (1983) Weibull diameter distributions for mixed stands of western conifers. Can J For Res 13:85–88

Liu C, Zhang L, Davis CJ, Solomon DS, Gove JH (2002) A finite mixture model for characterizing the diameter distributions of mixed-species forest stands. For Sci 48:653–661

Liu C, Zhang SY, Lei Y, Newton PF, Zhang L (2004) Evaluation of three methods for predicting diameter distributions of black spruce (*Picea mariana*) plantations in central Canada. Can J For Res 34:2424–2432

Liu C, Beaulieu J, Prégent G, Zhang SY (2009) Applications and comparison of six methods for predicting parameters of the Weibull function in unthinned *Picea glauca* plantations. Scand J For Res 24:67–75

Lu J, Zhang L (2011) Modeling and prediction of tree height-diameter relationships using spatial autoregressive models. For Sci 57:252–264

Lynch TB, Murphy PA (1995) A compatible height prediction and projection system for individual trees in natural, even-aged shortleaf pine stands. For Sci 41:194–209

Magnussen S (1986) Diameter distributions in *Picea abies* described by the Weibull model. Scand J For Res 1:493–502

Maltamo M (1997) Comparing basal area diameter distributions estimated by tree species and for the entire growing stock in a mixed stand. Silva Fennica 31:53–65

Maltamo M, Kangas A (1998) Methods based on k-nearest neighbor regression in estimation of basal area diameter distribution. Can J For Res 28:1107–1115

Maltamo M, Puumalainen J, Päivinen R (1995) Comparison of beta and Weibull functions for modelling basal area diameter distribution in stands of *Pinus sylvestris* and *Picea abies*. Scand J For Res 10:284–295

Maltamo M, Kangas A, Uuttera J, Torniainen T, Saramäki J (2000) Comparison of percentile based prediction methods and the Weibull distribution in describing the diameter distribution of heterogeneous Scots pine stands. For Ecol Manage 133:263–274

Mateus A, Tomé M (2011) Modelling the diameter distribution of eucalyptus plantations with Johnson's S_B probability density function: parameters recovery from a compatible system of equations to predict stand variables. Ann For Sci 68:325–335

Matney TG, Farrar RM Jr (1992) A thinned/unthinned loblolly pine growth and yield simulator for planted cutover site-prepared land in the Mid-Gulf South. South J Appl For 16:70–75

Matney TG, Sullivan AD (1982) Compatible stand and stock tables for thinned and unthinned loblolly pine stands. For Sci 28:161–171

Matney TG, Ledbetter JR, Sullivan AD (1987) Diameter distribution yield systems for unthinned cutover site-prepared slash pine plantations in southern Mississippi. South J Appl For 11:32–36

McGee CE, Della-Bianca L (1967) Diameter distributions in natural yellow-poplar stands. USDA Forest Service, Southeastern Forest Experiment Station, Asheville, Research Paper SE-25

McTague JP, Bailey RL (1987) Compatible basal area and diameter distribution models for thinned loblolly pine plantations in Santa Catarina. Brazil For Sci 33:43–51

Mehtätalo L (2004) A longitudinal height-diameter model for Norway spruce in Finland. Can J For Res 34:131–140

Mehtätalo L, Gregoire TG, Burkhart HE (2008) Comparing strategies for modeling tree diameter percentiles from remeasured plots. Environmetrics 19:529–548

Meng Q, Cieszewski CJ, Strub MR, Borders BE (2009) Spatial regression modeling of tree height-diameter relationships. Can J For Res 39:2283–2293

Meyer HA (1940) A mathematical expression for height curves. J For 38:415–420

Mønness E (2011) The power-normal distribution: application to forest stands. Can J For Res 41:707–714

Nanang DM (1998) Suitability of the normal, log-normal and Weibull distributions for fitting diameter distributions of neem plantations in northern Ghana. For Ecol Manage 103:1–7

Nelson TC (1964) Diameter distribution and growth of loblolly pine. For Sci 10:105–114

Newton PF, Amponsah IG (2007) Comparative evaluation of five height-diameter models developed for black spruce and jack pine stand-types in terms of goodness-of-fit, lack-of-fit and predictive ability. For Ecol Manage 247:149–166

Newton PF, Lei Y, Zhang SY (2005) Stand-level diameter distribution yield model for black spruce plantations. For Ecol Manage 209:181–192

Nordhausen K, Nummi T (2007) Estimation of the diameter distribution of a stand marked for cutting using finite mixtures. Can J For Res 37:817–824

Nord-Larsen T, Cao QV (2006) A diameter distribution model for even-aged beech in Denmark. For Ecol Manage 231:218–225

Palahí M, Pukkala T, Blasco E (2007) Comparison of beta, Johnson's SB, Weibull and truncated Weibull functions for modeling the diameter distribution of forest stands in Catalonia (north-east Spain). Eur J For Res 126:563–571

Paulo JA, Tomé J, Tomé M (2011) Nonlinear fixed and random generalized height-diameter models for Portuguese cork oak stands. Ann For Sci 68:295–309

Pienaar LV, Harrison WM (1988) A stand table projection approach to yield prediction in unthinned even-aged stands. For Sci 34:804–808

Pienaar LV, Shiver BD (1981) Survival functions for site-prepared slash pine plantations in the flatwoods of Georgia and northern Florida. South J Appl For 5:59–62

Pienaar LV, Page HH, Rheney JW (1990) Yield prediction for mechanically site-prepared slash pine plantations. South J Appl For 14:104–109

Podlaski R (2010) Two-component mixture models for diameter distributions in mixed-species, two-age cohort stands. For Sci 56:379–390

Rennolls K, Wang M (2005) A new parameterization of Johnson's S_B distribution with application to fitting forest tree diameter data. Can J For Res 35:575–579

Reynolds MR Jr, Burk TE, Huang WC (1988) Goodness-of-fit tests and model selection procedures for diameter distribution models. For Sci 34:373–399

Rose CE Jr, Clutter ML, Shiver BD, Hall DB, Borders B (2004) A generalized methodology for developing whole-stand survival models. For Sci 50:686–695

Rupšus P, Petrauskas E (2010) The bivariate gompertz diffusion model for tree diameter and height distribution. For Sci 56:271–280

Russell MB, Amateis RL, Burkhart HE (2010) Implementing regional locale and thinning response in the loblolly pine height-diameter relationship. South J Appl For 34:21–27

Sarkkola S, Hökkä H, Laiho R, Päivänen J, Penttilä T (2005) Stand structural dynamics on drained peatlands dominated by Scots pine. For Ecol Manage 206:135–152

Schreuder HT, Hafley WL (1977) A useful bivariate distribution for describing stand structure of tree heights and diameters. Biometrics 33:471–478

Schreuder HT, Swank WT (1974) Coniferous stands characterized with the Weibull distribution. Can J For Res 4:518–523

Schreuder HT, Hafley WL, Bennett FA (1979) Yield prediction for unthinned natural slash pine stands. For Sci 25:25–30

Scolforo JRS, Tabai FCV, Grisi de Macedo RL, Acerbi WF Jr, Leandra de Assis A (2003) S_B distribution's accuracy to represent the diameter distribution of *Pinus taeda*, through five fitting methods. For Ecol Manage 175:489–496

Sharma M, Parton J (2007) Height-diameter equations for boreal tree species in Ontario using a mixed-effects modeling approach. For Ecol Manage 249:187–198

Shiver BD (1988) Sample sizes and estimation methods for the Weibull distribution for unthinned slash pine plantation diameter distributions. For Sci 34:809–814

Siipilehto J (1999) Improving the accuracy of predicted basal-area diameter distribution in advanced stands by determining stem number. Silva Fennica 33:281–301

Siipilehto J, Sakari S, Mehtätalo L (2007) Comparing regression estimation techniques when predicting diameter distributions of Scots pine on drained peatlands. Silva Fennica 41:333–349

Smalley GW, Bailey RL (1974a) Yield tables and stand structure for loblolly pine plantations in Tennessee, Alabama, and Georgia highlands. USDA Forest Service, Southern Forest Experiment Station, New Orleans, Research Paper SO-96

Smalley GW, Bailey RL (1974b) Yield tables and stand structure for shortleaf pine plantations in Tennessee, Alabama, and Georgia highlands. USDA Forest Service, Southern Forest Experiment Station, New Orleans, Research Paper SO-97

Soares P, Tomé M (2002) Height-diameter equation for first rotation eucalypt plantations in Portugal. For Ecol Manage 166:99–109

Somers GL, Oderwald RG, Harms WR, Langdon OG (1980) Predicting mortality with a Weibull distribution. For Sci 26:291–300

Stankova TV, Zlatanov TM (2010) Modeling diameter distribution of Austrian black pine (*Pinus nigra* Arn.) plantations: a comparison of the Weibull frequency distribution function and percentile-based projection methods. Eur J For Res 129:1169–1179

Staudhammer C, LeMay V (2000) Height prediction equations using diameter and stand density measures. For Chron 76:303–309

Stauffer HB (1979) A derivation for the Weibull distribution. J Theor Biol 81:55–63

Strub MR, Burkhart HE (1975) A class-interval-free method for obtaining expected yields from diameter distributions. For Sci 27:67–69

Sullivan AD, Clutter JL (1972) A simultaneous growth and yield model for loblolly pine. For Sci 18:76–86

Temesgen H, Hann DW, Monleon VJ (2007) Regional height-diameter equations for major tree species of Southwest Oregon. West J Appl For 22:213–219

Temesgen H, Monleon VJ, Hann DW (2008) Analysis and comparison of nonlinear tree height prediction strategies for Douglas-fir forests. Can J For Res 38:553–565

Tewari VP, Gadow Kv (1999) Modelling the relationship between tree diameters and height using S_{BB} distribution. For Ecol Manage 119:171–176

Tham Å (1988) Structure of mixed *Picea abies* (L.) Karst. and *Betula pendula* Roth and *Betula pubescens* Ehrh. stands in south and middle Sweden. Scand J For Res 3:355–370

Thomas V, Oliver RD, Lim K, Woods M (2008) LiDAR and Weibull modeling of diameter and basal area. For Chron 84:866–875

Trincado G, VanderSchaaf CL, Burkhart HE (2007) Regional mixed-effects height-diameter models for loblolly pine (*Pinus taeda* L.) plantations. Eur J For Res 126:253–262

Wang M, Upadhyay A, Zhang L (2010) Trivariate distribution modeling of tree diameter, height, and volume. For Sci 56:290–300

Weibull W (1951) A statistical distribution function of wide applicability. J Appl Mech 18:293–297

Woollons RC (1998) Even-aged stand mortality estimation through a two-step process. For Ecol Manage 105:189–195

Yang RC, Kozak A, Smith JHG (1978) The potential of Weibull-type functions as flexible growth curves. Can J For Res 8:424–431

Zanakis SH (1979) A simulation study of some simple estimators for the three parameter Weibull distribution. J Stat Comput Simul 9:101–116

Zarnoch SJ, Dell TR (1985) An evaluation of percentiles and maximum likelihood estimators of Weibull parameters. For Sci 31:260–268

Zarnoch SJ, Feduccia DP, Baldwin VC Jr, Dell TR (1991) Growth and yield predictions for thinned and unthinned slash pine plantations on cutover sites in the West Gulf region. USDA Forest Service, Southern Forest Experiment Station, New Orleans, Research Paper SO-264

Zasada M, Cieszewski CJ (2005) A finite distribution approach for characterizing tree diameter distributions by natural social class in pure even-aged Scots pine stands in Poland. For Ecol Manage 204:145–158

Zeide B, Zhang Y (2000) Diameter variability in loblolly pine plantations. For Ecol Manage 128:139–143

Zhang L, Packard KC, Liu C (2003) A comparison of estimation methods for fitting Weibull and Johnson's S_B distributions to mixed spruce-fir stands in northeastern North America. Can J For Res 33:1340–1347

Zhao D, Borders B, Wang M (2006) Survival model for fusiform rust infected loblolly pine plantations with and without mid-rotation understorey vegetation control. For Ecol Manage 235:232–239

Zhou B, McTague JP (1996) Comparison and evaluation of five methods of estimation of the Johnson system parameters. Can J For Res 26:928–935

Chapter 13
Size-Class Models for Even-Aged Stands

13.1 Defining Size Classes

Tree populations can be divided into groups, commonly called cohorts, based on species, diameter, height, or a combination of these and other characteristics. It is most common to form size classes on the basis of *dbh*. Size classes may be formed by specifying diameter classes of equal width (size-class cohorts), resulting in varying numbers per class. Alternatively, each cohort may be specified to contain the same number of trees (percentile cohorts) by varying the diameter class boundaries. Size-class cohorts are more commonly used, but there are examples of models that rely on percentiles cohorts.

Size-class models for even-aged stands typically generate future diameter distributions (stand tables) based on an initial measured diameter distribution. Fundamental to the stand projection process is the ability to estimate diameter growth and mortality by diameter class. Present and future stand tables can be converted to stock tables (volumes by size classes) and growth can be estimated. Various constraints are sometimes imposed on the stand table projection algorithm to insure consistency in behavior of the system.

Size-class models make use of information from inventories of trees by diameter classes and they avoid the necessity of assuming a particular distribution for tree diameters. Using size classes as the basic modeling unit provides a level of resolution between that of the whole-stand and individual-tree approaches.

13.2 Stand-Table Projection

Stand-table projection (STP) methods have been used to estimate forest growth for many different timber types. The procedure can be summarized as follows:

1. A present stand table showing numbers of trees in each *dbh* class is developed from a conventional inventory.

H.E. Burkhart and M. Tomé, *Modeling Forest Trees and Stands*, DOI 10.1007/978-90-481-3170-9_13,

2. Past periodic growth, by *dbh* classes, is determined from increment borings or from remeasurements of permanent sample plots.
3. Past diameter growth rates are applied to the present stand table to derive a future stand table (adjustments for mortality must be made).
4. Both present and future stand tables are converted to stock tables and periodic stand growth is obtained as the difference between the volume of the present stand and that of the future stand.

In instances where short-term projections are needed and reliable growth and yield models are not available, the general STP method can be applied. Classical stand table projection has many limitations, however, (e.g. assuming past growth rates by diameter class apply to the future, and difficulty with determining realistic mortality rates by diameter class without appropriate data from permanent plots), and models constructed from robust data sets are preferable. Size-class models, many of which are formalized extensions of the basic STP idea, have been developed and applied for growth and yield estimation in even-aged stands.

13.2.1 Stand-Table Projection Based on Change in Relative Basal Area

Pienaar and Harrison (1988), building on earlier work by Clutter and Jones (1980), developed a model to project an initial tree list or stand table in such a way that the future stand table will be consistent with the projected future survival and basal area per unit area. Assuming a list of tree diameters is available from sample plot measurements, the relative size of the ith tree was defined as $g_i/\bar{g}$ where g_i is the basal area of the ith tree and $\bar{g}$ is the mean basal area per tree. Using long-term remeasurement data from a slash pine spacing study, Pienaar and Harrison examined hypotheses concerning the change in relative tree sizes over time. If n trees survived from age t_1 to $t_2 > t_1$, and g_{1i} and g_{2i} were the basal areas of the ith survivor tree and $\bar{g}_1$ and $\bar{g}_2$ the mean basal area per tree of the n survivors at ages t_1 and t_2, respectively, then the relative size of the ith survivor was defined as $g_{1i}/\bar{g}_1$ and $g_{2i}/\bar{g}_2$ at ages t_1 and t_2, respectively, for $i = 1, 2, \ldots, n$.

Examination of the data from the slash pine spacing study showed that relative size of smaller than average-sized survivors decreased over time whereas the relative size of the larger trees increased over time. It was also evident that, for the same length of remeasurement period, the change in relative size decreased as age increased. In light of these trends the following model was formulated:

$$g_{2i}/\bar{g}_2 = (g_{1i}/\bar{g}_1)^{(t_2/t_1)^{\mathrm{b}}} \qquad (13.1)$$

where b is a parameter to be estimated from remeasurement data, and t_1 and t_2 are the initial and projection ages, respectively.

When an estimate of b is available, a projected stand table consistent with the observed or projected total basal area G_2 can be obtained as follows:

$$g_{2i} = G_2 \left(\frac{(g_{1i}/\bar{g}_1)^a n_i}{\sum_{i=1}^{k} (g_{1i}/\bar{g}_1)^a n_i} \right) \tag{13.2}$$

where $a = (t_2/t_1)^{\mathrm{b}}$, n_i is the number of survivors in the *i*th initial *dbh* class ($i = 1, 2, \ldots, k$), and g_{2i} the projected future total basal area of the n_i survivors.

Fitting Eq. 13.1 to the remeasurement data from slash pine resulted in:

$$g_{2i}/\bar{g}_2 = (g_{1i}/\bar{g}_1)^{(t_2/t_1)^{0.2333}}$$

To complete the stand table projection model a survival function (originally proposed by Clutter and Jones (1980) and shown as Eq. 12.22 in Chap. 12) and a basal area projection equation were fitted. When total survival and basal area projection equations are available, the stand table projection procedure proposed by Pienaar and Harrison requires that the predicted total mortality be identified in the initial stand table. It is assumed that the probability of a given tree dying during the projection interval is inversely proportional to its relative size. Given that a tree dies, it is then possible to compute a probability of its being of a particular relative size, and these probabilities are then used to allocate the observed or predicted total mortality to the initial tree list or stand table.

In a comparison of three yield projection systems (Weibull-distribution, percentiles-based, and Pienaar and Harrison's STP model), Borders and Patterson (1990) found that the stand table projection method of Pienaar and Harrison had less bias and greater precision when predicting stand and stock tables than the other methods. (Additional detail about Borders and Patterson's comparison is contained in Sect. 12.4.1.)

13.2.2 A Distribution-Independent Approach to Stand Table Projection

Tang et al. (1997) proposed an approach to projecting future stand diameter distributions based on current stand structure observed in forest inventory and future stand-level attributes predicted independently from a whole stand model. No theoretical probability density function was assumed for the empirical diameter distribution. Relationships between current and future stand distributions and stand-level attributes were established. The parameter recovery method was employed to derive the parameters in the tree survival function and diameter growth function. The growth function also included a stochastic component to mimic the differentiation of tree diameters over time.

The Lebesque-Stieltjes integral was applied by Tang et al. to derive a group of equations for the relationships between current and future stand diameter distributions and stand-level attributes. The parameters in the tree survival function and diameter growth function were recovered using these equations based on independent estimates of future stand mean diameter, quadratic mean diameter, and survival from a whole stand model. This disaggregation approach ensured that the resolutions at size-class distribution and/or individual tree levels were compatible with the stand-level aggregates. A stochastic error component, incorporated into the tree diameter growth function, mimicked the tree diameter differentiation process over time, and it improved prediction accuracy for future stand diameter distributions.

13.2.3 Stand Table Projection Algorithms that Incorporate a Diameter Growth Function

Bailey (1980) showed that by considering transformations of variables which preserve the functional form of the diameter distribution function, tree diameter growth models are implied. If d is the tree's diameter at the beginning of a growth period and Δd is the increment in diameter, then the model

$$\Delta d = (\beta_0 - d) + \beta_1 (d - \beta_3)^{\beta_2}$$

is implied by assuming the Weibull, lognormal, or generalized gamma distribution as a diameter distribution model. The exponential, normal, beta, or Johnson's *SB* distribution is implied for the special case of $\beta_2 = 1$. The relationship between diameter distribution functions and implied diameter growth has been applied by several developers of stand table projection models.

A stand table projection method based on existing estimates of future basal area and survival was derived by Nepal and Somers (1992). An observed stand table is first projected by applying either an existing diameter growth equation or a growth equation derived from appropriate diameter distributional assumptions. The stand table is adjusted by an algorithm that equates the future stand table to existing estimates of basal area and survival.

To apply the Nepal and Somers stand table projection algorithm, a stand table at time one is assumed available along with projection equations for future numbers of trees and basal area from a stand-level growth and yield model. An existing tree diameter growth equation can be applied or one can be derived by making distributional assumptions. Nepal and Somers assumed that the Weibull distribution would be an appropriate model for diameter at both points in time. The Weibull distribution and associated parameters can be written:

$$f(d) = \left[\exp - \left(\frac{d - a_i}{\mathrm{b}_i} \right)^{\mathrm{c}_i} \right] \left[\left(\frac{\mathrm{c}_i}{\mathrm{b}_i} \right) \left(\frac{d - a_i}{\mathrm{b}_i} \right)^{\mathrm{c}_i - 1} \right] \quad (13.3)$$

where

d = diameter at breast height
c_i = shape parameter at time i
b_i = scale parameter at time i
a_i = location parameter at time i

By assuming that trees do not change their relative sizes over time, the Weibull distribution at both periods implies the following individual diameter growth equation (Bailey 1980):

$$d_2 = a_2 - b_2\left(\frac{d_1 - a_1}{b_1}\right)^{\frac{c_1}{c_2}} \tag{13.4}$$

In the Nepal and Somers application the location parameter a_i was assumed to remain fixed at the lower endpoint of the lowest initial *dbh* class ($a_1 = a_2$).

Using data from permanent plots in naturally regenerated, even-aged longleaf pine stands, Nepal and Somers compared goodness-of-fit for diameter distributions produced with their proposed stand table projection model with those of a parameter recovery method using the Weibull function. In 75% of the cases the fit as measured by the Kolmogorov-Smirnov statistic was better for the proposed method than for the parameter recovery method. The proposed method also compared favorably with the procedures of Pienaar and Harrison (1988). Both methods (Nepal and Somers and Pienaar and Harrison) require an independent estimate of future survival and basal area.

Cao and Baldwin (1999) introduced an algorithm for stand table projection that produces estimates of stand basal area, numbers of trees per ha, and average tree diameter that are compatible with either observed or predicted values from growth and yield models. The Cao-Baldwin algorithm for stand table projection consists of three steps: (1) computing survival and allocating mortality, (2) deriving diameter growth for individual trees, and (3) adjusting projected diameters to match future average diameter and stand basal area.

After assuming that all mortality occurs at the beginning of the growth period, the surviving number of trees for the ith diameter class was predicted using the following survival function:

$$\hat{n}_{2i} = n_{1i}\left\{1 - \exp\left[b_1 - (d_i - d_{\min 1} + 1)\right]\right\} \tag{13.5}$$

where

n_{1i} = current number of trees per hectare in ith diameter class, $i = 1, 2, \ldots, p$
$\hat{n}_{2i}$ = future surviving number of trees per hectare in the ith diameter class,
d_i = midpoint diameter of the ith class,
$d_{\min 1}$ = midpoint of the current minimum diameter class, and
b_1 = coefficient to be determined.

Based on Eq. 13.5, more trees survive to the end of the growing period for large diameter classes as compared to small diameter classes, relative to the minimum diameter class. The coefficient b_1, which is negative, is calculated such that $\hat{n}_{2i}$ will sum to N_2, the total future surviving trees per hectare.

Since mortality rate is not evenly distributed across diameter classes, the diameter distribution will change after mortality. As a result, the stand attributes need to be updated. The current average diameter ($\bar{d}_1$) and basal area per hectare (G_1) after mortality are given by:

$$\bar{d}_1 = \frac{\Sigma_i \hat{n}_{2i} d_i}{N_2} \tag{13.6}$$

$$G_1 = \mathrm{k} \sum_i \hat{n}_{2i} d_i^2 \tag{13.7}$$

where $\mathrm{k} = \pi/40{,}000$ (to convert diameter in cm to area in square meters).

The minimum diameter was assumed to increase (i.e., all trees in the current minimum diameter class either die or move up to higher classes) when there is a sufficient shift in diameter distribution based on the arithmetic and quadratic mean diameters. Procedures for implied diameter growth followed those of Nepal and Somers (1992). Parameters of a Weibull distribution were recovered from $\bar{d}_1$ and G_1 to approximate the current diameter distribution of the stand immediately after mortality. Similarly, parameters of another Weibull distribution to characterize the future diameter distribution were recovered from $\bar{d}_2$ and G_2, the future average diameter and basal area per hectare, respectively.

After producing a stand table for the future stand, a constrained least squares procedure, similar to that employed by Matney and Farrar (1992) and Farrar and Matney (1994) (see Sect. 12.1.3.1), was used to adjust the future stand table. The final number of trees per hectare in the diameter class n_{2i} was calculated by minimizing

$$\sum_i (n_{2i} - \hat{n}_{2i})^2 \tag{13.8}$$

subject to the following constraints:

$$\sum n_{2i} = N_2 \tag{13.9a}$$

$$\sum n_{2i} d_i = N_2 \bar{d}_2 \tag{13.9b}$$

$$\sum n_{2i} d_i^2 = G_2/\mathrm{k} \tag{13.9c}$$

where the summation signs denote the sum over all diameter classes. This step can be interpreted as adjusting values of the stand table ($\hat{n}_{2i}$) to new values (n_{2i}) such that the three constraints (13.9a, 13.9b, and 13.9c) are met.

The above constrained least squares problem can be rewritten as:

$$\min.\sum(n_{2i}-\hat{n}_{2i})^2+2\lambda_1\left(\sum n_{2i}-N_2\right)+2\lambda_2\left(\sum n_{2i}d_i=N_2\bar{d}_2\right)$$
$$+2\lambda_3\left(\sum n_{2i}d_i^2-G_2/\mathrm{k}\right) \quad (13.10)$$

where Σ denotes the sum over all diameter classes (for values of i from 1 to p), p is the number of diameter classes, and λ_j's are Lagrangian multipliers.

Differentiating (13.10) with respect to n_{2i} and then setting the derivative equal to zero gives:

$$n_{2i}=\hat{n}_{2i}-\left(\lambda_1+\lambda_2 d_i+\lambda_3 d_i^2\right) \quad (13.11)$$

or

$$\lambda_1+\lambda_2 d_i+\lambda_3 d_i^2=\hat{n}_{2i}-n_{2i} \quad (13.12)$$

The Lagrangian multipliers (λ_j's) can be solved from the following system of three linear equations:

$$\lambda_1 p+\lambda_2\sum d_i+\lambda_3\sum d_i^2=\sum\hat{n}_{2i}-N_2 \quad (13.13a)$$

$$\lambda_1\sum d_i+\lambda_2\sum d_i^2+\lambda_3\sum d_i^3=\sum\hat{n}_{2i}d_i-N_2\bar{d}_2 \quad (13.13b)$$

$$\lambda_1\sum d_i^2+\lambda_2\sum d_i^3+\sum d_i^4=\sum\hat{n}_{2i}d_i^2-G_2/\mathrm{k} \quad (13.13c)$$

where Σ denotes the sum over all diameter classes (for values of i from 1 to p).

Cao and Baldwin (1999) used data on diameter distributions from direct-seeded stands of loblolly pine to compare the performance of their constrained least squares method of stand table adjustment to that of Nepal and Somers (1992) and to a Weibull distribution parameter-recovery approach. For the data used, the constrained least squares method provided the best goodness-of-fit statistics. Both stand-table projections methods (Cao-Baldwin and Nepal-Somers) were superior to the Weibull parameter-recovery method, which consistently ranked third based on all of the goodness-of-fit statistics (Kolgomorov-Smirnov, Chi-square, and an error index).

In an extension of work reported in Cao and Baldwin (1999), Cao (2007) presented a stand table projection system consisting of (1) computing survival of tree in each diameter class from a tree survival equation, (2) projecting diameters using a tree diameter growth equation, (3) reclassifying trees into new diameter classes, and (4) adjusting number of trees in each class to match total number of trees and basal area per hectare as predicted from the whole-stand model. Results indicated that incorporating the diameter growth equation, as contrasted

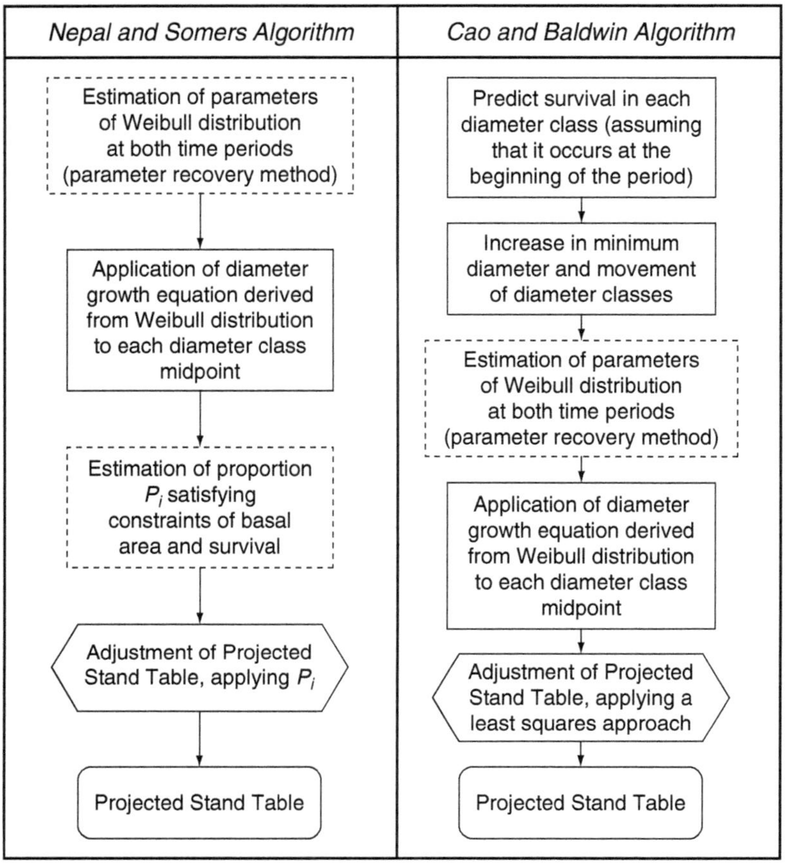

Fig. 13.1 Flowcharts showing the algorithms proposed by Nepal and Somers (1992) and Cao and Baldwin (1999) for projecting diameter distributions in even-aged forest stands (From Trincado et al. 2003)

to the method suggested by Nepal and Somers (1992) which relies on an implied diameter growth function, improved the projection of stand tables. The three adjusting methods (stand table projection, constrained least squares, and modified constrained least squares) produced comparable error indices, but the constrained least squares method consistently provided the best results when compared to the other approaches.

Trincado et al. (2003) compared the stand-table projection methods of Nepal and Somers (1992) and Cao and Baldwin (1999) using data from young eucalyptus plantations in Chile. (A flow chart of the two algorithms is displayed in Fig. 13.1.) The evaluation compared the observed and estimated diameter distributions for different projection intervals using the Kolmogorov-Smirnov test and an error index. Results showed that both methods are suitable for application in eucalypt plantations. However, the method proposed by Nepal and Somers proved to be more accurate, especially for projection periods of 4 years or more. Expected error and bias for observed and estimated total and merchantable volumes at the stand level were also evaluated. The observed error (mean absolute residual) and bias (mean residual) were relatively low for both methods; however, the Nepal-Somers approach exhibited greater accuracy in the estimation of both total and merchantable volume.

Allen et al. (2011) evaluated four methods for projecting stand tables of loblolly and slash pine plantations in East Texas, USA. The four methodologies included were: (1) Nepal and Somers (1992), (2) Cao and Baldwin (1999), (3) Cao (2007), and (4) a model that combines aspects of the Cao and Baldwin (1999) and Pienaar and Harrison (1988) algorithms. Validation of the stand table projection models, based on error indices computed by the method described by Reynolds et al. (1988), showed that the Nepal and Somers and Cao and Baldwin models consistently produced the lowest mean error index values for both species and for projection lengths varying from 3 to 21 years.

13.3 Percentile-Based Models

As an alternative to defining size classes as dbh class limits, a fixed number of cohort groups may be specified. Cohorts are groups of individual trees that are assumed to exhibit similar growth and, thus, are treated as single entities within the model. Percentile models consist of equations for predicting increment, mortality, and, in some cases, recruitment.

In the model developed by Clutter and Allison (1974) for radiata pine in New Zealand, an actual diameter distribution from a forest inventory can be entered. Alternatively, equations which predict average stand diameter, minimum stand diameter, and variance of the stand diameters from stand age and number of stems per unit area can be used to estimate parameters for a three-parameter Weibull distribution. The Weibull probability density function is divided into 25 classes of equal area. Thus, class 1 contains the smallest (by diameter) four percent of the trees in the stand, class 2 contains the next smallest four percent, etc. The median diameter in a class is assumed to represent all trees in the class. Total tree heights are predicted using the median diameter of each class and stand age and number of trees per unit area. Diameter and height values for the 25 cohorts are applied to estimate stand volume. Stand growth projection is carried out on a year-by-year basis, with predicted values per unit area being allocated among the diameter-based cohort groups. With diameter classes defined on the basis of relative size rather than in relation to fixed diameter limits, the problem of "class movement" is avoided.

Alder (1979) formulated a model for conifer plantations in East Africa by defining 10 classes based on fixed cumulative probability points of the diameter distribution. Ten median trees corresponding to the 5th, 15th, ... 95th percentiles of the cumulative tree diameter distribution are updated in the growth estimation process. Increment for a given diameter is predicted as a function of three main effects – site, age, and competitive status defined as the ratio of stand basal area to maximum basal area for the site and age of a subject stand. For a stand of given relative basal area, competitive status among individual trees is accounted for by the ratio of the diameter of a specified tree to the mean diameter of dominant trees.

13.4 Related Approaches

Size-class models have been successfully applied to a variety of even-aged forest types. In this approach the tree population is divided into a limited number of size classes (generally diameter classes). These size classes are projected forward to obtain future stand structure and yields. The size-class method is flexible, in that no assumptions about the underlying overall size-class distribution are required, and the approach is computationally efficient. When individual trees are followed through time, rather than grouping them into size classes, a distance-independent individual-tree modeling approach (Chap. 14) may be an attractive alternative.

Matrix models (Chap. 15) are another form of size-class model. Although matrix models could be applied with data from even-aged stands, this approach has seldom been taken, largely due to difficulties with modeling transition probabilities over time in even-aged structures. The matrix modeling approach has most commonly been applied to uneven-aged forests managed with a selection system of silviculture. Matrix models are presented and discussed in the chapter on growth and yield models for uneven-aged stands (Chap. 15).

References

Alder D (1979) A distance-independent tree model for exotic conifer plantations in East Africa. For Sci 25:59–71

Allen MG II, Coble DW, Cao QV, Yeiser J, Hung I-K (2011) A modified stand table projection growth model for unmanaged loblolly and slash pine plantations in East Texas. South J Appl For 35:115–122

Bailey RL (1980) Individual tree growth derived from diameter distribution models. For Sci 26:626–632

Borders BE, Patterson WD (1990) Projecting stand tables: a comparison of the Weibull diameter distribution method, a percentile-based projection method, and a basal area growth projection method. For Sci 36:413–424

Cao QV (2007) Incorporating whole-stand and individual-tree models in a stand-table projection system. For Sci 53:45–49

Cao QV, Baldwin VC Jr (1999) A new algorithm for stand table projection models. For Sci 45:506–511

Clutter JL, Allison BJ (1974) A growth and yield model for *Pinus radiata* in New Zealand. In: Fries J (ed) Growth models for tree and stand simulation. Royal College of Forestry, Stockholm, pp 136–160, Research Notes 30

Clutter JL, Jones EP (1980) Prediction of growth after thinning old-field slash pine plantations. USDA Forest Service, Southeastern Forest Experiment Station, Asheville, Research Paper SE-217

Farrar RM Jr, Matney TG (1994) A dual growth simulator for natural even-aged stands of longleaf pine in the South's East Gulf region. South J Appl For 18:147–155

Forss E, Gadow Kv, Saborowski J (1996) Growth models for unthinned *Acacia mangium* plantations in South Kalimantan. Indones J Trop For Sci 8:449–462

Knowe SA, Aherns GR, DeBell DS (1997) Comparison of diameter-distribution-prediction, stand-table-projection and individual-tree-growth modeling approaches for young red alder plantations. For Ecol Manage 98:49–60

Matney TG, Farrar RM Jr (1992) A thinned/unthinned loblolly pine growth and yield simulator for planted cutover site-prepared land in the Mid-Gulf South. South J Appl For 16:70–75

Nepal SK, Somers GL (1992) A generalized approach to stand table projection. For Sci 38:120–133

Pienaar LV, Harrison WM (1988) A stand table projection approach to yield prediction in unthinned even-aged stands. For Sci 34:804–808

Reynolds MR Jr, Burk TE, Huang WC (1988) Goodness-of-fit tests and model selection procedures for diameter distribution models. For Sci 34:373–399

Tang S, Wang Y, Zhang L, Meng C-H (1997) A diameter-independent approach to predicting stand diameter distribution. For Sci 43:491–500

Trincado VG, Quezada PR, Gadow Kv (2003) A comparison of two stand table projection methods for young *Eucalyptus nitens* (Maiden) plantations in Chile. For Ecol Manage 180:443–451

Chapter 14
Individual-Tree Models for Even-Aged Stands

14.1 Approach

Individual-tree models consist of a system of equations to simulate stand dynamics by incrementing each tree during a growth period in relation to its growing conditions. Tree growth, ingrowth and regeneration, and mortality are aggregated to provide estimates of stand growth and yield. Models based on individual-tree dynamics provide detailed information about stand development and structure, including the distribution of stand volume by size classes. Individual-tree models have inherent flexibility that permits modeling combinations of species mixtures and stand structures, management regimes and regeneration methods. Individual-tree models represent the highest level of abstraction and resolution in the suite of forest growth and yield models, and include feedback loops between stand structure and individual tree growth. This chapter focuses on individual-tree models developed for simulating development of even-aged stands; models for uneven-aged systems are described in Chap. 15.

14.2 Types of Individual-Tree Models

Individual-tree models may be divided into two classes based on whether or not tree locations are required tree attributes. Distance-independent (also sometimes called "position-independent", "location-independent", or "non-spatially-explicit") models usually project tree growth as a function of present size and stand-level variables such as age, site index, and stand density. Individual-tree locations are not specified when applying these models. Typically, distance-independent models consist of three basic components: (1) a diameter-growth equation, (2) a height-growth equation (or a height-diameter relationship to predict heights from *dbh* values), and (3) a mortality component. Mortality may be stochastically generated or it may be predicted as a function of growth rate and/or tree characteristics.

H.E. Burkhart and M. Tomé, *Modeling Forest Trees and Stands*,
DOI 10.1007/978-90-481-3170-9_14,

In the application of distance-dependent (which may be designated "position-dependent", "location-dependent", or "spatially-explicit") models, initial stand conditions are input or generated, and each tree is assigned a coordinate location. The growth of each tree is simulated as a function of its attributes, the site quality, and a measure of competition from neighboring trees. The competition index varies from model to model but in general is a function of the size of the subject tree and the size of and distance to competitors (Chap. 9). Estimated tree growth is often adjusted by a random component representing genetic and/or microsite variability. Survival is generally controlled stochastically as a function of competition and/or individual-tree attributes.

Yield estimates in individual-tree models are obtained by summing individual-tree volumes and multiplying by an appropriate expansion factor; growth is computed as the difference between successive yield estimates.

14.3 Growth Functions

Two approaches have been commonly used to predict growth in individual-tree models. In the first approach, tree increment is related to tree, stand, and site variables via regression analysis. A second approach involves establishing a growth potential, which is adjusted by a modifier or reduction factor based on a tree's competitive status and vigor. The potential times modifier formulation is fitted to tree measurement data using regression analysis techniques. Although both approaches have proven successful, the potential times modifier function approach has the feature of readily incorporating limits or bounds on growth relationships.

Diameter growth of individual trees can be expressed as diameter increment or basal area increment. In a study aimed at determining if it is preferable to estimate diameter growth from diameter or basal area increment measurements, West (1980) concluded that there is no practical difference between the two scales.

14.4 Distance-Dependent Models

Newnham (1964) developed the first spatially-explicit stand simulator based on growth of individual trees. Diameter increment of trees in planted stands of Douglas-fir was considered to be equal to that of open-grown trees as reduced by a measure of competition. Competition was assessed for each tree as the sum of "angles of intersection" of crowns of neighboring trees. Total heights were estimated through a regression equation using diameter at breast height and stand basal area as predictors. Newnham and Smith (1964) reported on the model's performance for Douglas-fir and lodgepole pine.

Since the work of Newnham, a number of distance-dependent, individual-tree models have been promulgated for even-aged, single species stands, including simulators developed by Arney (1974) and Mitchell (1975) for Douglas-fir, Hegyi (1974) for jack pine, Burkhart et al. (1987) for loblolly pine and Pukkala (1989a) for Scots pine, Faber (1991) for poplar, Soares and Tomé (2003) for eucalyptus plantations, and Perot et al. (2010) for mixed sessile oak-Scots pine stands. These models differ in detail, but all of them follow the general modeling principles and procedures first employed by Newnham.

Two well established models, one for plantations and another for mixed-species stands, will be used as typical examples of distance-dependent models.

14.4.1 Example Model Structure for Pine Plantations

A stand simulator for loblolly pine plantations, PTAEDA2, is, in many aspects, typical of distance-dependent, individual-tree models. The PTAEDA2 model (Burkhart et al. 1987) consist of two main subsystems – one dealing with the generation of an initial precompetitive stand and another with the growth and dynamics of that stand. Management subroutines added to the overall framework allow simulation of controlling hardwood competition levels, applying fertilizers, and implementing thinning. Input/output routines facilitate model operations.

A number of options are available for creating rectangular spatial patterns in PTAEDA2. Users may specify the distance between trees and between rows in a conventional manner allowing the program to compute the planted number of trees. Alternatively, the number of trees may be specified along with the ratio of planting distance to row width (e.g., 3:4, 1:2). If this ratio is omitted, square spacing is assumed. From this information, a simulation plot is generated and coordinate values are assigned to each of the planting locations. The juvenile stand is then advanced to an age of 8 years where intraspecific competition is assumed to begin. At this point, predicted juvenile mortality is assigned at random. Individual-tree dimensions are then generated for the residual stand. Tree *dbh* values are generated from a two-parameter Weibull distribution; the parameters of the Weibull distribution are estimated as functions of stand age, number of trees surviving, and average height of dominant and codominant trees at that age. Height is predicted for each tree from an equation that includes *dbh*, average height of the dominant stand, trees surviving, and age. Crown ratio for each tree is then calculated as a function of its total height, *dbh*, and age. After assigning dimensions to each tree, the competition effect of neighboring trees is calculated for each individual tree as:

$$CI_i = \sum_{j=1}^{n} \frac{d_j/d_i}{dist_{ij}} \tag{14.1}$$

where

CI_i = competition index of *i*th subject tree
n = number of competitors "in" with BAF 2.3 m^2 per ha sweep centered at *i*th tree

d_j = *dbh* of *j*th competitor
d_i = *dbh* of *i*th subject tree
$dist_{ij}$ = distance between subject tree *i* and *j*th competitor

After generation of the precompetitive stand, competition is evaluated and simulated trees are grown individually on an annual basis. In general, growth in height and diameter is assumed to follow some theoretical growth potential. An adjustment or reduction factor is applied to the potential increment based on a tree's competitive status and vigor, and a random component is then added representing microsite and/or genetic variability.

The potential height increment (Δh_{pot}) for each tree is the change in average height of the dominant and codominant trees, obtained as the first difference with respect to age of a site-index equation. A tree may grow more or less than this potential, depending on its individual attributes. Crown ratio is considered to be an expression of a tree's photosynthetic potential. It is used in conjunction with the competition index to compute an adjustment factor for height growth. The adjustment factor times the potential height growth (determined from a site-index equation) gives the estimated actual height growth for an individual tree with a given crown ratio and competition index. The final form of the height increment (Δh) equation that was fitted to measurement data was:

$$\Delta h = \Delta h_{pot}\left(\mathrm{b}_1 + \mathrm{b}_2 c_r^{\mathrm{b}_3} e^{(-\mathrm{b}_4 CI - \mathrm{b}_5 c_r)}\right) \tag{14.2}$$

where c_r is crown ratio and *CI* is competition index as computed by Eq. 14.1. Crown ratio appears two times in the height increment equation, thus allowing increasing growth with increasing crown ratio up to a maximum (which occurs at around $c_r = 0.6$ for Eq. 14.2 fitted to data from loblolly pine plantations) and then decreasing height increment thereafter. Assuming residual variability in height growth is normally distributed, a random component is added to the final growth determinations with variance equal to the residual mean square from the fitted regression.

The maximum *dbh* attainable for an individual tree of given height and age is considered to be equal to that of open-grown loblolly pines. An equation describing this relationship, developed from measurements of open-grown trees (Strub et al. 1975), is:

$$d_0 = \mathrm{b}_0 + \mathrm{b}_1 h + \mathrm{b}_2 t \tag{14.3}$$

where

d_0 = *dbh* of open-grown trees
h = total tree height
t = tree age

The first difference of Eq. 14.3 with respect to age represents maximum potential diameter increment.

$$\Delta d_{pot} = \mathrm{b}_1 \Delta h + \mathrm{b}_2 \tag{14.4}$$

where Δd_{pot} is the potential diameter increment and Δh is the observed height increment. This potential diameter increment is adjusted by a reduction factor that is a function of the tree's competition index and crown ratio, a measure of photosynthetic potential. The equation fitted to estimate diameter increment (Δd) was

$$\Delta d = \Delta d_{pot} \left(\mathrm{b}_1 c_r^{\mathrm{b}_2} \mathrm{e}^{-\mathrm{b}_3 CI} \right) \tag{14.5}$$

where c_r and CI remain as previously defined. In the case of diameter increment, crown ratio appears only once because growth in diameter is expected to increase monotonically as crown ratio increases. Finally a normally distributed random component is added to diameter-growth determinations with variance equal to the residual mean square from the fitted regression.

The probability that a tree remains alive in a given year is assumed to be a function of its competitive stress (CI) and vigor as measured by crown ratio. An equation for estimating "probability of survival" was developed using non-linear least squares for fitting the following form to dichotomous (0,1) data:

$$P_{live} = \mathrm{b}_1 c_r^{\mathrm{b}_2} \mathrm{e}^{-\mathrm{b}_3 CI^{\mathrm{b}_4}} \tag{14.6}$$

where P_{live} is the probability that a tree remains alive.

In PTAEDA2, survival probability is calculated for each live tree each year and used to determine annual mortality. P_{live} increases with increasing crown ratio and decreases with increasing competition. The calculated P_{live} is compared with a uniform random variate between zero and one. If P_{live} is less than this generated number, the tree is considered to have died.

The increment and mortality components, along with subroutines to simulate the effects of various levels of hardwood competition, thinning, and fertilization, were linked together in a computer program to simulate individual-tree growth and stand development. Figure 14.1 is a schematic diagram showing relationships between tree and stand components in this distance-dependent, individual-tree growth and yield model for loblolly pine plantations.

14.4.2 A Model for Complex Stand Structures

Because of their inherent flexibility, there has been considerable interest in developing spatially-explicit models for simulating multispecies, as well as pure, stands subject to a wide array of silvicultural inputs. A notable example of an individual-tree, distance-dependent stand modeling approach to encompass an array of stand structures is the SILVA model (Pretzsch et al. 2002).

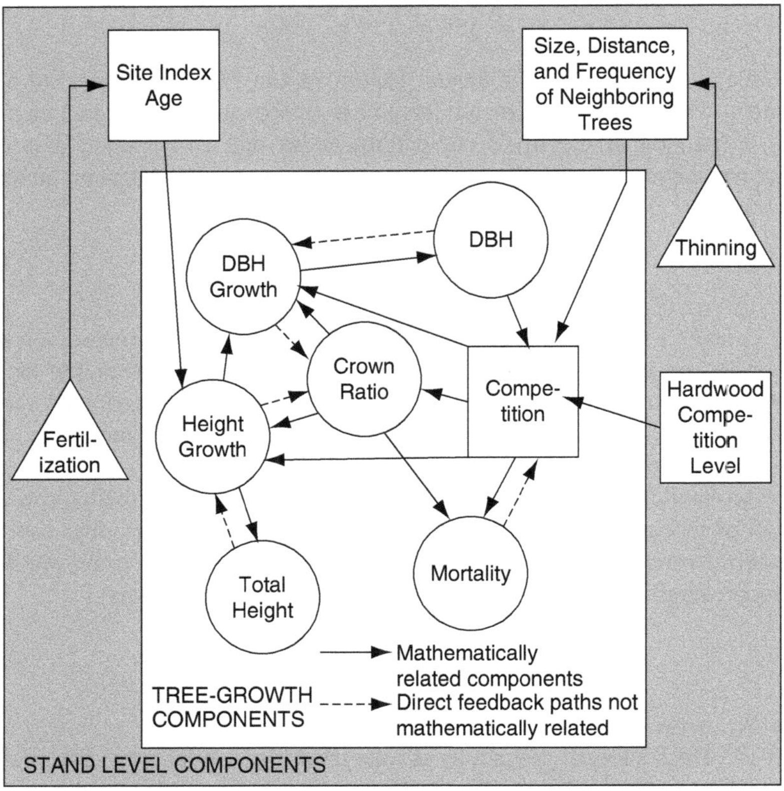

Fig. 14.1 Schematic diagram showing relationships between tree and stand components for an individual-tree, distance-dependent model of loblolly pine plantations (From Burkhart et al. 1987)

The SILVA model predicts growth in pure and mixed stands of any age composition, and it can be used to evaluate silvicultural alternatives at the stand level. Required model inputs include initial stand parameters, site variables, and selected silvicultural prescriptions.

In SILVA each tree is characterized by species, *dbh*, total height, height to crown base, crown diameter, and location coordinates. Species-specific crown models are used to represent three-dimensional crown shapes. The simulation time step with SILVA is 5 years, which corresponds with the measurement interval for the trial plots used in its development. For each growth cycle a three-dimensional competition analysis is computed to determine the degree of competition for each tree. Then preliminary tree growth is computed for use in the mortality module to determine if the individual is considered alive for the current simulation period. After removal of dead or harvested trees, competition indices and dimensional changes of each tree are recalculated.

The competition index is a geometrical competition measure calculated for the three-dimensional space surrounding a particular tree. After determining the degree of competition, natural mortality within the next simulation cycle is estimated. The mortality module calculates the survival probability from the dimensions of a tree and its estimated increase in basal area.

Thinnings are performed according to specifications defined by the user at the start of the simulation.

Site-dependent height growth potential is calculated for each tree; this potential is reduced to the expected height growth according to the individual tree's conditions defined by its competition index and crown dimensions. Tree diameter increment is also derived from potential growth; the potential diameter increment is used when estimating expected basal area growth.

New crown dimensions are calculated by estimating the height to the crown base and crown diameter from tree height and diameter.

Output from SILVA includes growth and yield information at the stand and tree level (such as stem number, basal area, timber volume, current and mean annual increment, and mean height). A stand visualization system allows for three-dimensional views of stand development, virtual walkthroughs and interactive thinning. Figure 14.2 provides an overview of the prediction algorithm employed by the growth simulator SILVA.

The SILVA model was originally parameterized with data from a large network of trial plots in Germany (Pretzsch et al. 2002). It was subsequently calibrated for Norway spruce and beech in Denmark by Brunner et al. (2006).

14.5 Generating Spatial Patterns

Spatially-explicit models can be used when projecting existing stands if the coordinate locations of trees are known or if the individual tree attributes can be assigned to points generated on a two-dimensional plane. Seldom are stem locations recorded in forestry practice, and most use of stem location-specific models results from simulations of various specified initial conditions and management treatments. Initiating a simulation scenario requires generating realistic spatial patterns and assigning initial values to the points. In the case of planted stands, the generation of locations for each planting space is straight forward (stochastic variation in planting spots can be incorporated, if desired). Mortality of planted seedlings prior to inter-tree competition can be assigned at random or in a clumped distribution, after which the simulation of stand dynamics can be projected forward.

Forest structure – spatial distribution, species diversity, and variation in tree dimensions – affects biodiversity and forest dynamics, growth and yield. Consequently a great deal of effort has been devoted to developing quantitative measures of forest structure (Tomppo 1986; Penttinen et al. 1992; Mateu et al. 1998; Kokkila et al. 2002; Pommerening 2002; Aguirre et al. 2003).

Quantitative spatial pattern descriptions, which are required for implementing distance-dependent growth and yield models, have been developed for a number of applications. Daniels (1978) examined spatial patterns in 5–12-year-old loblolly

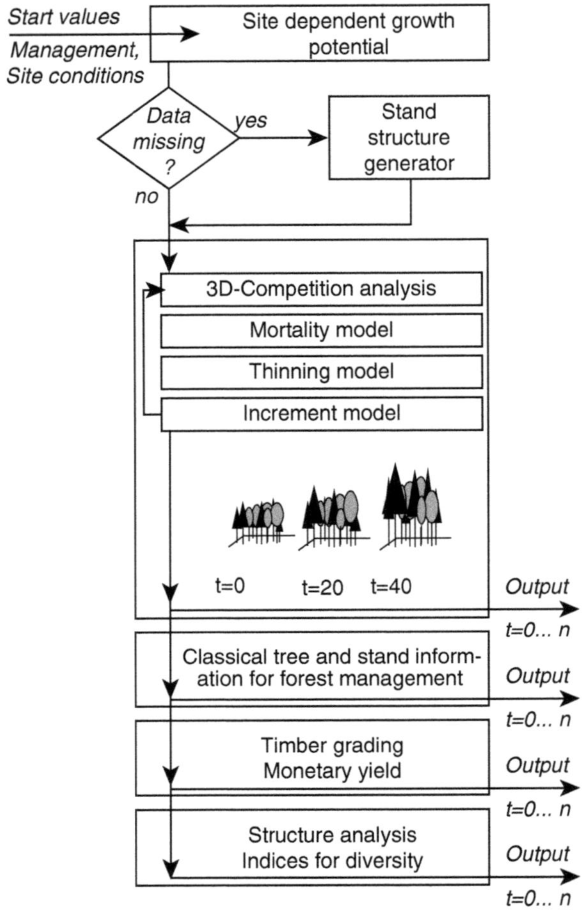

Fig. 14.2 Overview of the prediction algorithm in the growth simulator SILVA 2.2. Variable t denotes the simulation time in years, running from 0 to n (From Pretzsch 2009)

pine stands of seed origin using point-to-plant distance methods. A nonrandomness index indicated that clumping was prevalent in stands regenerated by seed. Observed distance frequencies from data collected in 40 stands were described by continuous distributions and a simulator was constructed to generate spatial patterns and individual tree sizes at age 10 for seeded loblolly pine (Daniels et al. 1979b). Tree growth and mortality were then incremented annually in the framework of the distance-dependent simulator of Daniels and Burkhart (1975).

Pommerening (2006) evaluated forest spatial structural indices for generating spatial patterns. A family of individual tree neighborhood-based indices, which are measures of small-scale variations in tree positions, species, and dimensions, was used. When expressed as frequency distributions, these indices provide information on spatial structure, which can be used to produce tree coordinates as input data for growth simulations or visualizations. In a subsequent paper, Pommerening and Stoyan (2008) presented an efficient method of synthesizing spatial tree point patterns from nearest neighbor summary statistics sampled in small circular subwindows. Data on nearest neighbor summary statistics can be feasibly obtained as part of forest surveys, thus providing information needed for generating spatial patterns for use in growth modeling and visualization.

Pretzsch (1997) described the stand structure generator STRUGEN that is used to produce stem coordinates for all trees at an initial time for use in the spatially explicit simulator SILVA. To generate stand structures, a two-dimensional homogeneous Poisson process is used as well as a set of two-dimensional distribution functions which determine mixture type and intermingling intensity of main and associated tree species. A distance function determines minimum distances between competing neighboring trees. The pattern produced is thus the result of a combination of an inhomogeneous Poisson process for generating mixture units and a so-called hard-core process for determining minimum distances between neighbors. Application of the STRUGEN generator requires information on diameters and species of all trees in an area to be simulated.

14.6 Controlling Plot Edge Bias

When implementing distance-dependent models it is generally necessary to account for stem spatial patterns beyond the simulation plot when computing competition index values for trees near the plot borders in order to avoid what is commonly termed "edge bias". Sample plots on which growth measurements are taken are typically rather small with a sizeable portion of the trees being at or close to the plot boundaries. Likewise, the simulation plot created in model applications may contain a relatively high proportion of border trees. Hence, a method for edge bias correction is required.

When fitting equations or simulating stand growth, one possibility is to include only the interior trees. This option may not be feasible for small plots and it has the disadvantage that much of the measurement data are not used. Edge-bias compensation methods fall into three general categories. The first, and most widely used method, consists of extending plots through translation or reflection. Both translation and reflection methods can create unrealistic periodicities in spatial patterns, but the problem is much more pronounced in the reflection approach (Fig. 9.8 illustrates translation and reflection). Spatial periodicities from translation, which typically involves shifts to all four sides as well as to the four corners of the simulation plot, generally do not create detectable biases in growth predictions. A second general approach is based on estimating competition outside the plot from surrounding trees inside the measurement or simulation plot. The third general approach involves generating spatial structure outside the plot based on stem-distance functions.

Monserud and Ek (1974) judged methods which involve shifting the simulation plot image to form a set of border plots best on the basis of likely bias reduction and the relative simplicity of introduced spatial pattern periodicity. Martin et al. (1977)

followed up with a study aimed at control of plot edge bias in forest simulation models with an approach termed the linear expansion method. When comparing the linear expansion method with plot image translation techniques they concluded that both methods behaved equally well on square plots 0.08 ha in size. However, as plot size diminished or its shape deviated from square, the linear expansion method provided greater accuracy and lower bias than translation techniques.

Radtke and Burkhart (1998) compared four methods of edge-bias compensation with the alternative of ignoring off-plot trees to determine their relative adequacy in modeling crown closure from individual tree crown measurements. By shrinking the size of experimental plots in a loblolly pine spacing trial, measurements of "off plot" data were available to compare the results from edge-bias compensation methods. Three edge-bias computation algorithms were found to perform equally well: translation, reflection via a reflecting line through the edge trees, and a random arrangement of interior trees around the plot.

Pommerening and Stoyan (2006) evaluated the performance of six different approaches to edge-correction: no correction, translation, reflection, buffer zone, and two nearest-neighbor methods. The performance of edge-correction methods depended strongly on the algorithm of the indices and the spatial pattern involved. In fact, they found that some edge-correction methods introduced more error than ignoring edge-bias entirely. The reflection method resulted in highly-biased values, and the authors recommended that it not be used. Translation, buffer zone, and a nearest-neighbor method were judged suitable methods to reduce error.

14.7 Distance-Independent Models

Distance-independent models are very similar in structure and operation to distance-dependent models, the primary difference being in how competition is represented. In the case of distance-independent models a measure of stand density may be included in the tree increment equations along with an indication of the relative competitive status of the individual tree such as the basal area of trees larger than the subject tree or the ratio of mean stand diameter to the diameter of the subject tree. The data required for developing and implementing individual tree increment and mortality equations that are not spatially explicit are generally more readily available and less expensive to acquire than spatially-explicit observations.

Distance-independent, individual-tree models typically consist of a diameter increment equation, a height increment or height-diameter function, and a mortality function. Increment may be predicted directly as a function of tree, stand, and site variables, or it may be estimated by a potential growth function multiplied by a modifier.

Tree diameter and height information is inserted in tree volume, weight, or taper equations to obtain estimates of tree contents; the tree values are summed and converted to unit area estimates. The tree increment and mortality equations, along with silvicultural response functions, are implemented in a stand simulator. Because individual tree values are summed to obtain unit area values, no assumption about or restrictions on the underlying dbh or height distributions are required.

Individual-tree models that are tree position independent have been developed for a variety of stand types. Arney (1985) reported on an individual-tree, distance-independent model that uses the potential growth times modifier functions approach for diameter and height development of even-aged Douglas-fir stands. Amateis et al. (1989) fitted diameter increment and survival equations for loblolly pine plantations using equation forms from the distance-dependent model of Daniels and Burkhart (1975) with the ratio of quadratic mean diameter to individual tree diameter being substituted for a spatially explicit competition index. Equations for basal area increment and height-diameter relationships were developed by Harrison et al. (1986) to predict growth of even-aged Appalachian mixed hardwood stands. Lynch et al. (1999) presented an individual tree growth and yield prediction system for even-aged, natural stands of shortleaf pine.

Complex stand structures (mixed species and uneven– as well as even-aged stands) have been modeled using an individual- tree, distance-independent framework. Andreassen and Tomter (2003) developed basal area growth models for Norway spruce, Scots pine, birch and other broadleaves in Norway. The candidate input variables included: tree size, competition index, site conditions, and stand variables. Huang and Titus (1995) used data from permanent sample plots to model periodic diameter increment of white spruce grown in boreal mixed-species stands in Alberta, Canada, as a function of tree diameter, total height, relative competitiveness of the tree in the stand, species composition, stand density, and site productivity. The Forest Vegetation Simulator (FVS), developed and maintained by the United States Forest Service, is a distance-independent individual-tree growth and yield platform that has been widely applied. Building on the framework laid out by Stage (1973), a large number of variants of FVS have been developed (Crookston and Dixon 2005) for timber types in North America as well as in other parts of the world. As an example Monserud and Sterba (1996) fitted a basal area increment model for individual trees growing in even- and uneven-aged forest stands in Austria. The Monserud-Sterba basal area increment model was patterned after the basal area function in FVS and described in detail by Wycoff (1990). Wycoff's basal area increment model for individual conifers in the Northern Rocky Mountains of the USA does not include site index or age because of the irregular stand structures found in mixed-conifer stands in the region.

As with distance-dependent models, two typical distance-independent models will be described to illustrate the structure of this approach.

14.7.1 *Example Model for Pure, Even-Aged Stands*

Lynch et al. (1999) developed a distance-independent individual-tree growth and yield system for even-aged shortleaf pine stands. The basic components of the model are individual-tree basal area growth and survival probability equations, plus compatible height-*dbh* and height-growth projection equations. The basal area growth model uses the potential-modifier approach in which a Chapman-Richards (Richards 1959) function, constrained by maximum tree size (Shifley and Brand 1984), represents potential tree growth. This potential is multiplied by a logistic function modifier. The modifier function (Murphy and Shelton 1996) is constrained to assume values between 0 and 1; it reduces potential growth on the basis of variables representing stand and tree attributes. Equation 14.7 was fitted to predict basal area growth of individual shortleaf pine trees in even-aged natural stands:

$$\Delta g_i = \frac{b_1 g_i^{b_2} - \left(b_1 g_i / g_{max}^{1-b_2}\right)}{1 + \exp(b_3 + b_4 G + b_5 t + b_6 R_i + b_7 g_i)} \tag{14.7}$$

where Δg_i is annual basal area growth of tree *i*; g_i is basal area of tree *i*; *t* is stand age; R_i is the ratio of quadratic mean stand diameter to the *dbh* of tree *i*; *G* is stand basal area; g_{max} is the maximum expected basal area for a shortleaf pine tree in managed stands; and $b_1, b_2, \ldots, b_7$ are coefficients estimated from data.

Individual-tree survival was predicted using a logistic model. The dependent variable was 1 for trees that were alive at both ends of the measurement interval and 0 for trees that were alive at the first measurement but dead prior to the second. Iteratively reweighted least squares was used to achieve homogeneity of variance. The weight was the inverse of the variance $P^t(1 - P^t)$ where *P* is the annual probability of survival predicted by the logistic model and t is the number of years in the measurement period. An annual mortality prediction equation was desired; consequently, because the remeasurement intervals were not the same for all plots, the logistic model was raised to a power equal to the number of years in the measurement interval when estimating the parameters. Model (14.8) was selected to predict probability of survival in natural shortleaf pine stands:

$$P_i = \frac{1}{1 + \exp\left[-\left(b_0 + \frac{b_1}{R_i} + b_2 G_s + b_3 h_{dom}\right)\right]} \tag{14.8}$$

where P_i is the probability of annual survival for tree *i*; h_{dom} is the average height of dominant and codominant trees; $b_0, b_1, \ldots, b_3$ are coefficients to be estimated; and other variables are as defined in Eq. 14.7.

Lynch and Murphy (1995) developed a compatible height prediction and projection system for individual shortleaf pine trees (see, also, Sect. 12.2). Their equation for predicting either current or future individual shortleaf pine tree heights is:

$$(h_i - h_d) = b_1 (h_{dom} - h_d)^{b_2} \exp(b_3 d_i{}^{b_4}) \tag{14.9}$$

where h_i is total height of tree *i*; h_d is height to *dbh*, d_i is *dbh* of tree *i*; and h_{dom} is as defined for Eq. 14.8.

Prediction of future tree heights may be more accurate if information from previous tree heights can be used as a predictor variable. Since the shortleaf pine data set contained two height measurements, it was possible to develop the following projection equation that can be used to predict future heights based on previously measured heights:

$$(h_{2i} - h_d) = (h_{1i} - h_d)\left(\frac{h_{dom2} - h_d}{h_{dom1} - h_d}\right)^{b_2} \exp\left[b_3\left(d_{2i}^{b_4} - d_{1i}^{b_4}\right)\right] \tag{14.10}$$

where h_{1i} and h_{2i} are time 1 and 2 total height of tree i; h_{dom1} and h_{dom2} are time 1 and 2 average total heights of dominant and codominant trees; and d_{1i} and d_{2i} are time 1 and 2 *dbh* values of tree i.

Equation 14.10 uses measured height at time 1, if available, to predict future heights at time 2, leading to better predictions than could be obtained with Eq. 14.9. Equations 14.9 and 14.10 are compatible in the sense that when Eq. 14.9 is used to generate a height at time 1, the predicted height at time 2 given by Eq. 14.10 is the same as would be predicted by Eq. 14.9.

A representative subsample of trees was selected for developing an individual tree crown ratio prediction model. The data set consisted of 3,132 shortleaf pine trees on which total height and height to live crown were measured. The crown ratio equation, which is a form previously used by Dyer and Burkhart (1987) and is constrained to give estimates between zero and one, had the form:

$$c_{ri} = 1 - \exp\left[-\left(b_0 + \frac{b_1}{t}\right)\left(\frac{d_i}{h_i}\right)^{b_2}\right] \tag{14.11}$$

where c_{ri} is crown ratio of tree i; b_0, b_1, and b_2 are parameters to be estimated; and other variables are as previously defined.

Equations 14.7, 14.8, 14.9, 14.10 and 14.11 plus the site index equation of Graney and Burkhart (1973) and the taper function of Farrar and Murphy (1987) were incorporated into a ShortleafPineStandSimulator (SLPSS) to simulate growth and yield of even-aged, naturally-regenerated shortleaf pine stands (Huebschmann et al. 1998). The basic input to the simulator consists of current stand conditions in the form of either a stand table (number of trees by *dbh* classes) or inventory data from field plots.

Each tree (or group of trees in a *dbh*-class interval) is grown on a year-by-year basis. Yearly basal area increment is estimated by Eq. 14.7 and the probability of survival by Eq. 14.8. A tree survives the year if its probability of survival exceeds the value of a uniformly distributed random number (restricted to the interval 0–1) generated for that tree.

Equation 14.9 or 14.10 is used to estimate total height for each tree, and Eq. 14.11 is applied to calculate the crown ratio. The height and crown ratio estimates determine which of Farrar and Murphy's (1987) taper functions is used to compute the tree's volume. The simulator is capable of conducting low or free thinning to specified levels of residual stand basal area. Other types of thinning can be accomplished by specifying a desired residual stand table.

14.7.2 A Distance-Independent Modeling Platform for Complex Stand Structures

The Forest Vegetation Simulator (FVS) had its beginning as the Prognosis Model for Stand Development (Stage 1973). The modeling structure of Prognosis did not include site index or age in order that stands of mixed species and irregular age structure could be projected, if desired. When the Prognosis Model was adopted as a common modeling platform by the United States Department of Agriculture, Forest Service, it was soon thereafter designated the Forest Vegetation Simulator or FVS. Geographic specific versions of FVS are called variants. Over 20 FVS variants have been developed for forested areas in the United States and for part of British Columbia, Canada (Crookston and Dixon 2005).

Stand development is simulated by predicting changes in dimensions of trees that compose the stand. Tree growth is predicted separately for large trees and for small trees. For large trees diameter increment is predicted, then height growth is predicted as a function of diameter increment and other variables. In the case of small trees, height growth is predicted first and diameter increment is estimated as a function of height growth and other variables.

Prediction of diameter increment is central to FVS. All facets of projections of large-tree development are dependent in part on diameter or diameter increment. The overall behavior of FVS is strongly influenced by the behavior of the diameter increment model and the subsequent use of dbh and dbh increment in the prediction of other tree attributes (Crookston and Dixon 2005).

Wycoff (1990) presented a detailed derivation of the diameter increment model that is at the core of most FVS variants. Individual-tree basal area increment is modeled as a function of tree size, site variables, and competition. Tree size is expressed by dbh; site effects are captured with habitat type, location, slope, aspect, and elevation. When modeling competition, the overall level of stand density is quantified by crown competition factor CCF (Sect. 8.6.2). Crown ratio was included for individual trees along with CCF to further reflect tree vigor, treatment history, and overall stand density. Finally, relative tree size was expressed in terms of basal area in trees larger than the subject tree.

Mortality predictions in FVS are intended to reflect competition-induced or typical mortality rates. Mortality from insects, pathogens, and fire are accounted for in extensions to FVS. Mortality from causes such as logging damage, animal damage, or wind storms can be simulated by user-specified commands.

A distinguishing feature of FVS is its ability to automatically calibrate internal models to reflect local deviations from the regional growth trends represented in the variant. If three or more tree records for a species have measured heights, the model parameters of the height-diameter function for that species are adjusted. When growth increment data are provided on five or more sample trees per species, parameters of the large-tree diameter increment model, the small tree height increment model, or both will also be scaled (Crookston and Dixon 2005).

Random effects are incorporated into FVS projections. A random component is assigned to the distribution of errors associated with the prediction of the logarithm of basal area increment. The effects of this variation propagate in highly nonlinear ways through most of the remaining components of FVS.

14.8 Annualized Growth Predictions from Periodic Measurements

Most individual-tree models are calibrated to produce annual increments of tree characteristics. The data available for estimating coefficients in the component equations, however, generally come from periodic measurements. Furthermore, the time between measurements may be variable. Hence, the modeller is faced with the problem of determining the most appropriate values for annual increments to use in equation fitting.

When the desired time interval for prediction is not the same as the interval at which the data used for equation fitting were collected, some kind of interpolation is required. Linearity assumptions, while convenient and easy to apply, are generally inconsistent with the growth function selected and can lead to biased growth projections. McDill and Amateis (1993) developed two interpolation methods based on the assumption that the estimated functional form provides useful information for making interpolations and that interpolation should be consistent with the functional form of the growth equation. An empirical application using tree height growth data indicated that the proposed interpolation methods provided better results than methods based on an assumption of linear growth.

Cao (2000) developed an iterative method for estimating annual diameter growth and survival for individual trees from periodic measurements. The iterative method out-performed the averaging (constant growth rate) approach for predicting tree survival, tree diameter growth, and stand basal area. The superior performance of the iterative method was attributed to its accounting for the variable rate of diameter growth and tree survival probability as functions of changing stand and tree attributes. These methods were subsequently extended and applied by Cao et al. (2002) when developing an individual tree growth modeling system for the loblolly pine-shortleaf pine forest type in Louisiana, USA. The Cao et al. (2002) growth system includes models for individual tree survival, diameter growth, height growth, and change in crown ratio. A multivariate extension of a two-step, model-based interpolation method was used to estimate parameters of annual tree growth equations based on measurements from 7-year growth periods.

Models of individual tree annual diameter and height growth, as well as annual mortality, of beech in Denmark were developed by Nord-Larsen (2006). An iterative method for continuously updating individual-tree and stand attributes using the hypothesized functional form of a system of difference equations was applied due to irregular measurement intervals in the data.

Weiskittel et al. (2007) used a variation of the iterative technique suggested by Cao (2000) to estimate annualized diameter and height growth equations for plantations of Douglas-fir, western hemlock, and red alder in the Pacific Northwest region of the United States. The inclusion of multi-level mixed effects improved the model fits, but the random effects were not closely related with physiographic features, climate variables, or soil properties.

Cao and Strub (2008) developed an approach to simultaneously estimate parameters of an annual tree growth model in which the sum of log-likelihood functions for tree survival and diameter growth was maximized. Four methods for acquiring interim values of stand density were evaluated with periodic measurement data from a loblolly pine seed source study. Updating attributes, in which individual tree values were summarized at the end of each year within the growth period to predict interim stand-level attributes, and predicting attributes, in which stand attributes were predicted annually using a stand-level model, performed better in predicting tree survival and diameter growth than did linear interpolation or using initial values of stand attributes at the beginning of the growth period.

An individual-tree growth model was developed by Crecente-Campo et al. (2010) with data from 54 permanent plots of Scots pine located in Galicia (northwestern Spain). The study involved two model fitting approaches, one assuming constant growth and mortality rates in the period between two consecutive inventories and another considering variable growth and mortality rates in the same period. Evaluation of the model via simulation of growth and mortality in the period between inventories showed that the variable growth rate approach provided slightly better results than the constant growth rate assumption.

Nunes et al. (2011) developed an annual individual tree survival and growth model for pure, even-aged stands of maritime pine in Portugal using a large data set containing irregularly time-spaced measurements. Two approaches were compared for modeling annual tree growth: direct estimation of future diameter or height using well-known growth functions formulated in difference form and estimation of diameter or height using a function in differential form for estimating the increment over an annual period. In both approaches the function parameters were related to tree and stand variables reflecting the competition status of individual trees. The second approach performed slightly better than the first.

14.9 Simultaneous Estimation of Model Component Equations

Typically, the component equations of individual-tree models have been fitted individually using ordinary least squares. However, because individual-tree growth models are based on multivariate attributes observed on the same individuals, the resulting set of growth equations can be considered a simultaneous system and simultaneous regression techniques can be considered for parameter estimation.

Hasenauer et al. (1998) compared results from fitting an individual tree basal area increment model, a height increment model, and a crown ratio model separately using ordinary least squares (OLS) and simultaneously by applying two-stage and three-stage least squares (2SLS, 3SLS). The general formulation of the three equations followed the approach of Wykoff (1990) and Monserud and Sterba (1996) by predicting y as a function of tree size (*SIZE*), competition (*COMP*), and site descriptors (*SITE*):

$$y = a + f_1(SIZE; \boldsymbol{b}) + f_2(COMP; \boldsymbol{c}) + f_3(SITE; \boldsymbol{d}) \tag{14.12}$$

with y a response variable that is typically a logarithmic transformation of some measure of increment or dimensional change, a the intercept, $\boldsymbol{b}$ the vector of coefficients for tree size variables, $\boldsymbol{c}$ the vector of coefficients for the competition variables, and $\boldsymbol{d}$ the vector of coefficients for the site variables. Site descriptors were restricted to basic topographic features elevation, slope, and aspect.

The specific model forms fitted were

(a) *Basal area increment*

$$\begin{aligned} y_1 &= a_1 + \mathrm{b}_{11} \ln d - \mathrm{b}_{12} d^2 + \mathrm{b}_{13} \ln C + \mathrm{c}_{11} G_{>di} - \mathrm{c}_{12} CCF \\ &\quad + \mathrm{d}_{11} ELEV^2 - \mathrm{d}_{12} SL^2 \end{aligned} \tag{14.13}$$

where $y_1 = \ln(\Delta g)$, Δg is 5-year basal area increment; d the diameter at breast height; $C = (1/c_r) - 1$ where c_r is the crown ratio; $G_{>di}$ the basal area of trees larger in diameter than the subject tree; *CCF* the crown competition factor; *ELEV* is elevation and *SL* is the tangent of the slope angle.

(b) *Height Increment*

The second model predicts the logarithm of height increment $y_2 = \ln(\Delta h)$:

$$\begin{aligned} y_2 &= a_2 + \mathrm{b}_{21} \ln d + \mathrm{b}_{22} h^2 + \mathrm{b}_{23} \ln C + \mathrm{c}_{21} G_{>di} + \mathrm{c}_{22} CCF \\ &\quad + \mathrm{d}_{21} ELEV^2 + \mathrm{d}_{22} SL \end{aligned} \tag{14.14}$$

where h is total tree height, $\Delta h = h_2 - h_1$, and others variables remain as defined for Eq. 14.13.

(c) *Crown Ratio*

$$\begin{aligned} y_3 &= a_3 + \mathrm{b}_{31} h/d + \mathrm{b}_{32} h + \mathrm{b}_{33} d^2 + \mathrm{c}_{31} G_{>di} + \mathrm{c}_{32} \ln CCF + \mathrm{d}_{31} ELEV \\ &\quad + \mathrm{d}_{32} SL^2 + \mathrm{d}_{33} SL \cos AZ + \mathrm{d}_{34} SL \sin AZ \end{aligned} \tag{14.15}$$

where $y_3 = \ln((1c_r) - 1)$, *h/d* is the height/diameter ratio, *AZ* is the azimuth, and all other parameters are as previously defined.

When an endogenous (dependent) variable appears on the right hand side of another equation in the system, multistage estimation techniques (such as 2SLS or 3SLS) are necessary to obtain parameter estimates that are consistent and asymptotically efficient. In the system of Eqs. 14.13, 14.14 and 14.15 an endogenous variable, crown ratio, appears as a predictor variable rendering OLS estimates biased.

Results from fitting the system (14.13, 14.14 and 14.15) using data from more than 7,500 Norway spruce trees indicated the presence of strong cross-equation correlations, especially between the basal area and height increment models. Correlations between basal area increment and crown ratio predictions were weak. Using 3SLS resulted in improved standard error of the estimates for the basal area increment model but the gain in precision for the height increment and crown ratio models was negligible.

Huang and Titus (1999) fitted a system of three interdependent, tree-level nonlinear equations for white spruce grown in boreal mixed-species stands in Alberta, Canada. The system consisted of equations to predict total tree height, periodic diameter increment and height increment. Because variables that were on the left-hand side of the equations also appeared on the right-hand side of the equations in the system, the system was estimated using nonlinear simultaneous techniques. Testing of cross-equation correlations indicated that the error terms of the related equations in the system were significantly correlated, suggesting that the parameter estimates obtained from simultaneous techniques are consistent and asymptotically more efficient than those obtained from ordinary least squares procedures applied to individual equations of the system. Parameter estimates obtained by applying nonlinear ordinary least squares, nonlinear two-stage least squares, nonlinear three stage least squares and seemingly unrelated nonlinear regression were, however, similar.

Compatible height and diameter increment models for lodgepole pine, trembling aspen, and white spruce were developed by Nunifu (2009). For each species the diameter and height growth models were considered as a system and seemingly unrelated regression was applied to account for cross-equation correlation. The growth model developed by Nunes et al. (2011) also involved simultaneous fitting of components. The individual tree diameter growth model and the survival probability model were fitted simultaneously using seemingly unrelated regression (SUR); parameters of the height function were obtained separately.

14.10 Incorporating Stochastic Components

The growth of individual trees differs by microsite, environment, spacing and genetic attributes. Deterministic growth functions attempt to capture the mean response to site, competition, and genetic inputs, but inevitably residual, unexplained variation will remain. Random or stochastic components are often incorporated in individual-tree models as a means of dealing with unexplained variation. Adding random components is expected to increase variation in the predicted tree sizes so that size attributes from models more nearly match what is observed in the field.

One common means of incorporating stochastic variation is to add a random component with mean zero and variance equal to the residual mean square error from equations fitted to diameter and height increment.

Stage and Wycoff (1993) developed a model of the stochastic process affecting individual-tree increments in diameter that included the serial correlation of errors. The sensitivity of projections in the Prognosis Model for Stand Development was tested by simulation. Results showed that incorporation of stochasticity into diameter increment predictions reduced bias in projected stand volumes and diameter ranges.

In a study of residual variation associated with tree diameter growth predictions in Scots pine and Norway spruce, Miina (1993) modeled the variation in residuals by four random parameters which described level, trends, autocorrelation and error variance of the time series of the residuals. The modeled residual variation can be added as a stochastic component to growth estimates to take into account the total variation in diameter growth. Simulations showed that adding the modeled stochastic variation to diameter growth resulted in more rapid differentiation of trees into diameter classes than in deterministic growth simulations. Addition of the stochastic residual variation to the diameter growth estimates did not, however, notably affect estimates of the volume growth of the stand.

In a review of stochastic structure and individual-tree growth models, Fox et al. (2001) pointed out that past applications have relied on deterministic predictions or have added an unstructured random component to predictions. They identified the important components of stochastic structure as spatial, temporal, and nested and described methodologies for incorporating stochastic structure in growth model estimation and prediction. Benefits from incorporation of stochastic structure include valid statistical inference, improved estimation efficiency, and more realistic and theoretically sound predictions (Fox et al. 2001).

14.11 Relating Predictions from Whole-Stand and Individual-Tree Models

Growth and yield models for forest management decision support range from relatively simple whole-stand models to models that aggregate individual tree values to obtain stand estimates. Due to accumulation of error in detailed tree-level models, the stand-level predictions are generally not as reliable as those from whole-stand models. On the other hand, stand-level models do not provide the level of detail about stand structure that is often required. Hence, there has been considerable interest in constraining individual-tree models with whole-stand characteristics to obtain well-behaved projections of overall values while maintaining stand detail.

Somers and Nepal (1994) linked stand-level and individual-tree models via an algorithm based on the assumption that stand-level estimates of basal area, trees per unit area, and average diameter are correct and that the projected individual tree values should be adjusted to equate to the stand-level values. A disaggregation function based on relative-size growth was developed by Zhang et al. (1993) for Douglas-fir to distribute stand volume growth to a list of individual trees. Qin and Cao (2006) used data from loblolly pine plot measurements to fit whole-stand and individual-tree equations. Outputs from the individual-tree model were then adjusted to match observed stand attributes (number of trees, basal area, and volume per ha) by four disaggregation methods: proportional yield, proportional growth, constrained least squares, and coefficient adjustment techniques were applied to disaggregate predicted stand growth among trees in a tree list. Results showed that, compared to the unadjusted individual-tree model, the adjusted model performed better in predicting stand attributes, while providing comparable predictions of tree diameter, height, and survival probability.

In an effort aimed at ensuring that individual tree growth models provide both precise tree-level and stand-level predictions, Zhang et al. (1997) developed a model based on conceptual relationships between basal area of each diameter class and the diameter increment in the same class. An algorithm estimating multiresponse regressions was used to fit the constrained regression equation system to data from loblolly pine plantations. The simultaneous equation system provided essentially the same individual tree increment estimates as the unconstrained model but superior estimates for basal area of each diameter class and for total stand basal area.

Cao (2006) described past approaches used to adjust or condition summation of individual-tree values to be consistent with whole-stand estimates. The first approach, called a disaggregative approach, involves adjusting predictions from individual-tree models to equal that predicted by a whole-stand model (e.g., Zhang et al. 1993; Somers and Nepal 1994; Qin and Cao 2006). In the second approach, multiresponse regression techniques are used to constrain an individual-tree model to optimize for tree and diameter class levels (e.g., Zhang et al. 1997). A third approach was developed in which an individual tree model was constrained by stand attributes so that the model is optimized at the tree and stand levels (Cao 2006). Fitting the three alternative approaches to data from a loblolly pine seed source study showed that the third approach (individual-tree/whole stand) produced results that were comparable to those from the same tree model constrained with number of trees and basal area in each diameter class. Both constrained tree models were slightly better than the unconstrained tree model in predicting tree and stand attributes. The disaggregative approach, in which outputs from the individual tree model were adjusted with stand growth predictions from a whole-stand model, provided the best predictions of tree- and stand-level survival and growth.

14.12 Comparisons of Growth and Yield Models with Varying Levels of Resolution

The primary difference between spatially-explicit and non-spatially explicit models is the measure of inter-tree competition employed. Consideration of the spatial arrangement of trees is inherent in certain forest management decisions (such as, effect of rectangularity of planting spacing on subsequent growth, impact of clumpiness following thinning on future yields, etc.). Spatial information is expensive to obtain and it is generally not available from inventory plots commonly used when implementing growth and yield models. Consequently, the additional value, in terms of predictive ability of spatially-explicit as opposed to non-spatially explicit models, has been a topic of considerable interest.

Pukkala (1989b) compared spatial and non-spatial growth models for even-aged stands of Scots pine. In the non-spatial models competition was described at the stand level only; the spatial models considered competitors within 5 m of the subject tree. The use of spatial predictors increased the coefficient of determination for estimation of the logarithm of diameter increment from 43% to 56% if past growth was not known and from 62% to 75% if past growth was known (the respective decreases in the standard error of estimate were 10% and 15% points).

Various comparisons of individual-tree and alternative model structures have been conducted. Daniels et al. (1979a) used an independent data set to evaluate and compare three models for predicting merchantable yields of loblolly pine plantations. Their analysis included a whole-stand, diameter-distribution, and individual-tree model. The observed number of trees and dominant height on the test plots were used and merchantable yield was estimated by each model. Deviations of estimated from observed yields reveals that (1) all three models provided accurate estimates; (2) all three models were free from prediction bias due to stand attributes; and (3) the whole-stand and diameter-distribution models exhibited greater precision than the individual-tree model.

A comparison of a distance-dependent, individual-tree and a diameter-class model for describing changes in stand density and structure of Lake States' northern hardwoods was carried out by Ek and Monserud (1979). Projections were made by each model for 5–26 years over a range of stand conditions and harvest treatments. Results from numerous performance tests and comparisons of actual and predicted diameter distributions, basal areas, and numbers of trees indicated that the individual-tree model was considerably more sensitive to harvest treatments and reproduction response than the diameter-class model.

Knowe et al. (1997) used data from a red alder planting spacing study to compare three modeling approaches. The diameter-distribution model was based on the Weibull function with percentile-based parameter recovery (Chap. 12), the stand-table-projection approach relied on changes in relative tree size (defined as the ratio of individual-tree basal area to average basal area per tree in the stand) over time (Chap. 13), and the distance-independent, individual-tree growth function was patterned after that used in the Prognosis model (Wycoff 1990). When these three approaches were used to predict stand structures and dynamics in plantings of 7–16 years of age with densities ranging from slightly less than 1,000 to nearly 14,000 trees/ha, the individual-tree model provided the best representations of observed diameter distributions at all planting densities, stand ages, and growth intervals.

14.13 Developing Growth and Yield Models with Consistency at Varying Levels of Resolution

Growth and yield models form a continuum of complexity and detail, or 'resolution', ranging from relatively simple whole-stand yield prediction equations to more complex simulation models of the development and interactions of individual trees in a stand. In choosing appropriate stand models for growth and yield estimation, users should consider the reliability of estimates, flexibility to reproduce desired management alternatives, ability to provide sufficient detail for decision making, and efficiency in providing this information. In forestry practice models of varying levels of resolution may be applied, and it is often important that numeric consistency among the estimates be exhibited.

Ritchie and Hann (1997a) evaluated the efficiency of six disaggregative methods and two individual-tree methods in terms of their ability to predict 5-year basal area increment for Douglas-fir stands in western Oregon, USA. In general, the individual-tree approach was superior to the disaggregative approach for prediction of both stand and tree growth.

Three approaches to characterizing the diameter distribution of a future stand were constructed by Qin et al. (2007). The first approach was the "parameter-recovery" method, which links a whole-stand model to a diameter-distribution model. Two additional approaches provided linkages between an individual-tree model and a diameter-distribution model. Tree survival and diameter-growth equations were applied to the tree list (the "tree-projection" method) or to the diameter distribution (the "distribution-projection" method) at the beginning of the growth period. When evaluated with data from a loblolly pine seed source study, all three methods produced similar results in terms of Reynolds et al.'s (1988) error indices, whereas the distribution-projection method outperformed the other two methods in predicting total and merchantable volumes per hectare. The analysis of Qin et al. (2007) demonstrated that the diameter-distribution model could be linked to either a whole-stand model or an individual-tree model with comparable success.

Numerous investigations have been conducted to elucidate specific relationships between growth and yield predictions at different levels of resolution. There has been, however, relatively little effort devoted to developing integrated systems of stand models of different levels of resolution that are related in an overall unified mathematical structure. An exception is the Forest Resources Evaluation Program, or FREP (Leary 1979), which was designed to include a general system of models to project forest growth from inventory records, at varying levels of resolution, for various stand types in the Great Lake States region of the United States. The FREP approach is based on a system of simultaneous-difference equations. In the simplest case one difference equation may be used for whole-stand growth estimates for stands of single-species composition. Mixed stands are modeled using two or more equations of the same form. The system was designed so that it could be expanded to include one equation for each tree to yield an individual-tree-level model. The difference equation employed uses a growth potential multiplied by a modifier to estimate realized individual or aggregated tree growth. When the size of and distance to neighbors are included in the growth modifier function the model becomes distance-dependent. Thus, the system provides a unified theoretical structure for models ranging from whole-stand models to individual-tree distance-dependent models. Although the FREP system was conceived as being capable of operating at multiple levels of resolution, in actual application the individual-tree distance-independent approach has been used almost exclusively.

Daniels and Burkhart (1988) developed an integrated system of stand models—ranging from distance dependent, individual-tree, to whole-stand levels—by applying a unified mathematical structure. The point density measure used in the distance-dependent model was area potentially available. The mean area per tree was estimated as the inverse of the number of trees per unit area, so that point density reduces to stand density and a distance-independent individual-tree model results. Trees were grouped in size classes to collapse the distance-independent tree level model in a size-class projection model. Continuing, the dimensions of the model were collapsed to an "average" tree to produce a stand-level projection model.

References

Adame P, Hynynen J, Cañellas I, del Río M (2008) Individual-tree diameter growth model for rebollo oak (*Quercus pyrenaica* Willd.) coppices. For Ecol Manage 255:1011–1022

Aguirre O, Hui GY, Gadow Kv, Jiménez J (2003) An analysis of spatial forest structure using neighborhood-based variables. For Ecol Manage 183:137–145

Alder D (1979) A distance-independent tree model for exotic conifer plantations in East Africa. For Sci 25:59–71

Amateis RL, Burkhart HE, Walsh TA (1989) Diameter increment and survival equations for loblolly pine trees growing in thinned and unthinned plantations on cutover, site-prepared lands. South J Appl For 13:170–174

Andreassen K, Tomter SM (2003) Basal area growth models for individual trees of Norway spruce, Scots pine, birch and other broadleaves in Norway. For Ecol Manage 180:11–24

Arney JD (1974) An individual tree model for stand simulation in Douglas-fir. In: Fries J (ed) Growth models for tree and stand simulation. Royal College of Forestry, Stockholm, pp 38–46, Research Notes 30

Arney JD (1985) A modeling strategy for the growth projection of managed stands. Can J For Res 15:511–518

Avila OB, Burkhart HE (1992) Modeling survival of loblolly pine trees in thinned and unthinned plantations. Can J For Res 22:1878–1882

Brunner A, Hahn K, Biber P, Skovsgaard JP (2006) Conversion of Norway spruce: a case study in Denmark based on silvicultural scenario modelling. In: Hasenauer H (ed) Sustainable forest management growth models for Europe. Springer, Berlin, pp 343–371

Burkhart HE, Farrar KD, Amateis RL, Daniels RF (1987) Simulation of individual tree growth and stand development in loblolly pine plantations on cutover, site-prepared areas. Virginia Polytechnic Institute and State University, Blacksburg, Pub. FWS-1–87

Christopher TA, Goodburn JM (2008) The effects of spatial patterns on the accuracy of Forest Vegetation Simulator (FVS) estimates of forest canopy cover. West J Appl For 23:5–11

Cao QV (2000) Prediction of annual diameter growth and survival for individual trees from periodic measurements. For Sci 46:127–131

Cao QV (2006) Predictions of individual-tree and whole-stand attributes for loblolly pine plantations. For Ecol Manage 236:342–347
Cao Q, Strub M (2008) Evaluation of four methods to estimate parameters of an annual tree survival and diameter growth model. For Sci 54:617–624
Cao QV, Li S, McDill ME (2002) Developing a system of annual tree growth equations for the loblolly pine-shortleaf pine type in Louisiana. Can J For Res 32:2051–2059
Crecente-Campo F, Soares P, Tomé M, Diéguez-Aranda U (2010) Modelling annual individual-tree growth and mortality of Scots pine with data obtained at irregular measurement intervals and containing missing observations. For Ecol Manage 260:1965–1974
Crookston NL, Dixon GE (2005) The forest vegetation simulator: a review of its structure, content, and applications. Comp Elec Agric 49:60–80
Daniels RF (1978) Spatial patterns and distance distributions in young seeded loblolly pine stands. For Sci 24:260–266
Daniels RF, Burkhart HE (1975) Simulation of individual tree growth and stand development in managed loblolly pine plantations. Virginia Polytechnic Institute and State University, Blacksburg, Pub. FWS-5–75
Daniels RF, Burkhart HE (1988) An integrated system of forest stand models. For Ecol Manage 23:159–177
Daniels RF, Burkhart HE, Strub MR (1979a) Yield estimates for loblolly pine plantations. J For 77:581–583, 586
Daniels RF, Burkhart HE, Spittle GD, Somers GL (1979b) Methods for modeling individual tree growth and stand development in seeded loblolly pine stands. Virginia Polytechnic Institute and State University, Blacksburg, Pub. FWS-1–79
Dennis B, Brown BE, Stage AR, Burkhart HE, Clark S (1985) Problems of modeling growth and yield of renewable resources. Am Stat 39:374–383
Dyer ME, Burkhart HE (1987) Compatible crown ratio and crown height models. Can J For Res 17:572–574
Ek AR, Monserud RA (1979) Performance and comparison of stand growth models based on individual tree and diameter-class growth. Can J For Res 9:231–244
Faber PJ (1991) A distance-dependent model of tree growth. For Ecol Manage 41:111–123
Farrar RM Jr, Murphy PA (1987) Taper functions for predicting product volumes in natural shortleaf pines. USDA Forest Service, Southern Forest Experiment Station, New Orleans, Research Paper SO-234
Fortin M, Bédard S, DeBlois J, Meunier S (2008) Accounting for error correlations in diameter increment modelling: a case study applied to northern hardwood stands in Quebec, Canada. Can J For Res 38:2274–2286
Fox JC, Ades PK, Bi H (2001) Stochastic structure and individual-tree growth models. For Ecol Manage 154:261–276
Gould PJ, Marshall DD, Harrington CA (2008) Prediction of growth and mortality of Oregon white oak in the Pacific Northwest. West J Appl For 23:26–33
Graney DL, Burkhart HE (1973) Polymorphic site index curves for shortleaf pine in the Ouachita Mountains. USDA Forest Service, Southern Forest Experiment Station, New Orleans, Research Paper SO-85
Hamilton DA (1990) Extending the range of applicability of an individual tree mortality model. Can J For Res 20:1212–1218
Harrison WC, Burk TE, Beck DE (1986) Individual tree basal area increment and total height equations for Appalachian mixed hardwoods after thinning. South J Appl For 10:99–104
Hasenauer H, Monserud RA, Gregoire TG (1998) Using simultaneous regression techniques with individual-tree growth models. For Sci 44:87–95
Hegyi F (1974) A simulation model for managing jack-pine stands. In: Fries J (ed) Growth models for tree and stand simulation. Royal College of Forestry, Stockholm, pp 74–90, Research Notes 30
Huang S, Titus SJ (1995) An individual tree diameter increment model for white spruce in Alberta. Can J For Res 25:1455–1465

Huang S, Titus SJ (1999) Estimating a system of nonlinear simultaneous individual tree models for white spruce in boreal mixed-species stands. Can J For Res 29:1805–1811

Huebschmann MM, Lynch TB, Murphy PA (1998) Shortleaf pine stand simulator: an even-aged natural shortleaf pine growth and yield model user's manual. Oklahoma State University Agricultural Experiment Station, Stillwater, Research Report P-967

Knowe SA, Aherns GR, DeBell DS (1997) Comparison of diameter-distribution-prediction, stand-table-projection, and individual-tree-growth modeling approaches for young red alder plantations. For Ecol Manage 98:49–60

Kokkila T, Mäkelä A, Nikinmaa E (2002) A method for generating stand structures using Gibbs marked point process. Silva Fennica 36:265–277

Leary RA (1979) Design. In: A generalized forest projection system applied to the Lake States region. USDA Forest Service, North Central Forest Experiment Station, St. Paul, pp 5–15, General Technical Report NC-49

Liu J, Burkhart HE (1994a) Spatial autocorrelation of diameter and height increment predictions from two stand simulators for loblolly pine. For Sci 40:349–356

Liu J, Burkhart HE (1994b) Spatial characteristics of diameter and total height in juvenile loblolly pine (*Pinus taeda* L.) plantations. For Sci 40:774–786

Lynch TB, Murphy PA (1995) A compatible height prediction and projection system for individual trees in natural, even-aged shortleaf pine stands. For Sci 41:194–209

Lynch TB, Hitch KL, Huebschmann MM, Murphy PA (1999) An individual-tree growth and yield prediction system for even-aged natural shortleaf pine forests. South J Appl For 23:203–211

Martin GL, Ek AR, Monserud RA (1977) Control of plot edge bias in forest stand growth simulation models. Can J For Res 7:100–105

Mateu J, Usó JL, Montes F (1998) The spatial pattern of a forest ecosystem. Ecol Mod 108: 163–174

McDill ME, Amateis RL (1993) Fitting discrete-time dynamic models having any time interval. For Sci 39:499–519

Mette M, Albrecht A, Ammer C, Biber P, Kohnle U, Pretzsch H (2009) Evaluation of the forest growth simulator SILVA on dominant trees in mature mixed silver fir-Norway spruce stands in south-west Germany. Ecol Mod 220:1670–1680

Miina J (1993) Residual variation in diameter growth in a stand of Scots pine and Norway spruce. For Ecol Manage 58:111–128

Mitchell KJ (1975) Dynamics and simulated yield of Douglas-fir. For Sci Monograph 17:39

Monserud RA, Ek AR (1974) Plot edge bias in forest stand growth simulation models. Can J For Res 4:419–423

Monserud RA, Sterba H (1996) A basal area increment model for individual trees growing in even- and uneven-aged forest stands in Austria. For Ecol Manage 80:57–80

Moore JA, Zhang L, Newberry JD (1994) Effects of intermediate silvicultural treatments on the distribution of within-stand growth. Can J For Res 24:398–404

Murphy PA, Shelton MG (1996) An individual-tree basal area growth model for loblolly pine stands. Can J For Res 26:327–331

Murphy PA, Lawson ER, Lynch TB (1992) Basal area and volume development of natural even-aged shortleaf pine stands in the Ouachita Mountains. South J Appl For 16:30–34

Nepal SK, Somers GL (1992) A generalized approach to stand table projection. For Sci 38: 120–133

Newnham RM (1964) The development of a stand model for Douglas fir. PhD thesis, Univ. of British Columbia. 201pp

Newnham RM, Smith JHG (1964) Development and testing of stand models for Douglas-fir and lodgepole pine. For Chron 40:494–502

Nord-Larsen T (2006) Modeling individual-tree growth from data with highly irregular measurement intervals. For Sci 52:198–208

Nunes L, Tomé J, Tomé M (2011) Prediction of annual tree growth and survival for thinned and unthinned even-aged maritime pine stands in Portugal from data with different time measurement intervals. For Ecol Manage 262:1491–1499

Nunifu TK (2009) Compatible diameter and height increment models for lodgepole pine, trembling aspen, and white spruce. Can J For Res 39:180–192

Penttinen A, Stoyan D, Henttonen HM (1992) Marked point processes in forest statistics. For Sci 38:806–824

Perot T, Goreaud F, Ginisty C, Dhôte J-F (2010) A model bridging distance-dependent and distance-independent tree models to simulate the growth of mixed forests. Ann For Sci 67:502p1–502p11

Pommerening A (2002) Approaches to quantifying forest structures. Forestry 75:305–324

Pommerening A (2006) Evaluating structural indices by reversing forest structural, analysis. For Ecol Manage 224:266–277

Pommerening A, Stoyan D (2006) Edge-correction needs in estimating indices of spatial forest structure. Can J For Res 36:1723–1739

Pommerening A, Stoyan D (2008) Reconstructing spatial tree point patterns from nearest neighbor summary statistics measured in small subwindows. Can J For Res 38:1110–1122

Pommerening A, LeMay V, Stoyan D (2011) Model-based analysis of the influence of ecological processes on forest point pattern formation – a case study. Ecol Mod 222:666–678

Pretzsch H (1997) Analysis and modeling of spatial stand structures. Methodological considerations based on mixed beech-larch stands in Lower Saxony. For Ecol Manage 97:237–253

Pretzsch H (2009) Forest dynamics, growth and yield. Springer, Berlin

Pretzsch H, Biber P, Ďurský J (2002) The single tree-based stand simulator SILVA: construction, application and evaluation. For Ecol Manage 162:3–21

Pukkala T (1988) Effect of spatial distribution of trees on the volume increment of a young Scots pine stand. Silva Fennica 22:1–17

Pukkala T (1989a) Prediction of tree diameter and height in a Scots pine stand as a function of the spatial pattern of trees. Silva Fennica 23:83–99

Pukkala T (1989b) Predicting diameter growth in even-aged Scots pine stands with a spatial and non-spatial model. Silva Fennica 23:101–116

Pukkala T, Kolström T (1987) Competition indices and the prediction of radial growth in Scots pine. Silva Fennica 21:55–67

Pukkala T, Kolström T (1991) Effect of spatial pattern of trees on the growth of a Norway spruce stand. A simulation model. Silva Fennica 25:117–131

Qin J, Cao QV (2006) Using disaggregation to link individual-tree and whole-stand growth models. Can J For Res 36:953–960

Qin J, Cao QV, Blouin DC (2007) Projection of a diameter distribution through time. Can J For Res 37:188–194

Radtke PJ, Burkhart HE (1998) A comparison of methods for edge-bias compensation. Can J For Res 28:942–945

Reed DD, Burkhart HE (1985) Spatial autocorrelation of individual tree characteristics in loblolly pine stands. For Sci 31:575–587

Reynolds MR Jr, Burk TE, Huang WC (1988) Goodness-of-fit tests and model selection procedures for diameter distribution models. For Sci 34:373–399

Richards FJ (1959) A flexible growth function for empirical use. J Exp Bot 10:290–300

Richie MW, Hann DW (1997a) Evaluation of individual-tree and disaggregative prediction methods for Douglas-fir stands in western Oregon. Can J For Res 27:207–216

Richie MW, Hann DW (1997b) Implications of disaggregation in forest growth and yield modeling. For Sci 43:223–233

Schröder J, Rodríguez Soalleiro R, Alonso Vega G (2002) An age-independent basal area increment model for maritime pine trees in northwestern Spain. For Ecol Manage 157:55–64

Shater Z, de-Miguel S, Kraid B, Pukkala T, Palahí M (2011) A growth and yield model for even-aged *Pinus brutia* Ten. stands in Syria. Ann For Sci 68:149–157

Shifley SR, Brand GJ (1984) Chapman-Richards growth function constrained for maximum tree size. For Sci 30:1066–1070

Soares P, Tomé M, Soares P (2003) GLOBTREE, an individual tree growth model for *Eucalyptus globulus* in Portugal. In: Amaro A, Reed D (eds) Modelling forest systems. CAB International, Wallingford, pp 97–110

Somers GL, Nepal SK (1994) Linking individual-tree and stand-level growth models. For Ecol Manage 69:233–243

Stage AR (1973) Prognosis model for stand development. USDA Forest Service, Intermountain Forest and Range Experiment Station, Ogden, Research Paper INT-137

Stage AR, Wycoff WR (1993) Calibrating a model of stochastic effects on diameter increment for individual-tree simulations of stand dynamics. For Sci 39:692–705

Stage AR, Wycoff WR (1998) Adapting distance-independent forest growth models to represent spatial variability: effects of sampling design on model coefficients. For Sci 44:224–238

Stauffer HB (1978) Aggregating points to fit Pielou's index of nonrandomness. Can J For Res 8:355–363

Sterba H, Monserud RA (1997) Applicability of the forest stand growth simulator PROGNAUS for the Austrian part of the Bohemian Massif. Ecol Mod 98:23–34

Strub MR, Vasey RB, Burkhart HE (1975) Comparison of diameter growth and crown competition factor in loblolly pine plantations. For Sci 21:427–431

Subedi N, Sharma M (2011) Individual-tree diameter growth models for black spruce and jack pine plantations in northern Ontario. For Ecol Manage 261:2140–2148

Tomppo E (1986) Models and methods for analysing spatial patterns of trees. Commun Ins For Fenn 138:1–65, Helsinki

Vettenranta J (1999) Distance-dependent models for predicting the development of mixed coniferous forests in Finland. Silva Fennica 33:51–72

Weiskittel AR, Garber SM, Johnson GP, Maguire DA, Monserud RA (2007) Annualized diameter and height growth equations for Pacific Northwest plantation-grown Douglas-fir, western hemlock, and red alder. For Ecol Manage 250:266–278

West PW (1980) Use of diameter increment and basal area increment in tree growth studies. Can J For Res 10:71–77

Wycoff WR (1990) A basal area increment model for individual conifers in the northern Rocky Mountains. For Sci 36:1077–1104

Zhang L, Moore JA, Newberry JD (1993) Disaggregating stand volume growth to individual trees. For Sci 39:295–308

Zhang S, Amateis RL, Burkhart HE (1997) Constraining individual tree diameter increment and survival models for loblolly pine plantations. For Sci 43:414–423

Chapter 15
Growth and Yield Models for Uneven-Aged Stands

15.1 Special Considerations for Modeling Uneven-Aged Stands

Some forests are managed with selective harvests applied on a particular cutting cycle. Thus a continuous forest cover is maintained, but the cover is composed of trees that differ markedly in age. Such forests, which are often composed of multiple species, are termed "uneven-aged" to distinguish them from forests managed on a rotation or even-aged basis.

Because age has nebulous meaning in the context of uneven-aged forests, it is not usable as a variable for growth and yield prediction purposes. Also, site quality assessment by the site index method is questionable because of initial suppression of advanced reproduction, especially for shade tolerant species, and because site index is an age-dependent variable. Hence models that include uneven-aged structures generally do not involve age or site index as predictor variables.

Models such as SILVA (Pretzsch et al. 2002) and FVS (Crookston and Dixon 2005) that were presented in the previous chapter were developed to accommodate pure and mixed stands of even- and uneven-aged structures. Another notable example is MOSES, a distance-dependent growth simulator that has been calibrated for all major tree species in Austria and Switzerland (Hasenauer et al. 2006). The MOSES model was designed to include forest types ranging from even-aged, pure stands with no treatment to uneven-aged mixed-species stands with thinning. Palahí et al. (2008) used different modeling approaches and predictors when developing an individual-tree-based simulator for growth of even- and uneven-aged *Pinus brutia* growing in pure or mixed stands in Greece. The model for even-aged forests consists of site index, diameter growth, height growth, and mortality components. For uneven-aged forests, a past growth index is used instead of site index.

The focus of this chapter, however, is on modeling methods that have been developed specifically for application to uneven-aged forests. The basic frameworks of whole-stand, diameter-distribution, size-class, and individual-tree modeling structures have been adapted to stand types that are age indeterminate. Matrix models,

H.E. Burkhart and M. Tomé, *Modeling Forest Trees and Stands*,
DOI 10.1007/978-90-481-3170-9_15,

which are a type of size-class model, are well suited to modeling uneven-aged forests; hence they are described in this chapter. An overview of growth and yield models for uneven-aged stands can be found in the review paper by Peng (2000).

The most common approaches to modeling uneven-aged forests include: (i) whole-stand models based on elapsed time from specified initial conditions, (ii) models that produce size class information from continuous distributions, (iii) size-class models that rely on discrete distributions (stand table projection and matrix models), and (iv) models that aggregate individual tree values for stand-level estimates.

15.2 Whole-Stand Models

15.2.1 Equations Based on Elapsed Time

Moser and Hall (1969) developed a volume growth-rate model for uneven-aged stands of mixed northern hardwoods by fitting a basal area growth equation and a stand volume function. Murphy and Farrar (1982a) applied the Moser-Hall approach to develop models for basal area and volume projection of uneven-aged loblolly-shortleaf pine stands. Basal area growth rate was described by Richards (1959) generalized form of Bertalanffy's differential equation, namely

$$dG/dt = \mathrm{n}G^{\mathrm{m}} - \mathrm{k}G \tag{15.1}$$

where

dG/dt = periodic annual basal area growth
G = average basal area during the period
n, m, k = parameters to be estimated

The stand volume function fitted was:

$$V = \mathrm{b}_0 G^{\mathrm{b}_1} \tag{15.2}$$

where

V = volume per unit area
G = basal area per unit area
b_0, b_1 = parameters to be estimated

Basal area at the end of the growth period can be estimated using the integrated form of (15.1):

$$G_t = \left\{\mathrm{n/k} - \left[\mathrm{n/k} - G_0^{(1-\mathrm{m})}\right] e^{-(1-\mathrm{m})\mathrm{k}t}\right\}^{1/(1-\mathrm{m})} \tag{15.3}$$

where G_t is the projected basal area, G_0 is the initial basal area, and t is elapsed time. Initial basal area, G_0, can be substituted in Eq. 15.2 to predict stand volume at the start of the period, and projected basal area, G_t, can be predicted from Eq. 15.3 and then used in Eq. 15.2 to estimate projected stand volume.

The use of elapsed time from specified initial conditions, as pioneered by Moser and Hall (1969), has been a highly useful construct in modeling uneven-aged forests. Moser (1972) further developed the approach by modeling the components of net growth for number of trees and basal area by a system of first-order, ordinary differential equations for an uneven-aged forest stand. An additional expansion of the approach to include the distribution of stand growth by both size classes and individual growth components for all-aged forest stands was presented by Moser (1974).

15.2.2 Whole-Stand Models with Stand-Table Information

Lynch and Moser (1986) developed a growth model for two species groups in a mixed stand. Future stand conditions for each species group are determined by integrating a system of differential equations which relate rates of change in per unit area values of basal area, sum of diameters, and numbers of trees to current amounts of basal area, sum of diameters, and numbers of trees. Solutions to the system at future times include estimates of basal area, sum of diameters, and number of trees for each species group. Parameter recovery (Chap. 12) was used with the Weibull distribution to develop predicted stand tables.

Building on the methods of Moser (1974) for using a system of differential equations to project uneven-aged stand dynamics, Lynch and Moser (1986) proposed the following system for stands with two species groups (softwood and hardwood):

$$dN_{k1}/dt = f_{k1}(N_1, N_2, S_1, S_2, G_1, G_2) \tag{15.4}$$

$$dS_{k1}/dt = \beta_{k1}(dN_{k1}/dt) \tag{15.5}$$

$$dG_{k1}/dt = \alpha_{k1}(dN_{k1}/dt) \tag{15.6}$$

$$dN_{k2}/dt = f_{k2}(N_1, N_2, S_1, S_2, G_1, G_2) \tag{15.7}$$

$$dS_{k2}/dt = g_{k2}(S_k N_k dN_{k2}/dt) \tag{15.8}$$

$$dG_{k2}/dt = z_{k2}(G_k N_k dN_{k2}/dt) \tag{15.9}$$

$$dS_{k3}/dt = g_{k3}(N_1, N_2, S_1, S_2, G_1, G_2) \tag{15.10}$$

$$dG_{k3}/dt = z_{k3}(N_1, N_2, S_1, S_2, G_1, G_2) \tag{15.11}$$

$$dN_k/dt = dN_{k1}/dt - dN_{k2}/dt \tag{15.12}$$

$$dS_k/dt = dS_{k1}/dt - dS_{k2}/dt + dS_{k3}/dt \tag{15.13}$$

$$dG_k/dt = dG_{k1}/dt - dG_{k2}/dt + dG_{k3}/dt \tag{15.14}$$

where

$k = 1$ if species group is softwood and 2 if species group is hardwood,
$i = 1$ if ingrowth, 2 if mortality, 3 if survivor growth,
N_{ki} = number of species k trees of type i,
S_{ki} = sum of diameters on species k trees of type i,
G_{ki} = basal area on species k trees of type i,
N_k = number of species k trees,
S_k = sum of diameters on species k trees,
G_k = basal area on species k trees.

The system of Eqs. 15.4, 15.5, 15.6, 15.7, 15.8, 15.9, 15.10, 15.11, 15.12, 15.13, and 15.14 is designed to relate stand conditions at a particular time (initial conditions) to rates of change in these conditions. Ingrowth, mortality, and survivor growth rates are used to determine net change rates. Since the equations in the system do not depend explicitly on age, this system could be used to model even-aged as well as uneven-aged stands.

A solution to the system of Eqs. 15.4, 15.5, 15.6, 15.7, 15.8, 15.9, 15.10, 15.11, 15.12, 15.13, and 15.14 can be used to obtain the parameters of a two parameter probability density function representing the diameter distribution for each species group. Basal area and sum of diameters per unit area are related to diameter distribution parameters through the following equations:

$$S_k = N_k\left[E(d\,|b_k, c_k)\right] \tag{15.15}$$

$$G_k = cN_k\left[E(d^2\,|b_k, c_k)\right] \tag{15.16}$$

where

d = individual tree diameter at breast height, (b_k, c_k) are diameter distribution parameters,
$E(d\,|b_k, c_k)$ = the expected value of individual tree diameter at breast height,
$E(d^2\,|b_k, c_k)$ = the expected value of the square of individual tree diameter at breast height.

If predictions for S_k and G_k are obtained by solving Eqs. 15.4, 15.5, 15.6, 15.7, 15.8, 15.9, 15.10, 15.11, 15.12, 15.13, and 15.14 at a particular time, Eqs. 15.15 and 15.16 can be solved simultaneously for the diameter distribution parameters. These parameters can be used to obtain predicted stand tables. Unit area values for number of trees, sum of diameters, and basal area for each species group are used as initial conditions for the system of Eqs. 15.4, 15.5, 15.6, 15.7, 15.8, 15.9, 15.10, 15.11, 15.12, 15.13, and 15.14. Integration of the system to some subsequent time provides predicted number of trees, basal area, and sum of diameters for each species group that can be used in Eqs. 15.15 and 15.16 to obtain predicted diameter distribution parameters, and subsequently, stand tables. In their application, Lynch and Moser (1986) obtained predicted diameter distribution parameter for each of two species groups by using the left truncated Weibull distribution. The authors stated that their approach may be thought of as a method of predicting future plot attributes from their present values with the added capability of characterizing diameter distributions from the means of certain attributes.

Khatouri and Dennis (1990) used the general structure proposed by Lynch and Moser (1986) to develop a model to predict growth and yield in uneven-aged cedar stands in the Ajdir Forest in Morocco. The model predicts yield from differential equations and diameter distributions. Rates of change of ingrowth, mortality, and survivor growth are related to stand conditions. Numerical integration gives growth and yield projections through time. Predicted stand tables are produced by estimating Weibull distribution parameters from the results of the system of stand-level equations.

15.3 Diameter Distribution Approach

Diameter distributions in uneven-aged forests typically follow a reverse J-shaped curve in which the numbers of trees decrease as diameter increases (Fig. 15.1). The exponential function (Moser 1976) and the doubly truncated exponential probability

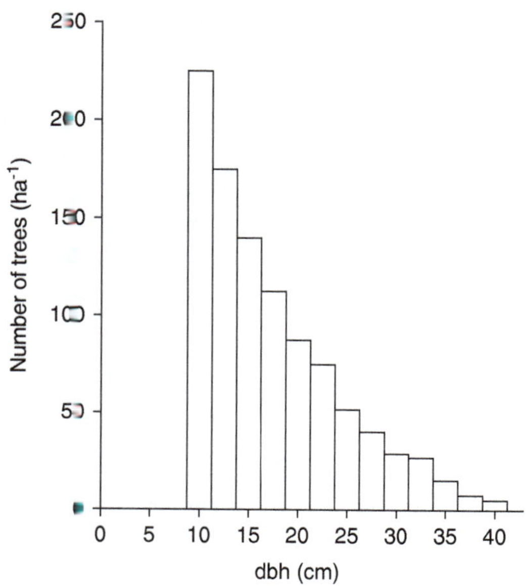

Fig. 15.1 Typical *dbh* distribution for regular, uneven-aged stands (Adapted from Avery and Burkhart 2002)

density function (Murphy and Farrar 1982b) have been proposed to specify a diameter distribution for a given stand density level. The Weibull function has also been found highly useful for describing diameter distributions in uneven-aged stands because it assumes a reverse J-shape when the shape-parameter, c, is less than 1 (for $c = 1$, the negative exponential results).

While stand-table values are sometimes produced as auxiliary information to whole-stand predictions, the continuous distribution approach focuses on development of the stand table itself for yield estimation. Hyink and Moser (1983) generalized the diameter distribution approach (Chap. 12) for projecting yields and stand structure in even-aged stand to include stands of indeterminate age. Attention was focused on developing parameter prediction models (PPM) and parameter recovery models (PRM) for both even- and uneven-aged situations. Parameter prediction models consist of forecasting future number of trees and the associated values of the parameters of a probability density function (pdf) describing the diameter distribution of those trees. Parameter recovery models use the future values of particular yield attributes directly to compute implied parameter values of the pdf characterizing the underlying diameter distribution.

Hyink and Moser (1983) developed methodology for calculating yield attributes as a function of tree diameter at any specified time, t, by generalizing the work of Strub and Burkhart (1975) as:

$$y_{ij} = N_t \int_{D_{lj}}^{D_{uj}} g_i(x) f_x(x;\theta_t) dx \tag{15.17}$$

where

$f_x(x;\theta_t)$ = a probability density function (pdf), describing the diameter distribution,
x = tree diameter at breast height (*dbh*),
θ_t = a vector of length m, containing the parameters of the pdf specifying the diameter distribution at time t,
N_t = total number of trees per unit area at time, t, distributed by $f_x(x;\theta_t)$,
$g_i(x)$ = any yield attribute that is a function of tree *dbh*, indexed by i
D_{uj} = upper *dbh* limit of *j*th size class,
D_{lj} = lower *dbh* limit of the *j*th size class, and
y_{ij} = per unit area value of the yield attribute given by the *i*th function of tree *dbh* in the *j*th size class, $(D_j : D_{lj} < D_j \leq D_{uj})$.

If N is replaced by

$$1/\int_{D_{lj}}^{D_{uj}} f_x(x;\theta_t) dx$$

in Eq. 15.17, the y_{ij}'s represent mean tree values.

Assuming that the diameter distribution is adequately characterized by $f_x(x;\theta_t)$, use of Eq. 15.17 with any given $g_i(x)$ requires only knowledge of N_t and θ_t. In the case of applications of the PPM to even-aged stands, number of trees surviving at age t (years since stand establishment) is specified and the parameters of the pdf used to describe frequencies by *dbh* are predicted as a function of stand age, site index (or height of the dominant stand), and trees surviving. Hyink and Moser (1983) extended the PPM to uneven-aged stands. The three-parameter Weibull function with location parameter (a) fixed was chosen as the diameter distribution model. Then the average annual change in numbers of trees (*dN*/*dN*) and in the Weibull parameters ($d\theta/dt$) were approximated from remeasurement data. Linear and nonlinear least squares techniques were used to estimate coefficients for the differential equations required by the model.

Hyink and Moser (1983) also developed a generalized parameter recovery model by noting that when omitting the time subscript, *t*, Eq. 15.17 may be rewritten as

$$h_{ij}(N_j,\theta) = N\int_{D_{lj}}^{D_{uj}} f_x(x;\theta)dx \frac{\int_{D_{lj}}^{D_{uj}} g_x(x) f_x(x;\theta)dx}{\int_{D_{lj}}^{D_{uj}} f_x(x;\theta)dx} = y_{ij} \qquad (15.18)$$

Let $N_j = N\int_{D_{lj}}^{D_{uj}} f_x(x;\theta)dx$ for any $h_{ij}(N_j,\theta)$ i.e., N_j is the number of stems per unit area associated with the corresponding y_{ij}. If the *j*th class interval is coincident with the full range of $f_x(x;\theta)dx$ then $N_j = N$. The reminder of Eq. 15.18 is a general formula for the expected value of any $g_i(x)$ given $f_x(x;\theta)$ for any diameter class interval, D_j.

Given N_j, y_{ij} and $k = 1$ values of θ for a particular $h_{ij}(N_j,\theta)$, one may solve for the *k*th value of θ,

$$\theta_k = F_k(y_{ij}, N_j, \theta)$$

If one further assumes that a distinct (y_{ij}, N_j) pair is known for each of the $kh_{ij}(N_j,\theta)$'s such that no $h_{ij}(N_j,\theta)$ is a linear combination of any other, a system of equations may be formed, such as,

$$\theta_1 = F_1\left[(y_{ij}, N_j)_1, \theta\right]$$
$$\theta_2 = F_2\left[(y_{ij}, N_j)_2, \theta\right]$$
$$\vdots$$
$$\theta_k = F_k\left[(y_{ij}, N_j)_k, \theta\right]$$

that results in a unique solution for θ, given $[(y_{ij}, N_j)_m, m = 1, k]$. By assuming that the diameter distribution underlying the $[(y_{ij}, N_j)_m, m = 1, k]$ is adequately characterized by $f_x(x; \theta)$, the system represents a general model for the "recovery" of diameter distribution parameters (Sect. 12.1.3) from stand-average attributes.

A distribution-based growth and yield model for uneven-aged loblolly-shortleaf pine stands was developed by Murphy and Farrar (1988). Stand-level equations for merchantable basal area, sawtimber basal area, quadratic mean diameter, and maximum tree diameter were derived first. Parameters for the doubly truncated Weibull distribution were obtained by the "parameter recovery" technique by using estimates of maximum diameter, quadratic mean diameter, and the sawtimber-merchantable basal area ratio. Compatibility for basal area estimates was maintained between the stand-level estimates and those produced from the Weibull distribution.

15.4 Size-Class Models

15.4.1 Stand-Table Projection Equations

In lieu of continuous distributions, discrete size classes can be projected using a generalized stand table projection procedure. (Sect. 13.2 contains a general description of the stand table projection concept.) Ek (1974) presented equations to predict periodic ingrowth, mortality, and survivor growth, by diameter classes, in northern hardwood stands. Net 5-year change (i.e., the change from t_0, the time of initial measurement, to t_1, the time of final measurement, where t_1 is 5 years after t_0) in the number of trees in a diameter class Δn_c was defined as

$$\Delta n_c = \text{stand ingrowth} - \text{mortality} - \text{upgrowth} + \text{ingrowth}$$

The component equations of this generalized stand-table projection model are

$$n_{is} = \mathrm{b}_0 N^{\mathrm{b}_1} e^{(\mathrm{b}_2 G^{\mathrm{b}_3} N^{-1})} \quad (15.19)$$

$$n_m = \mathrm{b}_0[(G/N)/(g_c/n_c)]^{\mathrm{b}_1} \quad (15.20)$$

$$n_u = \mathrm{b}_0 n_c^{\mathrm{b}_1} S[(g_c/n_c)/(G/N)]^{\mathrm{b}_2} e^{\mathrm{b}_3 G} \quad (15.21)$$

where

n_{is} = stand ingrowth (merchantable trees at t_1 that were nonmerchantable at t_0)
n_m = mortality (trees present in diameter class at t_0 but dead at t_1)
n_u = upgrowth (trees present in a diameter class at t_0 but growing into next larger diameter class at t_1)
n_i = ingrowth (upgrowth from next lower measured diameter class)

N = number of trees per unit area in the stand
G = stand basal area per unit area
n_c = number of trees in a specified diameter class
g_c = basal area of trees in a specified diameter class
S = site index (dominant height at index age)

Equations 15.19, 15.20, and 15.21 can be used to project observed or hypothetical stand tables which can then be converted to stock tables. Because of the growth rates involved for northern hardwood stands, ingrowth n_{is} would be added only to the smallest merchantable class (15-cm *dbh* class) when 5-cm groupings are used. Diameter-class ingrowth, n_i, is equal to the upgrowth n_u computed from the next lower diameter class (e.g., upgrowth computed for 20-cm class is ingrowth to 25-cm class). For projections longer than 5 years, a new stand table must be constructed at 5-year intervals. The new stand table is prepared by adding ingrowth trees from smaller size classes to the number of survivors in a class that did not move to the next larger class. Basal areas can be computed by using the class midpoint diameters. The new stand table then serves as the initial conditions for the next 5-year projection. Volumes can be computed by applying an appropriate size-class volume equation, such as

$$V_m = b_1 g_c^{b_2} S^{b_3} G^{b_4}$$

where V_m is merchantable volume in a specified diameter class, the b_i's are constants to be estimated from data, and the other variables remain as previously defined.

15.4.2 Matrix Model Approach

The stand table projection (STP) method can be formalized and implemented by computation of a transition matrix. In the simplest form of transition matrix, a sequence of events, each with a finite number of possible outcomes, is represented by probabilities. It is assumed that the probability of transition from a given state to any given outcome is dependent only on the value of the preceding state and that the transition probabilities are stationary (i.e., do not change) over time. The outcomes, called states, are finite, mutually exclusive, and exhaustive. This special case of projection by use of the type of transition matrix just described is called a Markov chain, and the assumptions on which it is based coincide with those of classic stand table projection methodology (Sect. 13.2) that has been widely applied in forestry.

The difference between STP and the Markov chain approach is that instead of collecting increment core data from the forest of interest to derive diameter movement data, observations from growth plots in a timber type of interest are used to estimate transition probabilities. As with the inventory-based STP approach, information on ingrowth and mortality is needed to complete a Markov chain stand projection system.

Three variations, which differ in the underlying assumptions invoked when developing the transition matrix, will be recognized: Markov chains, Usher matrices, and generalized transition matrices.

15.4.2.1 Markov Chains

When constructing a Markov model for stand table projection one assumes that at any time the system can be represented by a finite number of states. Further assumptions are that the probability of movement to a given future state depends only upon the current state and transition probabilities do not change over time. These assumptions are restrictive but are tenable for short-term projection of uneven-aged forests, especially if a regular cutting cycle is imposed to maintain a relatively constant forest structure.

Trees of any given initial diameter in the stand table must remain in the same class, grow into another class, die, or be cut. The Markov chain in forestry applications has two types of states, transient and absorbing. Transient states are finite in duration; diameter classes are transient states because all trees must eventually grow into a larger class, die, or be harvested. The mortality and harvest states are considered absorbing states since trees that enter these states can not leave them. Probabilities of movement can be expressed as a matrix (**M**) to predict change during a single time interval:

$$\mathbf{V}_1 = \mathbf{M}\mathbf{V}_0$$

The probabilities for multiple time steps can be captured as

$$\mathbf{V}_n = \mathbf{M}^n\mathbf{V}_0$$

where **M** represents the Markov matrix containing the probabilities of movement, n represents the number of steps, and $\mathbf{V}_0$ and $\mathbf{V}_n$ are vectors describing the initial and final states respectively.

Two primary assumptions are required. First, the *markov assumption* requires that the probability of any event occurring depends only on the initial state (in the case of STP, this implies that the probability of remaining, moving up, dieing or becoming part of a harvest cut depends only on tree *dbh*). Second, the *stationarity assumption* requires that the movement probabilities do not change over time.

Bruner and Moser (1973) used data from permanent growth plots in uneven-aged mixed hardwood stands in Winconsin, USA, to apply a Markov chain approach to diameter distribution projection. The transition matrix involved 25 states: 23 *dbh* classes and one class each for mortality and harvest. This study showed that number of survivor trees was accurately predicted, but predictions of diameter distributions, number of mortality trees, and number of harvested trees were less accurate. Discrepancies between observed and predicted values were larger for projections beyond one period. The authors conjectured that the loss of predictive ability over longer time periods may be attributable to failure to satisfy the stationarity assumption. An examination of 19 years of remeasurement data showed that the transition probabilities were fairly constant for diameter classes with a high initial frequency of trees. This was not true, however, for diameter classes with a small initial number of trees which suggests that accurate predictions require good estimates of transition probabilities, which, in turn, depend on a sufficiently large number of trees in the sample data.

15.4.2.2 Usher Matrices

Leslie (1945, 1948) pioneered the use of matrix models for studying animal populations where the animals are grouped into age classes. Usher (1966, 1969) adapted the Leslie approach for applications to selection-managed forests by using tree size rather than age classes. Tree diameters are more readily obtained and more generally available for uneven-aged forests than ages. In addition, the Usher approach involves choosing a time interval and class width so that during a projection period trees can grow no more than one class; these restrictions allow for a substantial reduction in the number of parameters to be estimated.

Markov models contain only transition probabilities, but recruitment can be predicted by employing non-zero values in the top row of the matrix. These values, termed *fecundity*, reflect the number of offspring for each individual in the corresponding cell of the state vector. Fecundity values in the matrix allow recruitment to vary according to the presence of trees in various size classes. Thus, the matrix may be reduced to four vectors: growth (trees move into the next class or remain in their present class), fecundity (recruitment), mortality, and harvest.

Most forestry applications of transition matrices for stand table projection have been built on the Usher approach as opposed to a strictly Markov process because of the increased flexibility and efficiency that the Usher model offers.

15.4.2.3 Generalized Matrices

The Buongiorno and Michie Model

Buongiorno and Michie (1980) modified the Usher model approach to construct a model for uneven-aged forest management which would describe growth more accurately. Modifications included making ingrowth only partially dependent on harvest and allowing for ingrowth to respond to changes in stand density and diameter distribution. Consequently, stands could grow at an increasing, constant, or decreasing rate, depending upon their attributes. Because the Buongiorno and Michie construct has been widely adopted and applied for transition matrix modeling of uneven-aged forests, it will be described in some detail here.

Trees in a stand are divided into a finite number, n, of diameter classes. The expected number of living trees within each size class at a specific time, t, is denoted by $y_{1t}, y_{2t}, \ldots, y_{nt}$. Therefore the entire stand of living trees is represented at time t by the column vector

$$\mathrm{y}_t = [y_{it}]\, i = 1, \ldots, n$$

During a specific growth period θ the trees in a given diameter class i may remain in the same class or advance to a larger size class. They may also die during the interval θ, or they may be harvested. The number of trees harvested from diameter class i during the interval θ is denoted by h_{it}, and is represented by the column vector

$$\mathrm{h}_t = [h_{it}]\, i = 1, \ldots, n.$$

Furthermore, let a_i denote the probability that a live tree in size class i at time t which is not harvested during the interval θ will be alive and in the same size class at time $t + \theta$. Also, let b_i denote the probability that a live tree in size class $i-1$ at time t which is not harvested during the interval θ will be alive and in size class i at time $t + \theta$. Finally, I_t designates the expected ingrowth, i.e., the expected number of trees entering the smallest size class during the interval θ. The projected stand at time $t + \theta$ may then be entirely determined from the attributes at time t, the harvest during θ, and the ingrowth during θ by the n equations:

$$\begin{aligned} y_{1t+\theta} &= I_t + a_1(y_{1t} - h_{1t}) \\ y_{2t+\theta} &= \mathrm{b}_2(y_{1t} - h_{1t}) + a_2(y_{2t} - h_{2t}) \\ &\ldots \\ y_{nt+\theta} &= \mathrm{b}_n(y_{n-1t} - h_{n-1t}) + a_n(y_{nt} - h_{nt}) \end{aligned} \tag{15.22}$$

The ingrowth function incorporated by Buongiorno and Michie assumes that abundance of ingrowth is inversely related to stand basal area and that, for a given basal area, is directly related to number of trees (i.e., ingrowth is favored by stands of small trees). These considerations led to an expected ingrowth function of the form:

$$I_t = \beta_0 + \beta_1 \sum_{i=1}^{n} g_i(y_{it} - h_{it}) + \beta_2 \sum_{i=1}^{n} (y_{it} - h_{it}) \tag{15.23}$$

with $I_t \geq 0$. Where g_i is the basal area of the tree of average diameter in size class i, while β_0, β_1 and β_2 are constraints which are expected to be, respectively, positive, negative, and positive. Using (15.23) as the expression for I_t leads to a new expression for the number of trees in the smallest size class as a function of the number of trees in all size classes and of the harvest.

$$y_{1t+\theta} = \beta_0 f_1(y_{1t} - h_{1t}) + \cdots + f_n(y_{nt} - h_{nt}) \tag{15.24}$$

where

$$f_1 = a_1 + \beta_1 g_1 + \beta_2$$
$$f_i = \beta_1 g_i + \beta_2 \ \text{ for } i > 1$$

The final model takes then the form

$$y_{t+\theta} = \mathbf{G}(y_t - h_t) + \mathrm{c} \tag{15.25}$$

where **G** and **c** are respectively a matrix and a column of constant coefficients:

$$\mathbf{G} = \begin{bmatrix} f_1 & f_2 & \cdots & f_n \\ b_2 & a_2 & & \\ & b_3 & a_3 & \\ & & \cdots & \\ & & b_n & a_n \end{bmatrix}, \mathbf{c} = \begin{bmatrix} \beta_0 \\ 0 \\ 0 \\ \vdots \\ 0 \end{bmatrix} \tag{15.26}$$

The differences between model (15.25) and Usher's model were enumerated by Buongiorno and Michie. First, in (15.25) harvest is represented by a variable vector, h_t, instead of being represented by the coefficients of matrix **G**. This allows for the harvest strategy to vary, instead of being always a specific fraction of the stock in each diameter class. Second, in (15.25), part of the mortality is lost, even if the stand is harvested. Third, and most important, in model (15.25), ingrowth is set to a constant β_0, modified by coefficients in the first row of **G** which reflect the changes in ingrowth due to changes in stand structure.

Data from permanent growth plots in northern hardwoods stand in Wisconsin and Michgan, USA, were used to estimate the elements of **G** and **c** denoted in (15.26). Trees were grouped into seven diameter classes. Estimation of the probabilities a and b in (15.22) consisted of simple proportions because the data set used indicated for each plot the number of trees in each diameter class which between two successive inventories either remained in the same diameter class, moved up one class, were harvested, or were lost to mortality. The resulting matrix of a and b coefficients, as identified in (15.22), was

$$\begin{bmatrix} 0.72 & & & & & & \\ 0.23 & 0.70 & & & & & \\ & 0.26 & 0.67 & & & & \\ & & 0.30 & 0.65 & & & \\ & & & 0.30 & 0.66 & & \\ & & & & 0.30 & 0.81 & \\ & & & & & 0.19 & 0.86 \end{bmatrix} \tag{15.27}$$

The ingrowth Eq. 15.23 was estimated by linear regression from data on ingrowth, number of trees in each size class and harvested trees, with the following results:

$$I_t = 109.0 - 9.65 \sum_{i=1}^{7} g_i (y_{it} - h_{it}) + 0.27 \sum_{i=1}^{7} (y_{it} - h_{it}) \tag{15.28}$$

Although the coefficient of determination was small (0.15), all coefficients were highly significant.

While ingrowth appears to be a highly random process there is a systematic and predictable feedback of stand conditions on it, which may be altered by harvest. This feedback process is represented by the first row in the matrix **G** and the vector **c** which can be computed from (15.27) and (15.28) using Eq. 15.24. The resulting estimated matrices are:

$$\mathbf{G} = \begin{bmatrix} 0.81 & -0.043 & -0.22 & -0.43 & -0.69 & -0.98 & -1.3 \\ 0.23 & 0.70 & 0 & 0 & 0 & 0 & 0 \\ 0 & 0.26 & 0.67 & 0 & 0 & 0 & 0 \\ 0 & 0 & 0.30 & 0.65 & 0 & 0 & 0 \\ 0 & 0 & 0 & 0.30 & 0.66 & 0 & 0 \\ 0 & 0 & 0 & 0 & 0.30 & 0.81 & 0 \\ 0 & 0 & 0 & 0 & 0 & 0.19 & 0.86 \end{bmatrix}, \quad \mathbf{c} = \begin{bmatrix} 109.0 \\ 0 \\ 0 \\ 0 \\ 0 \\ 0 \\ 0 \end{bmatrix} \tag{15.29}$$

In matrix **G** in (15.29) the top row represents fecundity, the main diagonal indicates trees remaining in a diameter class, and the second diagonal shows upgrowth. In vector **c**, 109.0 represents the average ingrowth observed (trees per ha per 5 years) for the data used.

The bi-diagonal Usher matrix approach of Buongiorno and Michie (1980) has been applied to a variety of forest types, including oak-hickory stands in the USA (Michie and McCandless 1986), natural stands of dipterocarp trees in Indonesia (Mendoza and Setyarso 1986), a tropical rain forest of Nigeria (Osho 1991), and a subtropical forest in southern Brazil (Spathelf and Durlo 2001).

Estimating Transition Probabilities

When applying the matrix modeling approach, analysts must adopt a method for estimating the transition probabilities. Michie and Buongiorno (1984) evaluated four methods for computing the coefficients of their matrix model (Buongiorno and Michie 1980). Applying ordinary least squares (OLS) to each equation and use of seemingly unrelated regression (SUR) techniques produced biased results. Constraining transition probabilities and mortality rates for a particular diameter class to add to unity alleviated the bias found in the OLS and SUR alternatives. However, the best method consisted of using the individual tree data directly to determine the proportion of trees which stayed in the diameter class they were in at the beginning of the growth period and the proportion which grew into the next size class.

Lowell and Mitchell (1987) proposed that logistic regression be used to simultaneously estimate growth and mortality. A logistic model to predict the probability that a tree will attain a future diameter class can be specified. Given the initial diameter distribution of a forest stand, the future diameter distribution can be projected by estimating the proportion of the stems in each diameter class which attain a specified future diameter and the proportion which fails to achieve at least zero growth (i.e. mortality). The authors used permanent plot data for an oak-hickory forest in Missouri, USA, to calibrate such a logistic model. Validation results indicated that the model performs satisfactorily (i.e. estimates are unbiased) for individual trees over a 5-year prediction period, and for stand characteristics over 5-, 10-, 15-, and 20-year predictions, although precision declines as the prediction period lengthens.

Projecting for Fractional Time Intervals

A drawback of the matrix modeling approach is that projections can only be made in integer multiples of the measurement interval for the plot data. However, Harrison and Michie (1985) presented a matrix factorization approach whereby a one-year matrix ($\mathbf{M}_1$) may be estimated from a n-year matrix ($\mathbf{M}_n$) such that $\mathbf{M}_1^n \approx \mathbf{M}_n$. In a case example, the approximated 1-year matrix reasonably replicated the original matrix model (of the bi-diagonal, Usher matrix form of Buongiorno and Michie 1980) in terms of numbers of trees and basal area growth.

Relating Transition Probabilities to Stand Density

The stationarity assumption is not always tenable when developing matrices of transition probabilities. In an application of the bi-diagonal Usher matrix modeling approach for the forests of the Jura Mountains in France, Buongiorno et al. (1995) found that the probabilities of transition between size classes were affected strongly by the basal area of the stand. Consequently they developed an upgrowth matrix which depends on the stand basal area after the cut. Virgilietti and Buongiorno (1997) noted that the probabilities of transition (trees moving from one size to the next) was affected by stand density and tree size when modeling forests in the Italian Alps. Regression equations fitted to data showed a negative effect of basal area on the upgrowth rate of spruce and fir and a significant negative relationship between transition probabilities and the size of larches. However, the coefficients of determination (R^2) for the fitted equations were low, and in the final model the transition probabilities were set at their mean for spruce, fir and beech, and at their mean for a given diameter for larch.

FIBER – A Two-Stage Model

Solomon et al. (1986) developed a model called FIBER based on the matrix approach with dynamic transition probabilities for mixed species stands with frequent harvests in the New England region of the United States. Periodic remeasurements were available for trees classified by species and diameter class. A diameter distribution for each species at time t was written in matrix notation as a column vector:

$$\mathbf{y}_t = (y_{1t}, y_{2t}, \ldots, y_{nt})'$$

where y_{it} denotes the number of trees prior to harvest in the ith diameter class at time t.

The number of trees that were harvested from the ith diameter class at time t was denoted by h_{it}; therefore, the set of harvests at time t can be written as:

$$\mathbf{h}_t = (h_{1t}, h_{2t}, \ldots, h_{nt})'$$

To project the growth and mortality of trees within a stand, a_{it} denoted the proportion of survivor trees in the ith diameter class at time t that remain in diameter class i at time $t+k$. The proportion of survivor trees in the ith diameter class at time t that are in the $(i+1)$th diameter class at time $t+k$ was denoted by b_{it}, and c_{it} denoted the proportion of survivor trees in the ith diameter class at time t in the $(i+2)$th diameter class at time $t+k$. The proportion of trees in diameter class i that die by time $t+k$ was denoted by m_{it}. Ingrowth, the number of live trees that grow into the smallest diameter class during the time interval from t to $t+k$, was denoted by I_{t+k}. The ingrowth was expressed as a function of residual basal area (G_R) at time t by the following form:

$$I_{t+k} = \alpha_0 + \alpha_1 G_{Rt} = \alpha_0 + \alpha_1 \sum_{i=1}^{n} g_i (y_{it} - h_{it}) \tag{15.30}$$

where g_i is the basal area of the tree of average diameter in diameter class i and α_0 and α_1 are regression coefficeients to be determined.

The model FIBER was developed as a two-stage model. Tree growth and mortality were related to tree size and density, both before and after thinning, using information from growth studies in New England. Therefore, the proportion of trees remaining in a diameter class (a_{it}), the proportions of trees growing out of a diameter class (b_{it}, c_{it}) and the mortality rate (m_{it}) are also related to tree size and stand density. These proportions are assumed to have a multinomial distribution. The first stage consists of a set of four simultaneous linear regression equations that are used to estimate the proportions from independent stand variables. Specifically, the equations in FIBER are:

$$a_{it} = \beta_{01} + \beta_{11} G_{It} + \beta_{21} G_{Rt} + \beta_{31} d_i$$

$$b_{it} = \beta_{02} + \beta_{12} G_{It} + \beta_{22} G_{Rt} + \beta_{32} d_i$$

$$c_{it} = \beta_{03} + \beta_{13} G_{It} + \beta_{23} G_{Rt} + \beta_{33} d_i$$

$$m_{it} = \beta_{04} + \beta_{14} G_{It} + \beta_{24} G_{Rt} + \beta_{34} d_i \tag{15.31}$$

where β_{jk} is the regression coefficient to be estimated, G_{It} is the stand basal area prior to harvest at time t, G_{Rt} is the stand basal area after harvest at time t, and d_i is the midpoint of the diameter class i.

In the second stage, the diameter distribution of the stand is projected from time t to time $t+k$. The number of trees in the first or smallest class at time $t+k$ is the number of trees that stay in the class from time t plus the ingrowth. Therefore,

$$y_{1,t+k} = a_{1t}(y_{1t} - h_{1t}) + I_{t-k} = \alpha_0 + (a_{1t} + \alpha_1 g_1)(y_{1t} - h_{1t}) + \alpha_1 \sum_{i=2}^{n} g_i (y_{it} - h_{it})$$

The number of trees in the second class at time $t+k$ is the number of trees that moved from the first class at time t into the second class at time $t+k$ plus the number of trees that stayed in the second class:

$$y_{2,t+k} = b_{1t}(y_{1t} - h_{1t}) + a_{2t}(y_{2t} - h_{2t})$$

Similarly, for $i \geq 3$,

$$y_{i,t+k} = c_{i-2,t}(y_{i-2,t} - h_{i-2,t}) + b_{i-1,t}(y_{i-1,t} - h_{i-1,t}) + a_{it}(y_{it} - h_{it})$$

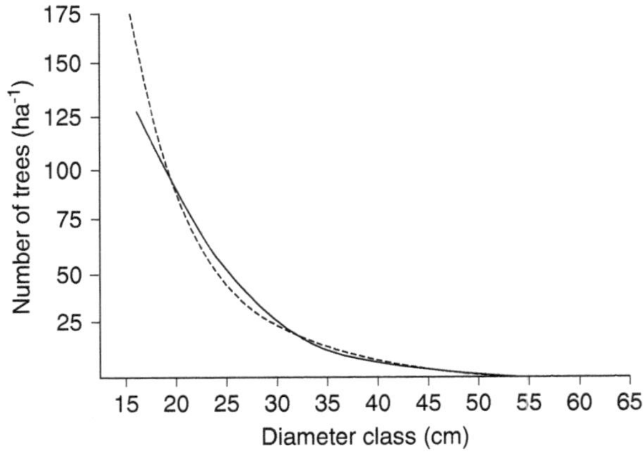

Fig. 15.2 Actual (-) and predicted (—) average diameter distributions after 15 years for softwood stands managed at a residual density of 27.5 m^2/ha (From Solomon et al. 1986)

Hence, the second stage of the model can be written in matrix form:

$$\begin{bmatrix} y_{1,t+k} \\ y_{2,t+k} \\ y_{3,t+k} \\ y_{4,t+k} \\ \cdot \\ \cdot \\ \cdot \\ y_{n,t+k} \end{bmatrix} = \begin{bmatrix} f_{1t} & f_{2t} & f_{3t} & f_{4t} & & \cdot & \cdot & f_{nt} \\ b_{1t} & a_{2t} & 0 & 0 & & \cdot & \cdot & 0 \\ c_{1t} & b_{2t} & a_{3t} & 0 & & \cdot & \cdot & 0 \\ 0 & c_{2t} & b_{3t} & a_{4t} & & \cdot & \cdot & 0 \\ \cdot & \cdot & \cdot & \cdot & & \cdot & \cdot & \cdot \\ \cdot & \cdot & \cdot & \cdot & & \cdot & \cdot & \cdot \\ \cdot & \cdot & \cdot & \cdot & & \cdot & \cdot & \cdot \\ 0 & 0 & 0 & 0 & \cdots & c_{n-2,t} & b_{n-1,t} & a_{nt} \end{bmatrix} \times \begin{bmatrix} y_{1t} - h_{1t} \\ y_{2t} - h_{2t} \\ y_{3t} - h_{3t} \\ y_{4t} - h_{4t} \\ \cdot \\ \cdot \\ \cdot \\ y_{nt} - h_{nt} \end{bmatrix} + \begin{bmatrix} \alpha_0 \\ 0 \\ 0 \\ 0 \\ \cdot \\ \cdot \\ \cdot \\ 0 \end{bmatrix} \tag{15.32}$$

where $f_{jt} = \begin{cases} a_{1t} + \alpha_1 g_1, & \text{when } j = 1 \\ \alpha_1 g_j, & \text{when } j = 2, 3, \ldots, n \end{cases}$

Using matrix notation (15.32) can be written more compactly as follows:

$$y_{t+k} = G_t(y_t - h_t) + A \tag{15.33}$$

Given the initial diameter distribution and parameter estimates, (15.31), the diameter distribution at any subsequent time can be calculated by successive application of (15.33). This model overcomes the stationarity assumption because new probabilities are computed prior to each projection step.

The FIBER model has been applied to a variety of species and management practices found in New England. Figure 15.2 shows actual and predicted average diameter distributions after 15 years for softwood stands managed at a residual density of 27.5 m^{2}ha^{-1}. Mengel and Roise (1990) used a similar approach to that of Solomon et al. (1986) to develop a diameter-class matrix model for bottomland hardwood stands in the southern United States.

Combinations with Other Modeling Approaches

Transition probabilities were related to stand density by Pukkala and Kolström (1988) when simulating the development of Norway spruce stands in Finland. An individual-tree growth model was used to generate stands with different densities. Next regression analysis was used to relate the transition probabilities calculated for different diameter classes from the growth model to stand density.

In another application that combined different modeling approaches, Picard et al. (2002) developed a matrix model for smaller trees and an individual-based component for trees above a threshold diameter. The combined model utilized the simplicity and efficiency of matrix modeling approaches where tree frequencies are high (as in small diameter classes in uneven-aged forests) and the strengths of individual-tree models for high-valued trees occurring at lesser frequencies (as in large diameter classes in uneven-aged forests). The individual-based component for trees above the threshold diameter was built from data so as to ensure continuity with the matrix component.

15.5 Individual-Tree Models

Because of their inherent flexibility in handling a wide variety of stand structures, individual tree models have been found useful for characterizing uneven-aged forests. Models that do not incorporate age and site index as predictors have been developed to accommodate age indeterminate stands and have been applied to even-aged stands as well. Some representative examples of such models are described in the previous chapter. Peng (2000) tabulated a chronology of selected growth and yield models for uneven-aged stands for the pre-1960s period to the time of publication of his review paper. In this section an example of a distance-dependent model that was developed to be applicable to forests of mixed species and uneven age is briefly described. The general structure of a distance-independent model formulated specifically for uneven-aged stands is then presented.

15.5.1 A Distance-Dependent Approach

The FOREST model published by Ek and Monserud (1974) is a spatially-explicit simulator for mixed species, uneven-aged stands (it can also be applied in even-aged monocultures).

Input for FOREST, a distance-dependent model, is a set of tree coordinates and associated tree characteristics (e.g., height, diameter, age, clear bole length, and species). Tree coordinates and tree characteristics may also be generated by the simulator. Each tree is then "grown" for a number of projection periods based on potential growth functions, modified by an index of competition. The competition index is based on the assessment of relative tree size, crowding, and shade tolerance. Mortality results when the probability of survival for a stem falls below a threshold value, which is dependent on the competitive status of a tree. In any "year" of the simulation, optional reproduction routines may be called to allow for regeneration by seed and sprout production of the overstory. Silvicultural treatments, including site alteration, cutting, or pruning operations, may also be implemented as the stand develops. Output of the model is in the form of periodic stand tables with yield and mortality for various products.

15.5.2 A Distance-Independent Model

Pukkala et al. (2009) developed a distance-independent diameter growth model, a height prediction model, a survival model, and an ingrowth model for uneven-sized forest stands in Finland. They described the stands as uneven-sized rather than uneven-aged because information on tree ages was not available. Regardless, the modeling principles and concepts applied fall within the context of growth and yield models for uneven-aged stands.

Data were available from two long-term experiments, a set of temporary sample plots, and sample plots from the third National Forest Inventory of Finland. Models were fitted for a range of growing sites and tree species. Additional detail, with coefficients for the fitted equations, is provided by Pukkala et al. (2009).

The potential predictors of diameter increment included tree size, competition (stand basal area and basal area in larger trees), forest site type, and temperature sum (sum of degree days over 5°C). The ages of trees and stands were not used as predictors since stand age is not defined for uneven-aged stands, and tree ages are seldom measured in inventories of uneven-sized stands. Therefore, it was assumed that, in a given stand, trees of a specified size and with a certain level of competition will grow similarly regardless of their age. Due to the omission of age, site index based on age and dominant height was not included.

The model for the 5-year diameter increment was of the following form:

$$\ln(\Delta d) = \mathrm{a}_1 + \mathrm{a}_2 G_{>spruce} + \mathrm{a}_3 G_{>other} + \mathrm{a}_4 \ln G + \mathrm{a}_5 \sqrt{d} + \mathrm{a}_6 d^2 + \mathrm{a}_7 MT + \mathrm{a}_8 VT + \mathrm{a}_9 CT + \mathrm{a}_{10} CIT + \mathrm{a}_{11} \ln TS \quad (15.34)$$

where Δd is the 5-year over-bark diameter increment (cm), d is the diameter at breast height (cm), G is the total basal area of trees larger than 5 cm in *dbh* ($\mathrm{m}^2\mathrm{ha}^{-1}$) and TS is the temperature sum (degree days); MT, VT, CT and CIT are indicator variables which indicate whether the site type is *Myrtillus* (MT), *Vaccinium* (VT), *Calluna* (CT) or *Cladonia* and poorer (CIT). In each stand, only one indicator variable equals one while the others are zeroes. If all indicator variables are zeroes, the model predicts the growth on *Oxalis-Myrtillus* type. $G_>$ is the basal area of trees larger than the subject tree (m^2ha^{-1}), which was computed separately for spruce ($G_{>spruce}$) and other tree species ($G_{>other}$).$G_>$ describes the competitive position of a tree within a stand.

In height modeling, variables were selected to describe tree size and site. The height model was a modification of the Hossfeld function:

$$h = \frac{a_1 + a_2MT^+ + a_3VT + a_4CT + a_5CIT}{1 + (b_1/d) + (b_2/d^2)} \tag{15.35}$$

where h is tree height (m) and d is the diameter at breast height (cm). MT^+ is an indicator variable, which equals 1 if the site type is *Myrtillus* or better (otherwise $MT^+ = 0$).

Survival was modeled with a logistic function which guarantees that the predicted probability of survival will be between zero and one. The survival model was as follows:

$$p_6 = \frac{1}{1 + \exp\left[-(a_1 + a_2\sqrt{d} + a_3 \ln G + a_4G_{>spruce} + a_5G_>)\right]} \tag{15.36}$$

where p_6 is the probability of survival for the following 6-year period. The 6-year survival probability can be converted into a 5-year probability as: $p_5 = p_6^{5/6}$.

The ingrowth limit was taken as 5 cm. Thus, the ingrowth model predicts the number of trees that will exceed the 5-cm *dbh* limit during the following 5-year period. The model form was:

$$\ln(N_I + 1) = a_1 + a_2\sqrt{G} + a_3 \ln G + a_4\sqrt{N_{spruce}} + a_5\sqrt{N_{other}} + a_6MT^- \tag{15.37}$$

where N_I is the number of ingrowth trees per hectare, N_{other} is the number of non-spruce trees ($dbh > 5$ cm) per hectare, and N_{spruce} is number of spruce trees per hectare ($dbh > 5$). MT^- is an indicator variable for sites which are *Myrtillus* type or poorer. G is the basal area of trees larger than 5 cm *dbh* (m^2ha^{-1}).

The model for the diameter of ingrowth predicts the mean diameter of the ingrowth trees at the end of the 5-year period. The model was as follows:

$$\ln(\bar{d}) = a_1 + a_2 \ln G + a_3MT + a_4VT^- \tag{15.38}$$

where $\bar{d}$ is the mean diameter of ingrowth trees at the end of the 5-year period. VT^- is an indicator variable which equals 1 if the site type is *Vaccinium* or poorer. The model predicts smaller diameters for stands with high basal area, i.e., increasing stand basal area decreases the diameter growth of ingrowth.

References

Avery TE, Burkhart HE (2002) Forest measurements, 5th edn. McGraw-Hill, New York

Bruner HD, Moser JW Jr (1973) A Markov chain approach to the prediction of diameter distributions in uneven-aged forest stands. Can J For Res 3:409–417

Buongiorno J, Michie BR (1980) A matrix model of uneven-aged forest management. For Sci 26:609–625

Buongiorno J, Peyron JL, Hollier F, Bruciamacchie M (1995) Growth and management of mixed-species, uneven-aged forests in the French Jura: implications for economic returns and tree diversity. For Sci 41:397–429

Crookston NL, Dixon GE (2005) The forest vegetation simulator: a review of its structure, content, and applications. Comp Electron Agric 49:60–80

Ek AR (1974) Nonlinear models for stand table projection in northern hardwood stands. Can J For Res 4:23–27

Ek AR, Monserud RA (1974) FOREST: a computer model for simulating the growth and reproduction of mixed species forest stands. University of Wisconsin, Madison, Research Paper R2635

Harrison TP, Michie BR (1985) A generalized approach to the use of matrix growth models. For Sci 31:850–856

Hasenauer H, Kindermann G, Steinmetz P (2006) The tree growth model MOSES 3.0. In: Hasenauer H (ed) Sustainable forest management growth models for Europe. Springer, Berlin/Heidelberg, pp 64–70

Hyink DM, Moser JW Jr (1983) A generalized framework for projecting forest yield and stand structure using diameter distributions. For Sci 29:85–95

Khatouri M, Dennis B (1990) Growth-and-yield model for uneven-aged *Cedrus atlantica* stands in Morocco. For Ecol Manage 36:253–266

Leslie PH (1945) On the use of matrices in certain population mathematics. Biometrika 33:183–212

Leslie PH (1948) Some further notes on the use of matrices in population mathematics. Biometrika 35:213–245

Lowell KE, Mitchell RJ (1987) Stand growth projection: simultaneous estimation of growth and mortality using a single probabilistic function. Can J For Res 17:1466–1470

Lynch TB, Moser JW Jr (1986) A growth model of mixed species stands. For Sci 32:697–706

McTague JP, Stansfield WF (1994) Stand and tree dynamics of uneven-aged ponderosa pine. For Sci 40:289–302

Mendoza GA, Setyarso A (1986) A transition matrix forest growth model for evaluating alternative harvesting schemes in Indonesia. For Ecol Manage 15:219–228

Mengel DL, Roise JP (1990) A diameter-class matrix model for southeastern U.S. Coastal plain bottomland hardwood stands. South J Appl For 14:189–195

Michie BR, Buongiorno J (1984) Estimation of a matrix model of forest growth from re-measured permanent plots. For Ecol Manage 8:127–135

Michie BR, McCandless FD (1986) A matrix model of oak-hickory stand management and valuing forest land. For Sci 32:759–768

Moser JW Jr (1972) Dynamics of an uneven-aged forest stand. For Sci 18:184–191

Moser JW Jr (1974) A system of equations for the components of forest growth. In: Fries J (ed) Growth models for tree and stand simulation. Royal College of Forestry, Stockholm, pp 260–287, Research Notes 30

Moser JW Jr (1976) Specification of density for the inverse j-shaped diameter distribution. For Sci 22:177–180

Moser JW Jr, Hall OF (1969) Deriving growth and yield functions for uneven-aged forest stands. For Sci 15:183–188

Murphy PA, Farrar RM (1982a) Interim models for basal area and volume projection of uneven-aged loblolly-shortleaf pine stands. South J Appl For 6:115–119

Murphy PA, Farrar RM (1982b) Calculation of theoretical uneven-aged stand structures with the exponential distribution. For Sci 28:105–109
Murphy PA, Farrar RM Jr (1983) Sawtimber volume predictions for uneven-aged loblolly-shortleaf pine stands on average sites. South J Appl For 7:45–49
Murphy PA, Farrar RM Jr (1988) A framework for stand structure projection of uneven-aged loblolly-shortleaf pine stands. For Sci 34:321–332
Osho JSA (1991) Matrix model for tree population projection in a tropical rain forest of south-western Nigeria. Ecol Mod 59:247–255
Palahí M, Pukkala T, Kasimiadis D, Poirazidis K, Papageorgiou AC (2008) Modelling site quality and individual-tree growth in pure and mixed *Pinus brutia* stands in north-east Greece. Ann For Sci 65:501p1–501p14
Peden LM, Williams JS, Frayer WE (1973) A Markov model for stand projection. For Sci 19:303–314
Peng C (2000) Growth and yield models for uneven-aged stands: past, present and future. For Ecol Manage 132:259–279
Perot T, Goreaud F, Ginisty C, Dhôte J-F (2010) A model bridging distance-dependent and distance-independent tree models to simulate the growth of mixed forests. Ann For Sci 67:502p1–502p11
Picard N, Bar-Hen A, Franc A (2002) Modeling forest dynamics with a combined matrix/individual-based model. For Sci 48:643–652
Pretzsch H, Biber P, Ďurský J (2002) The single tree-based stand simulator SILVA: construction, application and evaluation. For Ecol Manage 162:3–21
Pukkala T, Kolström T (1988) Simulation of the development of Norway spruce stands using a transition matrix. For Ecol Manage 25:255–267
Pukkala T, Lähde E, Laiho O (2009) Growth and yield models for uneven-sized forest stands in Finland. For Ecol Manage 258:207–216
Pukkala T, Lähde E, Laiho O (2011) Using optimization for fitting individual-tree growth models for uneven-aged stands. Eur J For Res 130:829–839
Richards FJ (1959) A flexible growth function for empirical use. J Exp Bot 10:290–300
Roberts MR, Hruska AJ (1986) Predicting diameter distributions: a test of the stationary Markov model. Can J For Res 16:130–135
Sallnäs O (1990) A matrix growth model of the Swedish forest. Studia Forestalia Suecica 183, 23 pp
Solomon DS, Hosmer RA, Hayslett HT Jr (1986) A two-stage matrix model for predicting growth of forest stands in the Northeast. Can J For Res 16:521–528
Spathelf P, Durlo MA (2001) Transition matrix for modeling the dynamics of a subtropical seminatural forest in southern Brazil. For Ecol Manage 151:139–149
Strub MR, Burkhart HE (1975) A class-interval-free method for obtaining expected yields from diameter distributions. For Sci 27:67–69
Temesgen H, Gadow Kv (2004) Generalized height-diameter models – an application for major tree species in complex stands of interior British Columbia. Eur J For Res 123:45–51
Temesgen H, Mitchell SJ (2005) An individual-tree mortality model for complex stands of southeastern British Columbia. West J Appl For 20:101–109
Trasobares A, Pukkala T, Miina J (2004) Growth and yield model for uneven-aged mixtures of *Pinus sylvestris* L. and *Pinus nigra* Arn. in Catalonia, north-east Spain. Ann For Sci 61:9–24
Usher MB (1966) A matrix approach to the management of renewable resources, with special reference to selection forests. J Appl Ecol 3:355–367
Usher MB (1969) A matrix model for forest management. Biometrics 25:309–315
Virgilietti P, Buongiorno J (1997) Modeling forest growth with management data: a matrix approach for the Italian Alps. Silva Fennica 31:27–42

Chapter 16
Modeling Response to Silvicultural Treatments

16.1 Need to Model Response to Silvicultural Treatments

Silvicultural practices such as thinning, control of competing vegetation, application of fertilizers, and use of genetically improved planting stock are commonly applied when wood production is the principal management goal. Growth and yield models that allow for varying management inputs are essential for making informed decisions about which treatments to apply at what times and at what levels. General model structures for even-aged stands are described in Chapters 10, 11, 12, 13 and 14. While model parameters may be estimated using data from plots that have received specific treatments, forest managers frequently wish to consider a wide range of treatment options.

Silvicultural treatments are often-times applied in combination. At the time of thinning, for example, vegetation control and fertilizer application might also be implemented. Studies that encompass all combinations of treatments that might be of interest do not exist, nor is it feasible to install such studies. The data available will generally be for one specific silvicultural practice, or perhaps a combination of two. Thus, the inclusion of silvicultural treatment effects is a modeling problem that goes far beyond fitting equations to data.

This chapter focuses on approaches for modeling growth response to silvicultural practices – primarily thinning, vegetation control, fertilizer applications, and genetic improvement – in even-aged structures. The impact of silvicultural practices on wood characteristics is touched upon in the chapter that follows. Most of the past work on modeling response to silvicultural inputs has been on mid rotation (that is between crown closure and final harvest) treatments. However, many important silvicultural decisions are also made at stand establishment.

The site preparation method used, initial spacing, extent of competition control and other cultural treatments during the stand establishment phase greatly influence subsequent stand development and the type and timing of subsequent silvicultural interventions. Thus, we will first consider modeling the impacts of cultural treat-

H.E. Burkhart and M. Tomé, *Modeling Forest Trees and Stands*,
DOI 10.1007/978-90-481-3170-9_16,

ments during the stand establishment phase and then address methods for predicting response to mid rotation silvicultural treatments.

16.2 Modeling Response of Juvenile Stands

Spacing trials have provided valuable data for modeling juvenile stand development especially for trials where measurements began at an early age and were acquired at frequent intervals. Zhang et al. (1996) used data from a loblolly pine spacing trial to estimate juvenile tree growth in diameter, height and crown ratio prediction. Analyses indicated that effects of stand density on diameter and crown ratio development became significant soon after planting. Stand density also affected height growth, but to a much lesser extent than diameter growth.

Peracca and O'Hara (2008) studied relationships between growing space per tree and growth components for giant sequoia, ponderosa pine, and Douglas-fir in the Sierra Nevada region of the USA. Relationships between growing space and tree height, diameter, and percentage of live crown all showed increasing trends as growing space per tree increased. Hybrid poplar plantations at two locations in southern British Columbia, Canada, were examined by Johnstone (2008) 9 years after planting to determine the effects of plantation spacing on tree and stand growth. Spacing had a direct, significant effect on individual-tree characteristics; however, for a given amount of growing space per tree, rectangularity (within row to between row spacing ratio) had no significant effect on individual-tree diameter, height or total tree volume.

Models of the juvenile phase of stand development are relatively rare, but there are some notable exceptions. Belli and Ek (1988) published a framework of prediction equations for growth and survival of red pine and white spruce during the first 5 years after planting. Their model incorporated data synthesized from published reports of planting experiments in the Great Lakes region of the United States. The authors characterized their model as a "first step" in modeling the brief, but crucial, establishment phase of planted conifers.

Management of competing vegetation is a standard practice for ensuring successful establishment of conifer plantations in the Pacific Northwest region of the United States. To be effective, vegetation treatments must be applied at young ages. Accordingly, models for response to vegetation management treatments were developed for juvenile Douglas-fir stands (Knowe et al. 1992, 1997b; Knowe 1994a, b; Knowe and Stein 1995). The salient aspects of models for juvenile Douglas-fir will be briefly described here. These components of young-stand growth were structured so that they could be interfaced with growth models for more mature stands, thus enabling managers to make rotation-age projections of the effects of treatments during the establishment/juvenile stage.

Knowe et al. (1992) published a parameter recovery procedure for the Weibull distribution, based on diameter percentiles and modified it to incorporate the effects of interfering vegetation in young Douglas-fir plantations. Four percentiles (0, 25th,

50th, 95th) of the cumulative probability distribution were predicted as functions of quadratic mean diameter and age.

In a subsequent analysis, Knowe (1994a) incorporated the effect of inter-specific competition and vegetation management treatments in stand table projection models for Douglas-fir saplings. A projection equation was developed for relative tree size, defined as the ratio of individual-tree diameter to quadratic mean diameter. An additional equation was developed to project quadratic mean diameter in order that individual-tree diameters could be projected from relative size. The stand table projection system performed similarly to the diameter distribution prediction system (Knowe et al. 1992) based on the Weibull function.

Height-age and height-diameter functions (Knowe 1994b) that may be used in conjunction with diameter distribution or stand table projection models were fitted to data from young Douglas-fir plantations. Analysis of height growth patterns for dominant trees indicated significant differences between total vegetation control treatment and operational release treatments or no treatment. Different height-diameter curve shapes were associated with total vegetation control and the operational release and no treatments.

Knowe and Stein (1995) developed prediction models based on the Weibull distribution function and stand-table projection models using changes in relative diameter for 2- to 10-year-old Douglas-fir plantations. Both modeling approaches incorporated the effects of site preparation, animal protection, and competing vegetation. Equations were derived for predicting survival, height growth of dominant trees, height-diameter relationships and the development of woody vegetation over time to facilitate estimating stand structure and dynamics after various site-preparation and animal-protection treatment conditions.

Cover-projection models based on algebraic difference formulations of an exponential-power function to describe shrub recovery and development patterns after clearcutting and site preparation were elaborated by Knowe et al. (1997b). The effect of six treatments on shrub growth patterns was investigated by incorporating indicator variables into the rate and shape parameters of the models.

Variations of variables and equations developed for young plantations of Douglas-fir (Knowe et al. 1992, 1997b; Knowe 1994a, b; Knowe and Stein 1995) were integrated to predict current tree size distributions and project stand dynamics of planted Douglas-fir with hardwood competitors in the coast ranges of the Pacific Northwest, USA (Knowe et al. 2005). Stand-level equations were included for the following components: dominant height and survival projection of planted Douglas-fir; basal area of planted Douglas-fir; hardwood basal area projection; diameter distribution prediction function for planted Douglas-fir; and height, crown width and height to crown base prediction equations for individual planted Douglas-fir trees. Prediction equations provide estimates of current size and stand structure based on stand characteristics and age, while projection equations provide estimates of future size based on current size and future stand characteristics.

Ritchie and Hamann (2006), noting the need to model the dynamics of competing vegetation, published growth equations for competing vegetation in young plantations. Their growth equations were developed to be consistent with individual tree model architecture. Response variables were height increment, basal diameter increment, and crown width. Three common competing shrub and three competing hardwood species were included in the analysis. Fit statistics for hardwoods were generally much better than those obtained for shrub species. The equations were developed for use in an individual-tree plant growth model for young plantations in southern Oregon and Northern California, USA, where the primary planted species are ponderosa pine and Douglas-fir.

Mason et al. (1997) used data from 27 site preparation experiments to construct a model that predicts growth, survival and size class distributions for radiata pine in New Zealand with respect to altitude, weed control, cultivation, fertilization, and initial stocking during the first 5 years after planting. The question of linking the model with existing growth and yield models for older crops was discussed and a theoretical structure was proposed that clarified assumptions required if the models are jointly used to evaluate establishment practices throughout a complete rotation. Subsequently, Mason (2001) added effects of initial seedling diameter and plant handling to the overall structure of the model of juvenile growth and survival of radiata pine.

Westfall et al. (2004) developed a system of equations to simulate growth of loblolly pine before the onset of intraspecific competition. Treatment response functions were included for various site preparation, herbaceous weed control, and fertilization practices. These functions modify the baseline model predictions to simulate the effects of treatments on tree growth and stand development. This system was incorporated into a distance-dependent growth-and-yield simulator to make growth projections from time of planting through rotation age for intensively managed stands of loblolly pine in the southeastern United States.

16.3 Frameworks for Modeling Stand Level Response

16.3.1 Response Functions

Continued progress in genetic improvement and site-specific silvicultural prescriptions necessitates perpetual adjustment of existing yield prediction models to reflect responses to these practices. Pienaar and Rheney (1995) proposed a methodology for modeling stand level growth and yield response to silvicultural treatments. The core of their proposed system is a dominant height growth model that consists of a treatment response term to accommodate different response patterns. The model of height growth response is then used as a vehicle for modeling the resultant effects on stand basal area and volume production.

Various silvicultural practices may have differential effects on height growth over the life of a stand. In addition to the absolute magnitude of the effect, some treatments may have only a relatively brief effect compared to a baseline or standard treatment, while others may have a longer lasting effect. Pienaar and Rheney (1995) developed the following height growth model that is capable of reflecting such differential effects of silvicultural practices:

$$h_{dom} = a_0(1 - e^{-a_1 t})^{a_2} + b_1 t_{st} e^{-b_2 t_{st}} \quad (16.1)$$

where h_{dom} is dominant height, t is plantation age, t_{st} is years since the treatment was applied, and a_0, a_1, a_2, b_1 and b_2 are parameters that define a particular growth curve. The expected treatment response is explicitly represented in the model in terms of readily interpretable parameters, rather than implicitly by attempting to relate the parameters of the basic height growth model (a_0, a_1, a_2) to different silvicultural treatments. When no additional treatment is applied, the second term in Eq. 16.1 is zero, so that the first term represents the baseline, or standard height growth for a given site. The second term represents the cumulative effect of an additional treatment on dominant height over time. Such a treatment can be applied at the time of stand establishment (in which case $t_{st} = t$) or at some later time. In either case the parameters b_1 and b_2 determine the magnitude and pattern of the response. When different levels of a treatment are involved, one or both of these parameters are considered to be a function of the treatment level.

Model (16.1) provided an accurate description of the average dominant height growth response of slash pine plantations that received a variety of different silvicultural treatments. With a satisfactory description of the height growth response to silvicultural treatment, attention was turned to accounting for the basal area and volume growth response in the context of existing stand level basal area and volume prediction models. A stand level basal area prediction model (Pienaar and Rheney 1993) that includes average dominant height as a predictor, in addition to age and trees per unit area, did not adequately account for the effect of all treatments on basal area growth and accordingly an adjustment term was added. For a stand level volume prediction model that includes both average dominant height and basal area as predictor variables, in addition to age and trees per unit area, no additional adjustment factors were required for the treatments evaluated in Pienaar and Rheney's analyses.

Snowdon (2002) described the concepts of Type 1 and Type 2 responses to silvicultural treatments. Type 1 responses are those that advance the stage of stand development but do not change the inherent productivity of the site. An example would be weed control. Type 2 responses occur from treatments which result in a long-term change in site properties. For example, correction of a phosphorus nutrient deficiency.

Schumacher-type yield and projection models that incorporate Type 1 and Type 2 concepts were developed and tested with data from field experiments and compared with the model of Pienaar and Rheney (1995). The model formulation of Snowdon (2002) performed better than the Pienaar-Rheney model in yield form but the Pienaar-Rheney model tended to have superior performance in projection form. The consequences of assuming Type 1 and Type 2 responses for experimental design, optimal management regimes, model development, and subsequent applications are discussed by Snowdon (2002).

16.3.2 Distributing Stand Growth Response to Individual Trees

Modeling response to intermediate silvicultural treatments has generally focused on whole-stand or individual-tree relationships. An alternative approach, consisting of distributing within-stand basal area growth following silvicultural treatment using a relative size-relative growth (*RSG*) function, was investigated by Moore et al. (1994). Relative growth was defined as the ratio of individual tree basal area growth to stand total basal area growth on a unit area, and relative size was defined similarly, giving:

$$\frac{i_g}{i_G} = b_0 + b_1\frac{g}{G} + b_2\left(\frac{g}{G}\right)^2$$

where i_g is tree basal area growth, i_G is stand basal area growth, g is tree basal area and G is stand basal area. The parameters b_0, b_1 and b_2 were expressed as linear functions stand density, quadratic mean *dbh* and the coefficient of variation of tree basal area distribution.

Results showed that the *RSG* function performed well for distributing stand basal area growth to individual trees following silvicultural treatments in even-aged Douglas-fir stands in the Inland Northwest United States. Thinning and fertilization treatments did not change the relationship between relative tree basal area growth and relative tree basal area and did not alter the relationship between average tree size, stand density and structure. Hence, the authors concluded that there is no need to develop treatment specific *RSG* functions.

When applying the methodology of Moore et al. (1994) the absolute growth effects of silvicultural treatments must first be estimated at the stand level. Growth is then subsequently distributed to a list of individual trees using a *RSG* function.

16.4 Modeling Response to Selected Silvicultural Treatments

Studies of tree and stand response to silvicultural practices are often focused on a single cultural treatment. Hence it may be necessary to model response to given treatments separately and then to integrate these response functions into an overall model structure. This approach may prove satisfactory for estimating response to specific inputs, but the modeling of interactions among treatments remains problematic. In the sections that follow, an overview of methods for estimating growth response to thinning, vegetation control, fertilizer application, and genetic improvement is presented.

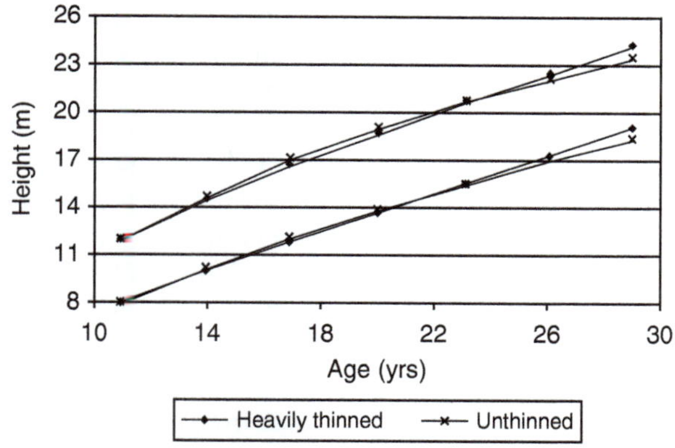

Fig. 16.1 Dominant height curves generated for heavily thinned and unthinned stands of loblolly pine with a thinning intensity of 50% basal area removed and a thinning age of 11 years (From Sharma et al. 2006)

16.4.1 Thinning

Thinning alters stand density, mean tree size, and stand structure. Tree and stand response to thinning depends on species, site variables, stand age and the intensity and type of thinning imposed. All aspects of tree and stand growth dynamics are affected by an abrupt modification of stand density, with some characteristics being more affected than others. Here we present a brief overview of selected reports on modeling specific aspects of tree and stand response to thinning.

16.4.1.1 Dominant Height

Sharma et al. (2006) used data from a thinning study installed in loblolly pine plantations across the southeastern United States to evaluate thinning impact on height growth of dominant and codominant trees. Height growth was reduced initially by thinning but was increased after 3 years following thinning. The average total height of dominant and codominant trees in thinned stands exceeded its counterpart in unthinned stands after 12 years following treatment. Initial growth response to thinning was less at older stand ages than at younger ages. A model was constructed to characterize the development of height in thinned and unthinned stands. The model reflects the initial suppression of dominant and codominant height growth followed by acceleration as a result of thinning. However, the impact of thinning on dominant height development was relatively minor (Fig. 16.1), and, for many practical applications, it can be ignored.

16.4.1.2 Basal Area

Pienaar (1979) proposed a procedure by which the predicted basal area growth and yield in thinned plantations is derived from unthinned plantations of the same age, site index, and number of stems per hectare as remain in the thinned plantation immediately after thinning. An estimate of growth after thinning is obtained by adjusting the projected growth of the unthinned plantation to allow for the degree of suppression that existed in the thinned plantation, relative to the unthinned counterpart. Different thinning intensities and thinning methods can be accommodated within the proposed general formulation.

Further analyses of basal area projection for thinned and unthinned pine plantations were published by Pienaar and Shiver (1984, 1986) and Pienaar et al. (1985). In their paper published in 1986, Pienaar and Shiver further generalized stand-level basal area and basal area growth equations for unthinned as well as thinned plantations.

Long term remeasurement data from both thinned and unthinned slash pine plots in South Africa were used in the analyses. Pienaar and Shiver (1986) observed that the magnitude of the difference in basal area between thinned and unthinned stands of the same age, trees per unit area, and average dominant height depends on the age when the thinning occurred and the intensity of the thinning. Thus, the following general prediction equation was proposed to accommodate both unthinned and thinned plantations:

$$\ln G = b_0 + b_1 \frac{1}{t} + b_2 \ln N + b_3 \ln h_{dom} + b_4 \frac{\ln N}{t} + b_5 \frac{\ln h_{dom}}{t} + b_6 \frac{N_t t_t}{N_{at} t} \tag{16.2}$$

where

t = plantation age at last thinning
N = present number of trees per unit area
N_t = number of trees removed in last thinning
N_{at} = number of trees remaining after last thinning
G = basal area per unit area
t = plantation age
h_{dom} = average dominant height

The term $b_6(N_t t_t)/(N_{at} t)$ modifies the basal area of unthinned plantations of given age, stems per unit area, and average dominant height to predict the basal area for comparable thinned plantations. In the nonlogarithmic form of the prediction equation, it is a multiplicative modifier theoretically between 0 and 1. For any given age, t, the earlier a thinning of given intensity (N_t/N_{at}) occurs, the larger (closer to 1) the modifier will be. If thinnings of different intensities occur the same time ago, so that (t_t/t) and N_{at} are the same, then the modifier will be larger for the less intensive thinning.

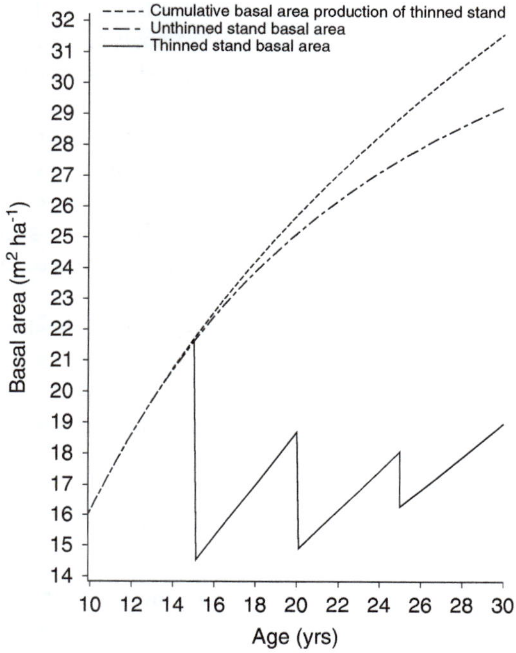

Fig. 16.2 Comparisons of predicted basal area and basal-area growth for slash pine plantations with an initial stocking of 16 m^2 ha^{-1} at age 20 years (From Bailey and Ware 1983)

A basal area projection equation was derived from the prediction Eq. 16.2:

$$\ln G_2 = \ln G_1 + b_1\left(\frac{1}{t_2}-\frac{1}{t_1}\right) + b_2(t_2 - t_1) + b_3\left(1-\frac{t_1}{t_2}\right) + b_4\left(\frac{1}{t_2^2}-\frac{1}{t_1 t_2}\right) + b_5 \ln N_1\left(\frac{1}{t_2}-\frac{1}{t_1}\right) + b_6 \ln h_{doml}\left(\frac{1}{t_2}-\frac{1}{t_1}\right) + b_7\left(\frac{N_t t_t}{N_{at} t_2}-\frac{N_t t_t}{N_{at} t_1}\right) \quad (16.3)$$

Bailey and Ware (1983) developed a thinning index based on the ratio of quadratic mean diameter of trees removed in thinning to the quadratic mean diameter of all trees before thinning in order to reflect kind and level of thinning employed. The ratio was transformed into an indexing variable. The new variable, or thinning index, was conditioned to increase from negative to positive as thinnings progress from removing large trees to removing small trees. Further, the index is equal to zero when no thinning is imposed or when thinning does not affect stand mean tree diameter, as in row thinning. The thinning index resulted in improved predictions when incorporated as a multiplier in a basal-area projection model. A comparison of cumulative basal area production of thinned stands and basal area development of unthinned stands is shown in Fig. 16.2.

Hasenauer et al. (1997) investigated basal area growth in loblolly pine plantations after thinning and developed a generalized equation for projecting stand basal area development for a range of thinning treatments, site conditions and stand ages at time of thinning. Data from a region-wide thinning study installed across the southeastern United States were analyzed. Each installation of the study consisted of a control plot, a lightly thinned plot and a heavily thinned plot. All thinning was from below; analyses were conducted after four re-measurements, taken at 3-year intervals, were completed. The results of the study indicated that basal area of thinned plots approaches that of their unthinned counterparts. Based on these findings, Hasenauer and co-authors developed a basal area projection equation that includes a thinning response factor and accounts for the effects of competing hardwood vegetation by including basal area of the hardwood component in the total stand basal area variable.

With the assumption that basal area of the thinned plots may eventually converge toward, but not exceed, that of the unthinned plots, using height to describe the time trend, and considering hardwood competition effects, Hasenauer et al. (1997) proposed the following model for projecting basal area for loblolly pine per hectare, Gp_2:

$$Gp_2 = Gp_1^{(h_{dom1}/h_{dom2})} \exp\left[(Gp_1/Gt_1)^{b_0} b_1 S^{b_2} TR \left(1 - \frac{h_{dom1}}{h_{dom2}}\right)\right] \qquad (16.4)$$

where Gp_1 is the basal area of the pines at the beginning of the growing period, Gt_1 is the total basal area including hardwood trees, and h_{dom1} is the corresponding dominant height; h_{dom2} is the dominant height at the end of the growing period, *S* is the site index, and *TR* is the thinning response variable, which is defined as

$$TR = \left(\frac{Gp_{at}}{Gp_{bt}}\right)^{b_3(h_{domt}/h_{dom2})} \qquad (16.5)$$

The ratio between Gp_{at}, the pine basal area after thinning, and Gp_{bt}, the pine basal area before thinning, indicates the relative thinning intensity. h_{domt} is the dominant height at the time of thinning and h_{dom2} remains as previously defined. Depending on the parameter b_3, *TR* (Eq. 16.5) has the following effect within Eq. 16.4: if b_3 is zero (i.e. not significant), the basal area development of the thinned plots is the same as that of the unthinned counterparts with the same initial basal area. If b_3 is positive, the basal area of the thinned plots diverges from that of the unthinned plots, and if b_3 is negative, the basal area of the thinned plots converges toward that of the unthinned plots. The thinning effect itself decreases with increasing time since thinning because h_{dom2} of Eqs. 16.4 and 16.5 increases.

In unthinned stands *TR* equals 1. Thus (16.4) assumes that, depending on the dominant height and the site index, the maximum possible stand basal area development occurs in unthinned stands and will reach an upper asymptote because the tree height increment of taller trees tends towards zero. The maximum basal area of thinned stands will not exceed the maximum basal area of unthinned plots. In thinned stands, however, the thinning response function, *TR*, will always be greater than 1 (with b_3 negative) and gives the rate at which the basal area of the thinned plots converges toward that of the unthinned plots. Figure 16.3 provides the basal area development following thinning using the mean statistics at plot establishment.

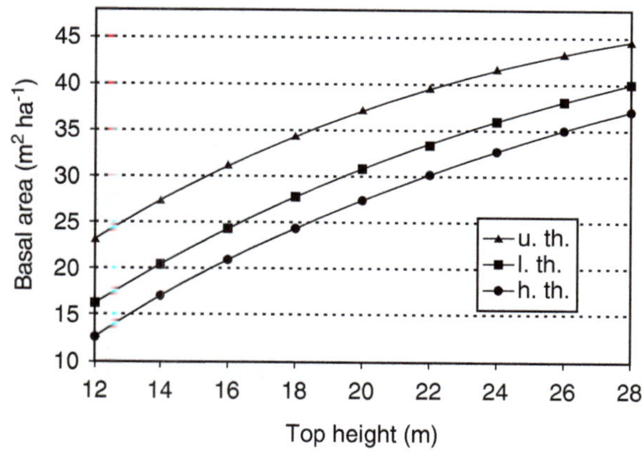

Fig. 16.3 Basal area vs. top height development for unthinned (*u.th.*), lightly thinned (*l.th.*), and heavily thinned (*h.th.*) plots in loblolly pine plantations. The starting values used are the mean statistics at plot establishment (From Hasenauer et al. 1997)

16.4.1.3 Survival

Avila and Burkhart (1992) fitted logistic functions separately to data from thinned and unthinned loblolly pine plots to estimate the probability of survival of individual trees. The fitted function, with distance-dependent and distance-independent measures of competition in conjunction with crown ratio, can be used in individual-tree based growth and yield simulators.

Stand-level survival functions have been fitted to observations from thinned stands. For example, Clutter and Jones (1980) fitted the following function to remeasured plots in slash pine plantations in the Coastal Plain regions of the USA:

$$N_2 = \left[N_1^{a_1} + a_2\left(t_2^{a_3} - t_1^{a_3}\right)\right]^{1/a_1} \tag{16.6}$$

where N_2 is trees per unit area at age t_2, N_1 is trees per unit area at age t_1, and $t_2 > t_1$. When estimating survival in thinned loblolly pine plantations, Lemin and Burkhart (1983) found Eq. 16.6 best among the alternatives evaluated. In both instances (slash pine and loblolly pine) the stands were thinned from below and site index, thinning intensity, type of thinning, and age at time of thinning were not explicitly included in the model.

In an analysis with data from the same study as that used by Clutter and Jones, Bailey et al. (1985) developed a survival function for thinned and unthinned stands that directly incorporated a measure of thinning, age at the time of thinning, and site index. The model of Bailey et al. (1985) is of the form:

$$N_2 = N_1\left(\frac{t_2}{t_1}\right)^{b_1} e^{\left[(b_0 + b_2 S)(t_2 - t_1) + b_3 \frac{dg_{rt}}{dg_{bt}} I_{t22.5}\left(\frac{1/t_2 - 1/t_1}{t_t}\right)\right]} \tag{16.7}$$

where:

$I_{t22.5} = 1$ if $t_2 < 22.5$ year; $= 0$ if $t_2 > 22.5$
dg_{rt} = quadratic mean diameter of the trees removed in thinning

dg_{bt} = quadratic mean diameter of the whole stand before thinning
t_t = age of stand at last thinning
N_i = number of trees surviving per hectare at age i
t_i = stand age at time i
S = site index

16.4.1.4 Height-Diameter

Zhang et al. (1997) fitted Eq. 16.8 to data from unthinned, lightly thinned, and heavily thinned plots in loblolly pine plantations:

$$h = b_1 h_{dom}^{b_2} \, e^{\left[\frac{b_3}{t} + \left(\frac{1}{d} - \frac{1}{d_{\max}}\right)\left(b_4 + b_5 \frac{\ln N}{t}\right)\right]} \tag{16.8}$$

where h is the predicted total tree height for a tree with dbh equal to d in a stand of age t, in which maximum dbh equals $d_{\max}$ and number of trees per unit area equals N.

Three thinning response models were incorporated into model (16.8) to account for the effect of thinning. There was little difference in mean square error for the base Eq. 16.8 and equations fitted with a thinning response variable. Consequently, the authors concluded that the influence of thinning on height-diameter relationships can be explained by thinning effects on variables of average dominant height (h_{dom}), maximum stand diameter ($d_{\max}$), and number of trees per unit area (N).

In a subsequent analysis that utilized additional measurements of the study plots used by Zhang et al. (1997), Russell et al. (2010) found that including geographic coordinates in addition to a thinning response modifier improved predictions of individual tree heights as compared with a baseline model without these covariates.

16.4.1.5 Crown Measures

Studies have shown that reducing stand density through thinning slows the recession of height to the crown base in shade-intolerant conifers (Kramer 1966; Siemon et al. 1976). These results suggest that crowns develop differently for thinned and unthinned stand conditions. With results on differential crown recession in mind, Short and Burkhart (1992) developed a function using percentage of basal area removed to express thinning effect in a crown height recession model. Their results indicated that including a thinning variable improved the fit of the model and produced a substantial increase in accuracy for predictions from the model.

Liu et al. (1995) developed a thinning response function that includes thinning intensity, stand age at time of thinning, and elapsed time since thinning. Biologically, there should be no immediate response at the time of thinning. Instead, response to thinning should begin at zero and increase to some maximum as the crowns of the residual trees respond to the extra growing space and additional sunlight.

Then, as the stand again closes, the response should diminish and approach that of an unthinned condition. With these considerations in mind, the following thinning response function was derived:

$$TR = \left(\frac{G_{at}}{G_{bt}}\right)^{\frac{-(t-t_t)^2+k(t-t_t)}{t^2}} \tag{16.9}$$

where:

TR = thinning response
t = stand age
t_t = age of stand at time of thinning
G_{at} = basal area after thinning
G_{bt} = basal area before thinning
k = duration parameter for the thinning effect

The duration of thinning response (in years) is determined by the value of the duration parameter, k. The first derivative of the exponential part of the Eq. 16.9 with respect to $t - t_t$, the time elapsed since thinning, indicates that the maximum thinning response will occur at

$$\frac{kt_t}{k + 2t_t}$$

years after thinning. Thus, age of maximum response depends on age of the stand at time of thinning and k. Using Eq. 16.9, an allometric crown ratio model was specified:

$$c_r = 1 - \left(\frac{G_{at}}{G_{bt}}\right)^{\frac{r\left[-(t-t_t)^2+k(t-t_t)\right]}{t^2}} e^{[-(b_o+b_1/t)d/h]} \tag{16.10}$$

where c_r is crown ratio, d is dbh, h is total tree height, r is a rate parameter, and other variables are as previously defined. The rate parameter, r, is dimensionless and along with G_{at}/G_{bt}, t and t_t defines the shape of the response function.

In a similar way, Liu et al. (1995) specified a crown increment equation by substituting Eq. 16.9 for the original Short and Burkhart (1992) thinning response function, giving:

$$\Delta h_{cb} = b_0\left(\frac{G_{at}}{G_{bt}}\right)^{\frac{r\left[-(t-t_t)^2+k(t-t_t)\right]}{t^2}} h_{dom}^{b_1} e^{(b_2c_r^{0.5}+b_3CI+b_4t)} \tag{16.11}$$

where Δh_{cb} is increment in height to the base of the crown, CI is a competition index, and all other variables remain as previously defined.

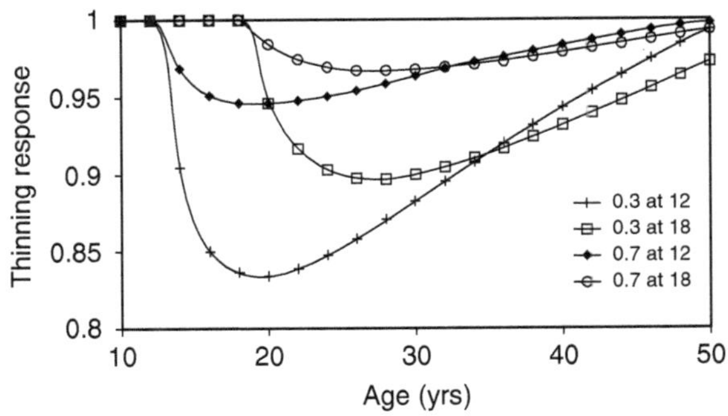

Fig. 16.4 Effects of thinning on crown ratio as a function of thinning intensity (I, defined as the ratio of basal area after to before thinning), time of thinning, and time elapsed since thinning (From Liu et al. 1995)

Eqs. 16.10 and 16.11 were fitted using data from a region-wide set of thinning plots in loblolly pine plantations located across the southeastern United States. Each location of the thinning study consisted of a control plot and a lightly-thinned and a heavily-thinned plot (thinnings were from below). The thinning response rate parameter, r, was positive indicating that trees in heavily thinned plantations will have less crown-height recession or larger crown ratio than trees in lightly thinned or unthinned stands. Likewise, the thinning response duration parameter, k, was also positive ensuring that thinning impact gradually increases after thinning to some level and then generally decreases. Figure 16.4 shows the behavior of the thinning response function (16.9) with regard to crown ratio for two hypothetical stands thinned at age 12 with after-to-before thinning basal area ratios of 0.3 and 0.7 and for two stands thinned at age 18 with the same 0.3 and 0.7 thinning intensities. Response to other silvicultural treatments, such as fertilizer applications, also exhibit behavior that can be modeled via a rate and duration parameter. Function (16.9) might serve as a more general response to silvicultural treatment expression.

In their analysis of crown response to thinning, Liu et al. (1995) found that the allometric model approach (16.10) and the increment function for height to base of crown (16.11) gave approximately equivalent results.

16.4.1.6 Stem Profile/Volume

The form exponent was used by Tassissa and Burkhart (1998) to evaluate thinning effects on stem form of loblolly pine trees. Any solid of revolution can be generated by rotating a curve of the form

$$y = k\sqrt{x^r}$$

around the X-axis where y is radius, x is height or length, k is a constant relating the rate of change in radius with height (length) and r is the form exponent. As the form exponent changes, different solids are generated. When r is 1 a paraboloid is obtained; when r is 2 a cone; when r is 3 a neiloid.

Thinning significantly increased the form exponent particularly in the lower bole leading to a more neiloid form. A stem profile model that accounts for changes in stem form due to thinning effects was developed by incorporating a variation of the thinning modifier function of Liu et al. (1995).

In an analysis of the same loblolly pine data as that used by Tasissa and Burkhart (1998), Tasissa et al. (1997) found that total stem volume and implied taper functions were significantly different for trees from thinned and unthinned stands.

16.4.1.7 Product Proportions

In addition to increasing diameter growth and average tree size, thinning treatments are also imposed to remove defective stems and thus to improve tree quality in the residual stand. To accurately reflect product distributions and stand values, growth and yield models must account for both number of trees by size classes and proportions of trees in each class that qualify for conversion to specified products. Burkhart and Bredenkamp (1989) modeled the proportion of trees in three product classes (pulpwood, sawtimber, peelers) in thinned and unthinned loblolly pine plantations using the Chapman-Richards function. The function was constrained to go through zero when *dbh* was at the lower limit for specified products and to exhibit an upper asymptote of one:

$$P_d = \left(1 - e^{b_1(d - d_{lower})}\right)^{b_2}$$

where:

P_d = proportion of trees with *dbh* d having a specified product classification
d = *dbh* class
d_{lower} = lower *dbh* limit of product size class
b_1, b_2 = parameters to be estimated

In a later analysis of plot data from the same thinning study as that analysed by Burkhart and Bredenkamp (1989), Amateis and Burkhart (2005) applied proportional odds modeling methods to develop equations for predicting product proportions (pulpwood, sawtimber, peelers) from tree *dbh*, stand basal area before and after thinning (thinning intensity), and number of thinnings imposed. Thinning had its greatest impact on the product distribution in the larger diameter classes. The more intensive thinning treatments removed almost all the lower valued (pulpwood) trees from the larger diameter classes. In the intermediate diameter range, comprised of a mix of pulpwood- and sawtimber-sized trees, the thinning intensity had little influence on the product distribution (Fig. 16.5).

The system of product classification developed by Amateis and Burkhart (2005) can be incorporated into individual tree growth and yield models to predict the

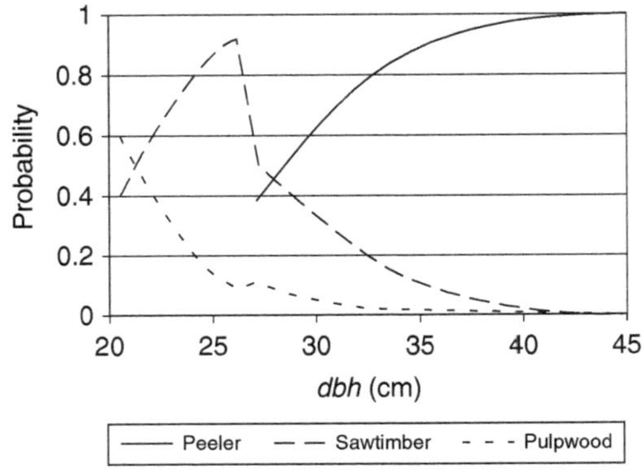

Fig. 16.5 Predicted product proportions (peeler, sawtimber, and pulpwood) by *dbh* for a loblolly pine plantation thinned once from 30 to 18 m^2 ha^{-1} (From Amateis and Burkhart 2005)

probability that a particular tree qualifies for a certain product category. The system can also be used with diameter distribution models to estimate the proportion of trees in each of the product categories identified for specified diameter classes.

16.4.1.8 Incorporating Thinning Response into Stand Simulators

Westfall and Burkhart (2001) used data from a long-term thinning study in loblolly pine to develop thinning response variables that were subsequently incorporated into a distance-dependent, individual-tree model, PTAEDA2 (see Sect. 14.4.1 for a description of the model structure). Height increments and mortality components needed no additional refinement to account for response to effects of thinning. The diameter increment and crown ratio components, however, could not account for thinning effects in their original form and thinning response functions were added to these equations. Results showed significant improvements in predictive ability when a thinning response function was added to the diameter increment model. There was no significant improvement in crown ratio prediction due to adding a thinning response function.

While incorporating thinning response variables can improve predictive ability, Westfall and Burkhart (2001) cautioned that the thinning response modifications can result in unanticipated model behavior when incorporated into stand simulators. Improvements in predictive ability exhibited by a given equation outside a stand simulator do not necessarily equate to a similar level of performance when incorporated into the overall stand model. Interrelationships among various components within stand simulators such as PTAEDA2 are complex and changes to any of the equations can have a large effect on the behavior of the other components. The best combination of growth equations for simulating stand dynamics cannot be fully determined by evaluating individual equations. Candidate equations should be evaluated through analysis of systems output and the appropriate combination chosen that provides the best predictive result for the attributes of primary interest.

16.4.2 Vegetation Control

Unwanted vegetation in commercial forests competes with the crop trees for space, water, nutrients and light. The benefit to crop plants from controlling weeds is well recognized and there have been numerous studies aimed at assessing growth response to competition control. Wagner et al. (2006) reviewed results from 60 of the longest-term studies on vegetation management in Canada, the United States, Brazil, South Africa, New Zealand and Australia. A majority of the studies reported substantial gains (30–500% increases in wood volume) from the most effective vegetation treatments. More recently, McCarthy et al. (2011) summarized the state of forest vegetation management practices in Europe.

Many vegetation management studies involve a limited number of experimental treatments (often just with and without weed control). From a growth modeling standpoint, response to various levels of weed control is needed. Further, a preponderance of the research of growth response to competition control has been aimed at light-demanding pioneer species. The emphasis in North America has been on loblolly pine in the southeastern USA and on conifers (especially Douglas-fir) in the west (Walstad and Kuch 1987). We draw extensively on modeling work from these regions to illustrate principles and methods that are applicable to estimating crop tree response to weed control.

A model developed by Burkhart and Sprinz (1984) predicts survival and growth and yield of unthinned loblolly pine plantations with varying levels of hardwood competition in the main canopy. The approach taken to modeling hardwood competition effects on yield was to regard values observed in plantings on essentially weed-free former agricultural land as upper limits and to compute reduction factors based on the level of hardwood competition measured in a region-wide set of plots in plantations established on cutover, site-prepared areas. Of the stand components examined (i) height-age development (ii) height diameter equations, (iii) individual tree volume relationships, (iv) diameter distributions, and (v) survival relationships, the major impact of competing hardwoods was a reduction in pine survival and diameter growth. An equation was estimated to relate pine survival to percent of basal area in the main canopy. Moments of the Weibull distribution were related to hardwood competition level such that the variance of the *dbh* distribution of pines remains constant regardless of the amount of hardwood competition but the mean diameter and mean squared diameter (and thus basal area) are reduced with increasing levels of hardwood in the main canopy. With pine survival and diameter development related to percent of total basal area in hardwood in the main canopy, the behavior of stand composition in terms of pine and hardwood competition needed to be considered. Data on percent basal area composition of pine and hardwood were available for ages 11 and 24 from measurements in a site preparation study. The fitted relationship to estimate percent basal area in hardwoods at age 24 to percent at age 11 gave a slope coefficient of 0.97 (not significantly different from 1.0), hence, the composition by basal area, after crown closure, was assumed constant. This assumption can hold only for relatively short time periods.

Other analyses have found composition of basal area of the main canopy to be a useful predictor (Smith and Hafley 1987; Knowe 1992a), but the projection of stand basal area composition through time remains a relatively uncertain component in the models.

A loblolly pine simulator that accounts for hardwood competitors was developed by Smith and Hafley (1987). Their model is based on a bivariate distribution of height and diameter in which the minimum and modal height and diameter are adjusted for amount of hardwood basal area. The proportion of total stand basal area that is composed of hardwoods is allowed to vary over time.

Knowe (1992a) fitted basal area and diameter distribution models for loblolly pine plantations with hardwood competition in the Piedmont and Upper Coastal Plain of Alabama, Georgia, and South Carolina in the USA. A model for pine basal area yield was developed to account for the effects of varying hardwood levels. Percentiles of the diameter distribution needed to recover the parameters of a Weibull function were predicted from stand characteristics and proportion of hardwoods defined by the ratio of hardwood basal area to total basal area per unit area.

The basal area yield model employed was:

$$G_p = b_0 h_{dom}^{b_1} e^{(b_2 R_s + b_3 G_h / G)} \qquad (16.12)$$

where:

G_p = pine basal area per unit area
h_{dom} = mean height of the dominant and codominant pines
R_s = relative spacing (mean distance between trees divided by h_{dom})
G_h = hardwood basal area
G = total basal area

Figure 16.6 shows basal area yield for site index 20 m (base age 25 years) for ages 10–30 years and with hardwood percentage of basal area of 0, 10, 20 and 30.

A parameter recovery procedure based on the 0th (minimum *dbh*) 25th, 50th, and 95th percentiles of the diameter distributions was used to estimate diameter distributions for pine plantations with varying levels of hardwood competition (Fig. 16.7). Diameter distributions predicted by procedures developed by Knowe (1992a) were compared with those for comparable stand conditions from the models of Burkhart and Sprinz (1984) and Smith and Hafley (1987). Figure 16.8 illustrates the range in shapes exhibited by the three models for diameter distributions of plantations of site index 20 m (25-year base) with 0% and 20% basal area in hardwood species.

In an analysis focused on predicting the impact of inter-specific competition in young loblolly pine stands, Knowe (1992b) found very little difference among three types of models to predict the effects of hardwoods and total vegetation control on quadratic mean diameter. The percentile-based parameter recovery system for the three-parameter Weibull function included direct and indirect effects of

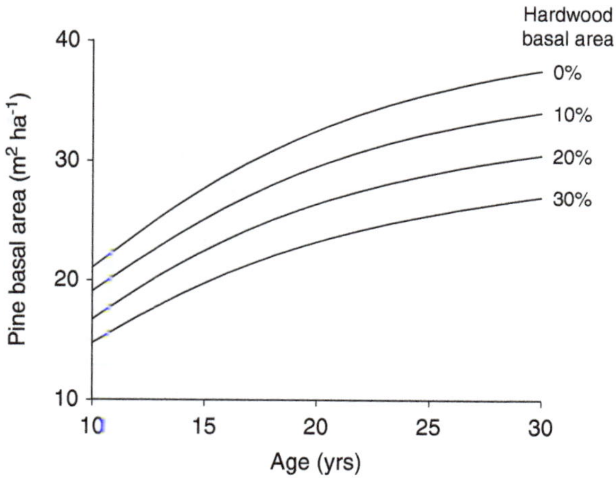

Fig. 16.6 Predicted effect of hardwoods on planted loblolly pine basal area for site index 20 m. Hardwood competition is expressed as the ratio of hardwood basal area to total basal area in the stand (From Knowe 1992a)

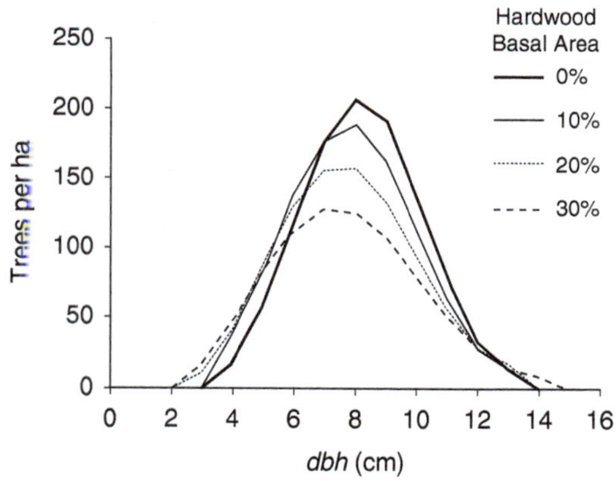

Fig. 16.7 Predicted effect of increasing proportion of hardwood basal area on loblolly pine diameter distributions for age 25 and site index 20 m (base age 25) (From Knowe 1992a)

competition. The effects of increasing inter-specific competition included skewing the distributions toward smaller diameters, decreasing the variation in diameters, and increasing the coefficient of variation in diameters.

Shiver and Brister (1996) used data from yield plots in naturally regenerated, even-aged loblolly pine stands in the Georgia Piedmont, USA, to fit the following function to estimate pine volume per unit area (V_p) for stands with varying amounts of hardwood competition:

$$V_p = b_0 G^{b_1} h_{dom}^{b_2} N_p^{b_3} e^{b_4 G_h / G} \tag{16.13}$$

where G is total stand basal area (pine and hardwood), h_{dom} is height of dominant and codominant loblolly pine trees, N_p is number of pine trees per unit area, and G_h is basal area of hardwoods.

Volume by product classes was determined by utilizing the model introduced by Amateis et al. (1986) to predict the merchantable portion of total stand volume from

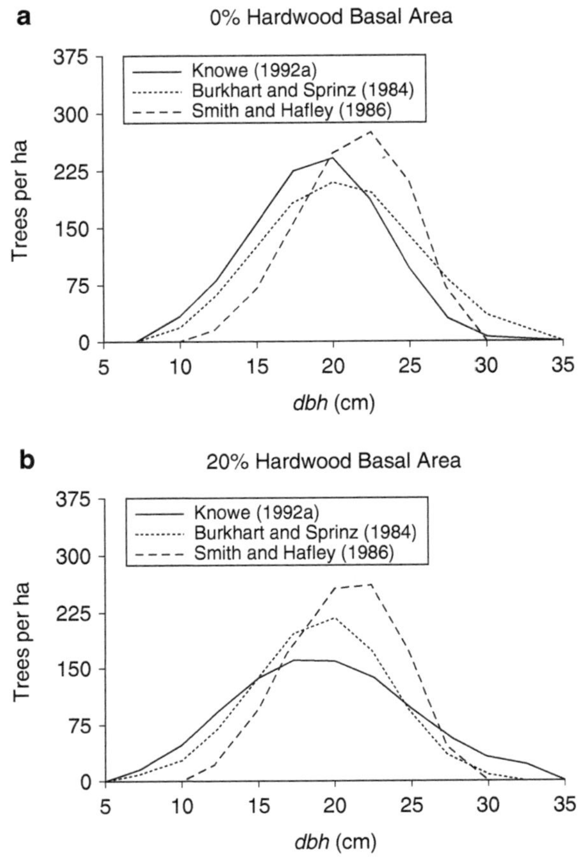

Fig. 16.8 Comparison of the predicted effect of hardwood competition on loblolly pine diameter distributions for age 25 and site index 20 m (base age 25) obtained for three published models. (**a**) 0% hardwood basal area. (**b**) 20% hardwood basal area. (From Knowe 1992a)

threshold diameter (d_T), quadratic mean diameter ($\bar{d}_g$), merchantable top diameter limit to (d_t), and trees per unit area, namely:

$$V_m = V e^{b_1\left(\frac{d_t}{\bar{d}_g}\right)+b_2 N^{b_2}\left(\frac{d_T}{\bar{d}_g}\right)^{b_3}} \tag{16.14}$$

where V_m is merchantable pine volume per unit area for trees with $dbh \geq d_T$ to top diameter limit of d_t.

Using 5-year remeasurement data from the same plots as those originally analyzed by Shiver and Brister, Martin and Brister (1999) developed a system of growth equations to project yield over time that accounts for hardwood competition. Their system allows for an increase in the proportion of hardwood basal area over time. The projected pine basal area and trees per hectare are adjusted to account for increased hardwood levels.

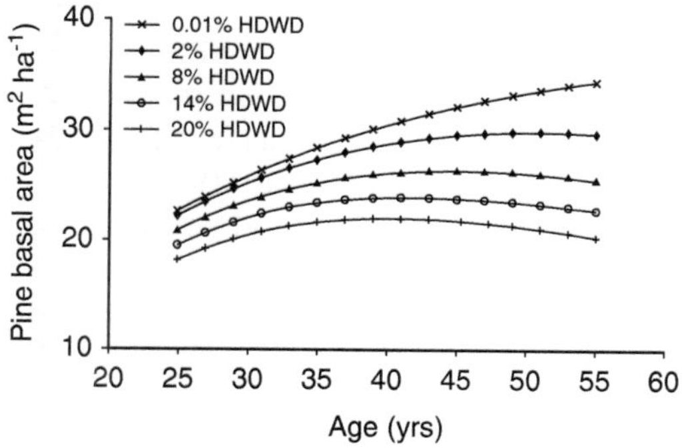

Fig. 16.9 Projected natural-stand loblolly pine basal area over time starting with $23 m^2 ha^{-1}$ of total basal area at five different initial hardwood percentages utilizing Eqs. 16.15 and 16.16 (From Martin and Brister 1999).

The following equation form was fitted to project the proportion of hardwood basal area from t_1 to t_2:

$$\frac{G_{h2}}{G_2} = \left(\frac{G_{h1}}{G_1}\right)^{(t_1/t_2)^{b_1}} \tag{16.15}$$

where G_{hi}/G_i = proportion of hardwood basal area at age t_i.

Pine basal area was projected using an equation form with two components; the first is a basal area projection equation and the second is a multiplicative reduction factor that reduces the pine basal area according to the projected proportion of hardwood basal area in the stand at age t_2:

$$G_{p2} = \left[G_{p1}^{(t_1/t_2)} e^{b_1(1-t_1/t_2)}\right]\left(1 - \frac{G_{h2}}{G_2}\right)^{(1-t_1/t_2)} \quad 0 < \frac{G_{h2}}{G_2} < 1 \tag{16.16}$$

where G_{pi} is pine basal area per unit area at age t_i. Using Eq. 16.16, basal area growth curves were generated for selected values of G_{p1}, G_{h2}/G_2, t_1 and t_2 (Fig. 16.9).

Martin and Brister (1999) also formulated a function to project number of trees per unit area for stands with varying levels of hardwood basal area:

$$N_{p2} = \left[N_{\min} + (N_{p1} - N_{\min})e^{b_1(t_2-t_1)}\right]\left(1 - \frac{G_{h2}}{G_2}\right)^{(1-t_1/t_2)} \tag{16.17}$$

where N_{pi} is pine trees per unit area at age t_i and $N_{\min}$ is defined by an arbitrarily fixed lower bound for number of trees per unit area.

The yield equations of Shiver and Brister (1996), which were fitted to the initial plot measurements, were found to be accurate for stands with a low proportion of hardwoods but as the proportion of hardwoods increased above 0.3 the yield functions over estimated pine volume. Hence, Martin and Brister (1999) fitted a new yield function (16.18):

$$V_p = b_0 G^{b_1} h_{dom}^{b_2} N_p^{b_3} \tag{16.18}$$

where variables remain as defined previously. Equation 16.18 does not explicitly account for the proportion of hardwoods in the stand, but rather stand composition is accounted for in the estimation of projected pine basal area and trees per unit area (16.16 and 16.17). Merchantable volumes can be obtained by utilizing equations published in Shiver and Brister (1996).

The Stand Prognosis Model structure (Stage 1973), which is now designated FVS or Forest Vegetation Simulator (Sect. 14.7.2), can be used to calculate expected yield for mixed-conifer forests in the Inland Northwest of the USA for management regimes that include vegetation management. Stage and Boyd (1987) provide an overview, with examples, of how the FVS system might be applied to simulate vegetation management practices in the Inland Northwest which include site preparation, early conifer release from shrub and herb competition, and cleaning and thinning to favor particular conifer species while controlling stand density.

Because of the importance of individual tree height development in models of conifer stands, Salas et al. (2008) developed and evaluated an individual-tree height growth model for Douglas-fir in the Inland Northwest. The model predicts growth for all tree sizes continuously, rather than requiring a transition from models of juvenile to mature growth phases. Effects of overstory and understory vegetative competition on height growth are included. The model requires attained height rather than tree age as a predictor variable, thus avoiding the necessity of specifying site index. Site effects are expressed as a function of ecological habitat type, elevation, aspect, and slope.

The model of Salas et al. (2008) is intended for incorporation into stand models with an individual-tree level of organization. The Richards growth model with a power transformation

$$\frac{dh^c}{dt} = b(a^c - h^c) \tag{16.19}$$

was analytically integrated to obtain a yield equation (i.e. cumulative growth). Integrating (16.19) between t_0 and t_1 gives height at t_1, that is:

$$h_1 = a\left[1 - (1 - (h_0/h_1)^c)e^{-b(t_1-t_0)}\right]^{1/c} \tag{16.20}$$

where a, b and c are parameters to be estimated, with a being the upper asymptote, b a rate parameter and c a shape parameter.

Salas et al. (2008) introduced competition effects on height growth by modeling the b parameter in (16.20) as a function of overstory and understory vegetation. The modified growth parameter denoted by b' has the expression:

$$b' = b_0\left(b_1 - \frac{1}{1 + e^{b_2}}\right)$$

$$b_2 = b_{20} + b_{21}\frac{C_{ovs0}}{\sqrt{h_0}} + \frac{h_{uns0}}{h_0^2}b_{23}C_{uns0} \tag{16.21}$$

where b_0 is a parameter related to the maximum growth rate, C_{ovs} is a variable that represents the overstory competition (a stand density measure such as basal area per unit area), h_{uns} is the average height of the understory and C_{uns} is a measure of the amount of competing vegetation (e.g. understory cover or understory crown volume). The 0 subscript of the variables means that those are measured at the beginning of the period. Parameters to be estimated are b_0, b_1, b_{20}, b_{21} and b_{23}.

Due to the hierarchical nature of the data available, Salas et al. (2008) used a mixed-effects approach to estimating model parameters. The complexity of the model and diversity of the data sets used precluded estimating optimum values of all parameters simultaneously. Consequently, the required parameters were estimated in stages. Results from the fitted model showed that both overstory and understory density affect height growth.

In a review of modeling applied to vegetation management, Mason and Dzierzon (2006) discussed models that range in resolution from single yield equations to complex representations of processes affecting growth and competition for light, water, and nutrients. Six generic model forms were identified. Models were further categorized by whether they focus on weed population dynamics or on processes directly affecting growth of crop plants. The need persists for models that accurately forecast forest crop yields for a wide variety of sites and vegetation management practices.

16.4.3 Fertilizer Applications

Fertilizer applications are an important silvicultural tool for increasing tree growth and accelerating stand development. Growth and yield projections that accurately reflect response to fertilizer treatments are required when determining prescriptions for nutrient amendments during various stages of stand development.

Duzan et al. (1982) related response of loblolly pine plantations in the Coastal Plain and Piedmont regions of the southeastern USA to stand conditions. Volume increment equations were developed using basal area, site index and fertilizer treatment as predictor variables. Response to fertilization was calculated as the difference between growth estimates for control and fertilized stands at given levels of basal area and site index.

A simple model was formulated by Ballard (1984) for predicting Douglas-fir growth response to nitrogen fertilizer applications in western North America. Cumulative volume growth response, resulting from fertilizer application, was predicted by:

$$R = kt_f A_N S_c S_s S$$

where k is a constant, t_f is time since fertilizer treatment; A_N is amount of fertilizer nutrient applied, S_c is stand composition, S_s is a stocking factor, and S is site quality as expressed by site index. The model includes information on amount of nutrient applied, the temporal aspect of response to fertilizer, and indicates that response is conditional on stand characteristics and site variables.

Shafii et al. (1990) developed individual-tree diameter growth models for quantifying within-stand response to nitrogen fertilization of grand fir and Douglas-fir in northern Idaho, USA. Because the Prognosis model (Stage 1973) is widely used in the Inland Empire region, Shafii et al. (1990) first specified a Prognosis-type diameter increment model as follows:

$$\ln(i_{du}^2) = b_0 + I_{hab} + I_{trt} + I_{sp} + b_1(sl\cos(asp)) + b_2(sl\sin(asp)) + b_3 sl + b_4 sl^2 + b_5 el + b_6 el^2 + b_7 c_r + b_8 c_r^2 + b_9 CCF_{>d} + b_{10}\frac{d}{\bar{d}} + b_{11} I_{trt}\ln d + b_{12} G I_{trt} \qquad (16.22)$$

where:

i_{du} = inside-bark diameter growth
b_0 = constant term representing the overall regression intercept
I_{hab} = dummy variable representing habitat type
I_{trt} = dummy variable representing treatment type (control, fertilized, thinned, thinned and fertilized)
I_{sp} = dummy variable representing species (grand fir, Douglas-fir)
sl = stand slope percent
asp = stand aspect
el = stand elevation
c_r = individual-tree percent live crown
$CCF_{>d}$ = crown competition factor in trees larger than the subject tree
d = initial *dbh* of subject tree
$\bar{d}$ = average stand diameter
G = initial stand basal area
$b_1 - b_{12}$ = regression coefficients

Model (16.22) includes three categorical variables expressing the differential effects (direction of shift in the intercept as well as change in the slope) of habitat type, treatment and species on diameter growth. The site factors included in the model, i.e. slope, aspect, habitat type, and elevation, are the same as those given in the Prognosis diameter-increment model specification.

Since site index is a commonly used method for estimating site quality, Shafii et al. (1990) formulated a second diameter-increment model by replacing the seven site-dependent terms in (16.22) with site index. Using site index as the measure of site quality makes the model more parsimonious, eliminates collinearity among the site factors, and reduces the potential for ill conditioning between these and other regression variables in the model.

The diameter-increment model with site index as a predictor took the form:

$$\ln\left(i_{du}^2\right) = b_0 + I_{trt} + I_{sp} + b_1 S + b_2 c_r + b_3 c_r^2 + b_4 CCF_{>d} + b_5 \frac{d}{\bar{d}} + b_6 I_{trt} \ln d + b_7 G I_{trt} \tag{16.23}$$

where S is grand fir site index and all the other terms are as previously defined.

Regression results for models (16.22) and (16.23) were similar over the specified growth periods. All slope coefficients associated with the continuous variables were significant and had comparable standard errors for all three growth periods. The R^2 values associated with (16.22) for 14-, 10-, and 5-year growth periods were 0.71, 0.72, and 0.69, respectively. For Eq. 16.23 the R^2 values were 0.70, 0.70 and 0.68 for the three growth periods. Residuals from both models showed no bias when displayed by all tree, density, and competition variables as well as the predicted value of diameter growth.

The individual-tree growth response analysis conducted by Shafii et al. (1990) showed that nitrogen fertilizer changes the distribution of diameter increment across tree size classes within a stand. Larger trees showed more growth response to nitrogen treatments than smaller trees. Carlson et al. (2008) reported similar results for response of loblolly pine in the southeastern USA to mid-rotation fertilizer treatments of nitrogen and phosphorous. Both absolute growth response and relative growth response of individual trees were greater for the larger trees.

Hynynen (1993) reported on individual-tree models for predicting tree basal area growth response following nitrogen fertilization in middle-aged, managed stands of Scots pine in southern Finland. Data from unfertilized control plots were used to develop a reference model for basal area growth. Growth response of fertilized trees was calculated as the difference between observed growth and predicted reference growth. The temporal distribution of tree basal area growth response was modeled using the Weibull function. Parameters of the Weibull distribution were expressed as a function of stand characteristics.

Diameter and height growth models for fertilized loblolly pine were developed by Hynynen et al. (1998) using data from mid-rotation plantations throughout the southeastern United States. Tree growth in fertilized stands was predicted with a reference growth model multiplied by an equation predicting the relative growth response following fertilization. The temporal distribution of growth responses was modeled by the Weibull function. Information about dose, nutrients, and time elapsed since treatment is needed to predict the relative growth response following fertilization. The resultant equations for growth response are compatible with individual-tree based model structures.

The strategy adopted by Hynynen et al. (1998) in modeling tree diameter growth, i_d, following fertilizer application was to incorporate the effects of fertilization into a base growth model. Thus, a single growth model was developed using data from both non-fertilized and fertilized stands. The following multiplicative model form was chosen as the basic model structure:

$$i_{dF} = i_d R_F$$

where i_{dF} is the increment in diameter with incorporation of the effect of fertilization. The first part of the model (i_d) predicts tree growth without fertilization (reference growth). It includes the effects of site as well as tree and stand characteristics.

The second part of the model (R_F) includes the direct fertilization effects on tree growth. It predicts the relative growth responses following fertilization. Thus, R_F is a multiplier which is applied to reference growth to predict the growth of fertilized trees. The R_F function was developed in such a way that it can be used in conjunction with reference growth models other than the one parameterized for the overall model of Hynynen et al. (1998).

Information on the duration and distribution of growth response in mid-rotation loblolly pine stands following fertilization (e.g. Ballard 1981) indicates that the response increases with a peak response occurring during the first 4 year after fertilization. There is a fairly rapid decline in response following the peak. The Weibull function was applied in modeling the temporal distribution of the response to the fertilizer treatments in loblolly pine. Hynynen et al. (1998) modeled tree height growth as the product of potential height growth multiplied by a modifier function. Increment of stand dominant height was assumed to represent the stand-level potential height increment. Height growth of an individual tree was regarded to be either smaller or greater than dominant height increment depending on the tree's competitive status and vigor. In modeling height growth response following fertilization, a similar strategy to that applied in developing the diameter growth model was adopted.

The fertilizer response functions (R_F) for diameter and dominant height increment account for direct fertilization response; these response functions are conditioned to be greater than or equal to one. The R_F functions, which predict growth response due to fertilizer application, will not, however, by themselves provide estimates of growth increase due to nutrient amendments. Growth response is also affected by the predicted reference growth. Because fertilization changes the patterns of stand development, e.g., the development of stand basal area, it has a strong effect on the reference growth prediction. The magnitude and duration of the total absolute growth response can only be obtained by simulating the stand development using the reference growth model and the fertilization response function together.

Although the fertilization response function is always ≥ 1, it is possible to obtain negative absolute growth responses in the simulation of individual tree development. After the fertilization response has diminished only the reference growth model affects growth prediction. Actual growth of a fertilized tree can then be either greater or smaller than what it would be if the tree had not been fertilized, depending on relative tree size and stand basal area. Therefore, the models can take into account the effects of changed patterns of stand development in fertilized stands during long-term simulations (Hynynen et al. 1998).

The temporal pattern of the growth response to fertilization indicated that stand basal area growth peaks at around 2 years after fertilization. Thereafter, the response starts to decrease and levels off around 8 years after fertilization (Fig. 16.10a).

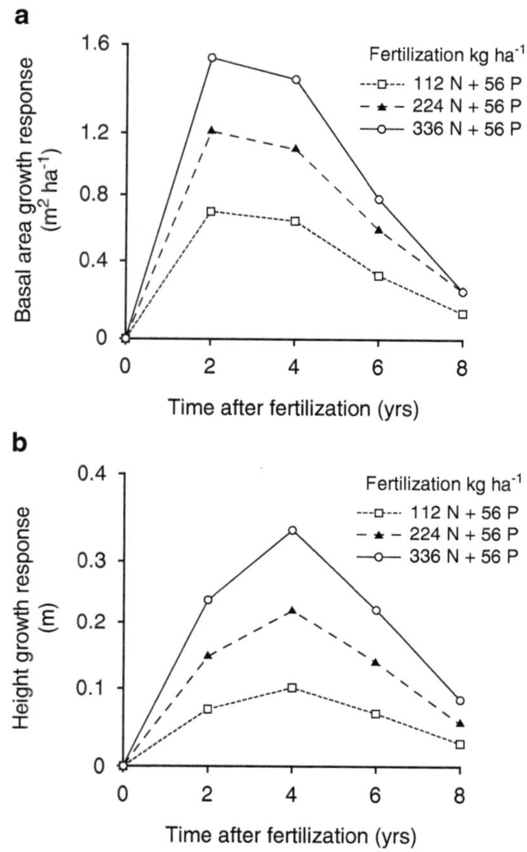

Fig. 16.10 Predicted response in loblolly pine stand basal area growth (**a**), and stand dominant height increment (**b**) with varying fertilizer treatments (From Hynynen et al. 1998)

For stand dominant height, growth response reaches its maximum level somewhat later than basal area, around 2–4 years after fertilization (Fig. 16.10b).

The magnitude of the growth response is strongly affected by the dose and the elements added. Fertilization with phosphorous and nitrogen results in much greater response than nitrogen applications alone (Fig. 16.11). In the fertilized stands, phosphorous application gave significant growth response only when added with nitrogen; the models will predict no growth response after fertilization with only phosphorous application. The increase in phosphorous doses from 28 to 56 kg ha^{-1} had no significant effect on the response. Consequently, the effect of phosphorous was included in the models using a categorical variable.

Bailey et al. (1989) used data from a regional fertilization study of mid-rotation slash pine plantations in the USA to fit prediction equations for basal area and trees per unit area, stand dominant height, diameter distributions and individual tree heights. Prediction equations include nitrogen and phosphorous fertilization rates and soils groups as predictor variables. The integrated system of equations allows calculation of expected yields by diameter class.

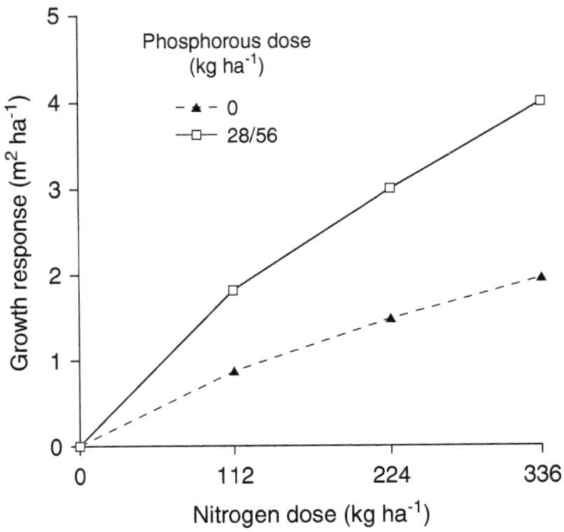

Fig. 16.11 Effect of dose and type of fertilizer on loblolly pine stand basal area growth response (From Hynynen et al. 1998)

The model of Bailey et al. (1989) was based on measurement data from early fertilizer trials in slash pine. Subsequent trials on intensively prepared sites showed greater growth response than what was exhibited in the plots analyzed by Bailey et al. (1989). Accordingly, Martin et al. (1999) developed an updated system of equations for slash pine plantations in the lower Coastal Plain fertilized at mid-rotation that express the combined effects of soil group and nitrogen (N) and phosphorous (P) fertilization on survival, basal area growth and yield, dominant height growth, and the stand diameter distribution. A diameter growth model that accepts an initial diameter distribution (or tree list) provides the ability to predict future diameter distributions. Predictor variables include combinations of three mid-rotation fertilizer treatments (no fertilizer, N only, N and P).

Data from a fertilizer response study in loblolly pine plantations at different sites in the southeastern United States were used by Amateis et al. (2000) to develop response models for dominant height and basal area following mid-rotation nitrogen (N) and phosphorous (P) fertilization. Nonlinear regression models developed from the data predict total cumulative response as a function of the interaction of N and P application rates, drainage class of the site, stand conditions when fertilized, and time since fertilization. Stand variables that were found to be significant predictors of response included site index, age, basal area, number of surviving trees, and dominant height at fertilization. The response models can be applied in conjunction with baseline models developed with data from unfertilized stands to estimate volume response to mid-rotation fertilizer applications. For dominant height, Amateis et al. (2000) applied the general response model

$$h_{domFt} - h_{domt} = a_0 t_t^{a_1} e^{a_2 t_t} \tag{16.24}$$

where h_{domFt} and h_{domt} are the dominant height for fertilized and unfertilized conditions, respectively, at time t following treatment; t_t is the time since treatment, in years; a_0, a_1 and a_2 are parameters. Equation 16.24 combines a power function and an exponential function, is conditioned to be zero when t_t is zero, and increases to a maximum value and then diminishes asymptotically toward zero. The shape of the response curve is strongly influenced by a_2 and the maximum response and the time to maximum response depend on the values of a_0, a_1, and a_2.

Nutrient factors found to affect the magnitude of response to fertilization included the amount of *N* and *P* applied. Stand conditions at time of fertilization that affected response included site index, age, dominant height, and number of trees. In addition, somewhat poorly, poorly, and very poorly, drained sites exhibited a greater dominant height response than moderately well, well, and somewhat excessively drained sites. Therefore, the parameter b_0 was defined as a function of the amount of *N* and the particular stand and site characteristics at fertilization. After including the drainage class, the model to be fitted was:

$$a_0 = (1 - e^{a_{01} A_N})h_{dom}^{a_{02}}\, t^{a_{03}} S^{a_{04}} N^{a_{05}} + a_6 I_{drain} \tag{16.25}$$

where A_N is the amount of nitrogen applied; h_{dom} is the dominant height at fertilization; *t* is stand age at fertilization, in years; *S* is site index; *N* is number of trees per unit area at fertilization. I_{drain} is 1 if the site is somewhat poorly, poorly, or very poorly drained, 0 otherwise, and $a_{01} - a_{06}$ are parameters to be estimated.

Since phosphorous (*P*) affects both the magnitude of the response and the time to maximum response, a_2 was defined as:

$$a_2 = a_{21} - a_{22} \ln(1 + A_P) \tag{16.26}$$

where A_P is the amount of phosphorous applied, and a_{21} and a_{22} are parameters. If a_{21} is negative and a_{22} is positive, then Eq. 16.26 implies that increasing amounts of *P* will increase the magnitude of the response and the time to maximum response. The logarithm transformation ensures that the relative effect diminishes with increasing amounts of *P*. Equations 16.25 and 16.26 were incorporated into Equation 16.24 to give:

$$h_{domFt} - h_{domt} = \left[(1 - e^{a_{01} A_N})h_{dom}^{a_{02}} t^{a_{03}} S^{a_{04}} N^{a_{05}} + a_6 I_{drain}\right] t_t^{a} e^{(a_{21} + a_{22} \ln(1 + A_P))t_t} \tag{16.27}$$

For modeling basal area response Amateis et al. (2000) found the combined power-exponential model of Eq. 16.24 was also suitable:

$$G_{Ft} - G_t = b_0 t_t^{b_1} e^{b_2 t_t} \tag{16.28}$$

where G_{Ft} and G_t are the basal area for fertilized and unfertilized conditions, respectively, at time *t* following treatment; b_0, b_1, and b_2 are parameters, and all other variables are as previously defined. As with dominant height, $b_0$was made a function of the amount of N and of stand characteristics at fertilization, which were significant predictors of response:

$$b_0 = \left(1 - e^{b_{01} A_N}\right) h_{dom}^{b_{02}} N^{b_{03}} G^{b_{04}} \tag{16.29}$$

In general, the data indicated that the amount of P has a positive effect on both the magnitude and duration of basal area response. However, very poorly drained sites were found to be responsive only to N and somewhat excessively, well, and moderately well drained sites exhibited an even greater response than average to the combined N and P treatments. Therefore, the amount of P and drainage class were included in the basal area response model in the function of b_2:

$$b_2 = b_{21} + (b_{22} + b_{23} I_{drain1} + b_{24} I_{drain2}) \ln(1 + A_P) \tag{16.30}$$

where $I_{drain1} = 1$ for very poorly drained sites, 0 otherwise; $I_{drain2} = 1$ for somewhat excessively, well and moderately well drained sites, 0 otherwise; and $b_{21} - b_{24}$ are parameters.

Combining Eqs. 16.28, 16.29 and 16.30 defines the final basal area response function:

$$G_{Ft} - G_t = \left[(1 - e^{b_{01} A_N}) h_{dom}^{b_{02}} N^{b_{03}} G^{b_{04}} t_t^{b1}\right] e^{(b_{21} + (b_{22} + b_{23} I_{drain1} + b_{24} I_{drain2}) \ln(1 + A_P)) t_t} \tag{16.31}$$

The equations fitted by Amateis et al. (2000) indicate that: (1) additional amounts of N will increase response but at a decreasing rate, (2) adding N and P together has a synergistic effect producing a greater response than N alone, (3) the magnitude of the response increases to some maximum after application and then diminishes, and (4) the magnitude and duration of the response is determined by the amount of N and P applied and the drainage class of the site.

Stand and site conditions at fertilization were also found to affect the response to treatment. For dominant height response, site index, dominant height, age, and number of stems at time of fertilization influenced the magnitude of response. Poorly drained sites were found to have a greater height response than other sites. For basal area, important stand characteristics were dominant height, number of stems, and basal area at fertilization. Very poorly drained sites had little basal area response to a combined N and P fertilizer treatment, whereas well-drained sites exhibited a greater than normal response to N and P. The specific response after fertilization for any particular stand results from a complex interaction of nutrients applied, rates of application, stand and site condition at time of application, and years since application (Amateis et al. 2000).

16.4.4 Genetic Improvement

Forecasting yields of stands established with improved trees requires that genetic effects be incorporated into growth and yield models. Typically, the genetic gain information available is for individual trees measured at young ages. The challenge, then, is to predict rotation-age yields on a per unit area basis by appropriately incorporating information from young trees. In addition, the process of identifying and selecting genetically superior trees is ongoing and continuous; by the time rotation-age yields are available for given genotypes, it is likely that those genotypes will have been supplanted by more advanced stock. Hence, the quantification of genetic improvement effects on forest stand dynamics, growth and yield has been and remains one of the most vexing problems of forest modeling.

Buford and Burkhart (1987) developed and tested a series of hypotheses concerning stand dynamics and growth patterns in loblolly pine plantations of improved stock relative to plantations of unimproved stock. Data available were from three sources: (1) a loblolly pine seed source study measured to age 25, (2) a 15-year-old block-plot open-pollinated progeny test, and (3) temporary plots in 34 operational plantings of improved loblolly pine stock. Results from this study indicated that at the seed source and family levels: (i) the shape of the height-age curve is influenced by the site, but the level is dependent on the genetic material planted, and (ii) the shape of the height-diameter relationship at a given age is dependent on the site and initial density, whereas the level is affected by genetics and is directly related to the dominant height of the seed source or family at that age. These results indicated differences in development among seed sources and first generation open-pollinated families on a given site can be represented by altering the level (site index) of the height-age equation.

In the analysis of Buford and Burkhart (1987), the logarithm of height-reciprocal of age model was used to describe dominant height development. Sprinz et al. (1989) reported differences in shape as well as level when the Chapman-Richards form of a height-age model was fitted to data from a loblolly pine source study in Arkansas, USA.

An approach for predicting genetic gains in loblolly pine using data from progeny tests with height-age models and stand simulations was presented by Knowe and Foster (1989). When the Chapman-Richards model was fitted to periodic remeasurements of height for a block-plot, open-pollinated progeny test, significant differences among families were detected in asymptote and rate parameters but not for the shape parameter. Differences in survival, height, and diameter were combined to examine trends in volume production associated with families. Predicted volume for each family was simulated by applying a bivariate distribution model of height and diameter (Hafley and Buford 1985) and using a family-specific height-age curve. Since height, diameter, and survival functions in the stand simulation model are inter-related, a change in one component such as the height-age curve would affect the other components.

Adams et al. (2006) examined differences in survival, diameter, dominant height, and stem profile among eight open-pollinated families and a commercial check of loblolly pine. Measurements from three spacings obtained over 17 years were evaluated for family-specific effects when applying a growth and yield model. Actual stand volume at age 17 years was compared with the volume predicted from age nine measurements using the stand models without modification and after modifications for family differences. Use of family-specific stem profile and site index equations in combination with density effects on survival and diameter prediction provided the best estimates of future stand volumes. Adams et al. (2006) concluded that incorporation of genetic effects in growth and yield models should focus on quantifying differences in family response to competition (i.e. ability to survive and growth at varying levels of stand density).

A stand simulation model (Prognosis model version 5.2) was used by Rehfeldt et al. (1991) to project yields per unit area from the progenies of western white pine selections made in year 7. Results of simulations suggested that the amount of increase in yield depends on the stand density, rotation length, and manner by which genetic gains accrue beyond year 25. Hamilton and Rehfeldt (1994) expanded the methods described by Rehfeldt et al. (1991) and provided a more general approach to the use of individual tree growth models for estimating anticipated stand-level gains in yield. The approach proposed by Hamilton and Rehfeldt (1994) consists of four steps:

1. Use growth data available from tree improvement plots to calibrate an existing model for the observed performance of unimproved stock and for anticipated performance of genetically improved stock.
2. Formulate assumptions and hypotheses about how genetic gains estimated from measures of early growth rates of individual trees will be expressed throughout the stand development process.
3. Run the calibrated model for the desired time interval (rotation) with and without modifications for genetic gain.
4. Evaluate the sensitivity of growth prediction to changes in assumptions and hypotheses about genetic gain and about stand development for improved stock.

Hamilton and Rehfeldt used a distance-independent individual-tree model (version 6.0 of the Inland Empire variant of the Stand Prognosis model) in their analyses, but they pointed out that their basic approach could be applied with any individual tree growth model that permits modification of growth rates of individual components. The Prognosis model has four basic growth components: basal area increment, height increment, mortality, and change in crown ratio. The model also includes a set of growth modifiers that facilitate simulation of the effects of changes in individual tree mortality rates and in individual tree growth rates for either height or basal area. Calibration of the Prognosis model for unimproved and improved stock of ponderosa pine was accomplished with growth multipliers. In addition to the cautions on use of growth multipliers given by Hamilton and Rehfeldt (1994), Hamilton (1994) provides a detailed description of the types of multipliers incorporated in the Prognosis model and limitations on their use.

Because of interdependence of model components, a change in height or basal area increment will affect the other. Problems may arise in selecting the proper set of multipliers or in determining appropriate values to assign to selected multipliers. If multipliers are used incorrectly, they can produce stand development patterns that are not realistic.

Data from radiata pine seed lots of varying genetic quality planted in block-plot genetic gain trials at 10 locations in New Zealand were analyzed by Carson et al. (1999). Permanent sample plots were measured annually for growth from ages 6–8 years since planting to ages 15–17 (mid-rotation). Seedlots from first generation open-pollinated seed orchards and a mix of crosses that all involved the top-performing parent were taller, larger in mean diameter, and with larger basal area and stem volume than seedlots originating from mild mass selection in harvested stands. The observed growth increases were quantified as changes in the rate of growth from that predicted by extant growth models in order to account for tree size and stocking differences. Seedlots from first-generation seed orchards and crosses of the top clone grew faster in height. Functions for basal area and stocking showed disproportionally larger changes than the baseline growth models, thus implying that basal area growth must be taken into account in order to obtain accurate predictions of gain in stand volume. The observed increases in growth rates were incorporated into stand growth models as "genetic-gain multipliers" in order to extrapolate predictions of growth of genetically improved seedlots beyond the sites, silvicultural regimes, and seedlots represented in the genetic gain trials.

Carson et al. (1999) calculated differences in growth among different genetic stocks using existing growth models (described by García 1984, 1994; Goulding 1994). The growth models developed for radiata pine in New Zealand include as state variables dominant height (h_{dom}), basal area (G), and number of stems per hectare (N):

$$\begin{aligned} \frac{dh_{dom}}{dt} &= f_1(h_{dom}) \\ \frac{dG}{dt} &= f_2(h_{dom}, G, N) \\ \frac{dN}{dt} &= f_3(h_{dom}, G, N) \end{aligned} \qquad (16.32)$$

Differences in growth rates with genetic improvement for both height and basal area were allowed to vary since the processes of growth are different for height and diameter (shoot-tip extension from apical meristem versus radial growth from cambial activity), and they are affected differently by intertree competition. Because of the form of the models, the basal area and stocking equations are difficult to separate; thus change in growth rate estimated for the basal area equation was assumed to be the same as that for the tree stocking equation. This assumption implied that when trees die as a result of competition from neighboring trees, an increase in the rate of basal area growth accelerates mortality at the same rate (Carson et al. 1999).

Although estimation of separate multipliers for basal area and stocking would be difficult to achieve with the structure of model (16.32), Carson et al. (1999) noted that multipliers could be similarly added to the last two equations by assuming a common multiplier for basal area and stems per hectare. A model incorporating genetic effects could be constructed by multiplying each of the growth-rate equations of (16.32) by a genetic-gain multiplier (m_i), where m_1 is the height equation multiplier, m_2 is the basal area equation multiplier, m_3 is the stocking equation multiplier, and $m_2 = m_3$. The multipliers would represent the relative growth improvement for the various state variables. An alternative and slightly different formulation was used because it was easier to implement, but the authors noted that the difference between the two approaches is unlikely to be of practical importance.

Xie and Yanchuk (2003) described the procedures used in British Columbia, Canada, for predicting breeding values of parents, estimating genetic worth of orchard seedlots, and projecting yields in genetically improved stock. Breeding value is a measure of the genetic quality of an individual as a parent. In British Columbia, breeding value for growth potential is expressed as percent gain of stem volume over the unimproved population at a designated rotation age.

Genetic worth represents the average level of genetic gain expected for the trait of concern at a designated rotation age when a seedlot is used for reforestation. Currently, the genetic worth of a seedlot is estimated in British Columbia by the mean breeding value of all the parents.

The yield of a genetically improved plantation is projected by incorporating the genetic worth of the seedlot into the existing growth model developed using extensive data from managed unimproved stands. This approach takes account of the stand dynamics determined by site conditions and silivicultural regimes and the declining nature of expected gain over time because of imperfect age-age genetic correlation.

In the procedure described by Xie and Yanchuk (2003), genetic worth is translated into an increase in site index at the designated rotation age; site index is then increased accordingly when projecting stand yields with a model designated TIPSY (Table Interpolation Program for Stand Yields). Xie and Yanchuk reasoned that since selection for timber production of commercially important conifers in British Columbia has been primarily based on height, the height-diameter or height-volume relations of the selected trees would not be expected to differ significantly from the unimproved stocks under the same stand conditions. Increases in height from genetic improvement are, thus, converted to volume gain using the height-volume relationship observed in unimproved trees. Accurate prediction of height gain is then the key for accurately projecting volume gains of genetically improved stocks. Similar assumptions, as noted previously, have been applied when modeling volume gains in open-pollinated southern pine in the USA, where the emphasis in genetic selection has also been on height development.

By contrast, Carson et al. (1999) found that the change in the rate of increase of basal area was much greater than the change in the rate of increase in dominant height, making increases in basal area and volume growth much greater than would

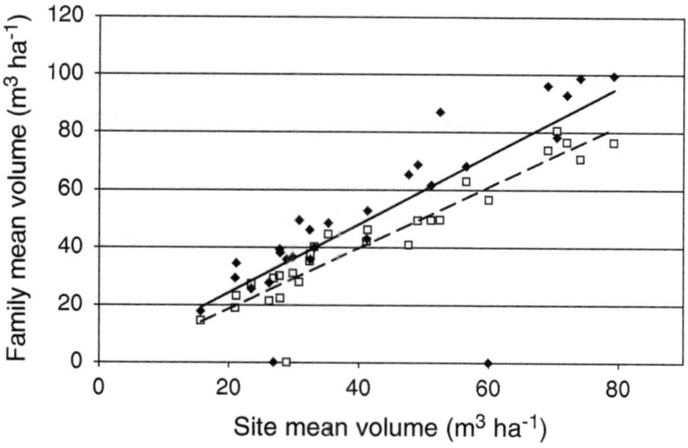

Fig. 16.12 Response of a highly productive family (*solid line* and *diamonds*) compared to an average family (*dashed line* and *open boxes*) across a range of sites at age 8 years. While the relative volume differences are the same across sites, the absolute volume differences are higher on the more productive sites (From McKeand et al. 2006)

be predicted from increasing site index alone. It is important to note, however, that diameter has been a main selection trait for improved growth potential of radiata pine in New Zealand.

Gould et al. (2008) developed methods to calculate genetic gain multipliers for use in individual tree models that predict periodic height and diameter growth of coast Douglas-fir for the Pacific Northwest region of the USA. Nonlinear mixed-effect models were initially developed to predict the average growth of trees in all families, which, taken together, represented unimproved populations. Phenotypic differences in growth rates were then calculated at the family level. Differences among families in height and diameter growth rates were analyzed using methods from quantitative genetics and raw phenotypic values. Because gain in total height and diameter at age 10 years is commonly available genetic information for improved Douglas-fir, equations were developed to predict genetic-gain multipliers from family breeding values for these traits. Incorporating multipliers in growth projection reduced the mean-square error of predicted growth of selected families.

Selection criteria for tree improvement generally consider much more that just volume traits. Stem straightness, form, branch size and branch angles, disease resistance, and wood properties usually enter into selection decisions as well. Consequently, modifications to growth and yield model predictions of product volumes and values may require adjustments of several components.

In a review of the performance of improved genotypes of loblolly pine planted across different soils and climates and with varying silvicultural inputs in the southeastern United States, McKeand et al. (2006) showed that open-pollinated (OP) families typically display a high degree of rank stability for productivity and quality traits (Fig. 16.12). With only a few exceptions families were generally stable in performance across all sites within a climatic zone. The authors conjectured that as tree improvement progresses towards deployment of more intensively selected genotypes and less genetically diverse sibling families or clones, consideration of genotype by environment (G × E) interactions may become more important. However, evidence was presented from numerous trials with full-sib families and clones demonstrating that G × E interaction for growth and other traits is no more significant than for OP families. Accordingly, McKeand et al. (2006) concluded that at present and for the foreseeable future, G × E interaction does not appear to be a major factor for the majority of deployed genetic sources under most silvicultural systems.

References

Adams JP, Matney TG, Land SB, Belli KL, Duzan HW Jr (2006) Incorporating genetic parameters into a loblolly pine growth-and-yield model. Can J For Res 36:1959–1967

Amateis RL (2000) Modeling response to thinning in loblolly pine plantations. South J Appl For 24:17–22

Amateis RL, Burkhart HE (2005) The influence of thinning on the proportion of peeler, sawtimber, and pulpwood trees in loblolly pine plantations. South J Appl For 29:158–162

Amateis RL, Burkhart HE, Burk TE (1986) A ratio approach to predicting merchantable yields of unthinned loblolly pine plantations. For Sci 32:287–296

Amateis RL, Burkhart HE, Liu J (1997) Modeling survival in juvenile and mature loblolly pine plantations. For Ecol Manage 90:51–58

Amateis RL, Liu J, Ducey MJ, Allen HL (2000) Modeling response to midrotation nitrogen and phosphorus fertilization in loblolly pine plantations. South J Appl For 24:207–212

Avila OB, Burkhart HE (1992) Modeling survival of loblolly pine trees in thinned and unthinned plantations. Can J For Res 22:1878–1882

Bailey RL, Ware KD (1983) Compatible basal-area growth and yield model for thinned and unthinned stands. Can J For Res 13:563–571

Bailey RL, Borders BE, Ware KD, Jones EP Jr (1985) A compatible model relating slash pine plantation survival to density, age, site index, and type and intensity of thinning. For Sci 31:180–189

Bailey RL, Burgan TM, Jokela EJ (1989) Fertilized midrotation-aged slash pine plantations – stand structure and yield prediction models. South J Appl For 13:76–80

Baldwin VC Jr, Feduccia DP, Haywood JD (1989) Postthinning growth and yield of row-thinned and selectively thinned loblolly and slash pine plantations. Can J For Res 19:247–256

Ballard R (1981) Optimum nitrogen rates for fertilization of loblolly pine plantations. South J Appl For 5:212–216

Ballard TM (1984) A simple model for predicting stand volume growth response to fertilizer application. Can J For Res 14:661–665

Belli KL, Ek AR (1988) Growth and survival modeling for planted conifers in the Great Lakes region. For Sci 34:458–473

Borders BE, Bailey RL (2001) Loblolly pine – pushing the limits of growth. South J Appl For 25:69–74

Buford MA (1986) Height-diameter relationships at age 15 in loblolly pine seed sources. For Sci 32:812–818

Buford MA (1989) Mean stem size and total volume development of various loblolly pine seed sources planted at one location. Can J For Res 19:396–400

Buford MA, Burkhart HE (1987) Genetic improvement effects on growth and yield of loblolly pine plantations. For Sci 33:707–724

Burkhart HE, Bredenkamp BV (1989) Product-class proportions for thinned and unthinned loblolly pine plantations. South J Appl For 13:192–195

Burkhart HE, Sprinz PT (1984) A model for assessing hardwood competition effects on yields of loblolly pine plantations. Virginia Polytechnic Institute and State University, Blacksburg, Pub. FWS-3-84

Carlson CA, Burkhart HE, Allen HL, Fox TR (2008) Absolute and relative changes in tree growth rates and changes to the stand diameter distribution of *Pinus taeda* as a result of midrotation fertilizer applications. Can J For Res 38:2063–2071

Carson SD, Hayes JD (1998) Diameter and height distributions in genetically improved *Pinus radiata*. Can J For Res 28:248–258

Carson SD, Garcia O, Hayes JD (1999) Realized gain and prediction of yield with genetically improved *Pinus radiata* in New Zealand. For Sci 45:186–200

Clason TR (1993) Hardwood competition reduces loblolly pine plantation productivity. Can J For Res 23:2133–2140

Clutter JL, Jones EP Jr (1980) Prediction of growth after thinning in old-field slash pine plantations. USDA Forest Service, Southeastern Forest Experiment Station, Asheville. Research Paper SE-217

Duzan HW Jr, Allen HL, Ballard R (1982) Predicting fertilizer response in established loblolly pine plantations with basal area and site index. South J Appl For 6:15–19

Euler Fv, Baradat P, Lemoine B, Lemoine B (1992) Effects of plantation density and spacing on competitive interactions among half-sib families of maritime pine. Can J For Res 22:482–489

Field RC, Clutter JL, Jones EP Jr (1978) Predicting thinning volumes for pine plantations. South J Appl For 2:59–61

Fox TR, Allen HL, Albaugh TJ, Rubilar R, Carlson CA (2007) Tree nutrition and forest fertilization of pine plantations in the southern United States. South J Appl For 31:5–11

García O (1984) New class of growth models for even-aged stands: *Pinus radiata* in Golden Downs Forest. N Z J For Sci 14:65–88

García O (1994) The state-space approach in growth modelling. Can J For Res 24:1894–1903

Gent JA, Allen HL, Campbell RG, Wells CG (1986) Magnitude, duration, and economic analysis of loblolly pine growth response following bedding and phosphorous fertilization. South J Appl For 10:124–128

Glover GR, Zutter BR (1993) Loblolly pine and mixed hardwood stand dynamics for 27 years following chemical, mechanical, and manual site preparation. Can J For Res 23:2126–2132

Gould PJ, Marshall DD (2010) Incorporation of genetic gain into growth projections of Douglas-fir using ORGANON and the forest vegetation simulator. West J Appl For 25:55–61

Gould P, Johnson R, Marshall D, Johnson G (2008) Estimation of genetic-gain multipliers for modeling Douglas-fir height and diameter growth. For Sci 54:588–596

Goulding CJ (1994) Development of growth models for *Pinus radiata* in New Zealand – experience with management and process models. For Ecol Manage 69:331–343

Hafley WL, Buford MA (1985) A bivariate model for growth and yield prediction. For Sci 31: 237–247

Hamilton DA Jr (1994) Uses and abuses of multipliers in the Stand Prognosis Model. USDA Forest Service, Intermountain Research Station, Ogden. General Technical Report INT-GTR-310

Hamilton DA Jr, Rehfeldt GE (1994) Using individual tree growth projection models to estimate stand-level gains attributable to genetically improved stock. For Ecol Manage 68:189–207

Harrington CA, Reukema DL (1983) Initial shock and long-term stand development following thinning in a Douglas-fir plantation. For Sci 29:33–46

Hasenauer H, Burkhart HE, Amateis RL (1997) Basal area development in thinned and unthinned loblolly pine plantations. Can J For Res 27:265–271

Haywood JD (2005) Influence of precommercial thinning and fertilization on total stem volume and lower stem form of loblolly pine. South J Appl For 29:215–220

Heald RC, Barrett TM (1999) Effects of planting density on early growth of giant sequoia (*Sequoiadendron giganteum*). West J Appl For 14:65–72

Heath LS, Chappell HN (1989) Growth response to fertilization in young Douglas-fir stands. West J Appl For 4:116–119
Hodge GR, White TL, Powell GL, de Souza SM (1989) Predicted genetic gains from one generation of slash pine tree improvement. South J Appl For 13:51–56
Hynynen J (1993) Modelling tree basal area growth response after nitrogen fertilization. In: IUFRO Proceedings Modelling Stand Response to Silvicultural Practices. Virginia Polytechnic Institute and State University, Blacksburg, pp 61–72, Pub. FWS-1-93
Hynynen J, Burkhart HE, Allen HL (1998) Modeling tree growth in fertilized midrotation loblolly pine plantations. For Ecol Manage 107:213–229
Hytönen J, Jylhä P (2011) Long-term response of weed control intensity on Scots pine survival, growth and nutrition on former arable land. Eur J For Res 130:91–98
Jack SB, Stone EL, Swindel BF (1988) Stem form changes in unthinned slash and loblolly pine stands following midrotation fertilization. South J Appl For 12:90–97
Johnstone W (2008) The effects of initial spacing and rectangularity on the early growth of hybrid poplar. West J Appl For 23:189–196
Knowe SA (1992a) Basal area and diameter distribution models for loblolly pine plantations with hardwood competition in the Piedmont and Upper Coastal Plain. South J Appl For 16:93–98
Knowe SA (1992b) Predicting the impact of interspecific competition in young loblolly pine plantations with diameter distribution models. For Ecol Manage 55:65–82
Knowe SA (1994a) Incorporating the effects of interspecific competition and vegetation management treatments in stand table projection models for Douglas-fir saplings. For Ecol Manage 67:87–99
Knowe SA (1994b) Effect of competition control treatments on height-age and height-diameter relationships in young Douglas-fir plantations. For Ecol Manage 67:101–111
Knowe SA, Foster GS (1989) Application of growth models for simulating genetic gain of loblolly pine. For Sci 35:211–228
Knowe SA, Stein WI (1995) Predicting the effects of site preparation and protection on development of young Douglas-fir plantations. Can J For Res 25:1538–1547
Knowe SA, Harrington TB, Shula RG (1992) Incorporating the effects of interspecific competition and vegetation management treatments in diameter distribution models for Douglas-fir saplings. Can J For Res 22:1255–1262
Knowe SA, Aherns GR, DeBell DS (1997a) Comparison of diameter-distribution-prediction, stand-table-projection, and individual-tree-growth modeling approaches for young red alder plantations. For Ecol Manage 98:49–60
Knowe SA, Stein WI, Shainsky LJ (1997b) Predicting growth response of shrubs to clear-cutting and site preparation in coastal Oregon forests. Can J For Res 27:217–226
Knowe SA, Radosevich SR, Shula RG (2005) Basal area and diameter distribution prediction equations for young Douglas-fir plantations with hardwood competition: coast ranges. West J Appl For 20:73–93
Knowe SA, Foster GS, Rousseau RJ, Nance WL (1994) Eastern cottonwood clonal mixing study: predicted diameter distributions. Can J For Res 24:405–414
Kramer H (1966) Crown development in conifer stands in Scotland as influenced by initial spacing and subsequent thinning treatment. Forestry 39:40–58
Lemin RC Jr, Burkhart HE (1983) Predicting mortality after thinning in old-field loblolly pine plantations. South J Appl For 7:20–23
Liu J, Burkhart HE, Amateis RL (1995) Projecting crown measures for loblolly pine trees using a generalized thinning response function. For Sci 41:43–53
Lowery RF, Lambeth CC, Endo M, Kane M (1993) Vegetation management in tropical forest plantations. Can J For Res 23:2006–2014
Lynch TB, Will RE, Hennessey TC, Heinemann RA, Holeman RT (2010) Relationships among diameter at breast height and loblolly pine attributes from local and nonlocal seed sources near the western edge of the natural range of loblolly pine. South J Appl For 34:149–153
Martin SW, Brister GH (1999) A growth and yield model incorporating hardwood competition for natural loblolly pine stands in the Georgia Piedmont. South J Appl For 23:179–185

Martin SW, Bailey RL, Jokela EJ (1999) Growth and yield predictions for lower Coastal Plain slash pine plantations fertilized at mid-rotation. South J Appl For 23:39–45

Mason EG (2001) A model of the juvenile growth and survival of *Pinus radiata* D. Don—adding the effects of initial seedling diameter and plant handling. New For 22:133–158

Mason EG, Dzierzon H (2006) Applications of modeling to vegetation management. Can J For Res 36:2505–2514

Mason EG, Whyte AGD, Woollons RC, Richardson B (1997) A model of the early growth of juvenile radiata pine in the Central North Island of New Zealand: links with older models and rotation-length analyses of the effects of site preparation. For Ecol Manage 97:187–195

McCarthy N, Bentsen NS, Willoughby I, Balandier P (2011) The state of forest vegetation management in Europe in the 21st century. Eur J For Res 130:7–16

McKeand SE, Jokela EJ, Huber DA, Byram TD, Allen HL, Li B, Mullin TJ (2006) Performance of improved genotypes of loblolly pine across different soils, climates, and silvicultural inputs. For Ecol Manage 227:178–184

McTague JP, Bailey RL (1987) Compatible basal area and diameter distribution models for thinned loblolly pine plantations in Santa Catarina. Brazil For Sci 33:43–51

Miller RE, Tarrant RF (1983) Long-term growth response of Douglas-fir to ammonium nitrate fertilizer. For Sci 29:127–137

Miller JH, Zutter BR, Newbold RA, Edwards MB, Zedaker SM (2003) Stand dynamics and plant associates of loblolly pine plantations to midrotation after early intensive vegetation management – a Southeastern United States regional study. South J Appl For 27:221–236

Moore JA, Mika PG, VanderPloeg JA (1991) Nitrogen fertilizer response of Rocky Mountain Douglas-fir by geographic area across the Inland Northwest. West J Appl For 6:94–98

Moore JA, Zhang L, Newberry JD (1994) Effects of intermediate silvicultural treatments on the distribution of within-stand growth. Can J For Res 24:398–404

Newton PF (2003) Systematic review of yield responses of four North American conifers to forest tree improvement practices. For Ecol Manage 172:29–51

Newton M, Cole EC (2006) Use of growth curve derivatives to illustrate acceleration and deceleration of growth in young plantations under variable competition. Can J For Res 36:2515–2522

Peracca GG, O'Hara KI (2008) Effects of growing space on growth for 20-year-old giant sequoia, ponderosa pine, and Douglas-fir in the Sierra Nevada. West J Appl For 23:156–165

Pienaar LV (1979) An approximation of basal area growth after thinning based on growth in unthinned plantations. For Sci 25:223–232

Pienaar LV, Rheney JW (1993) Yield prediction for mechanically site-prepared slash pine plantations in the Southeastern Coastal Plain. South J Appl For 17:163–173

Pienaar LV, Rheney JW (1995) Modeling stand level growth and yield response to silvicultural treatments. For Sci 41:629–638

Pienaar LV, Shiver BD (1984) An analysis and models of basal area growth in 45-year-old unthinned and thinned slash pine plantation plots. For Sci 30:933–942

Pienaar LV, Shiver BD (1986) Basal area prediction and projection equations for pine plantations. For Sci 32:626–633

Pienaar LV, Shiver BD, Grider GE (1985) Predicting basal area growth in thinned slash pine plantations. For Sci 31:731–741

Powers MD, Palik BJ, Bradford JB, Fraver S, Webster CR (2010) Thinning method and intensity influence long-term mortality trends in a red pine forest. For Ecol Manage 260:1138–1148

Rathbun LC, LeMay V, Smith N (2011) Diameter growth models for mixed-species stands of Coastal British Columbia including thinning and fertilization effects. Ecol Mod 222: 2234–2248

Rehfeldt GE, Wycoff WR, Hoff RJ, Steinhoff RJ (1991) Genetic gains in growth and simulated yield of *Pinus monticola*. For Sci 37:326–342

Richardson B (1993) Vegetation management practices in plantation forests of Australia and New Zealand. Can J For Res 23:1989–2005

Richardson B, Watt MS, Mason EG, Kriticos DJ (2006) Advances in modeling and decision support systems for vegetation management in young forest plantations. Forestry 79:29–42

Ritchie MW, Hamann JD (2006) Modeling dynamics of competing vegetation in young conifer plantations of northern California and southern Oregon, USA. Can J For Res 36:2523–2532

Russell MB, Amateis RL, Burkhart HE (2010) Implementing regional locale and thinning response in the loblolly pine height-diameter relationship. South J Appl For 34:21–27

Sabatia CO, Burkhart HE (2012) Competition among loblolly pine trees: does genetic variability of the trees in a stand matter? For Ecol Manage 263:122–130

Salas C, Stage AR, Robison AP (2008) Modeling effects of overstory density and competing vegetation on tree height growth. For Sci 54:107–122

Scott W, Meade R, Leon R, Hyink D, Miller R (1998) Planting density and tree-size relations in coast Douglas-fir. Can J For Res 28:74–78

Shafii B, Moore JA, Olson JR (1989) Effects of nitrogen fertilization on growth of grand fir and Douglas-fir stands in northern Idaho. West J Appl For 4:54–57

Shafii B, Moore JA, Newberry JD (1990) Individual-tree diameter growth models for quantifying within-stand response to nitrogen fertilization. Can J For Res 20:1149–1155

Sharma M, Smith M, Burkhart HE, Amateis RL (2006) Modeling the impact of thinning on height development of dominant and codominant loblolly pine trees. Ann For Sci 63:349–354

Shiver BD, Brister GH (1996) Effect of hardwoods and pine density on natural loblolly pine yields and product distribution. South J Appl For 20:99–102

Shiver BD, Rheney JW, Oppenheimer MJ (1990) Site-preparation method and early cultural treatments affect growth of flatwoods slash pine plantations. South J Appl For 14:183–188

Short EA III, Burkhart HE (1992) Predicting crown-height increment for thinned and unthinned loblolly pine plantations. For Sci 38:594–610

Siemon GR, Wood GB, Forrest WG (1976) Effect of thinning on crown structure in radiata pine. N Z J For Sci 6:57–66

Smith WD, Hafley WL, Smith WD, Hafley WL (1987) Simulating the effect of hardwood encroachment on loblolly pine plantations. In: Proceedings of the fourth biennial Southern silvicultural research conference. USDA Forest Service, Southeastern Forest Experiment Station, Asheville, pp 180–186. General Technical Report SE-42

Snowdon P (2002) Modeling Type 1 and Type 2 growth responses in plantations after application of fertilizer or other silvicultural treatments. For Ecol Manage 163:229–244

Sprinz PT, Burkhart HE (1987) Relationships between tree crown, stem, and stand characteristics in unthinned loblolly pine plantations. Can J For Res 17:534–538

Sprinz PT, Talbert CB, Strub MR (1989) Height-age trends from an Arkansas seed source study. For Sci 35:677–691

St Clair JB, Mandel NL, Jayawickrama KJS (2004) Early realized genetic gains for coastal Douglas-fir in the northern Oregon Cascades. West J Appl For 19:195–201

Stage AR (1973) Prognosis model for stand development. USDA Forest Service, Intermountain Forest and Range Experiment Station, Ogden, Research Paper INT-137

Stage AR, Boyd RJ Jr (1987) Evaluation of growth and yield responses to vegetation management of the mixed-conifer forests in the Inland Northwest. In: Management FV, Walstad JD, Kuch PJ (eds) Forest vegetation management for conifer production. Wiley, New York, pp 295–324

Staudhammer CL, Jokela EJ, Martin TA (2009) Competition dynamics in pure- versus mixed-family stands of loblolly and slash pine in the southeastern United States. Can J For Res 39:396–409

Stegemoeller KA, Chappell HN (1990) Growth response of unthinned and thinned Douglas-fir stands to single and multiple applications of nitrogen. Can J For Res 20:343–349

Stovall JP, Carlson CA, Seiler JR, Fox TR, Yanez MA (2011) Growth and stem quality responses to fertilizer application by 21 loblolly pine clones in the Virginia Piedmont. For Ecol Manage 261:362–372

Tasissa G, Burkhart HE (1997) Modeling thinning effects on ring width distribution in loblolly pine (*Pinus taeda*). Can J For Res 27:1291–1301

Tasissa G, Burkhart HE (1998) An application of mixed effects analysis to modeling thinning effects on stem profile of loblolly pine. For Ecol Mange 103:87–101
Tasissa G, Burkhart HE, Amateis RL (1997) Volume and taper equations for thinned and unthinned loblolly pine trees in cutover, site-prepared plantations. South J Appl For 21:146–152
Wagner RG (1993) Research directions to advance forest vegetation management in North America. Can J For Res 23:2317–2327
Wagner RG, Little KM, Richardson B, McNabb K (2006) The role of vegetation management for enhancing productivity of the world's forests. Forestry 79:57–79
Walstad JD, Kuch PJ (eds) (1987) Forest vegetation management for conifer production. Wiley, New York
Watt MS, Kimberley MO, Richardson B, Whitehead D, Mason EG (2004) Testing a juvenile tree growth model sensitive to competition from weeds, using *Pinus radiata* at two contrasting sites in New Zealand. Can J For Res 34:1985–1992
Weng YH, Kershaw J, Tosh K, Adams G, Fullarton MS (2008) Height-diameter relationships for jack pine seedlots of different genetic improvement levels. Silvae Gen 57:276–282
Weng YH, Park Y, Simpson D, Tosh K, Fullarton M (2010) Tree improvement effects on tree size distributions for *Picea glauca* and *Picea mariana* in New Brunswick. Scand J For Res 25:10–20
Westfall JA, Burkhart HE (2001) Incorporating thinning response into a loblolly pine stand simulator. South J Appl For 25:159–164
Westfall JA, Burkhart HE, Allen HL (2004) Young stand growth modeling for intensively-managed loblolly pine plantations in southeastern US. For Sci 50:823–835
Will R, Hennessey T, Lynch T, Holeman R, Heinemann R (2010) Effects of planting density and seed source on loblolly pine stands in southeastern Oklahoma. For Sci 56:437–443
Woodruff DR, Bond BJ, Ritchie GA, Scott W (2002) Effects of stand density on the growth of young Douglas-fir trees. Can J For Res 32:420–427
Xie C-Y, Yanchuk AD (2003) Breeding values of parental trees, genetic worth of seed orchard seedlots, and yields of improved stocks in British Columbia. West J Appl For 18:88–100
Xiong JS, Isik F, McKeand SE, Whetten RW (2010) Genetic variation of stem forking in loblolly pine. For Sci 56:429–436
Yu S, Chambers JL, Tang Z, Barnett JP (2003) Crown characteristics of juvenile loblolly pine 6 years after application of thinning and fertilizer. For Ecol Manage 180:345–352
Zhang L, Moore JA, Newberry JD (1993a) Disaggregating stand volume growth to individual trees. For Sci 39:295–308
Zhang L, Moore JA, Newberry JD (1993b) A whole-stand growth and yield model for interior Douglas-fir. West J Appl For 8:120–125
Zhang S, Burkhart HE, Amateis RL (1996) Modeling individual tree growth for juvenile loblolly pine plantations. For Ecol Manage 89:157–172
Zhang S, Burkhart HE, Amateis RL (1997) The influence of thinning on tree height and diameter relationships in loblolly pine plantations. South J Appl For 21:199–205
Zhang J, Oliver WW, Powers RF (2005) Long-term effects of thinning and fertilization on growth of red fir in northeastern California. Can J For Res 35:1285–1293
Zhao D, Kane M, Borders BE (2011) Growth responses to planting density and management intensity in loblolly pine plantations in the southeastern USA Lower Coastal Plain. Ann For Sci 68:625–635

Chapter 17
Modeling Wood Characteristics

17.1 Need for Information on Wood Characteristics

In addition to affecting tree growth and stand productivity, silvicultural treatments also affect characteristics of the wood produced. To fully evaluate the feasibility of various management alternatives, quantitative information on both wood quantity and quality is required. Wood quality is a term that is commonly used to denote wood characteristics that can be quantified and used to evaluate suitability for specified end uses. Modeling systems that encompass both volume growth and wood characteristic responses to silvicultural inputs allow managers to better understand interconnections between quantity and quality and thus to refine silvicultural practices for achieving maximum value.

Variation in wood properties within the stem has major implications for processing and utilizing wood for specific products. The patterns of variation differ among species, sites, individual genotypes, and for specified wood characteristics. Variation within the stem occurs from a pith to bark and along the stem for a given ring number. In pines within tree variation in wood properties is often very pronounced, with pith-to-bark trends being particularly marked (Zobel and van Buijtenen 1989). The strong radial variation within the stem can be related to ring age (number of rings from pith). However, variation in wood properties along the stem for a given ring age, while generally less pronounced than the pith-to-bark variation, can also be important. These considerations led Burdon et al. (2004) to advocate joint use of two separate concepts, corewood versus outerwood in the radial direction and juvenile versus mature wood in the vertical direction, the latter relating to the botanical concepts of juvenility and maturity.

Numerous wood characteristics affect wood quality, but those that are most commonly quantified include size of the juvenile wood core, specific gravity, the number, size and location of knots. A brief overview of modeling methods that have been employed for quantifying wood characteristics is given; examples where these types of equations have been incorporated into growth and yield simulators are also provided.

H.E. Burkhart and M. Tomé, *Modeling Forest Trees and Stands*,
DOI 10.1007/978-90-481-3170-9_17,

17.2 Juvenile Wood

17.2.1 Characteristics of Juvenile Wood

Juvenile wood is that formed in the central core of the stem (Fig. 17.1). This portion is also commonly referred to as corewood or crown-formed wood. The zone of juvenile wood extends outward from the pith; wood characteristics change rapidly in successively older growth rings and then eventually level off (Fig. 17.2). All tree species produce juvenile wood, but differences in properties of juvenile and mature wood tend to be more pronounced in conifers – especially in pine – and hence much effort has been devoted to modeling the transition from juvenile to mature wood in commercially important species such as radiata pine and the US southern pines. The transition is not abrupt but rather is gradual. A radial cross-section typically consists of three zones: a core or zone of crown-formed wood, a zone of transition wood, and a zone of mature wood. (Figs. 17.1 and 17.2).

Because juvenile wood has inferior characteristics (e.g. lower percentage of summerwood, lower specific gravity, shorter tracheids with larger fibril angles) for many uses, one must estimate where the transition from juvenile to mature wood

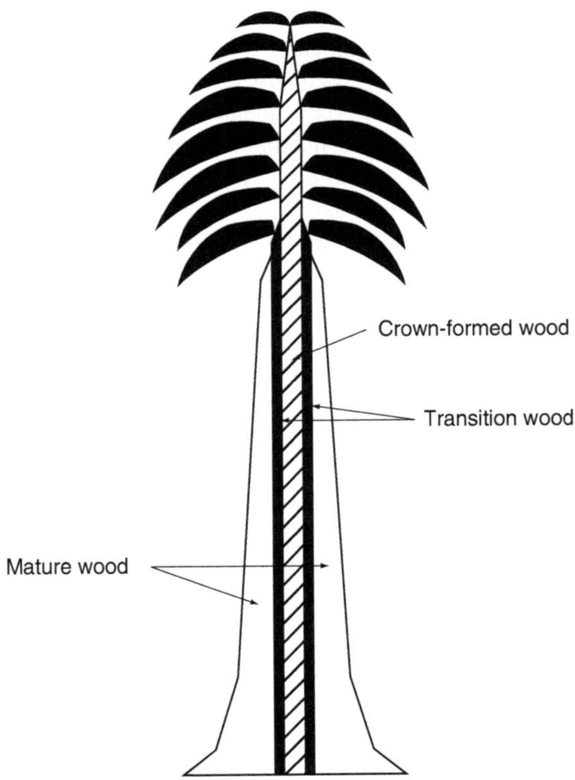

Fig. 17.1 Schematic diagram of core of crown-formed wood surrounded by band of transition wood and mature wood in standing tree (From Clark and Saucier 1991)

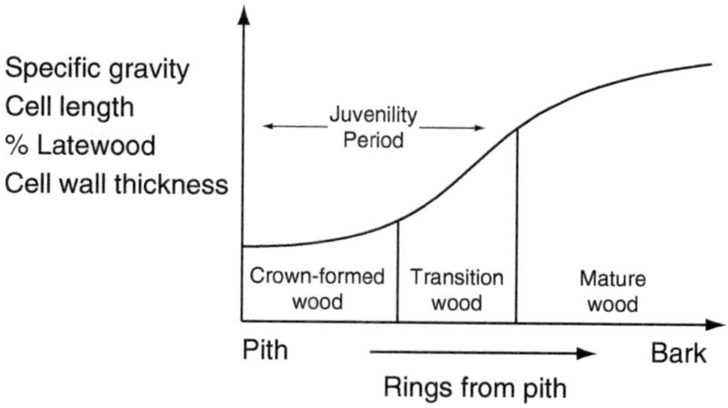

Fig. 17.2 General relationship of radial change in wood properties with age from pith and pattern of maturation (Adapted from Clark and Saucier 1989)

occurs in order to develop management regimens that limit the size of the juvenile wood core. While the transition is largely a function of tree-ring age, initial planting density strongly affects the size of the juvenile wood core that is produced.

Zobel and Sprague (1998) published an extensive review of juvenile wood in forest trees. A report on juvenile wood in coniferous trees was released by Thörnqvist (1993), and a synopsis of formation and properties of juvenile wood in US southern pines was compiled by Larson et al. (2001).

17.2.2 Estimating Juvenile-Mature Wood Demarcation

Knowing the age of transition from juvenile to mature wood characteristics and having information on diameter growth allows computation of the size of the juvenile core and the proportion of stand volume that consists of juvenile wood. Hence, a great deal of effort has been focused on establishing this boundary. The task is complicated because the transition is gradual, not abrupt, and the inherent variability is large. Because of the lack of a clearly defined border between juvenile and mature wood, visual inspection of data plots has sometimes been judged most appropriate (e.g. Bendtsen and Senft 1986; Clark and Saucier 1989). Several investigators, however, have successfully applied statistical modeling for estimating the demarcation between the juvenile and mature zones (Abdel-Gadir and Krahmer 1993; Tasissa and Burkhart 1998a; Sauter et al. 1999; Koubaa et al. 2005; Clark et al. 2006; Mora et al. 2007). Typically a radial profile at *dbh* is used in the analyses, but profiles at other stem heights have been used as well. The most commonly used variable for analyzing the transition is specific gravity, but other wood characteristics have also been included in some analyses.

Bendtsen and Senft (1986) determined the age of demarcation between juvenile and mature wood for plantation-grown eastern cottonwood and loblolly pine.

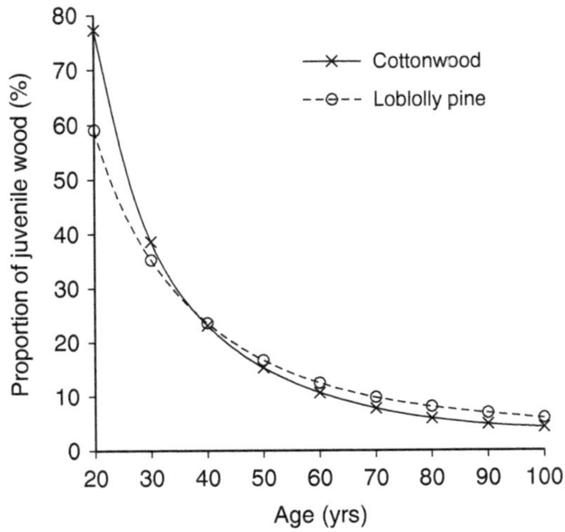

Fig. 17.3 Projected proportion of juvenile wood at height of 1.8 m in cottonwood and loblolly pine at ages 20–100 years (After Bendtsen and Senft 1986)

Several wood properties, including modulus of rupture (MOR), modulus of elasticity (MOE), compression strength, specific gravity, cell length, and fibril angle were analyzed using samples taken at a height of 1.8 m. None of the quantitative analysis methods considered produced acceptable results due to the gradual transition and high degree of variably in the data and the authors resorted to visual inspection of graphs to determine the age of demarcation. The demarcation between juvenile and mature wood was better defined in pine than in cottonwood and better defined in mechanical properties (MOR, MOE, compression strength) and in specific gravity than in cell length and fibril angle. The proportion of juvenile wood expected at various rotation ages was estimated to decrease rapidly for both pine and cottonwood, however, with juvenile wood constituting around 60% of the stem volume for pine at 20 years and 80% for cottonwood; the proportion declined to 24% for both species at age 40 (Fig. 17.3).

The influence of initial planting density on the age of transition from juvenile to mature wood of loblolly and slash pine was reported by Clark and Saucier (1989). Wood specific gravity of growth increments at breast height showed that planting density did not significantly affect the age of transition but it did affect the diameter of the juvenile core. Average diameter of the juvenile core at breast height ranged from 10 cm in slash pine planted at 1.8 × 1.8 m spacing to 16 cm in trees planted at 4.6 × 4.6 m. In loblolly pine, the average juvenile core diameter ranged from 13 cm in trees planted at 1.8 × 1.8 m to 20 cm in trees planted 3.6 × 3.6 m. The results of Clark and Saucier further showed that trees producing large annual diameter increments after converting to mature wood will contain proportionally less juvenile wood than slower growing trees (Fig. 17.4). And, in a separate but related phase of the study, the authors showed that the difference in the period of juvenility between slash and loblolly pine was not related to species but rather due to geographic location.

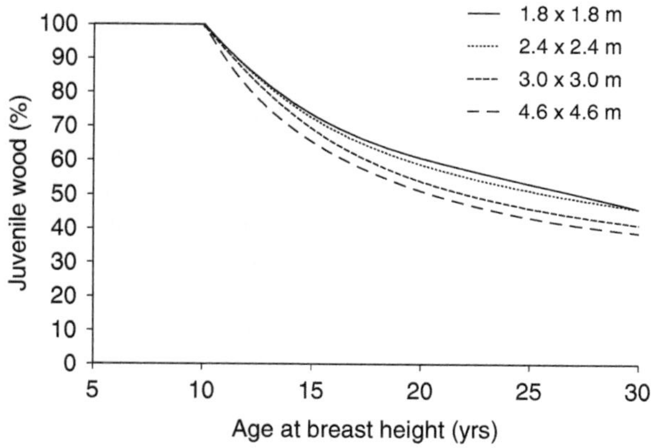

Fig. 17.4 Influence of spacing and age on proportion of tree basal area in juvenile wood of slash pine planted in the Upper Coastal Plain of Georgia (From Clark and Saucier 1989)

Abdel-Gadir and Krahmer (1993) applied piece-wise linear regression of specific gravity values to determine the demarcation point between juvenile and mature wood for Douglas-fir. The regression of density on age was assumed to follow one linear relation in the juvenile-wood stage but another in the mature-wood stage. The age of demarcation between juvenile and mature wood was estimated by minimizing the overall residual sum of squares with respect to the regression coefficients and the change point.

Specific gravity data from loblolly pine trees in a region-wide thinning study in southern United States were used by Tasissa and Burkhart (1998a) to determine age of demarcation between juvenile and mature wood. Iterative solutions in which the join point of two regressions is specified as an integer and segmented regression where the join point is estimated using nonlinear estimation techniques produced similar results after rounding the estimated age of demarcation from the segmented fit to an integer value. Demarcation ages determined at heights of 0.2, 1.4, 2.6, and 5.0 m were similar, indicating that within-tree variation in the age of demarcation is minor and that using measurements from a single height, such as breast height, should generally be adequate when quantifying the proportion of juvenile wood in loblolly pine. Thinning did not significantly affect the age of demarcation. There were differences in the age of demarcation across physiographic regions.

Sauter et al. (1999) applied segmented models to estimate the cambial age of juvenile-mature wood transition in samples taken at 4 m stem height from Scots pine trees in southwest Germany. The transition was estimated using latewood density profiles. The time series nature of the density data was considered by using generalized nonlinear regression and restricted likelihood regression procedures. A quadratic-linear fit showed the transition at cambial age of about 22 years with a standard deviation of 5–7 years.

Clark et al. (2006) estimated the age of transition from juvenile to mature wood in loblolly pine trees based on ring specific gravity, proportion of annual ring in latewood, and ring average microfibril angle. Threshold and segmented regression modeling approaches using measurement data collected at 1.3 m stem height showed that the age of transition differed by wood property and varied among geographic regions.

Koubaa et al. (2005) used polynomials to model radial patterns of several intra-ring traits of black spruce plantation-grown trees in Canada and to estimate the transition age from juvenile to mature wood. A modified logistic function was adopted by Mora et al. (2007) for modeling specific gravity profiles in loblolly pine trees in the southeastern USA. The function was used for demarcating corewood, transitional, and outerwood zones. No effects of silvicultural treatments (site preparation, fertilization, weed control) on the demarcation points were noted; however, the diameter of the juvenile core was increased as a result of treatments. A geographical trend for demarcation was observed with the northern sites requiring more time to reach a plateau in specific gravity than the southern sites.

17.3 Importance of Specific Gravity

Specific gravity is the most widely used criterion for evaluating quality of wood and its strength properties. In addition to modeling profiles of specific gravity to define the juvenile wood core, wood density is also important for predicting mechanical properties of wood and pulp yields. Although specific gravity has great utility as a predictor of the quality and strength properties of wood, it is highly variable by species, within and among trees of a given species, edaphic factors, climatic conditions, silvicultural treatments, and genetic stock. Hence, modeling wood specific gravity has presented significant challenges and it continues to command considerable attention from modelers.

Initial efforts were aimed at estimating average tree specific gravity from easily measured variables. With the advent of x-ray densitometry techniques (Cown and Clement 1983), determination of ring densities for large samples became feasible.

Specific gravity is the ratio of the weight of wood to the weight of an equal volume of water at a standard temperature; it is dimensionless and is usually expressed on an ovendry weight and green volume basis. A comprehensive compilation of published work on quantifying tree specific gravity was published by Zobel and van Buijtenen (1989). Consequently, the primary focus of this section is on more recent efforts to model specific gravity and to determine the extent to which it varies by environmental and silvicultural influences.

17.3.1 Models for Estimating Wood Specific Gravity

Because of inherent variability in specific gravity, and hence relatively low predictability, a number of past studies have correlated density of wood samples (typically increment cores taken at breast height) with stem specific gravity.

As an example, Burkhart and Beckwith (1970) related the specific gravity of a single circumferentially random increment core taken bark to pith at breast height to stem values to three top diameters for plantation-grown loblolly pine in the Piedmont region of Georgia, USA. Simple linear regressions of stem specific gravity on increment core specific gravity produced r^2 values 0.41, 0.45, and 0.47 for the stem to top diameters of 10, 8, and 0 cm, respectively. Because of the time and expense involved in taking wood samples, simple linear regressions were also fitted relating stem specific gravity to the reciprocal of age. The r^2 values for these regressions were, respectively, 0.28, 0.35, and 0.39 for the stem to top diameters of 10, 8, and 0 cm.

Repola (2006) used linear mixed-effects techniques to model wood density vertically along the stem of Scots pine, Norway spruce, and birch trees located on mineral soil sites in southern Finland. The constructed model can be calibrated for any stem using one or more density measurements at a freely chosen height; the calibrated model can then be used to determine the average wood density of a stem.

Data from loblolly pine plantations in the southeastern United States were used by Tasissa and Burkhart (1998b) to develop a ring specific gravity prediction model. Predictor variables included ring position, tree, stand and site attributes. The general framework of the specific gravity prediction model considered was:

$$\textit{Specific gravity} = f\ (\textit{tree, stand, site, and ring attributes})$$

where

tree attributes included *dbh*, total height, crown size, competition index
stand attributes included basal area per unit area, relative spacing, relative density, number of trees per unit area, quadratic mean diameter
site attributes included site index, and
ring attributes included physiological age, relative vertical position within a tree.

Much of the variation in ring specific gravity was due to within-tree variation; consequently, stand and site factors accounted for a limited proportion of variation in ring specific gravity. As the within-tree observations used in the study tended to be correlated, a direct covariance modeling approach was used to account for the correlation among observations. The selected model was:

$$R_{sg} = \mathrm{b}_0 + \mathrm{b}_1 \ln t_{pa} + \mathrm{b}_2(h_i/h) + \mathrm{b}_3 P_{lw} + \mathrm{b}_4 R_w + \mathrm{b}_5 CI$$

where

R_{sg} is ring specific gravity
t_{pa} is ring physiological age (the number of rings from the pith)
(h_i/h) is relative height (height above ground divided by total tree height)
P_{lw} is percent latewood within a ring (latewood width divided by ring width)
R_w is ring width
CI is competition index (quadratic mean diameter of the stand divided by tree *dbh*)
b_i $(i = 0,1, \ldots ,5)$ are regression coefficients

17.3.2 Impacts of Silviculture and Site on Specific Gravity

Results from Tasissa and Burkhart's (1998b) analyses showed that thinning treatments (30% and 50% of basal area removed; and 0% removed, or control) in loblolly pine did not alter the proportions of the annual ring in earlywood or latewood and thus did not affect ring specific gravity. There was, however, significant regional variation in average ring specific gravity. In an analysis of the impact of commercial thinning on selected wood properties of jack pine in eastern Canada, Schneider et al. (2008) reported that thinning increased ring width but thinning intensity did not influence ring density. Further, ring density followed the same pattern as percentage of latewood in which cambial age, relative height, and ring width were found to be important effects.

The impact of various silvicultural practices on wood specific gravity has been evaluated, including effects of initial spacing of red pine (Larocque and Marshall 1995) and of Japanese larch (Fujimoto and Koga 2009). Analyses of data from thinned and fertilized Scots pine stands in northern Sweden (Mörling 2002) and of loblolly pine stands in the coastal plain of North Carolina, USA (Antony et al. 2009) showed that fertilization greatly increased radial growth but did not significantly affect specific gravity. Nyakuengama et al. (2002) found that the responses in ring width and density of radiata pine to fertilizer application varied with fertilizer type and site; however, growth gains did not result in a marked change in wood density. Using data from slash and loblolly pine in the lower coastal plain of Georgia and Florida, Love-Myers et al. (2009) observed that fertilization treatments had similar short-term effects on the specific gravity of slash and loblolly pines, particularly in latewood, but the trees returned to a specific gravity pattern consistent with unfertilized trees within 2 or 3 years.

In general, silvicultural treatments do not result in changes in wood density unless the earlywood/latewood proportions are altered. Geographic variability in specific gravity, however, is very pronounced for some species. Using annual ring specific gravity of breast height increment cores from 3,957 trees across the natural range of loblolly pine, Jordan et al. (2008) showed that specific gravity increased significantly with age and varied significantly among physiographic regions. Antony et al. (2010) developed a three segment quadratic model and a semiparametric model to explain the vertical and regional variation in specific gravity of loblolly pine; maps showing regional variation in disk specific gravity at a specified height were then generated.

17.3.3 Relating Specific Gravity to Pulp Yields and Mechanical Properties

Increased wood density results in higher pulp yields and specific gravity is positively correlated with key wood mechanical properties of stiffness (modulus of elasticity, MOE) and strength (modulus of rupture, MOR). Specific gravity equations, coupled with volumetric measures, can be applied to determine yields in terms of dry weight or weight of actual wood substance, which is a much better indicator of pulp yields than wood volume or green weight (Burkhart and Beckwith 1970).

Empirical models relating MOE and MOR to specific gravity have been published by Pearson and Gilmore (1971, 1980), Zhang (1994, 1997), McAlister et al. (2000), Biblis et al. (2004), Schneider et al. (2008), and others.

Zhang (1994, 1997) observed that simple linear regressions of MOE and MOR on specific gravity explained approximately the same amount of variation as a curvilinear regression of the form $y = \alpha X^{\beta}$ where y represents MOE or MOR and X is specific gravity. While it is possible to estimate strength and stiffness of wood from specific gravity alone, many of the fitted models explained 50% or less of the variation. Because MOE and MOR are also affected by other factors, such as knots, slope of grain, and ring widths, additional predictors are often included. Biblis et al. (2004) found that MOR and MOE of lumber from 40-year-old loblolly pine plantations were significantly influenced by specific gravity, ring width, and percent latewood. In an analysis of wood properties of jack pine, Schneider et al. (2008) determined that wood stiffness (MOE) and strength (MOR) increased with cambial age and specific gravity and that MOR was also affected by ring width.

17.4 Modeling Ring Widths

Uniformity of growth rate affects wood structure and density variation both within and between growth rings (Bowyer et al. 2007). Ring widths are not uniform along the tree bole and the distribution is related to silvicultural treatments (Farrar 1961; Larson 1963). Hence, modeling of ring width distribution for various management treatments is an important aspect of quantifying wood quality for specified product objectives.

Tree ring width distribution can be determined by incrementing tree dbh, total height and other predictors if required and substituting into a tree taper function to describe the inside-bark stem profile each year. These annual stem profiles provide a distribution of the ring-width distribution throughout the bole. Alternatively, tree ring width distributions can be modeled directly.

Tasissa and Burkhart (1997) developed a ring width prediction model for loblolly pine that accounts for position in the tree, tree size, stand and site factors, as well as thinning effects. Thinning significantly increased ring width over most of the tree bole and its effects persisted over the 12 years since thinning was imposed. Significant regional variation in average ring width was also evident. Ring widths varied considerably within trees and among trees; the fitted model accounted for about 50% of the total variation.

The general framework for the ring-width prediction model of Tasissa and Burkhart was:

$$R_w = f \quad (\text{tree, stand, site, ring attributes})$$

Fig. 17.5 Thinning effects on average ring width by height of loblolly pine (From Tasissa and Burkhart 1997)

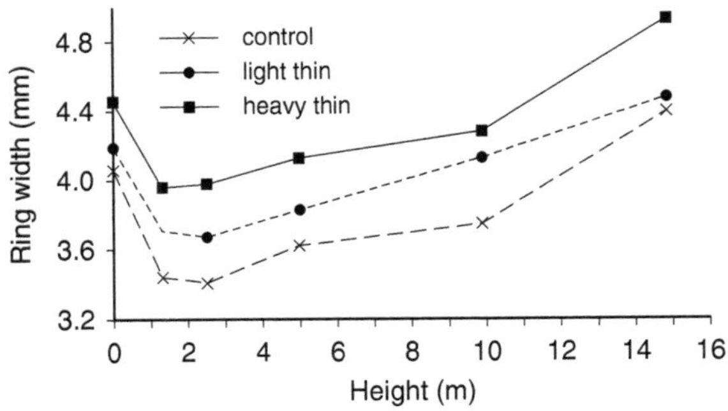

Empirical models that included various factors (variables) that have been shown to affect ring width distribution along the stem were evaluated, and the following expression was judged best overall:

$$R_w = \mathrm{b}_0 + \mathrm{b}_1(h_i/h) + \mathrm{b}_2 R_{tpa} + \mathrm{b}_3(h_i/h)R_{tpa} + \mathrm{b}_4(d/h) + \mathrm{b}_5 S + \mathrm{b}_6 CI$$

where

R_w is ring width
(h_i/h) is relative height where h_i is height above ground to the point of measurement
h is total tree height
R_{tpa} is relative physiological age (ring physiological age/disk physiological age)
d/h is a surrogate for stem taper
S is site index
CI is the stand-level competition index (quadratic mean diameter divided by tree *dbh*)
b_i (i = 0, … ,6) are regression coefficients

Since thinning significantly affects ring width, it was necessary to account for thinning effects to improve ring width prediction. A number of alternative thinning response functions were evaluated. Among the various specifications considered, a modification of the Liu et al. (1995) thinning variable appeared reasonable in that it enables accounting for the thinning intensity and time elapsed since thinning. The model with effect of thinning was:

$$R_w = \mathrm{b}_0 + \mathrm{b}_1(h_i/h) + \mathrm{b}_2 R_{tpa} + \mathrm{b}_3(h_i/h)R_{tpa} + \mathrm{b}_4(d/h) + \mathrm{b}_5 S + \mathrm{b}_6 CI + \mathrm{b}_7 THIN$$

where *THIN* is a thinning response function and other symbols are as defined previously.

Thinning substantially increased ring width over the stem of loblolly pine trees (Fig. 17.5). Following thinning, ring width first increased in the lower bole and then proceeded up the stem. The magnitude of increase in ring width was not constant over the entire bole; hence, basing diameter growth increases to response at dbh alone would not be adequate for many purposes.

17.5 Modeling Branches and Knots

In addition to specific gravity, the size and frequency of knots is central to assessing wood quality for structural timber. Knots greatly affect both appearance and strength; their occurrence is a primary factor when determining log and lumber grades.

Harris et al. (1975) noted that development of large knots in radiata pine could be compensated for by an increase in specific gravity. They calculated that bending resistance can be maintained when knot diameters increase 60% if there is an accompanying increase of 10% in wood density. Silvicultural practices influence branch, and hence knot, size. Branch development is influenced by initial planting spacing, timing and intensity of thinning, and by fertilizer applications. Pruning is sometimes applied to increase the amount of clear wood produced, and branch size and angle have been used as selection criteria in tree breeding programs.

17.5.1 Number, Size and Location of Branches

Colin and Houllier (1991) developed a model of the vertical trend of maximum branch diameter of Norway spruce in France. Their fitted equation consisted of a segmented polynomial with a join point at the height of the largest branch diameter of each sample tree. Subsequently, they published equations to describe the main characteristics of Norway spruce crowns from usual tree measurements (Colin and Houllier 1992). In particular, branch size, angle, number, and vertical position were predicted as a function of the tree diameter at breast height, total height, age and position along the stem.

Data from thinning experiments in Scots pine in Finland were used to construct models for predicting vertical trends of branch angle and branch diameter along the stem (Mäkinen and Colin 1998). Stand density measures were significant variables in the models; however, they could be excluded without loss of accuracy if variables describing tree dimensions were included as predictors. Relative crown length and stem diameter were adequate tree-level variables for describing branch characteristics. In a follow-up analysis of branchiness of Scots pine that used the same sample tree data, Mäkinen and Colin (1999) constructed models for (i) number of new branches on the stem apex, (ii) probability of a branch being alive, and (iii) proportion of the actual number of dead branches out of the predicted initial number of branches on the whorls below the crown base. The number of new branches was closely related to the height increment of the current year; the probability of a branch being alive was correlated with its age, tree age, height/dbh ratio, and its relative diameter within the branch whorl. Indices describing stand density and spatial arrangement of neighboring trees around the sample tree reduced the residuals slightly. Meredieu et al. (1998) further developed the models of Colin and Houllier by taking into account the within-tree correlation of branch characteristics when analyzing data from Corsican pine in France.

Mäkinen and Mäkelä (2003) developed models of cross-sectional area of branches in Scots pine at different heights within the live crown. The models were constructed to be applicable as submodels in growth simulators in order to predict the effects of silvicultural treatments on wood quality. Models were fitted for predicting cross-sectional area of the largest and smallest live branches in a whorl, relative cross-sectional area distribution on the basis of the largest branch, and cross-sectional area branches and its variance. Random variation of the dependent variables was partitioned into variance components at the stand, plot, tree, whorl, and branch level.

Using data from Finland and Sweden, Mäkinen et al. (2003b) predicted branch characteristics of Norway spruce from simple stand and tree measurements. Models were constructed to predict (i) crown ratio, (ii) the self-pruning ratio, (iii) number of living branches in a whorl, (iv) total number of branches in a whorl, (v) diameter of the thickest living branch of a whorl, (vi) diameters of smaller living branches of a whorl, and (vii) branch angle. Including stand-level variables as regressors in the models improved performance slightly. Similar procedures were applied by Mäkinen et al. (2003a) to develop models of branch characteristics along the stem of silver birch. The resultant models provide a framework for quantifying wood quality and evaluating application of silvicultural treatments to control branch development.

Spacing trials provide valuable information on the effects of stand density on branch development because extremes of density are included. Mäkinen and Hein (2006) and Hein et al. (2007), when analyzing data from three spacing experiments in Norway spruce in southwestern Germany, found that the most pronounced effect of stand density was an increase in branch diameter with a decrease in numbers of trees per unit area. At a density of 350 trees per ha, the maximum branch diameter of all sample trees exceeded the diameter limit of quality class B in the European quality standards for round wood. Benjamin et al. (2009b) used a combination of nonlinear multi-level mixed effects and generalized nonlinear modeling techniques to develop a series of equations to predict size and number of knots with respect to vertical location in black spruce trees from a spacing trial in Thunder Bay, Ontario, Canada. As growing space available for a given tree increased, crown size and branch size (and thus knot size) generally increased. Results of the analyses indicated that black spruce knot properties can largely be accounted for by tree bole and crown size variables.

In an analysis of several data sets of branch measurements from Douglas-fir trees, Weiskittel et al. (2007) found significant effects due to fertilization, thinning, pre-commercial thinning, vegetation control, and foliar disease (Swiss needle cast) for key crown structural attributes. Maximum branch size and total non-foliated crown radii were sensitive to the various stand factors, whereas no treatment effects were found for the number of branches within an annual segment or branch angle. When the data sets were combined and used to develop a single prediction equation, treatment effects were largely accounted for by changes in bole and crown size. In general, crown structural attributes were readily predicted from tree diameter at breast height, total height, and height to the crown base. Because specific combinations of these variables reflect the silvicultural regime under which the tree was grown, and because the allometric relationships between these tree-level variables and crown structural attributes were not greatly altered by the treatments, Weiskittel et al. concluded that general empirical equations can be expected to perform adequately for many purposes when integrated into models of forest growth and yield.

Kershaw et al. (2009) compared fixed- and mixed-effects nonlinear models for modeling vertical distribution of maximum knot size in black spruce. Mixed models are well suited for modeling knots because they are inherently nested at multiple levels including individual whorl, tree, and stand levels. Kershaw et al. derived six fixed-effects models of maximum knot size, ranging from a simple form to a more fully developed specification. Inclusion of random effects significantly improved all model fits; however, improvement in fit declined as the fixed-effects portion of the models incorporated variables related to trends observed in residual patterns. With the simple model form the nonlinear mixed-effects approach showed a reduction in root mean square error of 16% as compared to the generalized nonlinear model. In contrast, when a model form that accounted for primary covariates with a robust model form was used, the difference in root mean square error between nonlinear mixed-effects and generalized nonlinear models was approximately 6%. Although mixed-models are effective analytical tools for accounting for within- as well as between-subject correlation and for enhancing model flexibility, they are not necessarily a good substitute for robust model forms.

Most efforts to model branch locations have focused on location along the bole, with location around the bole being assigned by a random process. Because of the circular nature of branch azimuth data, conventional methods of statistical analysis cannot be applied. Circular statistics consist of converting observations to unit vectors or points on a unit circle. Doruska and Burkhart (1994) used circular statistics to examine branching patterns around the bole of loblolly pine trees. Analysis of branch azimuths on a whole-tree basis indicated that a uniform, or regular, distribution was appropriate. Circular correlation was applied to analyze rotational patterns within and between whorls, and a strong positive correlation was found for consecutive whorls of the same number of branches. In an analysis of the circular distribution of branches from plantation grown black spruce, Benjamin et al. (2009a) found that a random assignment (from a uniform distribution) of branches around the stem within each whorl was sufficient.

Most past studies have focused on modeling the vertical trend of external characteristics of branches located within the live crown. External characteristics of interest include branch diameter, angle, and length. While these types of models provide valuable information, they do not allow for the recovery of spatially explicit information or knot characteristics. Spatial information on internal branches (knots)

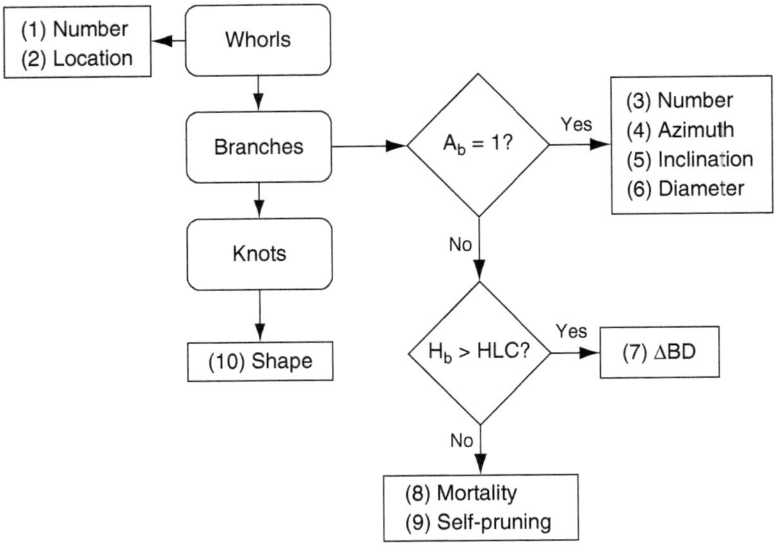

Fig. 17.6 Schematic diagram with the relationships between components, where A_b is branch age, h_b is branch height above ground, h_{cb} is height to the base of the live crown and ΔBD is the annual branch diameter increment (From Trincado and Burkhart 2009)

is essential for simulation of recovery of solid wood products by grade. Although limited in number, there are examples of forest-management-oriented simulation systems capable of modeling the complete process of branch dynamics. Two systems – one for radiata pine, the other for loblolly pine – will be briefly described.

Grace et al. (1999) presented the essential characteristics of a management-oriented system for modeling branch development in radiata pine. The model is based on a series of functions to predict branch initiation and development. These functions predict the number and location of branch clusters within each annual shoot; the number, orientation, and size of branches within each cluster; the change in branch diameter overtime; and branch mortality and stem cone occurrence.

Trincado and Burkhart (2009) reported on a model to represent the dynamics of branch and knot formation for loblolly pine, based on long-term measurements of crown dynamics from a spacing study and destructive analysis (whorl sections) of selected trees. The objectives were to (i) develop a framework to simulate the dynamics of first-order branches (initiation, growth, death, and self pruning), (ii) model internal stem structure over time in relation to branches contained within the stem (i.e., knots), and (iii) link the resulting equations and algorithms to an individual-tree growth and yield model. This integrated simulation system permits evaluation of the effects of silvicultural practices on sawlog quality and linkage with industrial conversion process (e.g., sawing simulation). A schematic diagram showing the relationships among components of the model is displayed as Fig. 17.6.

Information produced by the model of Trincado and Burkhart includes (i) trend of branch diameters along and around the stem, (ii) volume of knots (live and

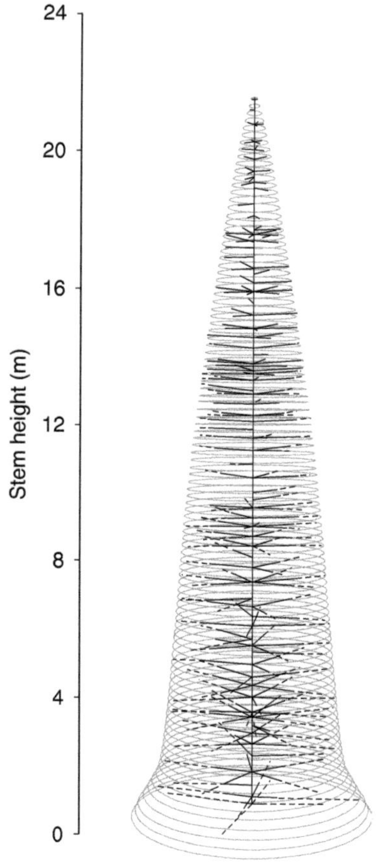

Fig. 17.7 Three-dimensional representation of internal branches (*knots*) of loblolly pine indicating live (*solid*) and dead (*dashed*) portions (Adapted from Trincado and Burkhart 2009)

dead portions), and (iii) spatial location, size and type (live and dead) of knots. Figure 17.7 shows a three-dimensional representation of internal branches, indicating live and dead portions, produced by the model for a representative tree.

17.5.2 Models of Knots

In past applications, internal branches (knots) have often been treated as cones, or, in some instances, the live portion was regarded as a cone and the dead portion as a cylinder. Stem dissection techniques have been applied to measure the internal, three-dimensional structure of knots within stems (e.g. Maguire and Hann 1987). Non-destructive methods involving computer tomography (or CT-scanning) and digital image analysis techniques have also been applied in a number of instances to develop models of internal knot properties (Moberg 1999, 2000, 2001, 2006; Pinto et al. 2003).

Moberg (2006) developed tree models whereby stem shape and internal knot structure of Scots pine and Norway spruce stems could be predicted using site, stand and tree variables. Knot properties included knot diameters, number of knots per whorl, sound knot length, and vertical knot inclination.

Trincado and Burkhart (2008) presented a model for representing the internal knot shape and structure for loblolly pine trees. Information on knot shape was recovered using a destructive analysis technique, the shape and structure (live/dead portion) of knots was characterized by a mathematical model, and analytical expressions were derived to predict the volume of individual knots. In their study the live portion was modeled using a simple one-parameter equation, and the dead portion was modeled assuming a cylindrical shape. It was assumed that knots (not external branches) produced in fast-growing plantations do not exhibit significant curvature. The general expression of the model was

$$r_{(l)}^2 = R^2\left(\frac{l}{L}\right)^{\beta} \quad 0 \leq l \leq L \tag{17.1}$$

where $r_{(l)}$ is the radius of the live portion of a knot at length l, R is the maximum radius of the live portion of a knot, L is the total length of the live portion of a knot, and β is the shape parameter. If $\beta = 1,2$, or 3, the geometrical shape generated is a paraboloid, cone, or neiloid, respectively. This equation meets two conditions: $r_{(l)} = 0$ when $l = 0$ and $r_{(l)} = R$ when $l = L$. Thus, if L and R are known, it is possible to estimate $r_{(l)}$ at any point. Additionally, the volume of this portion of a knot can be estimated by applying integral calculus:

$$V = \pi R^2 \int_0^L \left(\frac{l}{L}\right)^{\beta} dl \tag{17.2}$$

Based on model (17.2) analytical expressions were derived for estimating the volume of knots (live and dead portions) for three types of knot conditions on simulated trees: (i) live knots, (ii) nonoccluded dead knots, and (iii) occluded dead knots. Because multiple measurements were taken on each individual knot, the model was fitted using nonlinear mixed-effects modeling techniques. The within-individual errors were assumed homoscedastic, normally distributed, and uncorrelated with mean 0 and variance σ^2. Later, the random-effects parameter was related to other knot variables to incorporate covariates to explain additional between-individual variation in knot shape.

17.6 Incorporating Wood Quality Information into Growth and Yield Models

Forest management decision making involves using growth and yield models in an optimization context. To evaluate optimal economic returns, models of wood quality can be linked to growth and yield models through input variables that are output from the model of tree or stand development. Numerous equations for estimating tree or ring specific gravity, knot size and distribution, ring widths, and related characteristics that constitute wood quality have been published. This section provides examples of linking predictions of wood characteristics with growth and yield models in order to provide decision support for management on a value, rather than a volume, basis.

Maguire et al. (1991) incorporated wood quality variables into an individual-tree, distance- independent growth and yield model (designated ORGANON) that includes a crown development subcomponent. The model was developed for mixed conifer forests of southwestern Oregon, USA. Noting that crowns of conifers are dominated by annual whorls of branches, Maguire et al. obtained position of major branches or knots from height growth predictions. Mean maximum branch diameter was predicted as crown base receded past each whorl; branch diameter estimates were based on current depth of the whorl into the crown, tree diameter, stand relative density, and site index. Diameter of crown wood core was established as diameter inside bark of the stem at time of crown recession past each branch whorl. Simulations carried out to estimate harvested log distribution by branch size, whorl frequency, and crown wood indices for several thinning regimes showed good agreement with the limited field trial data available and with current understanding of stand and tree development. The authors concluded that individual-tree growth models that contain a crown recession component can be readily modified to predict crown wood core and indices of branch size.

Houllier et al. (1995) proposed a framework for analyzing the influence of silviculture and site quality on wood production of Norway spruce from both a quantitative and qualitative point of view (Fig. 17.8). The system consists of submodels that range from primary growth factors (site, stand density, and genetics) to quality of sawn boards. Tree and stand volume, stem taper, wood density, proportion of juvenile wood, and extent of knots were considered as the result of growth processes. Initially, a simple average tree growth model for pure, even-aged stands of Norway spruce in northeastern France was used to demonstrate how pieces of knowledge from silviculturalists, forest biometricians, and wood scientists can be brought together through computer simulation techniques.

Empirical ring-based models were developed by Ikonen et al. (2008) to predict the distribution of early wood percentage, wood density and fiber length along the stems of Scots pine and Norway spruce as affected by silvicultural treatments. Simulations by a process-based growth and yield model were used to analyze how thinning affects the growth and wood properties of Scots pine trees over a rotation as an average for the tree stem and also along the stem. The ring-based models for annual earlywood percentage (explained by ring width), wood density (a function of early wood percentage and cambial age) and fiber length (related to radial growth percentage and cambial age) predicted wood properties reasonably well at both an intra-ring and a cross-sectional level.

Mäkelä et al. (2010) reviewed methods for simulating wood quality in forest management models. Approaches were classified on the basis of their complexity,

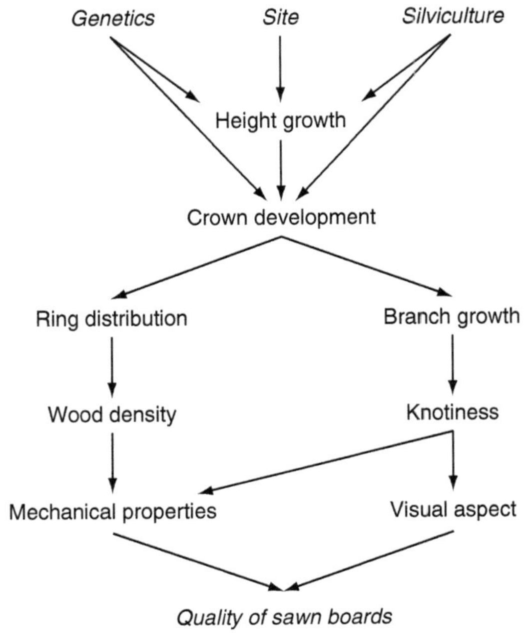

Fig. 17.8 General framework of a system for linking growth modeling to timber quality (From Houllier et al. 1995)

underlying principle and intended application. Examples were given of three types of dynamic models – empirical, hybrid, and mechanistic – applied for prediction of both sawn timber and fiber properties.

17.7 Linking Growth and Yield Models with Sawing Simulators

Output of log characteristics from growth and yield models can be linked to sawing simulators to obtain estimates of lumber yield by grade. These simulators allow for flexibility in studying different conversion rates from different types of mills and for varying product specifications.

The sawing simulation system called AUTOSAW published by Todoroki (1990) has been applied in a number of lumber recovery studies, including those of Barbour et al. (2003) and Weiskittel et al. (2006). A lumber recovery simulation system called Optitek (Forintek 1994) has been used in modeling lumber volume recovery in jack pine (Zhang and Tong 2005; Tong and Zhang 2006) and black spruce (Liu and Zhang 2005). One of the most comprehensive systems for linking silviculture to lumber yield and value is called SYLVER (Mitchell 1988; Di Lucca 1999). Developed in British Columbia, Canada, SYLVER integrates several subsystems to predict wood quality, product recovery, and financial return for various management regimes. An individual-tree model (TASS) provides tree height, diameter and knot size information that is passed to a log bucking program and then to a sawing simulator to cut logs into lumber using predefined cutting patterns.

Ikonen et al. (2003) demonstrated how tree stem properties can be linked to the properties of sawn timber through simulation of the structural growth of the tree and the sawing of the stem pieces subjected to quality grading. The growth and development of individual Scots pine trees was modeled in terms of the three-dimensional structure of the tree crown as determined by the influence of local light conditions on branch growth, which influences the distribution of the properties of the stem and wood such as stem form and knots. The three-dimensional distribution of these properties allows simulated cutting of the stem into logs and further into sawn pieces which can be graded based on quality. The part of the stem to be cut into logs is defined by the height of the stump and the minimum top diameter for a log acceptable for sawing. Stems are characterized by the number of logs obtainable for sawing, the length of the logs, the top and butt diameters of the logs, their volume and their taper. The wood is characterized primarily in terms of the occurrence of knots and their size and type (sound, dead). The sawing algorithm evaluates the knots in terms of diameter and type and calculates the grade and value of each piece. This procedure is repeated log by log, and a summary of the sawing is made for the whole stand, with a report on the total yield and value of the sawn timber obtained.

References

Abdel-Gadir A, Krahmer RL (1993) Estimating the age of demarcation of juvenile and mature wood in Douglas-fir. Wood Fiber Sci 25:242–249

Antony F, Schimleck LR, Daniels RF, Clark A III, Hall DB (2010) Modeling the longitudinal variation in wood specific gravity of planted loblolly pine (*Pinus taeda*) in the United States. Can J For Res 40:2439–2451

Antony F, Schimleck LR, Hall DB, Clark A III, Daniels RF (2011) Modeling the effect of midrotation fertilization of specific gravity of loblolly pine (*Pinus taeda* L.). For Sci 57:145–152

Antony F, Schimleck LR, Jordan L, Clark A, Daniels RF (2011) Effect of early age woody and herbaceous competition control on wood properties of loblolly pine. For Ecol Manage 262:1639–1647

Antony F, Jordan L, Daniels RF, Schimleck LR, Clark A III, Hall DB (2009) Effect of midrotation fertilization on growth and specific gravity of loblolly pine. Can J For Res 39:928–935

Aspinwall MJ, Li B, McKeand SE, Isik F, Gumpertz ML (2010) Prediction of whole-stem α-cellulose yield, lignin content, and wood density in juvenile and mature loblolly pine. South J Appl For 34:84–90

Bamber RK, Burley J (1983) The wood properties of radiata pine. Commonwealth Agricultural Bureaux, Slough

Barbour JR, Parry DL, Punches J, Forsman J, Ross R (2003) AUTOSAW simulations of lumber recovery for small-diameter Douglas-fir and ponderosa pine from southwestern Oregon. USDA Forest Service, Pacific Northwest Research Station, Portland, Research Note PNW-RN-543

Barnett JR, Jeronimidis G (eds) (2003) Wood quality and its biological basis. Blackwell Publishing/CRC Press, Oxford/Boca Raton

Beaulieu E, Schneider R, Berninger F, Ung C-H, Swift DE (2011) Modeling jack pine branch characteristics in Eastern Canada. For Ecol Manage 262:1748–1757

Bendtsen BA, Senft J (1986) Mechanical and anatomical properties in individual growth rings of plantation-grown eastern cottonwood and loblolly pine. Wood Fiber Sci 18:23–38

Benjamin JG, Chui YH, Kershaw JA Jr (2009a) Circular distribution of branches from plantation grown black spruce in Ontario. North J Appl For 26:15–20

Benjamin JG, Kershaw JA Jr, Weiskittel AR, Chui YH, Zhang SY (2009b) External knot size and frequency in black spruce trees from an initial spacing trial in Thunder Bay, Ontario. For Chron 85:618–624

Biblis E, Meldahl R, Pitt D, Carino HF (2004) Predicting flexural properties of dimension lumber from 40-year-old loblolly pine plantation stands. For Prod J 54(12):109–113

Bowyer JL, Shmulsky R, Haygreen JG (2007) Forest products and wood science: an introduction, 5th edn. Blackwell Publishing, Oxford

Burdon RD, Kibblewhite RP, Walker JCF, Megraw RA, Evans R, Cown DJ (2004) Juvenile versus mature wood: a new concept, orthogonal to corewood versus outerwood, with special reference to *Pinus radiata* and *P. taeda*. For Sci 50:399–415

Burkhart HE, Beckwith JR III (1970) Specific gravity prediction and dry-weight yield estimation. Tappi 53:603–604

Clark A III, Saucier JR (1989) Influence of initial planting density, geographic location, and species on juvenile wood formation in southern pine. For Prod J 39(7/8):42–48

Clark A III, Saucier JR (1991) Influence of planting density, intensive culture, geographic location, and species on juvenile wood formation in southern pine. Georgia Forest Research Paper 85, Georgia Forestry Commission

Clark A III, Daniels RF, Jordan L (2006) Juvenile/mature wood transition in loblolly pine as defined by annual ring specific gravity, proportion of latewood, and microfibril angle. Wood Fiber Sci 38:292–299

Colin F, Houllier F (1991) Branchiness of Norway spruce in north-eastern France: modeling vertical trends in maximum nodal branch size. Ann Sci For 48:679–693

Colin F, Houllier F (1992) Branchiness of Norway spruce in northeastern France: predicting the main crown characteristics from usual tree measurements. Ann Sci For 49:511–538

Cown DJ, Clement BC (1983) A wood densitometer using direct scanning with X-rays. Wood Sci Technol 17:91–99

Di Lucca CM (1999) TASS/SYLVER/TIPSY: systems for predicting the impact of silvicultural practices on yield, lumber value, economic return and other benefits. In: Stand density management conference: using the planning tools. Clear Lake Ltd, Edmonton, pp 7–16

Doruska PF, Burkhart HE (1994) Modeling the diameter and locational distribution of branches within the crowns of loblolly pine trees in unthinned plantations. Can J For Res 24:2362–2376

Farrar JL (1961) Longitudinal variation in the thickness of the annual ring. For Chron 37:323–330

Fayle DCF (1973) Patterns of annual xylem increment integrated by contour presentation. Can J For Res 3:105–111

Fayle DCF, MacDonald GB (1977) Growth and development of sugar maple as revealed by stem analysis. Can J For Res 7:526–536

Forintek Canada Corp (1994) Optitek: users's guide. Forintek Canada Corp, Sainte-Foy

Fujimoto T, Koga S (2009) An application of mixed-effects model to evaluate the effects of initial spacing on radial variation in wood density in Japanese larch (*Larix kaempferi*). J Wood Sci 56:7–14

Garber SM, Maguire DA (2005) Vertical trends in maximum branch diameter in two mixed-species spacing trials in the central Oregon Cascades. Can J For Res 35:295–307

Gardiner B, Leban J-M, Auty D, Simpson H (2011) Models for predicting wood density of British-grown Sitka spruce. Forestry 84:119–132

Gartner BL (2005) Assessing wood characteristics and wood quality in intensively managed plantations. J For 103:75–77

Grace JC, Pont D, Goulding CJ (1999) Modelling branch development for forest management. N Z J For Sci 29:391–408

Harris JM, James RN, Collins MJ (1975) Case for improving wood density in radiata pine. N Z J For Sci 5:347–354

Hein S, Mäkinen H, Yue C, Kohnle U (2007) Modelling branch characteristics of Norway spruce from wide spacings in Germany. For Ecol Manage 242:155–164

Houllier F, Leban JM, Colin F (1995) Linking growth modelling to timber quality assessment for Norway spruce. For Ecol Manage 74:91–102

Ikonen V-P, Kellomäki S, Peltola H (2003) Linking tree stem properties of Scots pine (*Pinus sylvestris* L.) to sawn timber properties through simulated sawing. For Ecol Manage 174:251–263

Ikonen V-P, Peltola H, Wilhelmsson L, Kilpeläinen A, Väisänen H, Nuutinen T, Kellomäki S (2008) Modelling the distribution of wood properties along the stems of Scots pine (*Pinus sylvestris* L.) and Norway spruce (*Picea abies* (L.) Karst.) as affected by silvicultural management. For Ecol Manage 256:1356–1371

Jordan L, Clark A III, Schimleck LR, Hall DB, Daniels RF (2008) Regional variation in wood specific gravity of planted loblolly pine in the United States. Can J For Res 38:698–710

Kellomäki S, Ikonen V-P, Peltola H, Kolström T (1999) Modelling the structural growth of Scots pine with implications for wood quality. Ecol Mod 122:117–134

Kershaw JA Jr, Benjamin JG, Weiskittel AR (2009) Approaches for modeling vertical distribution of maximum knot size in black spruce: a comparison of fixed- and mixed- effects nonlinear models. For Sci 55:230–237

Knapic S, Seppä IP, Usenius A, Pereira H (2011) Stem modeling and simulation of conversion of cork oak stems for quality wood products. Eur J For Res 130:745–751

Koubaa A, Isabel N, Zhang SY, Beaulieu J, Bousquet J (2005) Transition from juvenile to mature wood in black spruce (*Picea mariana* (Mill.) B.S.P.). Wood Fiber Sci 37:445–455

Larocque GR, Marshall PL (1995) Wood relative density development in red pine (*Pinus resinosa* Ait.) stands as affected by different initial spacings. For Sci 41:709–728

Larson PR (1963) Stem form development of forest trees. Forest Science Monograph No. 5

Larson PR, Kretschmann DE, Clark A III, Isebrands JG (2001) Formation and properties of juvenile wood in southern pines: a synopsis. USDA Forest Service, Forest Products Laboratory, Madison, General Technical Report FPL-GTR-129

Lasserre JP, Mason EG, Watt MS, Moore JR (2009) Influence of initial planting spacing and genotype on microfibril angle, wood density fibre properties and modulus of elasticity in *Pinus radiata* D. Don corewood. For Ecol Manage 258:1924–1931

Liu C, Zhang SY (2005) Models for predicting product recovery using selected tree characteristics of black spruce. Can J For Res 35:930–937

Liu J, Burkhart HE, Amateis RL (1995) Projecting crown measurements for loblolly pine trees using a generalized thinning response function. For Sci 41:43–53

Love-Myers KR, Clark A III, Schimleck LR, Jokela EJ, Daniels RF (2009) Specific gravity responses of slash and loblolly pine following mid-rotation fertilization. For Ecol Manage 257:2342–2349

Maguire DA, Hann DW (1987) A stem dissection technique for dating branch mortality and reconstructing past crown recession. For Sci 33:858–871

Maguire DA, Kershaw JA Jr, Hann DW (1991) Predicting the effects of silvicultural regime on branch size and crown wood core in Douglas-fir. For Sci 37:1409–1428

Mäkelä A, Mäkinen H (2003) Generating 3D sawlogs with a process-based growth model. For Ecol Manage 184:337–354

Mäkela A, Grace JC, Deckmyn G, Kantola A, Campioli M (2010) Simulating wood quality in forest management models. For Syst 19(SI):48–68

Mäkinen H, Colin F (1998) Predicting branch angle and branch diameter of Scots pine from usual tree measurements and stand structural information. Can J For Res 28:1686–1696

Mäkinen H, Colin F (1999) Predicting the number, death, and self-pruning of branches in Scots pine. Can J For Res 29:1225–1236

Mäkinen H, Hein S (2006) Effect of wide spacing on increment and branch properties of young Norway Spruce. Eur J For Res 125:239–248

Mäkinen H, Mäkelä A (2003) Predicting basal area of Scots pine branches. For Ecol Manage 179:351–362

Mäkinen H, Ojansuu R, Niemistö P (2003a) Predicting external branch characteristics of planted silver birch (*Betula pendula* Roth.) on the basis of routine stand and tree measurements. For Sci 49:301–317

Mäkinen H, Ojansuu S, Sairanen P, Yli-Kojola H (2003b) Predicting branch characteristics of Norway spruce (*Picea abies* (L.) Karst.) from simple stand and tree measurements. Forestry 76:525–546

McAlister RH, Powers HR, Pepper WD (2000) Mechanical properties of stemwood and limbwood of seed orchard loblolly pine. For Prod J 50(10):91–94

Megraw RA (1985) Wood quality factors in loblolly pine. Tappi Press, Atlanta, 88 pp

Meredieu C, Colin F, Hervé J-C (1998) Modelling branchiness of Corsican pine with mixed-effect models (*Pinus nigra* Arnold spp. *laricio* (Poiret) Maire). Ann Sci For 55:359–374

Miles PD, Smith WB (2009) Specific gravity and other properties of wood and bark for 156 tree species found in North America. USDA Forest Service, Northern Research Station, Newtown Square, Research Note NRS-38

Mitchell KJ (1988) SYLVER: modelling the impact of silviculture on yield, lumber value, and economic return. For Chron 64:127–131

Moberg L (1999) Variation in knot size of *Pinus sylvestris* in two initial spacing trials. Silva Fennica 33:131–144

Moberg L (2000) Models of internal knot diameter for *Pinus sylvestris*. Scand J For Res 15:177–187

Moberg L (2001) Models of internal knot properties for *Picea abies*. For Ecol Manage 147:123–138

Moberg L (2006) Predicting knot properties of *Picea abies* and *Pinus sylvestris* from generic tree descriptors. Scand J For Res 21:48–61

Mora CR, Allen HL, Daniels RF, Clark A III (2007) Modeling corewood-outerwood transition in loblolly pine using wood specific gravity. Can J For Res 37:999–1011

Mörling T (2002) Evaluation of annual ring width and ring density development following fertilisation and thinning of Scots pine. Ann For Sci 59:29–40

Mutz R, Guilley E, Sauter UH, Nepveu G (2004) Modelling juvenile-mature wood transition in Scots pine (*Pinus sylvestris* L.) using nonlinear mixed-effects models. Ann For Sci 61:831–841

Nyakuengama JG, Downes GM, Ng J (2002) Growth and wood density responses to later-age fertilizer application in *Pinus radiata*. IAWA J 23:431–448

Øvrum A, Vestøl GI, Høibø OA (2011) Modelling the effects of timber length, stand- and tree properties on grade yield of *Picea abies* timber. Scand J For Res 26:99–109

Pearson RG, Gilmore RC (1971) Characterization of the strength of juvenile wood of loblolly pine (*Pinus taeda* L.). For Prod J 21(1):23–31

Pearson RG, Gilmore RC (1980) Effect of fast growth rate on the mechanical properties of loblolly pine. For Prod J 30(5):47–54

Pinto I, Pereira H, Usenius A (2003) Analysis of log shape and internal knots in twenty Maritime pine (*Pinus pinaster* Ait.) stems based on visual scanning and computer aided reconstruction. Ann For Sci 60:137–144

Raymond CA, Joe B, Anderson DW, Watt DJ (2008) Effect of thinning on relationships between three measures of wood stiffness in *Pinus radiata*: standing trees vs. logs vs. short clear specimens. Can J For Res 38:2870–2879

Repola J (2006) Models for vertical wood density of Scots pine, Norway spruce and birch stems, and their application to determine average wood density. Silva Fennica 40:673–685

Sauter UH, Mutz R, Munro BD (1999) Determining juvenile-mature wood transition in Scots pine using latewood density. Wood Fiber Sci 31:416–425

Schneider R, Zhang SY, Swift DE, Bégin J, Lussier J-M (2008) Predicting selected wood properties of jack pine following commercial thinning. Can J For Res 38:2030–2043

Schultz EB, Matney TG (2006) An integrated growth and yield simulator for predicting loblolly pine dry weight pulp yields. Wood Fiber Sci 38:672–681

Tasissa G, Burkhart HE (1997) Modeling thinning effects on ring width distribution in loblolly pine (*Pinus taeda*). Can J For Res 27:1291–1301

Tasissa G, Burkhart HE (1998a) Juvenile-mature wood demarcation in loblolly pine trees. Wood Fiber Sci 30:119–127

Tasissa G, Burkhart HE (1998b) Modeling thinning effect on ring specific gravity of loblolly pine (*Pinus taeda* L.). For Sci 44:212–223

Thörnqvist T (1993) Juvenile wood in coniferous trees. Document D13:1993. Swedish Council for Building Research, Stockholm

Todaro L, Macchioni N (2011) Wood properties of young Douglas-fir in Southern Italy: results over a 12-year post-thinning period. Eur J For Res 130:251–261

Todoroki CL (1990) AUTOSAW system for sawing simulation. N Z J For Sci 20:332–348

Tong QJ, Zhang SY (2006) Modelling jack pine lumber value recovery in relation to tree characteristics using Optitek simulation. For Prod J 56(1):66–72

Trincado G, Burkhart HE (2008) A model of knot shape and volume in loblolly pine trees. Wood Fiber Sci 40:634–646

Trincado G, Burkhart HE (2009) A framework for modeling the dynamics of first-order branches and spatial distribution of knots in loblolly pine trees. Can J For Res 39:566–579

Watt MS, Zoric B (2010) Development of a model describing modulus of elasticity across environmental and stand density gradients in plantation-grown *Pinus radiata* within New Zealand. Can J For Res 40:1558–1566

Weiskittel AR, Maguire DA, Monserud RA, Rose R, Turnblom EC (2006) Intensive management influence on Douglas-fir stem form, branch characteristics, and simulated product recovery. N Z J For Sci 36:293–312

Weiskittel AR, Maguire DA, Monserud RA (2007) Modeling crown structural responses to competing vegetation control, thinning, fertilization, and Swiss needle cast in coastal Douglas-fir of the Pacific Northwest, USA. For Ecol Manage 245:96–109

Yang K-C (1994) Impact of spacing on width and basal area of juvenile and mature wood in *Picea mariana* and *Picea glauca*. Wood Fiber Sci 26:479–488

Zhang SY (1994) Mechanical properties in relation to specific gravity in 342 Chinese woods. Wood Fiber Sci 26:512–526

Zhang SY (1997) Wood specific gravity-mechanical property relationship at species level. Wood Sci Technol 31:181–191

Zhang SY, Tong QJ (2005) Modeling lumber recovery in relation to selected tree characteristics in jack pine using sawing simulator Optitek. Ann For Sci 62:219–228

Zobel BJ, Sprague JR (1998) Juvenile wood in forest trees. Springer, Berlin

Zobel BJ, van Buijtenen JP (1989) Wood variation its causes and control. Springer, Berlin

Chapter 18
Model Implementation and Evaluation

18.1 Model Implementation in Forest Simulators

A forest growth model is a simplification of the forest and its dynamics that allows forecasting of forest development under different management alternatives. State variables are used to describe the status of the forest at selected points in time. Forest growth models include a set of interrelated functions and algorithms that mimic the development of the forest. The functions may be growth functions, if they predict the changes of state variables over time, or functions that predict the value of a state variable from the value of other state variables. The previous chapters of this book described methodologies used to develop either growth or prediction functions for the stand and tree variables that are commonly part of forest growth models. Before a model can be used in practice, the set of functions developed to predict the changes in the state variables over time has to be implemented via a computer program that, when given the values of the state variables, uses the growth and prediction functions to forecast the dynamics of the state variables and, consequently, the forest dynamics. Each growth and/or prediction function is usually designated as a model component or module. Forest growth models also include other components such as algorithms to mimic silvicultural treatments that the user wants to consider during the planning horizon. These algorithms range from simple modifications of the values of some state variables to prediction functions for the values of the state variables after applying the treatment (see Chap. 16) or to more complex algorithms that take individual trees into consideration as is the case for the simulation of thinning in individual tree models.

Computer programs that implement growth and prediction functions, jointly with algorithms for the simulation of silvicultural treatments and user-friendly interfaces, are commonly known as forest simulators. Forest simulators operate at different spatial levels, from the stand to landscape or even larger regions. Here we emphasize implementation and evaluation of stand simulators.

When selecting a growth and yield model for application, forest managers should consider the reliability of the estimates, the flexibility to reproduce desired

H.E. Burkhart and M. Tomé, *Modeling Forest Trees and Stands*,
DOI 10.1007/978-90-481-3170-9_18,

management alternatives, the ability to provide sufficient detail for decision making, and the efficiency for providing this information. Buchman and Shifley (1983) provide a set of criteria to assist users in evaluating and selecting forest projection systems. Factors such as data requirements, computing capacity required, and cost all enter the evaluation, but the most important factor is the intended application. As noted by Buchman and Shifley, in practice, evaluation is often an iterative process of (1) defining the intended application, (2) specifying critical performance requirements, and (3) evaluating the capacity of candidate systems.

18.1.1 Input/Output

Many examples of input/output functions for forest stand simulators exist, but software for implementing a model for growth and yield of even-aged yellow-poplar stands (Knoebel et al. 1986) is representative of computer programs that are commonly developed and used in practice. The model can be operated at the stand level (Chap. 11) or it can be used to produce estimates by diameter classes by employing a continuous distribution function (Chap. 12). Full documentation of the model development, source code for the computer program for implementation, and model evaluation is published in Knoebel et al. (1986). (A summary of the underlying model structure is given in Chap. 12, Sect. 12.1.3.1.) Figure 18.1 provides a flow chart diagram of the yellow-poplar growth and yield model. A brief summary of the input/output functions follows.

Input data required by the program are:

- Age at the beginning of the projection period
- Age at the end of the projection period
- Site index
- Basal area at the beginning of the projection period
- Number of trees per unit area at the beginning of the projection period
- Number of previous thinning

Given the input data, the following stand attributes are computed:

- Average height of the dominant and codominant trees
- Minimum diameter (at breast height)
- Arithmetic mean diameter
- Quadratic mean diameter

If stand-level estimates are desired, they are computed at this point:

- Number of trees per unit area
- Basal area
- Volumes

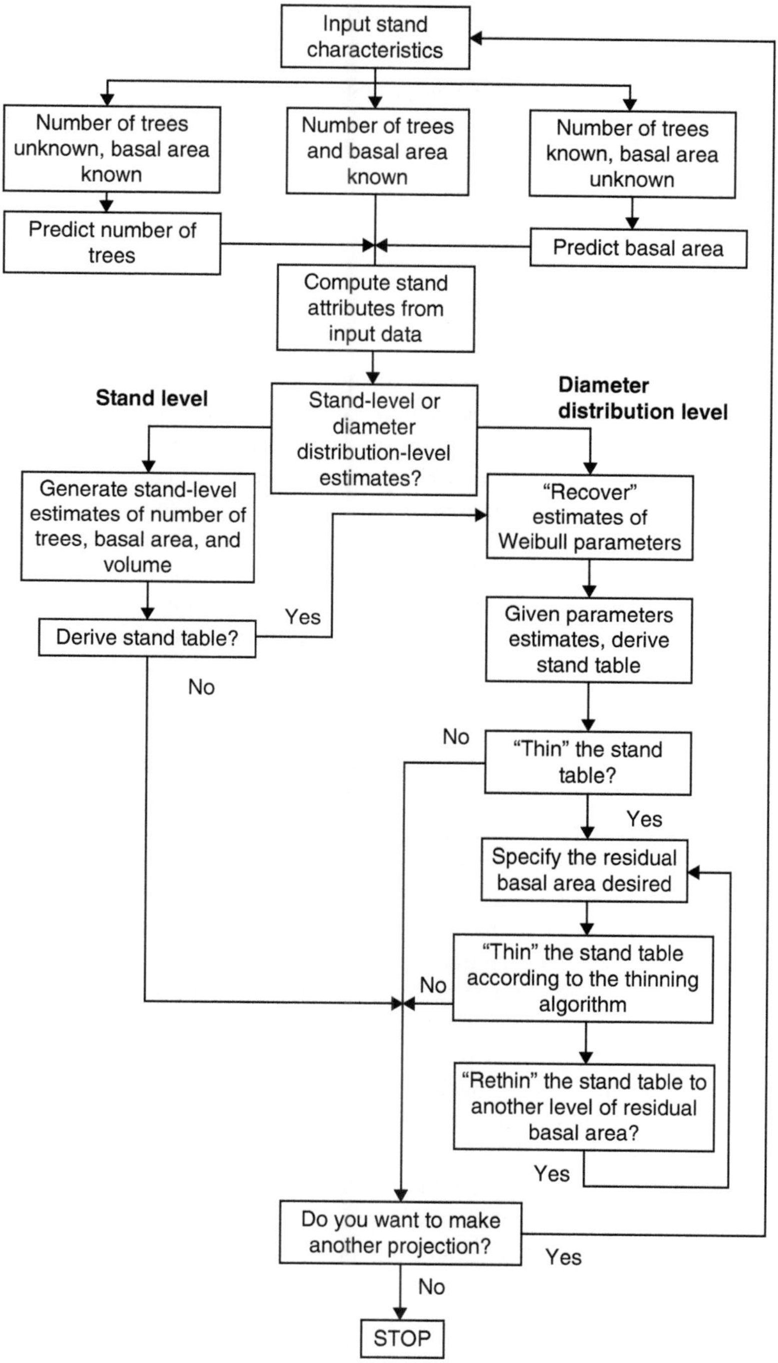

Fig. 18.1 Flow chart diagram of a computer program developed to provide stand level or diameter-distribution level estimates of growth and yield for yellow-poplar stands (Knoebel et al. 1986)

Once the stand-level attributes are generated and displayed, the user has the option to:

- Produce the corresponding stand/stock table
- Impose thinning
- Make another projection, or
- Terminate the growth and yield program

18.1.2 *Visualization*

While stand and stock tables provide adequate information for many types of decision analyses, a visual representation can enhance the usefulness of models for forest managers. With a visual image of the forest stand at various stages of development, managers can quickly grasp the need for, and consequences of, management actions such as thinning. A model that provides visual representations of forest development is valuable for presenting various management options to landowners, forest managers, and the general public. All aspects of different treatments (including the visual quality of stands produced under different management regimes) can be considered and integrated into the decision process.

Visual display can also enable stand modelers to gain new insights into the dynamics of forest stand development and to study the impact of various management treatments. Vast amounts of abstract data can be reduced to visual imagery to help scientists understand complex relationships more easily. Insights into stand development that may not be possible when using tables of numbers may become more apparent when visualization is employed. Although stand models are available for many forest types, scientific visualization capabilities are often not included. The addition of visualization capabilities has the potential for facilitating improvements in stand modeling, biological understanding of stand development, and forest management decision making.

Visualization software that can be linked to output from forest stand simulators is readily available (McGaughey 1998).[1] A list of tree diameters and heights (and sometimes crown measures) is passed from the growth and yield model to the visualization program. If the growth model is spatially explicit, tree attributes are represented at each individual's spatial coordinates. An example visualization of output produced by a distance-dependent growth model for loblolly pine is shown in Fig. 18.2. If spatial information is not included in the growth model, a spatial pattern can be assumed and the characteristics of trees in the list produced by the model can be distributed to the coordinate locations to produce an image of the stand.

[1]The USDA Forest Service stand visualization system (SVS) can be accessed at www.fs.fed.us/pnw/svs/.

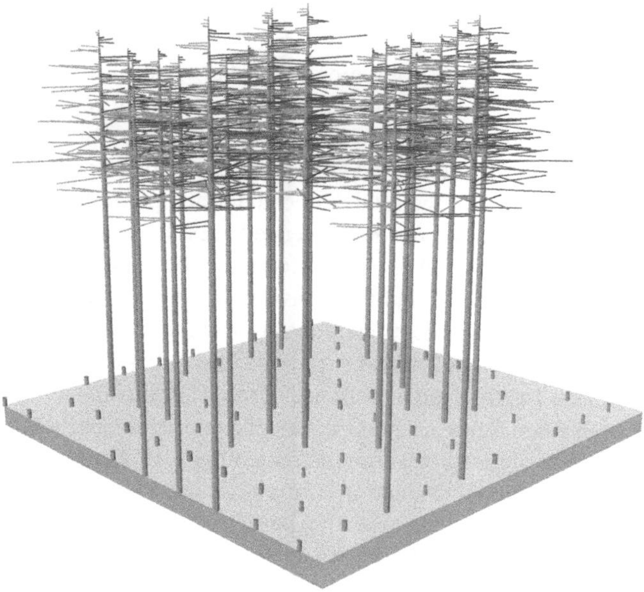

Fig. 18.2 Image of a loblolly pine stand created by linking the distance-dependent growth and yield model PTAEDA to the stand visualization system (*SVS*)

Visual imagery is a useful adjunct to numeric information produced by growth and yield models, but it cannot compensate for inherent weaknesses in the model itself. Regardless of the detail and attractiveness of output, users must carefully evaluate models for suitability for specific applications.

18.2 Model Evaluation

Growth and yield models are an essential tool for forest management planning. However, no matter how sophisticated the forest management planning method may be, the adequacy of the decisions depends to a great extent on the accuracy of the predictions provided by the models employed in planning. Any model is a simplification of reality and cannot be correct in every sense. Therefore each model should be evaluated and its limitations ascertained before it is adopted. Models used in decision support require a firm foundation in science and should produce predictions with quantified accuracy; furthermore they should be free of artifacts that confound choice of management alternatives (Stage 2003).

The terms model validation and model verification are sometimes used to indicate model evaluation. Validation involves a process to determine if a model performs at an acceptable level for its intended purpose; it is just one component of the larger task of model evaluation (Rykiel 1996). Van Horn (1971) defined validation as "the process of building an acceptable level of confidence that an inference about a simulated process is a correct or valid inference for the actual process". In this context model validation can never result in the acceptance of a model as right or wrong. It is instead a thorough analysis of model performance, including several procedures, that provides information that can be used to assess its adequacy for a particular use. Model evaluation should therefore provide as much information as possible about the model's behavior and predictive ability to allow users to decide if it is adequate for their intended uses. It should also reveal where future data collection and model revision efforts may be most useful (Vanclay and Skovsgaard 1997). Evaluation should be considered at every stage of model design and construction, when component functions are formulated and fitted to data, and when these components are assembled to provide the completed model. Most forest growth modelers are aware of the importance of model evaluation as a part of model construction, both when fitting each of the model components and when conducting an overall assessment of the model after implementing it via a computer program. In practice, however, additional evaluation is often needed when users are faced with selecting a model from those available for a particular objective.

Several authors have discussed the evaluation of models for natural systems. Some focused on the philosophical aspects of model validation (e.g. Reckhow and Chapra 1983; Oreskes et al. 1994), others applied a structured approach to model evaluation (e.g. Oderwald and Hans 1993; Soares et al. 1995; Janssen and Heuberger 1995; Vanclay and Skovsgaard 1997); in some cases, the proposed approach has been used to validate one or more existing models (e.g. Oderwald and Hans 1993; Soares et al. 1995; Tomé and Soares 1999).

Model evaluation involves several steps that are sometimes grouped in qualitative and quantitative tests (commonly designated model verification and validation, respectively). The first deals with a critical appraisal of model structure and biological realism while quantitative assessment implies comparisons of predictions with observations independent of those used to fit the model, usually including statistical evaluation of the magnitude of differences between the model and the real world. The importance of qualitative evaluation is stressed by some (e.g. Oderwald and Hans 1993; Soares et al. 1995) while statistical validation of the model has been dealt with by others (e.g. Reckhow and Chapra 1983; Mayer and Butler 1993; Power 1993).

The importance of including several procedures in model evaluation has been highlighted in the literature on forest growth models evaluation (*e.g.* Soares et al. 1995; Vanclay and Skovsgaard 1997), including qualitative as well as quantitative examinations of the model and going from theoretical to empirical issues.

According to Soares et al. (1995) and Vanclay and Skovsgaard (1997) validation procedures may be grouped as: (1) theoretical and biological aspects; (2) statistical properties; (3) characterization of model error; (4) statistical testing; (5) sensitivity analysis. When applying this framework to a practical problem, Tomé and Soares (1999) noted that some of the procedures apply mainly at the model building stage, as they are not problem-dependent, while others need to be implemented every time the model is going to be used for a different purpose, region or application.

The grouping of validation procedures proposed by Tomé and Soares (1999) was: (1) theoretical aspects of model building; (2) logic of model structure and biological aspects; (3) characterization of model errors. An elaboration on these three aspects follows.

18.2.1 Theoretical Aspects of Model Building

Theoretical aspects of model building include the analysis of the statistical properties of the model and sensitivity and uncertainty analyses. Modelers usually deal with these aspects of model evaluation, as they are not problem-dependent and they are highly technical. However, clear descriptions of the analytical methods employed also help users make informed choices regarding which models to adopt.

Analysis of statistical properties includes verification of the assumptions made when fitting statistical models. The nature of the error term (i.e. additive or multiplicative, independence of errors, etc.), errors associated with the independent variables, and the properties of parameter estimates in model functions are examples of issues that may need to be taken into account (for more details see Vanclay and Skovsgaard 1997). These analyses are usually made for each model component or set of components simultaneously fit to data.

Sensitivity analyses are conducted to determine those model components – parameters, initial values of state variables – which have the greatest influence on predictions (e.g. Liu et al. 1989; Botkin 1993; Jørgensen 1994). In practice the sensitivity analysis is carried out by changing the parameter or component and observing the corresponding effect on predicted outputs:

$$S_b = \frac{\partial \hat{y}}{\partial b}$$

where b is a parameter estimate. Sometimes sensitivity is expressed in relative terms:

$$Srel_b = \frac{\partial \hat{y}}{\partial b} \frac{b}{\hat{y}}$$

Results reveal parameters critical to model predictions, and parameters which may be redundant. Knowledge of sensitive parameters may guide applications (especially extrapolations) and the planning of model enhancements.

Gertner (1987, 1988) studied the estimation of the variance of growth and yield predictions and suggested the use of an error budget to evaluate the relative importance of each model component in the precision of the whole model.

Uncertainty analysis focuses on the impact of uncertainty in model inputs on its outputs (predictions). Uncertainty is commonly assessed using Monte Carlo methods, in which the model is run multiple times for combinations of different values of the parameters randomly selected from their distribution. The output of the model, for a certain input, is then a range of values instead of a unique prediction.

18.2.2 Logic of Model Structure and Biological Aspects

The analysis of the logic behind the model structure, including the model components, and of the compatibility of the model predictions with existing biological theories are usually referred to as qualitative evaluation. The model must be biologically realistic, agree with existing theories of forest growth and predict sensible responses to management actions. These aspects are not problem-dependent, but some of them are more relevant for certain applications than others. It is also important that, before employing a model as a decision aid tool, users gain knowledge of the model structure and the characteristics of each model component. Oderwald and Hans (1993) used the term "corroboration" to designate this type of analysis and listed the following items to be checked:

1. Do variables included in, and omitted from, the model agree with expectations?
2. Do the sign and magnitude of coefficients agree with expectations?
3. Are extrapolations outside the range of the development data reasonable?
4. Are transformations of model predictions reasonable (e.g. do model forecasts of future diameters also provide reasonable estimates of diameter increments, future volumes, mean increment curves, etc.)?
5. Are any contradictions present within the model?
6. Do derivatives, limits, maxima, minima, inflections, and the like, agree with expectations?

Figure 18.3 shows simulations of mean annual increment for different initial stand densities and site indices provided by the PTAEDA model. It can be seen that mean annual increment culminates earlier at higher densities and site indices, thus behaving according to biologic knowledge. Experienced foresters and other experts can be helpful in qualitative analysis, indicating areas where model predictions are contrary to expectations.

Matrix plots of simulated stand development trajectories showing a range of property-time and property-property relationships (Leary 1988, 1997) can offer useful insights into model behavior and provide an efficient way to reveal discrepancies in model predictions.

Simulations at extremes of stand conditions are often particularly revealing. Such simulations may encompass not only the upper and lower limits of site quality and stand density represented in the data, but also alternative stand structures (e.g. even- vs. uneven-aged, pure vs. mixed, thinned vs. unthinned, pruned vs. unpruned, etc.) (Vanclay and Skovsgaard 1997).

One important point raised by Soares et al. (1995) is that a model should not be rejected simply because it behaves in a counter-intuitive fashion; it may be our preconceptions that are wrong. Discrepancies should, however, cause us to critically reappraise the model, the data, and our preconceptions.

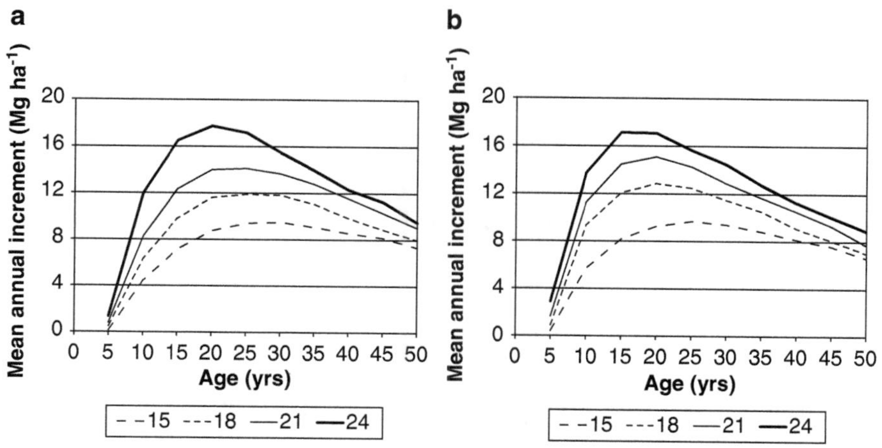

Fig. 18.3 Predicted mean annual increment for loblolly pine plantations from the PTAEDA model for site indices 15, 18, 21 and 24 m with 1,500 (**a**) and 3,000 (**b**) trees per ha at planting

18.2.3 Characterization of Model Errors

Characterization of model error refers to all the procedures for model evaluation that imply the comparison of model predictions with evaluation data. Knowing starting points of different stands (permanent plots or detailed continuous forest inventory) as well as the treatments that were applied during the time period described by the real data is provided as input to the model and the stand is simulated over that time period. Analysis of model error is based on the computation of prediction residuals or model errors, the differences between the observed and predicted values of all variables of interest. Error characterization may involve statistical testing or be based mainly on the computation of selected statistics and graphical analyses.

Model error characterization is clearly problem-dependent, as the evaluation data used should be in agreement with the region and conditions for which model predictions are needed.

The analysis of model error must be done for each of the model components as well as for the model as a whole. The analysis of different model components helps identify the main sources of the model error and also uncovers possible error compensation among model components that might hide real errors.

18.2.3.1 Evaluating Model Bias and Precision

Model error should be assessed in terms of two characteristics: bias and precision. The first refers to the deviation of the average of the model errors from zero and the second to the size of the model errors. Obviously both characteristics must be analyzed in order to assess model performance. Several statistics may be used to assess bias and precision (see Janssen and Heuberger 1995 for a complete list), but the most commonly used are the average model error to assess model bias and the mean absolute difference or mean squared error to assess model precision:

Model bias:

$$\bar{r}_p = \frac{\sum_{i=1}^{n} (y_i - \hat{y}_i)}{n} \quad \text{average model prediction error or average bias}$$

Model precision:

$$|\bar{r}_p| = \frac{\sum_{i=1}^{n} |y_i - \hat{y}_i|}{n} \quad \text{mean absolute prediction error}$$

$$mse = \frac{\sum_{i=1}^{n} (y_i - \hat{y}_i)^2}{n} \quad \text{mean squared prediction error}$$

where r_p is the prediction residual, y_i and $\hat{y}_i$ are, respectively, the observed and predicted values, and n is the number of observations in the evaluation data set.

The analysis of histograms of errors provides a good picture of model bias and precision, which may be complemented by quantiles of the distribution of residuals. The 5% and 95% percentiles are commonly chosen as they are not overly sensitive to extreme points in the data.

Plots of observed *versus* predicted values have often been used to characterize bias and precision. One simple statistic for assessing a model's performance is the correlation coefficient (r) between y and $\hat{y}$. Another common analysis is the linear regression of y on $\hat{y}$ to check whether the intercept (a) is near 0 and the slope (b) is near 1. Kobayashi and Salam (2000) and Gauch et al. (2003) argue that the correlation coefficient and the linear regression are not entirely satisfactory for model evaluation and suggest that the mean squared error (they identified it as mean squared deviation) is more informative. Gauch et al. (2003) proposed partitioning of *mse* into:

$$mse = sb + nu + lc$$

$$sb = \left(\bar{y} - \bar{\hat{y}}\right)^2 \quad \text{squared bias } (sb > 0 \text{ if b} = 1 \text{ and a} \neq 0)$$

$$nu = (1-b)^2 \frac{\sum_{i=1}^{n} \left(\hat{y} - \bar{\hat{y}}\right)^2}{n} \quad \text{non} - \text{unity slope } (nu > 0 \text{ if b} \neq 1)$$

$$lc = \left(1 - r^2\right) \frac{\sum_{i=1}^{n} (y - \bar{y})^2}{n} \qquad \text{lack of correlation } (lc > 0 \text{ if } r^2 \neq 1)$$

where symbols are as before.

These components have a simple geometric interpretation that reinforces the meaning of their names, as shown in Fig. 18.4. Departures from the ideal 1:1 line that corresponds to $mse = 0$ (shown on the top left of the figure) may occur due to translation, rotation or scatter (shown in the other three plots in the figure). Translation occurs when $b = 1$ but $a \neq 0$ and is indicated by a $sb > 0$. A slope different from 1 implies rotation and scatter is indicated by a low value of r^2. The resulting *mse* for any combination of these three problems is simply additive which makes this methodology particularly useful for model comparisons. In Fig. 18.5 models 2 and 3 have similar values of *mse*, but the sources of deviation are quite different; lack of correlation is prevalent in model 2 but all sources of deviation are roughly equal in model 3.

18.2.3.2 Computing Modeling Efficiency

Another useful technique is to compare predictions directly with observed data using a statistic analogous to R^2, and sometimes called modeling efficiency:

$$ef = 1 - \frac{\sum (y - \hat{y})^2}{\sum (y - \bar{y})^2}$$

where symbols are as before. This statistic provides a simple index of performance on a relative scale, where 1 indicates a 'perfect' fit, 0 reveals that the model is no better than a simple average, and negative values indicate a poor model indeed.

18.2.3.3 Looking for Tendencies in Model Error

One of the most efficient ways to examine model performance is to plot residuals or standardized residuals for all possible combinations of tree and stand variables to detect possible autocorrelation and other dependencies or systematic patterns. Figure 18.6 shows an example of the graphical analysis undertaken by Soares et al. (1995) when validating the PBRAVO model. Care is needed in interpreting these plots, as it is easy to focus on a few outliers and hard to see how many points are within the big "clouds". An analysis of the plots in Fig. 18.6 provides insight into the most important tendencies of model errors for the variables considered. For instance, the bias of volume estimations, which increases with stand volume, is immediately detected in both short term and long term predictions. It can be seen that the projection length has a rather small influence on bias or on the magnitude of the residuals but precision tends to deteriorate.

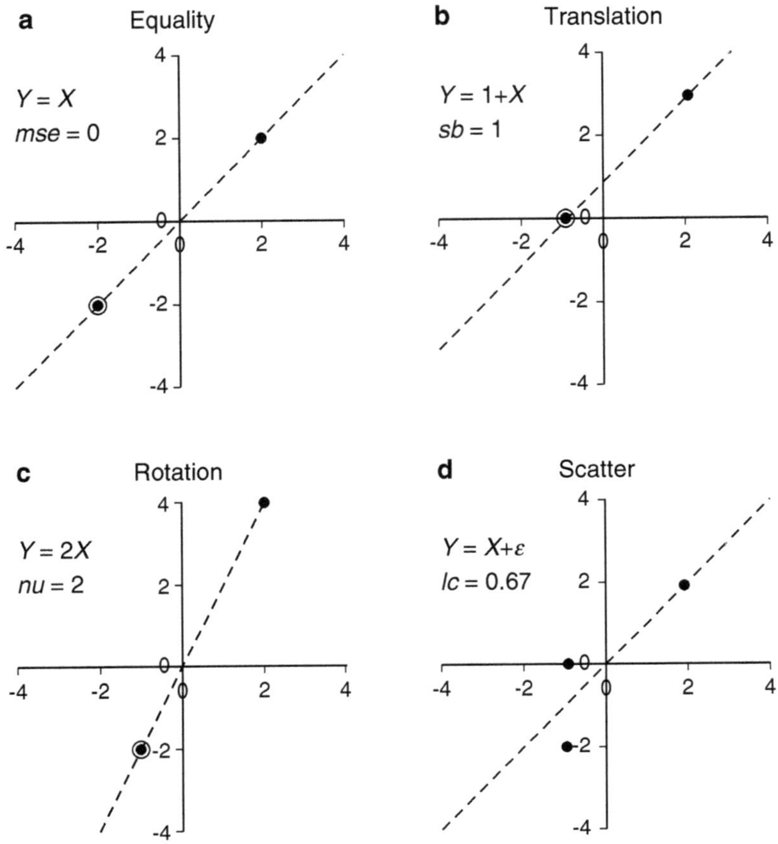

Fig. 18.4 Discrepancies between model predictions ($\hat{y}$) and real data (y). (**a**) Perfect equality between y and $\hat{y}$ leading to mse = 0; (**b**) Translation with $y = 1 + \hat{y}$, causing $sb = 1$; (**c**) Non-unity slope, in this case $y = 2\hat{y}$, causing $nu = 2$; (**d**) In this case lack of correlation is present, with $y = \varepsilon + \hat{y}$ and $lc = 0.67$ (Adapted from Gauch et al. 2003)

An efficient way to screen for major dependencies involves using multiple linear regression algorithms such as stepwise selection to find relationships between the residuals and site and stand variables. This procedure is an alternative to examining a large number of residual plots; it is not, however, a substitute for graphical inspection but rather serves as a way to highlight explanatory variables against which residuals should be plotted (Soares et al. 1995).

18.2.3.4 Statistical Tests

Characterization of model error may be complemented by several statistical tests that help to assess the magnitude of the deviations between model predictions and reality. The most common assessments include:

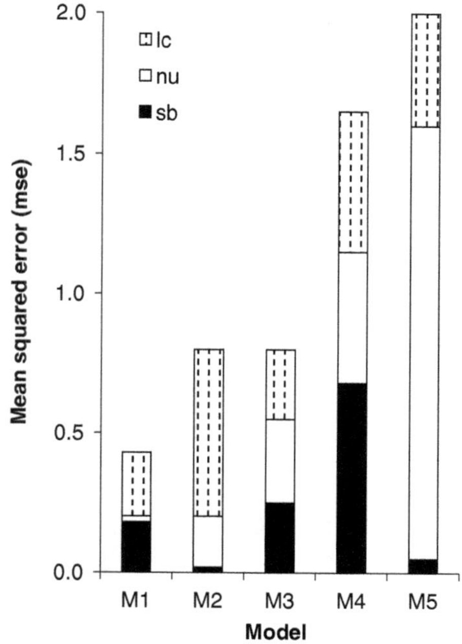

Fig. 18.5 Components (*sb* = squared bias; *nu* = non-unity slope; *lc* = lack of correlation) of mean squared error (*mse*) for five models. Models 2 and 3 show a similar *mse* but the analysis by components shows that their performance is clearly different (Adapted from Gauch et al. 2003)

1. bias and precision of the model and its components (test if mean prediction error is significantly different from zero (bias) and if the variance of prediction is larger than a specified critical value (precision));
2. goodness-of-fit of predicted size distributions (test if predicted size distributions significantly differ from observed values);
3. patterns in, and distribution of, residuals;
4. correlations over time and between components.

Many statistical tests of model performance have been suggested, but no single criterion can incorporate all aspects of model evaluation; it is desirable to use several simple tests to examine different facets of model behavior.

Yang et al. (2004) evaluated the usefulness of various statistical tests for forest biometric models. Ten tests, five parametric and five nonparametric, were selected for inclusion in the evaluation. Nine data sets were used to assess the behavior of 10 diagnostic tests. Results from this study demonstrated that the statistical tests have limited usefulness in model validation. None of the tests seemed to be generic enough to work well across a diverse range of models, data, assumptions, or constraints.

Given the results from their evaluation, Yang et al. (2004) recommended that analysts look at how well a model fits new, independent data rather than apply a statistical test to determine whether or not the model is good enough, because results will vary depending on the data, model types, study objectives, and the statistical test applied. Model validation is an attempt to judge whether or not a model is an

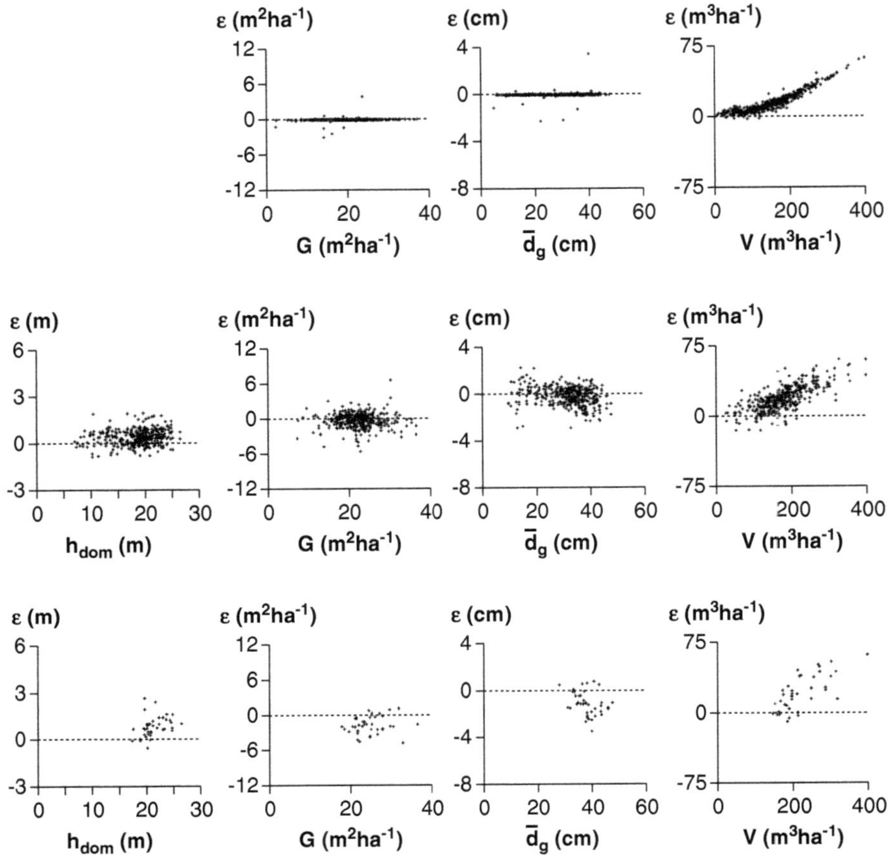

Fig. 18.6 Prediction residuals ($\varepsilon = y - \hat{y}$) versus observed dominant height (h_{dom}), stand basal area (G), quadratic mean diameter ($\bar{d}_g$) and standing volume (V) for various projection lengths (*top row*: no projection; *middle row*: 1–10 years; *bottom row*: 11–20 years)

acceptable representation of reality for some stated purpose. Statistical tests can be a valuable part of validation, but a composite process, rather than a simple pass-fail criterion of any particular test, is generally required.

18.2.4 Data for Model Validation

Quantitative evaluation of forest growth models requires comparison of predicted values with real data. To be rigorous the comparison must be made with a data set independent from the one used for model fitting. This poses the problem of acquiring an independent data set. When comparing models for an intended practical application, a new data set may purposely be gathered; however, this is generally not the case.

The need to base quantitative evaluation of models on an independent data set leads to the question of deciding whether two data sets are independent. Even if two data sets are taken at different time points or by different crews they are generally not totally independent. During model construction it is common to set aside a portion of the data for validation purposes (further details are given in Snee 1977; Myers 1990). This technique, known as cross-validation, has been questioned by several authors (including Yang et al. 2004; Cieszewski and Strub 2007) arguing that cross-validation of fitted models on so-called "independent data" is merely an example of the possible performance of the model with other data and of the magnitudes of the expected errors. Even if data splitting techniques have shortcomings, growth modelers often have to decide whether it is worthwhile splitting the data available into two subsets, one for developing and the other for testing of the model. This is not a trivial decision, especially when data are scarce. Setting some data aside may provide for a better test of the model, but it may also result in inferior parameter estimates (Vanclay and Skovsgaard 1997).

Double cross validation is a version of cross validation in which the total number of observations, n, is divided into k equal subsets. The model is fitted k times, each time setting aside one of the k subsets and using the observations not used in fitting to calculate the validation statistics. An extreme case occurs when $k = n$, so that each iteration of the model is derived from $n - 1$ observations, and the validation data set contains only one observation. This special case of double cross validation, widely used in model selection, results in what is known as the prediction sum of squares (PRESS) statistic (Myers 1990). One alternative is to form the subsets for the double cross validation according to some variable of the data such as plot age, geographic location, or other criteria that are relevant to a particular problem. Other re-sampling techniques, such as boot-strapping (Davison and Hinkley 1997), may also be used.

Kozak and Kozak (2003) studied the advantages of performing diverse types of cross-validation and double cross validation using seven data sets with two tree volume estimating models and a height–diameter model. The authors computed several fit and lack of fit statistics and concluded that cross validation by data splitting and double cross validation provide little, if any, additional information in the process of evaluating regression models over the fit of the models with the whole data set and corresponding computation of the fit and lack-of-fit statistics.

An independent sample was established by Burkhart (1971) to evaluate the accuracy and precision of a diameter distribution yield estimation technique (Bennett and Clutter 1968) for slash pine plantations. Analyses showed the yield estimates to be unbiased but not precise. Overall 80% of the estimated yields were within plus or minus 25% of the observed values; the plus and minus deviations were fairly well balanced and no trends over age, site index or stand density were detected. It was concluded that, for large samples, yields can be reliably predicted by the model tested.

Daniels et al. (1979) evaluated and compared three stand models (whole stand model of Burkhart et al. 1972; diameter distribution model of Burkhart and Strub 1974; individual tree model of Daniels and Burkhart 1975) with independent data on the basis of merchantable volume predictions for loblolly pine plantations. Analysis of deviations of estimated from observed yields revealed that all three models provided accurate estimates, and all were free from prediction bias due to stand characteristics. The whole stand and diameter distribution models were higher in precision than the individual tree simulation model. The authors opined that the selection of a model depends on the amount of stand detail desired and the management practices to be evaluated.

18.3 Applying Growth and Yield Models

When applying growth and yield models, it is assumed that stands are relatively homogenous with regard to independent variables (e.g. site index, stand density) used to predict stand values. If there is significant variation in predictors such as site index or stand density for a given tract, the area must be stratified into reasonably homogeneous stands and predictions made separately for each of these units to ensure accurate results.

Moeur and Ek (1981) used a distance-independent, individual-tree growth model to project changes in stand structure on aspen, red pine, and jack pine cover types in northern Minnesota, USA. Projections were made with 0.058-ha plots, plots aggregated within stands, and plots aggregated within cover types and compared with observed plot conditions. Plot by plot projections were most accurate in comparison with observed conditions, followed by stand and then cover-type aggregations.

Smith and Burkhart (1984) conducted a simulation study to assess the effect of sampling for predictor variable values on estimates of stand yield. Simple random and stratified random samples of stand density and tree height were drawn from simulated pine plantations containing nine strata. Sample outcomes were substituted into a yield function to produce estimates of mean yield. Stratifying by both site index and number of trees per unit area improved the precision of yield estimates by 2/3 over simple random samples. Site index was the more important of the two stratifying variables, but synergistic effects between the two attributes for stratification were found. Estimating mean yield with the average of the sample plot yields produced values with less bias and with smaller average error than using the alternative of applying the mean values of the independent variables to obtain an estimate of mean yield.

Users should also keep in mind that growth and yield predictions apply to net area. Forested tracts commonly have portions that are not occupied by forested cover; all nonproductive areas must be deducted from the overall tract area to avoid over estimates.

Finally, in most growth and yield models, no allowance is made for logging breakage or other losses during harvest; it is assumed that all material meeting minimum merchantability standards will be utilized. Adjustments must often be made in predicted values from models to approximate volumes that are likely to be realized under local harvesting and utilization conditions.

References

Bennett FA, Clutter JL (1968) Multiple-product yield estimates for unthinned slash pine plantations – pulpwood, sawtimber, gum. USDA Forest Service, Southeastern Forest Experiment Station, Asheville, Research Paper SE-35

Botkin DB (1993) Forest dynamics: an ecological model. Oxford University Press, New York

Buchman RG, Shifley SR (1983) Guide to evaluating forest growth projection systems. J For 81:232–234, 254

Burkhart HE (1971) Slash pine yield estimates based on diameter distribution: an evaluation. For Sci 17:452–453

Burkhart HE (1992) Scientific visualization for the study and use of forest stand simulators. Landsc Urb Plan 21:317–318

Burkhart HE, Strub MR (1974) A model for simulation of planted loblolly pine stands. In: Fries J (ed) Growth models for tree and stand simulation. Royal College of Forestry, Stockholm, pp 128–135, Research Notes 30

Burkhart HE, Parker RC, Strub MR, Oderwald RG (1972) Yields of old-field loblolly pine plantations. Virginia Polytechnic Institute and State University, Blacksburg, Pub. FWS-3-72

Cieszewski CJ, Strub M (2007) Parameter estimation of base-age invariant site index models: which data structure to use? A discussion. For Sci 53:552–555

Crow TR (1986) Comparing the performance of two northern hardwood growth projection systems. North J Appl For 3:28–32

Daniels RF, Burkhart HE (1975) Simulation of individual tree growth and stand development in managed loblolly pine plantations. Virginia Polytechnic Institute and State University, Blacksburg, Pub. FWS-5-75

Daniels RF, Burkhart HE, Strub MR (1979) Yield estimates for loblolly pine plantations. J For 77:581–583, 586

Davison AC, Hinkley DV (1997) Bootstrap methods and their application. Cambridge University Press, New York

Gauch HG Jr, Hwang JTG, Fick GW (2003) Model evaluation by comparison of model-based predictions and measured values. Agron J 95:1442–1446

Gertner G (1987) Approximating precision in simulation projections: an efficient alternative to Monte Carlo methods. For Sci 33:230–239

Gertner GZ (1988) Alternative methods for improving variance approximation of single tree growth and yield projections. In: Ek AR, Shifley SR, Burk TE (eds) Forest growth modelling and prediction. USDA Forest Service, North Central Forest Experiment Station, St. Paul, pp 739–746, General Technical Report NC-120

Huang S, Yang Y, Wang Y (2003) A critical look at procedures for validating growth and yield models. In: Amaro A, Reed D, Soares P (eds) Modelling forest systems. CAB International, Wallingford, pp 271–293

Janssen PHM, Heuberger PSC (1995) Calibration of process-oriented models. Ecol Mod 83:55–66

Jørgensen SE (1994) Fundamentals of ecological modelling. 2nd edn. Elsevier, Amsterdam

Knoebel BR, Burkhart HE, Beck DE (1986) A growth and yield model for thinned stands of yellow-poplar. For Sci Monogr 27:1–62

Kobayashi K, Salam MU (2000) Comparing simulated and measured values using mean squared deviation and its components. Agron J 92:345–352

Kozak A, Kozak R (2003) Does cross validation provide additional information in the evaluation of regression models? Can J For Res 33:976–987

Leary RA (1988) Some factors that will affect the next generation of forest growth models. In: Ek AR, Shifley SR, Burk TE (eds) Forest growth modelling and prediction. USDA Forest Service, North Central Forest Experiment Station, St. Paul, pp 22–32, General Technical Report NC-120

Leary RA (1997) Testing models of unthinned red pine plantation dynamics using a modified Bakuzis matrix of stand properties. Ecol Mod 98:35–46

Liu CM, Leuschner WA, Burkhart HE (1989) A production function analysis of loblolly pine yield equations. For Sci 35:775–788
Mayer DG, Butler DG (1993) Statistical validation. Ecol Mod 68:21–32
McCarter JB, Wilson JS, Baker PJ, Moffett JL, Oliver CD (1998) Landscape management through integration of existing tools and emerging technologies. J For 96(6):17–23
McGaughey RJ (1998) Techniques for visualizing the appearance of forestry operations. J For 96(6):9–14
Moeur M, Ek AR (1981) Plot, stand, and cover-type aggregation effects on projections with an individual tree based stand growth model. Can J For Res 11:309–315
Myers RH (1990) Classical and modern regression with applications, 2nd edn. Duxbury, Pacific Grove
Oderwald RG, Hans RP (1993) Corroborating models with model properties. For Ecol Manage 62:271–283
Oreskes N, Shrader-Frechette K, Belitz K (1994) Verification, validation, and confirmation of numerical models in the earth sciences. Science 263:641–646
Power M (1993) The predictive validation of ecological and environmental models. Ecol Mod 68:33–50
Rauscher HM, Young MJ, Webb CD, Robison DJ (2000) Testing the accuracy of growth and yield models for southern hardwood forests. South J Appl For 24:176–185
Reckhow KH, Chapra SC (1983) Confirmation of water quality models. Ecol Mod 20:113–133
Rykiel EJ Jr (1996) Testing ecological models: the meaning of validation. Ecol Mod 90:229–244
Smith JL, Burkhart HE (1984) A simulation study assessing the effect of sampling for predictor variable values on estimates of yield. Can J For Res 14:326–330
Snee RD (1977) Validation of regression models: methods and examples. Technometrics 19:415–428
Soares P, Tomé M, Skovsgaard JP, Vanclay JK (1995) Evaluating a growth model for forest management using continuous forest inventory data. For Ecol Manage 71:251–265
Stage AR (2003) How forest models are connected to reality: evaluation criteria for their use in decision support. Can J For Res 33:410–421
Sterba H, Monserud RA (1997) Applicability of the forest stand growth simulator PROGNAUS for the Austrian part of the Bohemian Massif. Ecol Mod 98:23–34
Stoltman AM, Radeloff VC, Mladenoff DJ (2004) Forest visualization for management and planning in Wisconsin. J For 102(4):7–13
Tomé M, Soares P (1999) A comparative evaluation of three growth models for eucalypt plantation management in coastal Portugal. In: Amaro A, Tomé M (eds) Empirical and process-based models for forest tree and stand growth simulation. Edições Salamandra/Novas Tecnologias, Lisboa, pp 517–533
Van Horn RL (1971) Validation of simulation models. Manag Sci 17:247–258
Vanclay JK, Skovsgaard JP (1997) Evaluating forest growth models. Ecol Mod 98:1–12
Wang X, Song B, Chen J, Crow TR, LaCroix JJ (2006) Challenges in visualizing forests and landscapes. J For 104:316–319
Yang Y, Monserud RA, Huang S (2004) An evaluation of diagnostic tests and their roles in validating forest biometric models. Can J For Res 34:619–629

Index

H.E. Burkhart and M. Tomé, *Modeling Forest Trees and Stands*,
DOI DOI 10.1007/978-90-481-3170-9,

www.ingramcontent.com/pod-product-compliance
Ingram Content Group UK Ltd.
Pitfield, Milton Keynes, MK11 3LW, UK
UKHW020130290926
15471UKWH00003B/400
* 9 7 8 9 0 4 8 1 3 1 6 9 3 *